Multiplatform E-Learning Systems and Technologies:
Mobile Devices for Ubiquitous ICT-Based Education

Tiong T. Goh
Victoria University of Wellington, New Zealand

Information Science REFERENCE

INFORMATION SCIENCE REFERENCE

Hershey · New York

Director of Editorial Content: Kristin Klinger
Senior Managing Editor: Jamie Snavely
Assistant Managing Editor: Michael Brehm
Publishing Assistant: Sean Woznicki
Cover Design: Lisa Tosheff
Printed at: Yurchak Printing Inc.

Published in the United States of America by
Information Science Reference (an imprint of IGI Global)
701 E. Chocolate Avenue
Hershey PA 17033
Tel: 717-533-8845
Fax: 717-533-8661
E-mail: cust@igi-global.com
Web site: http://www.igi-global.com/reference

Library of Congress Cataloging-in-Publication Data

Multiplatform e-learning systems and technologies : mobile devices for ubiquitous ICT-based education / Tiong T. Goh, editor.
p. cm.
Includes bibliographical references and index.
Summary: "This book addresses technical challenges, design frameworks, and development experiences that integrate multiple mobile devices into a single multiplatform e-learning systems"--Provided by publisher.
ISBN 978-1-60566-703-4 (hardcover) -- ISBN 978-1-60566-704-1 (ebook) 1. Educational technology. 2. Mobile communication systems in education. 3. Instructional systems--Design. I. Goh, Tiong T.
LB1028.3.M88 2009
371.33--dc22
2009002220

British Cataloguing in Publication Data
A Cataloguing in Publication record for this book is available from the British Library.

Table of Contents

Detailed Table of Contents

We propose a ubiquitous learning approach useful not only to acquire knowledge in the traditional educational meaning, but also to solve cross-environment everyday problems. By formalizing user request and profile through logic-based knowledge representation languages, a lightweight but semantically meaningful matchmaking process is executed in order to retrieve the most suitable learning resources. Standard formats for distribution of learning objects are extended in a backward-compatible way to support semantic annotations in our framework. The framework and algorithms are designed to be general purpose. Nevertheless, an application has been developed where the semantic-based Bluetooth/RFID discovery protocols devised in previous work, support users –equipped with an handheld device– to discover learning objects satisfying their needs in a given environment.

The aim of this chapter is to explore issues in effective system design to bring about pedagogically sound learning with mobile devices, including the emerging generation of new devices. We review pedagogical models and theories applicable to mobile learning (or m-learning) and ubiquitous learning (or u-learning, also sometimes called pervasive learning, or p-learning), consider the technological support available, and describe scenarios and case studies that exemplify the achievements and challenges for each paradigm. We will also consider possible abstractions that relate ways in which learners can work within varied pedagogical model(s) to make use of relevant supporting technologies, e.g., the notions of "personal learning workflows" and "group learning workflows."

In this work we present a structured method for automatically generating User Interfaces for e-learning environments. The method starts with a definition of the learning scenario where the different goals, jobs (professor-student/trainer-learner), and tasks are described and stored in a template. After, the description is mapped to FlowiXML, a learning process authoring tool, where graphically trainers or content designers draw the overall process. A learning process is viewed as a workflow and modeled using Petri net notation. From each step in the process model more details are added using user task models; user's activity interacting with a user interface is stored in such diagrams. Then, a transformational method for developing user interfaces of interactive information systems is used that starts from a task model and a domain model to progressively derive a final user interface. This method consists of three steps: deriving one or many abstract user interfaces from the task model, deriving one or many concrete user interfaces from each abstract interface, and producing the code of the final user interfaces corresponding to each concrete interface. The models and the transformations of these models are all expressed in UsiXML (User Interface eXtensible Markup Language) and maintained in a model repository that can be accessed by the suite of tools. Developing user interfaces in this way facilitates its automated generation over multiple computing platforms while maintaining portability and consistency between the multiple versions. Our approach is illustrated on an open Learning environment using a case study.

Mobile devices are becoming more and more commonplace across all walks of life from the workplace to leisure activities and even the classroom. Many schools shun the use of devices such as mobile phones in the classroom environment, but this will have to change as they become a more integral part of our daily lives. The ever increasing capabilities of these devices allow for opening up on new application domains. The ubiquitous use of mobile technology in the classroom may provide new and interesting ways for students to interact with subject matter. This chapter discusses the use of cross platform Bluetooth enabled mobile devices within the classroom setting to allow students to interact with subject matter in a new and interactive way using the ICT resources that are ever present in our daily lives.

This chapter presents research around pervasive and ubiquitous computing, particularly oriented in the field of human learning. We are studying several solutions to deliver content over a heterogeneous networks and devices. Converting and transmitting documents across electronic networks is not sufficient. We have to deal with contents and containers simultaneously. Related work in interface adaptation and plasticity (the capacity of a user interface to withstand variations of both the system physical characteristics and the environment while preserving usability) is presented and some examples of context-aware adaptation are exposed. We present an adaptive pervasive learning environment, based on contextual QR Codes, where information is presented to learner at the appropriate time and place, and according to a particular task. This learning environment is called PerZoovasive, where learning activities take place in a zoo and are meant to enhance classroom activities.

Building innovative m-learning systems can be challenging, because innovative technology is tied to innovative practice, and thus the design process needs to consider the social and professional context in which a technology is to be deployed. In this chapter we describe a methodology for co-design in m-learning, which includes stakeholders from the domain in the technology design team. Through a case study of a project to support nurses on placement, we show that co-design should be accompanied by co-deployment in order to manage the reception and eventual acceptance of new technology in a particular environment. We present both our co-design and co-deployment methodologies, and describe the techniques that are applicable at each stage.

A learner-adaptive self-learning environment has been developed in which both mobile phones and personal computers can be used as client terminals. The learner-adaptive function has been implemented using SCORM 2004 specifications. The specifications were extended to enable offline learning using mobile phones. Because the application-programming environment of mobile phones varies from carrier to carrier, a common content format was specified for the learning content and content-execution mechanisms were developed for each carrier's environment to maximize content-platform interoperability. The latest learning results achieved by using mobile phones were synchronized with the latest ones on the server-side sequencing engine so that the learner-adaptive function was available from personal

website. The pilot conducted utilised an SMS Prototype Tool Trigger that enabled dynamic information transfer between staff and students. Trigger facilitated live update reminders that assisted students to schedule their time and better organise themselves. Specifically, SMS technology was used to deliver physical class locations, availability and web addresses of iPod resources, important events, alerts for multimedia, examination schedules, and, assessment feedback by 'pushing' information to students. Trigger also provided students with pull access to study schedules and requirements. The aim of the test was to evaluate student response to the use of Trigger to improve the learning environment. The case study will identify student responses to the pilot and describe a current project that has extended the number of students participating in the study.

Educators strive to develop innovative teaching strategies to meet the expectations of digital natives that are accustomed to social networking environments. The Campus Connect project at Montclair State University provided an innovative mobile technology service, in order to meet these expectations. The program, which included a custom designed, high speed, rich media and GPS (location based services) capable cellular network as well as a rich array of cell phone based applications enabled students to customize their mobile phone for 24/7 access to the University's teaching and learning, information, and administrative resources. This chapter will describe the growth and evolution of the Campus Connect program and the applications that were frequented by the student population on mobile technology through this innovative program. In addition, a description of how these applications enhanced the learning environment will be provided as well as the changes the program underwent in order to best suit the demands of the changing population of students. Quantitative and qualitative survey results are offered to describe the student's reaction to using mobile technology in a learning environment as well as identify those applications that students utilized most often. Based on these results, recommendations for future iterations of the Campus Connect program will be provided, which can be used as a guide for administrators who may be contemplating comparable mobile technology programs at their institutions.

An important part of multiplatform or blended learning is designing learning environments that take full advantage of the relative strengths and weakness of the various platforms employed to meet learning objectives. The desktop has strengths that are conducive to immersive learning environments, whereas mobile devices excel in contextual learning and performance support roles. Blended learning then, is not merely porting the same content from one platform to another, but recognizing the need for unique implementations. This chapter will examine two general applications in which mobile learning takes advantage of the flexibility afforded by the platform. In the first case we will explore the possibilities

Foreword

I am pleased to write some words for this very timely collection. E-learning has come long way since its infusion in mainstream education. While advances are being made in a number of technological and pedagogical dimensions, ranging from innovations in semantic technologies to social networking through Web 2.0, significant growth in mobile device technologies in recent years has made it a real possibility to learn anytime and anywhere. This edited volume has not only recognized this aspect but has gone one step further in the realization of ubiquitous learning where mobile technology is not of one specific form. This has serious implications in terms of identifying ways for effective use of this technological advancement in education, particularly when the access to education is not limited to one type of devices. Even the same student is bound to use desktop computers and mobile phone to access and interact with educational content and activities at different times. In such as multiplatform learning environment, if we do not start looking critically at the pedagogical changes required to accommodate this, pedagogy will again be decades behind the technology.

This book is therefore a significant step towards making sure that pedagogy stays at par with technological advances. The chapters are divided into four sections, which logically flow from the theories and frameworks for supporting effective learning in mobile and ubiquitous learning environments, to the design and integration methodologies for various components in such environments. Next section focuses on various innovative tools that have been developed. Finally, the book concludes with a section on real examples of use cases.

One highlighting aspect is the abundance of focus on contextual, adaptive and personalized learning in this collection. With mobile and ubiquitous learning, the education is not any more a bulk process as it used to be in traditional classrooms, where individual student did not matter. Mobile and ubiquitous learning has really put the student at the center of learning, contributing significantly to constructivist learning paradigm.

Another breakthrough of mobile and ubiquitous learning is the awareness of situation that guides the learning process. Learning process does not follow a rigid path any more. Instead, it is now possible to customize and configure the content, activities and interaction to the real-time situation of the student. Mobile devices allow location awareness through technologies such as GPS and cellular base station reference, and environmental awareness through technologies such as RFID and QR Code. Chapters in this collection include these aspects at both research and implementation levels, making this collection useful for both researchers and practitioners.

There is a right balance between theory and practice in this book, covering pioneering innovations and well-proven applications of emerging technologies. Chapters focus on both pedagogy and technology, and therefore this collection should be useful for a wider community of researchers, early adopters and those who want to make sure that their teaching is informed by proven research. By having a stab at futuristic technological solutions, this collection will also serve as an archival reference for future generation of researchers by giving them insight of systematic technological advancements in education.

I am especially thrilled to comment on this collection, since the editor of this collection, Dr. Tiong Goh, is known to me for past many years, and I have witnessed his research capabilities and achievements, particularly those related to the multiplatform mobile adaptation framework he designed as part of his doctoral research to consider the factor of urgency in learning process. He has once again shown his prudence by putting together an excellent combination of high-quality research and implementation articles that would serve as stepping stone for others for years to come.

Kinshuk
Athabasca University, Canada

Kinshuk is Professor and Director of School of Computing and Information Systems at Athabasca University, Canada. He also holds iCORE/Xerox/Markin Industrial Research Chair in Adaptivity and Personalization in Informatics. Before moving to Canada in August 2006, Kinshuk worked at German National Research Centre for Information Technology as Postdoctoral Fellow, and at Massey University, New Zealand as Associate Professor of Information Systems and Director of Advanced Learning Technology Research Centre. He has been involved in large-scale research projects for adaptive and mobile learning environments and by 2008, he has published over 250 research papers in international refereed journals, conferences and book chapters. He is the Founding Chair of IEEE Technical Committee on Learning Technology and Editor of the SSCI indexed Journal of Educational Technology & Society (ISSN 1436-4522).

Preface

The objective of *Multiplatform E-Learning Systems and Technologies: Mobile Devices for Ubiquitous ICT-Based Education* is to provide researchers, e-learning adopters and practitioners with the most current information about various critical issues regarding design frameworks, the appropriate use of pedagogies, the adoption and extension of existing standards, the design of user interface and innovative tools, the adaptation, transformation and delivery of integrated learning content, the appropriate users' evaluation methodologies, and the scaffolding of existing mobile learning experiences toward multiplatform e-learning systems.

Multiplatform e-learning system is not mobile learning as per se, it is more comprehensive than mobile learning in various aspects. A simple definition of multiplatform e-learning systems is to regard learning systems that generate support and provide appropriate learning content concurrently to a proliferation of mobile devices such as wireless laptops, PDA, mobile phones, digital interactive TVs, iPhones, game consoles etc. In this context, an e-learning system that can support and engage learners through a multitude of access devices or objects is called a multiplatform e-learning system. Multiplatform e-learning system (sometime also known as a multi-device e-learning environment or as ubiquitous learning) is an emerging technology that opens a new research domain.

As in many new researches in emerging technologies, there are always challenges and risks involve in proposing such a book. However to engage and promote in a new research domain, the benefits and opportunities deriving from producing the book outweighed the risk. With the contributions from many outstanding authors and reviewers, and a small contribution from the editor, together we have managed to produce a first book in Multiplatform E-Learning Systems and Technologies for Ubiquitous ICT-Based Education. In the process of compiling the book, the term mobile learning is still widely used. It is hope that the publication of this book would accelerate the diffusion of multiplatform e-learning systems research into the main stream of ubiquitous ICT-based research and that researchers should begin thinking of a multitude of accessing platforms scenario instead of a single device.

In presenting *Multiplatform E-Learning Systems and Technologies: Mobile Devices for Ubiquitous ICT-Based Education*, from my engineering and information system background I am mindful about the strong need for theoretical foundation and practical usefulness. Therefore within this book, the frameworks and learning theories provide the foundation for development, design methodologies and integration provide the essential routes for successful implementation, innovative tools illustrate many alternative scenarios for engagement in ubiquitous ICT-based learning and various innovative case studies demonstrate validated learning experiences. This approach provides researchers and adopters the ease of understanding and absorption of critical knowledge and issues relevant to Multiplatform E-Learning Systems and Technologies for Ubiquitous ICT-Based Education.

The rest of this preface introduces the book in more details and position the situation and future challenges for Multiplatform E-Learning Systems research.

THE SITUATION AND FUTURE CHALLENGES

E-learning systems have now been adopted by many universities. With the advent of the Internet, the e-learning systems have been transformed into web based learning systems where content can be accessed beyond the conventional classroom and lab boundaries. In recent years, there has been a proliferation of devices capable of accessing the Internet, ranging from tablet PCs to mobile devices including telephones, smart phones and personal digital assistants (PDA), iPhones, game consoles and even appliances such as televisions, microwave ovens and refrigerators. Most of these devices are capable of accessing e-learning systems. However, till now, most popular e-learning systems such as Blackboard and WebCT are limited in delivering appropriate content to these proliferations of devices. Conventionally these e-learning systems are designed for personal computer usage. With the proliferation of access devices there is a need for these e-learning systems to extend their support to and provide appropriate content for these devices. Such an e-learning system that supports this multitude of devices is called a multiplatform e-learning system in this context. In this new context, many aspects differ from the original context. Firstly most of the mobile devices such as PDAs and smart phones are designed for telecommunication usage. These devices are characterized by a small screen, low memory, low power and distinctly different ways of interaction and navigation compared to desktop computers. When learners use these devices to access and interact with e-learning content instead of making a phone call, how do they feel about the interaction? What are the relevant factors that need to be considered even though the learning content may be identical? Furthermore, in this new context even if the access platform is similar to a desktop computer, other aspects may still differ from the original context. For instance, in most cases the connection bandwidth is likely to be lower than a school's local network. E-learning systems typically do not perform bandwidth estimation to make changes to content. The content remains the same irrespective of the changing context. In some cases the bandwidth might be too low for the delivery of multimedia content. Is it possible that some alternate ways of delivering content such as offline or plain text delivery could make learning more satisfying and useful than waiting endlessly for downloads? How might these alternative ways of delivering content influence learners? For instance, the affective components such as motivational factor of accessing the e-learning systems might be different at school than on the move. While on the move, different affective factors such as urgency may trigger students to explore the e-learning system. Can we extract and utilise learner's affective components? As teachers' help are not readily available on the move, a lot of self motivation is required. Would the present of urgency and absence of teachers' help influence learner?

While many attempts have been made to develop e-learning systems that can be accessed only through single devices such as desktop computers or certain mobile devices, multiplatform e-learning systems have not been well researched. With the escalating speed at which new mobile devices are being launched and wireless infrastructures are being developed, there is an ever-increasing need to acquire an understanding of the characteristics and learning experiences of multiplatform e-learning systems from various perspectives to support future Ubiquitous ICT-Based Education. It is hope that this book is timely to support such an endeavour.

OVERVIEW OF THE BOOK

This edited book is intended to address the latest development relevant to multiplatform e-learning systems and technologies for mobile devices and ubiquitous ICT-based education. It comprises contributions from leading researchers and practitioners all over the world in the field of multiplatform e-learning systems and technologies for mobile devices and ubiquitous ICT-based education. Inevitably this new environment provides both opportunities and challenges ahead. One aspect this book addresses is the integration and extension of current LMS towards multiplatform e-learning environment. For commercial and open source LMS providers, this provides an opportunity to extend the reach from single PC accessing device to various mobile devices. The practical nature of the book provides readers with real and proven knowledge and state-of-the art technologies on the design of multiplatform e-learning systems and technologies for mobile devices and ubiquitous ICT-based learning. Other aspects the book addresses are the frameworks in blended learning scenarios, heterogeneous pervasive environment, and designing with effective pedagogical systems.

Thus, this book presented extensive and yet critical issues relevant to the current technologies. Each chapter provide its research findings and briefly discuss on future research and how it will be useful in supporting multiplatform e-learning systems and technologies for mobile devices and ubiquitous ICT-based education.

The book is organized along four dimensions of theoretical and practical research. Section 1 addresses the framework and learning theory issues. This will help the reader to understand the foundation that support ubiquitous ICT-based learning. Section 2 addresses the design and integration issues. This will help the reader extend their theoretical understand into practical system. Section 3 addresses some practical tools that are useful to engage learners in various scenarios. This will help reader to consider various field learning activities. Section 4 provides various case studies. This will help reader understand the expected benefits and challenges should the reader intents to deploy similar technologies and systems.

The book has been organized into four sections of 20 chapters. A brief description of each of the chapters follows:

Section 1: Frameworks and Theories

In **Chapter 1**, Tim de Jong, Alba Fuertes, Tally Schmeits, Marcus Specht and Rob Koper describe a multi-platform extension of learning networks that provide mobile, contextualised learning content delivery and creation. They illustrate a possible extension to contextualise and more authentic forms of learning mediated by mobile devices. The chapter conclude with an outlook describing the components necessary to integrate multi-platform e-learning software in existing learning scenarios to achieve a larger-scale adaptation.

In **Chapter 2**, Michele Ruta, Floriano Scioscia, Simona Colucci, Eugenio Di Sciascio, Tommaso Di Noia, and Agnese Pinto propose a ubiquitous learning approach useful in acquiring knowledge in the traditional educational setting and capable of solving cross-environment everyday problems. The chapter introduces a lightweight and semantically meaningful matchmaking process to retrieve the most suitable learning resources. They proposed a generalised framework and algorithm and demonstrated with an application using semantic-based Bluetooth/RFID discovery protocols.

In **Chapter 3**, Wan Ng, Howard Nicholas, Seng Loke and Torab Torabi address the issues of effective learning system design for various mobile devices. They review various pedagogical models and theories applicable to mobile learning. Using scenarios and case studies they demonstrate various alternatives and challenges for each pedagogical model. A personal learning workflows and group learning workflows approach were proposed to work within varied pedagogical models.

In **Chapter 4**, Paul Hayes and Stephan Weibelzahl exploit text messaging for supporting learning in a variety of educational settings. This chapter demonstrates how the use of text messaging can contribute towards enhanced quality of learning. In particular the chapter focuses on the use of text messaging as a means of improving immediacy between instructors and students in third-level education. This chapter conclude with a discussion on the integration of text messaging for improving immediacy in Multiplatform E-Learning Systems.

In **Chapter 5**, Jonathan Bishop describes the Ecological Cognition Framework (ECF) that provides a thorough understanding of how actors respond to and influence their environment. Utilising the ECF the chapter shows that for an e-learning system to be an effective teacher it needs to be able to create five effects in the actors that use it. The effects are the belonging effect, the demonstration effect, the inspiration effect, the mobilisation effect, and the confirmation effect.

Section 2: Design and Integration

In **Chapter 6**, Juan Manuel González Calleros, Josefina Guerrero García, Jaime Muñoz Arteaga, Jean Vanderdonckt, and Francisco Javier Martínez Ruiz present a structured method for automatically generating User Interfaces for e-learning environments. Their method facilitates automated generation over multiple computing platforms while maintaining portability and consistency between the multiple versions. The method starts with a definition of the learning scenario where the different goals, jobs and tasks are described and stored in a template with the aid of FlowiXML, a learning process authoring tool and UsiXML, a User Interface eXtensible Markup Language tool.

In **Chapter 7**, Daniel C. Doolan, Tracey J. Mehigan, Sabin Tabirca, and Ian Pitt discuss the use of Bluetooth enabled mobile devices for cross platform application within the classroom setting to allow students to interact with subject matter in a new and interactive way using the ICT resources that present in our daily lives. The chapter provides an evaluation on the use of such cross platform learning applications and demonstrated that learning process is enhanced.

In **Chapter 8**, José Rouillard presents a solution to deliver content over a heterogeneous networks and devices. The chapter discusses work in interface adaptation and plasticity and illustrate examples of context-aware adaptation. In particular the chapter illustrates an adaptive pervasive learning environment that take place in a zoo. The system is based on contextual QR Codes, where information is presented to learner at the appropriate time and place, and according to a particular task.

In **Chapter 9**, David Millard, Yvonne Howard, Lester Gilbert, and Gary Wills describe a methodology for co-design in m-learning that includes stakeholders' inputs from the domain in the technology design team. The method emphasises ubiquitous learning design process that considers the social and professional context. A case study that supports nurses' placement illustrates the effectiveness of the co-design methodology.

In **Chapter 10**, Kiyoshi Nakabayashi presents a learner-adaptive self-learning environment for both mobile phones and personal computers. The learner-adaptive function has been implemented using SCORM 2004 specifications to enable offline learning using mobile phones. The functionality and usability of the system was evaluated and validated through two trial experiments.

In **Chapter 11**, Marc Alier Forment, María José Casany Guerrero, and Jordi Piguillem Poch analyze the complexities involved in the integration of Learning Management Systems (LMS) and ubiquitous learning. The chapter describes some standard interoperability architectures and related research and development projects that will allow better integration and interaction between the LMS and the m-learning applications. The chapter illustrates a case example with Moodbile that demonstrates a rich mobile client application with persistent storage capabilities and offline functionality.

Section 3:Innovative Tools

In **Chapter 12**, Dawn Woodgate, Danaë Stanton Fraser, Amanda Gower, Maxine Glancy, Andrew Gower, Alan Chamberlain, Teresa Dillon, David Crellin argue the relationships that exist between science education and public engagement in science, and 'formal' and 'informal' learning contexts. The chapter describes four case studies involving various mobile technologies, tools and platforms for ubiquitous ICT-based science-related learning inquires and activities.

In **Chapter 13**, Mattias Rost and Lars Erik Holmquist present a set of tools to support groups of students who are doing field studies. The tools include a wiki for gathering field notes and their group work material, a mobile application for capturing data (photo, video, audio, and text) and automatically uploading to the wiki, and a set of web tools which run on top of the wiki for increasing the awareness between students, and for browsing the captured data. The chapter describes the implementation of these tools and report on the experience.

In **Chapter 14**, Peter Byrne and Brendan Tangney present the design, implementation, and evaluation of a mobile learning application called the Stop-Motion Animation and Reviewing Tool (SMART). The application enables users to create animations on a mobile phone and is part of a larger generic suite of open-system software to facilitate the development of cross platform applications in the area of digital narrative production.

Section 4: Innovative Cases

In **Chapter 15**, Siu Cheung Kong presents a multiplatform e-learning system called the "Graphical Partitioning Model (GPM)" for learning knowledge of fraction equivalence. The chapter presents a case study on the use of the mobile version GPM for the learning of the targeted topic in a mobile technology supported environment. The case study reveals that there is a potential for the flexible use of the dual-version GPM to foster deep learning.

In **Chapter 16**, Kin-Choong Yow and Boon-Chong Seet describe a new platform for mobile and interactive learning between the professor and students during lectures. The new platform enables interactions through Multimedia Messaging Service (MMS) capable devices such as PDAs, Laptops, or Tablet PCs that are connected on the campus-wide Wireless LAN. The system enables and encourages response to questions or provides instance feedback on the lecture.

In **Chapter 17**, Joan Richardson and John Lenarcic describe a case study on the use of Short Message Service (SMS) to augment and support the provision of student administrative services. The system utilised SMS technology to deliver physical class locations, availability and web addresses of iPod resources, important events, alerts for multimedia, examination schedules, and, assessment feedback by 'pushing' information to students.

In **Chapter 18**, Patricia Kahn and Edward Chapel present a campus wide innovative mobile technology service program. The program, which included a custom designed, high speed, rich media and GPS (location based services) capable cellular network as well as a rich array of cell phone based applications enabled students to customize their mobile phone for 24/7 access to the University's teaching and learning, information, and administrative resources. The chapter describes how the applications enhanced the learning environment.

In **Chapter 19**, David Metcalf and David Rogers argue that an important part of multiplatform or blended learning is designing learning environments that take full advantage of the relative strengths and weakness of the various platforms employed to meet learning objectives. The chapter examines applications in which mobile learning takes advantage of the flexibility afforded by the platform. A case illustrates the possibilities presented by physical hyperlinks through the used of Near Field Communications, QR codes, and image recognition software.

Finally in **Chapter 20**, Shinichi Hisamatsu presents an interactive exhibition system for museums, which combines learning based on interactions with multiple physical objects and knowledge transmission. The system enable user to handle and look at an actual physical object and able to talk directly to the user. This "conversation" with the object as the user "grasps" and "feels" the object deepens the user's understanding of and interest in the object.

At the end of this book there is also a comprehensive index defining most of the terms that will be useful to reference for the exact meaning used by various authors in the book.

OPPORTUNITIES

Multiplatform E-Learning Systems and Technologies and Ubiquitous ICT-Based Education research is definitely in its early stage of research life cycle. There is a need to understand various aspects of the technologies pertaining to effective use of the technology to achieve satisfactory and effective learning outcomes. While the underlying technology may be ready for deployment there is indeed a lack of validated pedagogical theory to support effective design and development. This vacuum represents an opportunity for future research. There is also an opportunity to investigate various extensions of LMS to make appropriate interaction and delivery of existing content to various devices and new platforms. Indeed, a future volume may be necessary to continuously address the issues. Nevertheless, it is hope that this book will be a timely publication for both academics and practitioners who are interested in the design and development of future Multiplatform E-Learning Systems and Ubiquitous ICT-Based learning environments.

Tiong T. Goh
Victoria University of Wellington, New Zealand

Section 1
Frameworks and Theories

as well as, the integration of learning technologies in modern day education. In our opinion, a standardised, technology-supported process of installation, use, and integration would benefit a larger scale adoption of multi-platform learning systems and makes it possible to reuse and adapt existing learning designs in multiple learning contexts. Certainly, ease of use would lead to a greater enthusiasm to adopt new forms of education, which in its turn could increase the frequency of use. Therefore, we will illustrate a standardised process of creating authentic, blended, and ubiquitous learning scenarios and describe a technical infrastructure to help design these scenarios. More importantly, the technical infrastructure will provide generic interfaces and components that should ease the use with a range of devices and, furthermore, hide the technical details to reduce design complexity.

However, the design of an infrastructure for multi-platform, ubiquitous learning has to be grounded in theory. Consequently, in the next section, section two, we will first consider existing learning networks, the underpinning pedagogical theories, and how the pedagogical scenarios used could be extended with mobile devices. Section three describes an extension to learning networks to support blended learning with authentic real-world scenarios, which subsequently leads to technical requirements that will be described in section four. After that, a technical framework is considered in section five and illustrated using two learning scenarios in section six. Last, section seven provides our conclusions and an outlook to future developments.

Learning Networks

Learning networks (Koper, & Tattersall, 2004) are social software that support networks of lifelong learners, focusing at communities of self-directed learners. More importantly, they mean to exploit the heterogeneity of learners by creating communities where novices and experts can collaborate. Learning networks are founded on a combination of social-constructivist theories, more specifically, lifelong learning theories that integrate informal and formal learning approaches. Hence, to facilitate this integration, learning network software concentrates on supporting:

- Self-directed learning
- Learning in communities-of-practice
- Learning content creation, organisation and delivery

In the next subsections, we will shortly consider how learning network software supports these three settings, and see how learner support could be extended with mobile technology in a multi-platform e-learning system. In addition, we look at blended learning theory to extend current pedagogies in learning networks to include more authentic, real-world scenarios. After all, lifelong learning is learning anywhere and anytime and a supporting platform should ideally combine a variety of learning technologies to get the best out of each learning opportunity.

Self-Directed Learning

A lifelong learner is most often a self-directed learner (Brockett, & Hiemstra, 1991). Therefore, learning networks provide help for learners to self-organise their learning. A specific example of learner support are recommender systems that help learners deriving a learning path, a sequence of units of learning that would ultimately result in acquiring a learning goal (Drachsler, Hummel, & Koper, 2008). Another example of assistance for self-directed learners is assessment support that helps them position themselves on a learning path; i.e. which units of learning do they still need to carry out, and which ones they can skip (Kalz, et al., 2007). Furthermore, learning network software assists these learners to reflect (Schön, 1983; Schön, 1987) about their learning by preserving their growth in competency (Koper, & Tattersall,

2004). The learners *controlling* their own learning process is also specifically mentioned as a part of a task model for mobile learning presented in (Taylor, et al., 2006); thus, mobile learning could provide new ways of self-directed learning by facilitating learning content access nearly anyplace and anytime.

Learning in Communities-of-Practice

Next to self-directed learning, learning networks, as the name already states, support learner communities on a certain topic. The pedagogical theory underlying learning networks is mainly given by Wenger and Lave (Wenger, & Lave, 1991) who stressed the importance of knowledge acquisition in a cultural context and the integration in communities-of-practice. Bruner (Bruner, 1996) additionally states that learning should include social and cultural aspects. Hence, learning networks are social software for learning that provide several mechanisms to build, support, and maintain community processes in such communities-of-practice, among the most important are the following.

First, *collaboration:* Wenger and Lave (Wenger, & Lave, 1991) stated that learning requires collaboration, preferably in a heterogeneous group of learners, where novices can learn by interaction with experts. Communities in learning networks provide a central place for people to collaborate on joint learning tasks. Especially, these communities play an important role in finding appropriate peers to collaborate with and ideally lead to learners helping each other out.

Second, another important mechanism is technology-assisted *community reflection,* which allows a learner to find suitable learning peers, but also contrasts the learner's own experience to that of the community. Community-reflection makes it possible for learners to find experts to learn from, help out less experienced learners, or collaborate with learners that have similar backgrounds and are facing similar problems in their learning. For this reason, learning networks preserve a learner's action history, more specifically a record of their competence development, which can be used to position themselves in relation to others in the learning community. This is one type of social awareness which is aimed at sparking and maintaining active collaboration. Whereas learning networks provide technical assistance to raise social awareness, most of this assistance is meant to support web-based communities. In this sense, mobile technology could provide a link to real-world settings; an interesting approach being the BlueAware system presented in (Eagle & Pentland, 2005), which raises social awareness by notifying users when someone with similar interests is nearby.

Third, learning network software encourages *communication* between learners. Pask's conversation theory (Pask, 1975) states that learning occurs by using conversations to make knowledge (more) explicit. In addition, Wenger & Lave (Wenger & Lave, 1991) endorse the importance of communication by articulating that learning requires social interaction between peers. Moreover, according to Cognitive Flexibility Theory (Spiro, et al., 1992; Spiro, & Jehng, 1990), learning activities must provide multiple representations of content and support context-dependent knowledge. Especially, the theory identifies the importance of using interactive technology to support the learner in the learning process. The various opinions of learners represent multiple perspectives on learning content. Therefore, learning networks offer several communication channels between peers; this makes various forms of reflection possible, for example, learning by comments made by a peer, or learning by creating comments on knowledge created by another learner. One way mobile devices can extend the range of possibilities is by allowing communication between situated learners in an authentic learning situation and de-contextualised learners in a classroom or learning network (Terrenghi, Specht, & Moritz, 2004).

Learning Content Creation, Organisation and Delivery

In a review of new learning and teaching practices, Nesta Futurelab identified several pedagogical theories underpinning current learning technologies (Naismith, et al., 2004). One specific role of technology they found was assisting learners and teachers in coordinating learning and resources in learning activities. In learning networks, the coordination is mainly aimed at supporting self-directed learning and learning in communities of practice as we already have seen before. Next to that, learning network software makes available means to coordinate learning content creation, organisation, and delivery.

Learning content creation: constructivist theory (Bruner, 1966) brings forward learning as an active process, in which learners should construct new ideas or concepts based on their current knowledge. Moreover, learning has to take into account experiences and contexts that make the student willing and able to learn. Learning networks consist of learners that create their own learning content and provide that learning content to be used and improved by the community. Mobile and instant creation of learning content with associated context information, like for example GPS coordinates, has already been demonstrated in for example (De Jong, Specht, & Koper, 2007) and provides unique possibilities to add authentic learning content to learning communities.

Learning organisation: several pedagogical theories emphasise that instruction must be structured to be easily grasped by the student (Brockett, & Hiemstra, 1991; Bruner, 1966; Bruner, 1996). Furthermore, learning must not only be planned structured by a curriculum but also by the tasks and learning situations, and the interaction with the social environment of the learner (Wenger & Lave, 1991). Learning networks offer extensive support to organise learning based on units of learning, learning paths and pedagogical scenarios specified in IMS-LD (Drachsler, Hummel, & Koper, 2008; Koper, Olivier, & Anderson, 2003; Koper & Tattersall, 2005). Related to that, cognitive apprenticeship (Collins, Brown, & Newman, 1989) stresses the importance of structuring authentic learning processes to guide learners towards appropriate levels of knowledge by a constant process of contextualisation and de-contextualisation of knowledge. An interesting example providing learning organisation in a lifelong learning scenario that includes mobile devices is given in (Vavoula, & Sharples, 2002).

Learning content consumption: from a constructivist point of view knowledge is always contextualised, e.g. learning is always situated within its application and the community-of-practic (Mandl, Gruber, & Renkl, 1995). Furthermore, approaches like reflection in action and reflection about action describe the relevance of the context for enabling learning and self-reflection (Schön, 1983; Schön, 1987). While learning in learning networks is contextualised in the sense that it is situated in communities of practice, learning content is still mostly presented out of its situational context; i.e., the authentic context the knowledge needs to be applied in. An extension to contextualised mobile media could help to assist the learner in these authentic situations, by tailoring information delivery to an authentic learning context (Bardram, & Hansen, 2004; Klopfer, Squire, & Jenkins, 2002; Ogata, & Yano, 2004).

Blended Learning Scenarios

The integration of formal and lifelong learning approaches with informal learning activity support in learning networks is currently investigated in the TENCompetence project (Koper, 2005). While the learning networks in this project provide multi-platform access to learning content, and hence the possibility to implement blended learning scenarios, the project focuses at web-based and desktop delivery of learning content. With the recent uptake of mobile devices (Castells, et al., 2007), mobile information access has become more and more important. In addition, this new technology's impact on communication and learning in the younger generation is described as highly relevant for new forms of learning support (Green, & Hannon, 2007). However, the integration of mobile device technology and other new learning media with learning networks, such as smart phones, tablet PCs, smartboards, and gaming consoles, is mostly left out of scope. Moreover, the contextualisation of the learning content is limited. Since mobile devices offer unique possibilities for contextualised content creation and delivery, an extension with mobile devices would therefore offer the possibility to add real-world, context-specific learning scenarios in learning networks.

Several experts have indicated that learning should happen in relevant scenarios, situations, or contexts. Wenger and Lave (Wenger & Lave, 1991), for example, state that learning in a community-of-practice should use authentic tasks and learning situations, i.e., settings and applications that would normally involve the knowledge learnt. Sticht (Sticht, 1975), shares their emphasis in addressing the need to make learning relevant for the work context. Moreover, he states that the assessment of learning requires a context/content specific measurement. Related to that, Piaget (Paiget, 1970) highlights that learning should take place with activities or in situations that engage the learners and require adaptation. Teaching methods should be used that actively involve students and present challenges to the learner.

In a recent literature review of mobile contextualised software (De Jong, Specht, & Koper, 2008b), the authors made apparent that mobile devices have already been used to a large extent for social learning appliances. In particular, five application types of mobile social software for learning were exposed:

- Sharing content and knowledge
- Facilitate discussion and brainstorming
- Social awareness
- Guide communication
- Engagement and immersion

As we can see, the emphasis of mobile social software is quite similar to those of learning networks. A multi-platform learning system combining learning networks with mobile devices seems straightforward to create. In such a multi-platform approach to learning the benefits of both approaches would come together: on the one hand, self-directed learning and learning in communities-of-practice supported by the learning networks software. On the other hand, the learning content and learner interaction in learning networks can be extended with authentic, real-world creation, delivery, and interaction via mobile devices. In this way, blended learning scenarios could be created, integrating a range of technology, using the best technology to support a certain task in a certain situation or context: for instance, a mobile device to support on-the-spot learning in a field trip, or a smartboard to display learning content to a classroom full of learners.

A blended learning scenario that integrates mobile learning combines de-contextualisation and contextualisation of knowledge; theoretical knowledge learnt in a classroom setting could be transferred into practical knowledge in a real-world scenario. Moreover, through using context information, in combination with the creation or retrieval of learning content, several educational effects can be achieved:

- *Multiple perspectives on real-world objects*: by viewing and creating content in a real-world context, several opinions can be perceived and expressed, from which people can benefit through an indirect learning process (Efimova, & Fiedler, 2004).
- *Community-generated content* connected to relevant real-world objects and locations; an example for the effect and importance of self-generated contents in a learning community is presented in (Brandt, et

Table 1. A reference model for mobile social software

Content	Context	Information flow	Pedagogical model	Purpose
Documents Annotations Messages Notifications	Individuality Context Time Context Locations Context Environment or Activity Context Relations context	One-to-one One-to-many Many-to-one Many-to-many	Behaviourist Cognitive Constructivist Social Constructivist	Sharing Content and Knowledge Facilitate Discussion and Brainstorming Social Awareness Guide Communication Engagement and Immersion

al., 2002) about learning to operate medical devices.

- *Community interaction* and the creation of communities of interest around certain objects and locations, supporting contextualised learning (Wenger & Lave, 1991).
- *Different views on objects based on personal preferences.* Real-world objects can also be linked electronically to create relations between those objects and to create a so-called "internet of objects" (Mattern, 2004).
- *Recording of learning events;* allows for later reflection and eliciting of expert's knowledge, carried out in a work context during or shortly after the actual action performed (Schön, 1983; Schön, 1987).
- *Learning content tailored to a specific learning activity*; in the sense of cognitive apprenticeship (Collins, Brown, & Newman, 1989) the learner is guided towards appropriate levels of knowledge by a constant process of contextualisation and de-contextualisation of knowledge. Cognitive apprenticeship furthermore assumes this guidance takes place in an authentic learning situation.
- *Increasing motivation through active learning*, by actively involving the learner in the learning process, the learner involvement and motivation is increased. This as opposed to passive learning in a formal, classroom setting (Bruner, 1966).

Summarising, contextualised media enables the learner to create, retrieve, and use digital media in a relevant real-world context for notification, documentation, problem solving, reflection, communication and a variety of other learning activities. In the next sections, a technical extension of learning networks with contextualised mobile media will be laid out, to facilitate blended learning scenarios that combine social learning in learning networks with authentic scenarios in the real-world.

Extending Learning Networks with Contextualised Blended Learning Scenarios

In an earlier paper (De Jong, Specht, & Koper, 2008b) the authors have presented a reference model that can be used as a basis for future applications of mobile learning. The model will be used to extend the presented learning networks model to include context-aware mobile applications, which makes it possible to define contextualised blended learning scenarios in authentic settings. An overview of the reference model for mobile social software has been shown in Table 1, which combines each of the identified dimensions with its possible values.

The reference model describes the type of content that is used in contextualised learning tools, the context parameters taken into account for adaptation, the information flow, and on a higher level the main purpose and the underpinning pedagogical model.

- The content dimension describes the artefacts exchanged and shared by users, in an analysis of the literature the main types

of artefacts found were annotations, documents, messages, and notifications.
- The context dimension describes the context parameters taken into account for learning support. The main context dimensions identified are based on an operational definition of context by Zimmermann, et al. (Zimmermann, Lorenz, & Oppermann, 2007).
- The information flow classifies applications according to the number of entities in the system's information flow and the information distribution.
- The pedagogical paradigms and instructional models describe the main paradigm leading the design of contextualised media and the integration of media in real world contexts.
- The purpose describes applications according to the goals and methods of the system for enabling learning.

Using the reference model, mobile learning systems can be compared and classified by looking at the five dimensions; while one system could combine documents and annotations with locations context and a one-to-one information flow to support a learner in self-reflecting on the actions carried out in a specific location, another one with a many-to-many information flow would enable community-reflection for a group of learners. Thus, on the one hand, the reference model describes the manipulated knowledge resources, the context in which they are used, and the different flows of information. On the other hand, the higher level concepts of pedagogical model and purpose define how the content, context, and information flows are used and combined. The combinations of different values for each dimension lead to various forms of contextualised software with different purposes and different pedagogical underpinnings. Yet, the five dimensions should be seen as fairly independent. Despite the fact that they can be used to classify and derive applications of mobile learning, a specific combination of context, content, and information flow does not clearly specify the pedagogical model or purpose of the application. Still, some combinations of dimensions may be encountered more often than others for a certain pedagogical model or purpose. As an example, a system with a main purpose of *sharing content and knowledge* between its users, will most often use *documents* from the content dimension, *relations context* to describe social relations between the users, and a many-to-many information flow. Likewise, a social constructivist system like RAFT (Hine, Rentoul, & Specht, 2003), combines on-the-spot creation and delivery of documents with locations context, and messages between learners in a many-to-many information flow for increased engagement and immersion.

Learning network software is structured in four layers (Koper, 2005) that can be described using the dimensions content, information flow, and pedagogical model in the reference model described above. In addition then, the learning network model can be extended to include all aspects of the context dimension of the reference model. The four layers in a learning network can be mapped onto the reference model as follows:

- Knowledge Resources are reusable and self-contained pieces of learning content addressing a part of a larger course. These can consist of a variety of documents and annotations of the content dimensions.
- Units of Learning combine Knowledge Resources into Learning Designs that are underpinned by one of the pedagogical models of the reference model. The pedagogical scenarios are made up out of tasks and activities that can be described in a standard like IMS-LD (Koper, Olivier, & Anderson, 2003; Koper, & Tattersall, 2005). Learning designs furthermore can use the notifications of the content dimension to inform the learner about tasks and activities.

- Learning Communities consist of groups of learners interested in one specific topic and can be specified using individuality context, relations context, and the information flow between learners. Learners can communicate using the messages of the content dimension.
- A Learning Network is a collection of communities on a similar topic and can be fully described using the previous layers.

To be able to include authentic learning scenarios in the real-world would entail adding several additional context parameters to a learning network system and extending others to include more detailed information. Most notably, a learning network that includes learning in the real-world should be able to handle locations, time, and environmental or activity context. These three kinds of context can, together with the other forms, be combined to describe the learning situations (Dey, 2001) a learning scenario would take place in. For example, a history lesson could take into account certain historic locations that could be used to support field trips to those locations. More importantly, by defining more generic situations "in a restaurant", reusable scenarios can be defined that can be used to learn in a range of similar situations.

Technical Requirements for using Contextualised Media in Learning Networks

However, to make a seamless integration of learning networks with for mobile and contextualised technologies possible, the implementation of the software for the technologies should be based on existing standards and should additionally take into account the following requirements.

Multi-platform e-learning systems need to provide access to learning content from a wide range of devices, which requires a flexible technical infrastructure that is focused on *standardisation* and *reusability*. Technical standardisation will make the integration with existing learning management systems easier, and simplify the exchange of information between different devices and technologies. A client-server architecture adhering to existing web service standards is another kind of standardisation that will ease the interaction between heterogeneous devices and enable distributed technology (smartphones, iPods, desktops, smartboards) to communicate in a standardised and similar way.

All in all, standardisation is important because of information interchange between a variety of systems. In addition, standardisation makes the reuse of content easier. Next to the reuse of the learning content itself, pedagogical scenarios that integrate several situations, technologies, and learning theories should be written in reusable learning designs, specified in a standard like for example IMS-LD (Koper, Olivier, & Anderson, 2003; Koper, & Tattersall, 2005). *A modular server architecture*, in which new functionality can easily be added and integrated within existing learning designs, would increase this reusability.

Accessibility on different platforms calls for generic technical interfaces that make the system accessible from multiple clients. Additionally, accessibility requires adaptation of content to specific platforms; content created on one platform ideally should also be accessible using another. However, not all content is suitable to be displayed on all devices. Therefore, a technical framework supporting multi-platform learning approaches requires a certain flexibility providing *learning content filtering* and *learning content adaptation* to handle various formats and sources of learning content. The learning content should be specified in a device-independent XML format which can be easily translated to a standardised content mark-up language to be rendered for display on various devices.

In addition, *the independence of (mobile) client technology* is important because it allows for a more heterogeneous user group and to some extent circumvents the demands of rapidly changing/aging technology. The use of web-based content furthermore makes it possible to use *light-weight, easily portable clients* that integrate a web-browser to display the learning content and provide device-specific software to provide access to sensors. Next to this, specialised clients could be used for educational uses with a higher demand, when high performance is needed and the strengths of the technology should be exploited.

Finally, the multi-platform e-learning systems should be *easy to use*. This applies to the usability of the client software, but also to the integration of the technology in existing education. One way to realised the latter, is the use of tools aimed at a specific user groups. We propose at least two different user groups: first, one technical user group that manipulates and aggregates lower level information into higher level educational concepts. Second, we suggest an educational practitioner group that uses the educational concepts defined by the first group to create sound pedagogical scenarios. The design of a pedagogical scenario using multi-platform e-learning systems should be left educators, and therefore requires tools that operate on pedagogical concepts that those educators are familiar with. In any case, educators should not be bothered with technological details, and should work with higher level concepts and components designed by people with more technological knowledge.

Technical User Group

The technical user group creates higher level educational concepts for the educational practitioners. These concepts are created by defining aggregations of context information that has been acquired using the sensors. Moreover, certain actions can be defined using actuators.

Ideally, the technical user group would combine existing software components without writing any code. The creation of components should be a special case that only occurs rarely. Instead, the technical user group should be provided with two kinds of tools: (1) a visual aggregation tool that allows them to combine the components graphically, and (2) a rule-base architecture that makes it possible to define more complex component aggregations based on logic conditions about component inputs and outputs.

The technical user group uses the tools to specify both *situations* and *activities,* which can be used to define pedagogical scenarios. Situations are specified by an aggregation of context parameters and values and give the conditions in which a certain activity *can* or *should* take place. Conversely, activities specify certain actions or combinations of actions that should influence or drive learning (Koper, Olivier, & Anderson, 2003; Koper & Tattersall, 2005).

In a driving instruction scenario, a situation and activity could be defined as follows: to teach a student operating the vehicle not to drive too fast, a *situation* called "speeding" could be created that combines the two context *parameters* of time and location. Using the context *values* of these parameters the speed of a person can be calculated. Based on a *condition* defining the situation of "speeding", a decision can be made whether or not to carry out an *activity* that teaches the person what reaction is needed to prevent the person from driving too fast.

Educational Practitioner Group

An educational practitioner designs the pedagogical scenarios aimed at a specific learning content domain. Unlike, the technical user, an educational practitioner should not be bothered with technical details, like aggregations of sensor information and how to define situations on the basis of context parameters. Instead, an educational practitioner should be presented with known pedagogical and domain-specific concepts.

Pedagogical scenarios can be defined using learning designs that can be specified using standards as IMS-LD (Koper, Olivier, & Anderson,

2003; Koper & Tattersall, 2005). Learning designs use a combination of activities and learning content to create a variety of pedagogical scenarios. A lot of standardised activities are present within learning networks, among others the following examples:

- *the study* of learning theory,
- *on-spot content creation*; for example mobile content gathering,
- *community-reflection* on created content,
- *situation-specific* learning content delivery,
- *introduction to suitable learning peers,*
- *collaboration,*
- *discussion with peers.*

To create technology-mediated authentic learning scenarios, the situations in which these activities take place should be furthermore specified. In this case, three different conditions can take place. First, a situation could be pre-condition to an activity, thus, an activity will be sparked when a learner takes part in a situation. Second, a situation could be a post-condition that could be the result of an activity. Third, the situation can be monitored during an activity. By using this combination of activities, situations and learning content, complex learning scenarios can be created, two of which we will describe later.

Technical Framework

The requirements formulated in the previous section lead to the development of a modular client-service architecture which takes into account both the requirements for learning network software and a framework for contextualised media. In (De Jong, Specht, & Koper, 2008a) we developed a specification for a technical framework for contextualised media for learning, on the basis of the reference model that can be used as a guideline to implement contextualised learning networks. The framework for contextualised media described a layered software architecture that comprised of four layers, which enriches data step-by-step:

- The first and lowest layer collects the data captured by the client sensors and it acquires the electronic media created by the users. The data in this layer represents the simplest form of data, either *context information*, like for example a user location, or *electronic media*, as for example pictures created.
- The second layer groups the sensor data and electronic media into higher level concepts that can be used to represent real-world objects, locations, users or information attached them, like documents, annotations, user profiles, etcetera.
- The third layer provides the means to define activities, define application logic and processes, and combine the context metadata to take higher order decisions on the basis of semantically enriched data (from layer two). In this layer the educational processes based on the pedagogical paradigm in the reference model can be defined. Furthermore, the information flows and conditions for the delivery of content are defined here. Moreover, the adaptation to the user's personal preferences or physical objects the user interacts with, happen in the third layer.
- The fourth layer carries out actions and delivers the electronic media based upon the decisions that have been taken in layer three. This layer also chooses the correct actuator and suitable content for a certain situation. In short, the purpose of this layer is to carry out an action or change a real-world situation that is given by the last column.

Figure 1 illustrates the technical framework when these four layers are combined and integrated with learning networks software.

Figure 1. A technical framework integrating contextualised media and learning networks

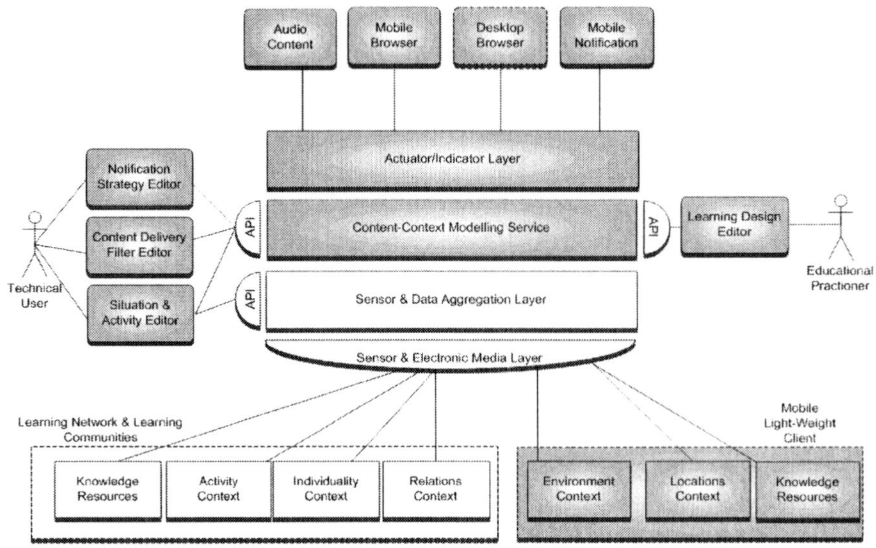

The learning network and learning communities are integrated via the sensor and electronic media layer on the one hand, and the actuator/indicator on the other. Knowledge Resources created in a learning community are delivered as electronic media to the first layer. Similarly, activities performed in the learning network, information about a learner, and social connections are described using activity, individuality and relations context, which is send as input to the sensor & electronic media layer. After the sensor data and content has been processed in the second layer, an action can be carried out on the basis of a decision made in the third layer, the content-context modeling service. The decision is made on the basis of a strategy defined by a technical user or an educational practitioner. As described in the requirements, the technical user defines several low-level strategies concerning the data aggregation, content filtering, and control logic. These strategies are created using the situation and activity editor, the content delivery filter editor, and the notification strategy editor. Alternatively, the education practitioner defines pedagogical scenarios in a learning design using a learning design editor. Once a decision for a suitable action in a certain context has been made, the fourth layer chooses an output channel, i.e. an actuator or indicator that can carry out the action to the learner in the learning network.

Application Scenarios of the Framework

The contextualised learning network software described could be used to carry out several mobile social learning scenarios, two of which we will provide in this section. The first example will describe a foreign language learning scenario, while the second will portray the benefits of blended learning scenarios in learning health and safety aspects in a real-world construction engineering scenario. The examples will illustrate how learning in learning networks can be combined with authentic, more informal, and formal classroom-based learning scenarios. We will concentrate on the use of mobile devices to support learning in context.

Foreign Language Learning

Language is a typical example of something that is widely used across contexts. Language learning takes place in different settings, for example, in a structured setting in an official language course in a educational institution, or a more unstructured, and common day-to-day setting in which language is acquired in a random manner. Additionally, the type of language learnt depends on the situation; some require informal daily speech, while other settings, i.e. business negotiations, require more formal language. Furthermore, language learning is addressed towards a certain community, most often a community of native or near-native practitioners, which uses a community-specific jargon (Petersen & Divitini, 2005). Especially, in an increasingly international world, acquiring this community-specific language becomes more and more important. Particularly, non-native speakers have a demand for just-in-time, situation-specific vocabulary to communicate in a more effective and efficient way.

This cross-context, situation-specific, community-based, and just-in-time nature makes language learning an interesting domain to explore and illustrate the possibilities and problems of a multi-platform e-learning system. In this sense, Petersen & Divitini (2005) have identified interesting community-based scenarios that include the use of mobile devices for learning (Petersen, & Divitini, 2005). More specifically, they emphasise that learning in communities is important because the students need to: (1) learn in an authentic cultural context where the local language is used, and (2) practice using the language with native speakers. In addition, we feel language learning would benefit from blended learning, combining de-contextualised theoretic language lessons, with contextualised authentic learning scenarios. An example of a de-contextualised language scenario is a structured online language learning course, much like the one taught at schools that train grammar, use vocabulary lists, and structured repetition. Conversely, contextualised scenarios would tailor vocabulary- and useful phrase lists to certain situations in daily-life. Paredes, et al. (2005) already demonstrated the context-aware language learning tool, LOCH, which assists learners in tasks that have to be solved by interacting with native speakers in the real-world (Paredes, et al., 2005). LOCH enables learners to directly get into contact with their teacher by using PDAs. The teacher can view the learner's locations and decide to give location-specific feedback. Moreover, the learners can create contextualised information like written annotations and pictures.

In a multi-platform learning network like the one we described, several connecting language learning scenarios can be implemented. A language learning network would include a variety of different learning communities each involved in learning a different language. Each community would consist of a heterogeneous group of native, near-native, and non-native language learners that create, possibly contextualised, multilingual learning content. The creation of learning content can furthermore be combined with community-reflection where more competent learners review the work done by novices. Furthermore, learners should be helped in finding appropriate (native) peers and a community-of-practice that would help them in their learning process; in this case, it would be interesting to couple native speakers that want to learn each other's languages. In any way, active use of a language by discussing with native peers would be an important part of language learning in learning networks.

Next to the community learning described above, language learning would also be beneficiary to self-directed learning processes, possibly mediated with mobile devices. The developed scenarios should allow for memorisation and repetition of language constructs, help to learn from errors by self-reflection on preserved learning history, and combine de-contextualised and contextualised knowledge that results in applying the knowledge learnt. Furthermore, the learning network software should help the self-directed learner in planning, structuring, self-monitoring, and evaluation of learning. Mobile devices could mediate these processes, for example by structured delivery of learning content for memorisation and repetition (Attewell & Webster, 2005). Another example is language learning by interaction with real-world objects. The objects are enriched with language learning content, for example a text message describing the object, or an audio fragment containing a useful phrase related to that object that can be accessed using a mobile device. Thus, the interaction with the objects and learning content in an authentic situation allows learners to learn a language. Furthermore, learners can create their own language learning content connected to objects (De Jong, Specht, & Koper, 2007).

Summarising, language learning in multi-platform learning networks include the following activities:

- Acquiring language on the move, tailored to specific situations,
- Active use of the language, by communication with native peers,
- Creation of learning content, either contextualised or de-contextualised,
- Commenting on peers,
- Discussion with peers,
- Memorisation and repetition,
- Planning, structuring, and self-monitoring of learning,
- Learning by interaction in the real-world.

In addition, countless situations could be defined that are used to contextualise the available language content, for instance standard situations as introducing yourself, ordering at a restaurant, bargaining in a shop, etcetera.

Learning Health and Safety Aspects in a Real-World Construction Engineering Scenario

In construction engineering, students have to learn how to apply the theoretical knowledge in the curriculum to real-world construction work scenarios. While currently most of the teaching is theoretical and classroom-based, students would benefit by actually seeing the principles applied in real construction work. Not only does such an exploration give students the opportunity to encounter real-world examples of knowledge applied, it also actively involves the students in the learning process and compel them to apply the theory just learnt (Bruner, 1966). This application scenario gives an example to get most benefit out of practical learning situations by mediating on-the-spot health and safety risks management learning with mobile devices.

The scenario is based on a Health and Safety Management course, which is part of the International Master in Construction Project Management taught at the Technical University of Catalonia (UPC). The aim of this course is to provide basic knowledge of health and safety (H&S) risks identification, H&S preventive measures and H&S regulations. Therefore, the course provides the know-how that will enable the future construction project managers to analyse and identify the H&S risks existing on a real construction site, in a clear, concise and comprehensive way and to choose the better and more efficient preventive measures to solve these risky situations. In order for students to build a better understanding of the concepts contained in the course, it is important that all the concepts exposed in the theoretical lessons can be recaptured by the students in real-world construction site scenarios, for instance, by using smart phones capable of displaying rich media content.

The course scenario is divided in three modules. First, in module one, the instructor exposes all the theoretical contents stressing the importance of the real-world construction examples and the use of digital contents, existing in repositories in the web, easily accessible for students. Second, module two aims at developing a workshop based on a real construction site. Students are provided with drawings of the current real state of the building. Then, the group of students (maximum 15 people) is moved to the construction site, where the H&S risk manager guides them through the site. Students are asked to identify the existing H&S risks, and the applied or missing preventive measures which they should draw on the provided drawings. Digital contents exposed by the instructor in the theoretical lessons can again be viewed by using the smart phones which allow the owners to access their work and improve their learning outside of a normal classroom context. Additionally, students are also encouraged to take pictures of the applied or missing preventive measures to be used in a reflective session afterwards. Last, module three is aimed at collecting and sharing all the students' reflections and observations using the drawings, pictures or videos recorded during the visit.

At the end of the course, students have gone through all the theoretical concepts related to H&S management, they have been at a real construction site where the theory has been applied, and finally they are asked to assume the role of the H&S risk manager checking the security of the site. Most of the learning process can be supported by multi-platform e-learning solutions.

In contrast to the language learning scenario, learning health and safety aspects in construction engineering mainly involves:

- Learning the theory: using pre-designed units of learning about the health and safety aspects.
- Contextualised content creation: the creation of GPS annotated pictures and other learning content describing the health and safety aspect on-site.
- A reflection session in the classroom afterwards discussing the created content to learn from each other's learning content.

The dissemination of the learning theory and the reflection session could be supported by the learning network software, while the contextualised content creation is typically done with mobile devices. Three different situations are found in this scenario: the pre-visit classroom-based session, the exploration of a real construction site, and the classroom reflection after the visit. These three situations can mainly be distinguished using location and time context information.

CONCLUSION AND OUTLOOK

In this chapter, we looked at extending learning networks to include more ubiquitous, lifelong learning scenarios. Especially, we emphasised on blended and authentic learning scenarios and provided a technical framework for contextualised learning in learning networks. Furthermore, we described some tools and scenarios to illustrate how a ubiquitous learning scenario could be designed and implemented. Based on the technical and scenario realisation, a number of conclusions can be made.

The adaptation of multi-platform e-learning systems will largely depend on the ease of use and an easy integration into current day education. Moreover, these systems should provide a clear surplus value to more traditional learning scenarios; especially, the learners should see the benefits of the technology. Thus, the success of multi-platform technology for learning, in our opinion, largely depends on the provision of pedagogical models for blended learning scenarios clearly indicating the benefits of the technology use to both educators as learners. However, the success of multi-platform learning might turn out to be more learner-driven. The recent uptake of mobile devices has made access to web-based learning content, personalised information, and social networks available nearly anywhere, anytime and anyplace (Castells, et al., 2007; Rheingold, 2002). Therefore, driven by the learner demand, the current web-based e-learning systems could gradually evolve, via a combination of web-based and mobile applications, into multi-platform systems providing learner, context, and device-specific learning content.

Especially, the increasing popularity of social software like for example flickr.com, youtube.com, already illustrates that a lot of potential learners are willing to create content that can be viewed and used by others. In addition, in software as facebook.com and twitter.com learners provide a lot of information about themselves, their social peers, and their current activities. Moreover, most of this social software already is accessible via mobile devices and thus increasingly used in real-world settings far away from the desktop computer. With learners enthusiastic to create learning content for themselves and others, providing detailed personalised information, and constantly communicating their activities, highly personalised, social learning communities can be derived that also provide support in real-world settings, for instance by introducing nearby learning peers to each other, communication between learners at home and on the move, or looking at similar learning settings encountered by others in the learning community. In this paper, we gave a possible path to derive such a mobile social learning platform.

ACKNOWLEDGMENT

A part of this chapter came about through research that was sponsored by the European Project MACE, which is funded by the European Commission's eContentPlus project, ECP 2005 EDU 038098.

REFERENCES

Attewell, J., & Webster, T. (2005). Engaging and supporting mobile learners. In J. Attewell & C. Savill-Smith (Eds.), *Mobile learning anytime everywhere* (pp. 15). Learning and Skills Development Agency.

Bardram, J. E., & Hansen, T. R. (2004). The AWARE architecture: supporting context-mediated social awareness in mobile cooperation. *Proceedings of ACM conference on Computer supported cooperative work* (pp. 192-201).

Brandt, E., Björgvinsson, E., Hillgren, P.-A., Bergqvist, V., & Emilson, M. (2002, October 19-23). PDA´s, Barcodes and Video-films for Continuous Learning at an Intensive Care Unit. *Proceedings of NordiCHI.*

Brickell, G., Herrington, J., & Harper, B. (2005). Technology-supported collaboration on field-based authentic tasks. *Proceedings of EdMedia 2005 World Conference on Educational Multimedia, Hypermedia and Telecommunications* (pp. 1961-1968), Norfolk, VA AACE.

Brockett, R. G., & Hiemstra, R. (1991). *Self-direction in adult learning: perspectives on theory, research and practice.* London: Routledge.

Bruner, J. (1966). *Toward a Theory of Instruction.* Cambridge, MA: Harvard University Press.

Bruner, J. (1996). *The Culture of Education.* Cambridge, MA: Harvard University Press.

Castells, M., Fernández-Ardèvol, M., Linchuan Qiu, J., & Sey, A. (2007). *Mobile Communication and Society: A Global Perspective.* Cambridge, Massachussets: MIT Press.

Collins, A., Brown, J. S., & Newman, S. E. (1989). Cognitive Apprenticeship: Teaching the Craft of Reading, Writing, and Mathematics. In L. B. Resnick (Ed.), *Knowing, Learning and Instruction* (pp. 453 - 494). Hillsdale N.J: Lawrence Erlbaum Associates.

De Jong, T., Specht, M., & Koper, R. (2007, October 17-19). *ContextBlogger: learning by blogging in the real world.* Paper presented at the conference ePortfolio 2007., Maastricht, The Netherlands.

De Jong, T., Specht, M., & Koper, R. (2008a). Contextualised media for learning. *Educational Technology & Society, 11*(2), 41–53.

De Jong, T., Specht, M., & Koper, R. (2008b). A reference model for mobile social software for learning. [IJCEELL]. *International Journal of Continuing Engineering Education and Life-long Learning, 18*(1), 118–138. doi:10.1504/IJCEELL.2008.016079

Dey, A. K. (2001). Understanding and Using Context. *Personal and Ubiquitous Computing, 5*(1).

Drachsler, H., Hummel, H., & Koper, R. (2008). Personal recommender systems for learners in lifelong learning: requirements, techniques and model. *International Journal of Learning Technology, 3*(4), 404–423. doi:10.1504/IJLT.2008.019376

Eagle, N., & Pentland, A. (2005). Social serendipity: mobilizing social software. *Proceedings of Pervasive Computing, IEEE* (pp. 28-34). April.

Efimova, L., & Fiedler, S. (2004, March 24-26). Learning Webs: Learning In Weblog Networks. *Proceedings of Web-based communities*, Lisbon, Portugal.

Green, H., & Hannon, C. (2007). *Young people are spending their time in a space which adults find difficult to supervise or understand.* London: Demos.

Hine, N., Rentoul, R., & Specht, M. (2003, May 19-20). Collaboration and roles in remote field trips. In Proceedings of MLEARN (London, United Kingdom: Learning and Skills Development Agency) 2003, 69–72.

Kalz, M., Van Bruggen, J., Rusmann, E., Giesbers, B., & Koper, R. (2007). Positioning of learners in learning networks with content-analysis, Metadata and Ontologies. *Interactive Learning Environments, 15*(2), 191–200. doi:10.1080/10494820701424585

Klopfer, E., Squire, K., & Jenkins, H. (2002, August). Environmental Detectives: PDAs as a window into a virtual simulated world. *Proceedings of IEEE International Workshop on Wireless and Mobile Technologies in Education (WMTE'02)* (pp. 95-98).

Koper, R. (2005). *TENCompetence: Building the European Network for Lifelong Competence Development* (Project Draft). Heerlen: Open Universiteit Nederland.

Koper, R., Olivier, B., & Anderson, T. (2003). *IMS Learning Design Information Model: IMS Global Learning Consortium.*

Koper, R., & Tattersall, C. (2004). New directions for lifelong learning using network technologies. *British Journal of Educational Technology, 35*(6), 689–700. doi:10.1111/j.1467-8535.2004.00427.x

Koper, R., & Tattersall, C. (2005). *Learning Design: A Handbook on Modelling and Delivering Networked Education and Training* (1st ed.). Heerlen: Springer-Verlag.

LOM. (2002). *Institute of Electrical and Electronics Engineers Learning Technology Standards Committee: IEEE standard for learning object metadata.* IEEE Standard 1484.12.1.

Mandl, H., Gruber, H., & Renkl, A. (1995). Situiertes Lernen in multimedialen Lernumgebungen. In L. J. Issing & P. Klimsa (Eds.), *Information und Lernen mit Multimedia* (pp. 167-178). Weinheim: Psychologie Verlags Union.

Mattern, F. (2004, November). Wireless Future: Ubiquitous Computing. *Proceedings of Wireless Congress*, Munich, Germany.

Naismith, L., Lonsdale, P., Vavoula, G., & Sharples, M. (2004). *Literature Review in Mobile Technologies and Learning* (Literature Review No. 11). University of Birmingham.

Ogata, H., & Yano, Y. (2004). Context-aware support for computer-supported ubiquitous learning. *The 2nd IEEE International Workshop on Wireless and Mobile Technologies in Education (WMTE '04)* (pp. 27-34).

Paiget, J. (1970). *The Science of Education and the Psychology of the Child.* New York: Grossman.

Paredes, R. G. J., Ogata, H., Saito, N. A., Yin, C., Yano, Y., Oishi, Y., & Ueda, T. (2005). LOCH: Supporting Informal Language Learning Outside the Classroom with Handhelds. *Proceedings of IEEE International Workshop on Wireless and Mobile Technologies in Education (WMTE '05)* (pp. 182-186).

Pask, G. (1975). *Conversation, Cognition, and Learning.* New York Elsevier.

Petersen, S. A., & Divitini, M. (2005). Language Learning: From Individual Learners to Communities. *Proceedings of IEEE International Workshop on Wireless and Mobile Technologies in Education (WMTE'05)* (pp. 169-173).

Rheingold, H. (2002). *Smart Mobs: The Next Social Revolution*: Perseus Books Group.

Schön, D. A. (1983). *The Reflective Practitioner: How Professionals think in Action.* London: Maurice Temple Smith.

Schön, D. A. (1987). *Educating the Reflective Practitioner.* San Francisco: Jossey-Bass.

Sharples, M. (2003). Disruptive Devices: Mobile Technology for Conversational Learning. *International Journal of Continuing Engineering Education and Lifelong Learning, 5/6*(12), 504–520.

Sharples, M. (2007). *Big Issues in Mobile Learning: Report of a workshop by the Kaleidoscope Network of Excellence Mobile Learning Initiative*: Learning Sciences Research Institute, University of Nottingham.

Sharples, M., Lonsdale, P., Meek, J., Rudman, P. D., & Vavoula, G. N. (2007). *An Evaluation of MyArtSpace: a Mobile Learning Service for School Museum Trips.* Paper presented at the conference mLearn 2007 conference, Melbourne.

Spiro, R. J., Feltovich, P. J., Jacobson, M. J., & Coulson, R. L. (1992). Cognitive flexibility, constructivism and hypertext: Random access instruction for advanced knowledge acquisition in ill-structured domains. In T. Duffy & D. Jonassen (Eds.), *Constructivism and the Technology of Instruction.* Hillsdale, NJ: Erlbaum.

Spiro, R. J., & Jehng, J. (1990). Cognitive flexibility and hypertext: Theory and technology for the non-linear and multidimensional traversal of complex subject matter. In D. Nix & R. Spiro (Eds.), *Cognition, Education, and Multimedia.* Hillsdale, NJ: Erlbaum.

Sticht, T. G. (1975). Applications of the audread model to reading evaluation and instruction. In L. Resnick & P. Weaver (Eds.), *Theory and Practice of Early Reading* (Vol. 1). Hillsdale, NJ: Erlbaum.

Taylor, J., Sharples, M., O'Malley, C., Vavoula, G., & Waycott, J. (2006). Towards a Task Model for Mobile Learning: a Dialectical Approach. *International Journal of Learning Technology, 2*(2/3), 138–158. doi:10.1504/IJLT.2006.010616

Terrenghi, L., Specht, M., & Moritz, S. (2004). RAFT. A role-based appliance for collaborative learning. *Proceedings of 2AD:Second International Conference on Appliance Design*, Bristol, UK: 2AD.

Vavoula, G. N., & Sharples, M. (2002). KLeOS: a personal, mobile, knowledge and learning organisation system. *Proceedings of IEEE International Workshop on Wireless and Mobile Technologies (WMTE '02)* (pp. 152-156).

Wenger, E., & Lave, J. (1991). *Situated Learning: Legitimate peripheral participation*. Cambridge, New York: Cambridge University Press.

Zimmermann, A., Lorenz, A., & Oppermann, R. (2007). An Operational Definition of Context. *Proceedings of Context 07*.

Chapter 2
A Knowledge-Based Framework for E-Learning in Heterogeneous Pervasive Environments

Michele Ruta
Politecnico di Bari, Italy

Floriano Scioscia
Politecnico di Bari, Italy

Simona Colucci
Politecnico di Bari, Italy

Eugenio Di Sciascio
Politecnico di Bari, Italy

Tommaso Di Noia
Politecnico di Bari, Italy

Agnese Pinto
D.O.O.M. s.r.l., Italy

ABSTRACT

We propose a ubiquitous learning approach useful not only to acquire knowledge in the traditional educational meaning, but also to solve cross-environment everyday problems. By formalizing user request and profile through logic-based knowledge representation languages, a lightweight but semantically meaningful matchmaking process is executed in order to retrieve the most suitable learning resources. Standard formats for distribution of learning objects is extended in a backward-compatible way to support semantic annotations in our framework. Framework and algorithms are absolutely general purpose, nevertheless an application has been developed where the semantic-based Bluetooth/RFID discovery protocols devised in previous work, support users –equipped with an handheld device– to discover learning objects satisfying their needs in a given environment.

DOI: 10.4018/978-1-60566-703-4.ch002

INTRODUCTION

Pervasive e-learning has been investigated in recent research because of its evolutionary impact on the definition of traditional e-learning: learning anytime, anyhow and anywhere. The main goal is to take full advantage from the possibility of performing the knowledge acquisition process also in case of lack of fixed infrastructures. Many studies recognize the independence of the learner's physical location and the availability of powerful learning devices as the main added value of electronic learning with respect to traditional approaches (Maurer, 1998). Hence, a full exploitation of ubiquitous computing technologies can deeply affect the most significant aspects of e-learning systems.

The main issues of the so-called mobile learning (m-learning) are identified and gathered in (Sharples, 2007). Beyond achieving benefits of electronic learning, m-learning allows a higher customization of the learning experience through adaptive techniques for content provisioning and organization. From this point of view, it is important to combine the usefulness of both e-learning approach and mobility technologies within a unified vision. Pervasive and Web-based technologies should be applied together in defining frameworks and guidelines to really allow a user to learn anywhere she is. The main challenge –or opportunity, we daresay– is to enable the knowledge acquisition across contexts and environments, rather than simply exploiting handheld devices for the fruition of learning contents. Hence, there is the need to move away from "adapting" activities and approaches designed for personal computers to mobile devices and contexts. On the contrary, a comprehensive approach should be outlined, taking into account:

- the complexity of mobile scenarios: the benefit of learning ubiquitously by using a portable device is balanced by the technological constraints of such devices (limited memory capacity, reduced computational capabilities, restricted battery power, small screen size, among others);
- the different dialectic relationship learners establish in those contexts with respect to wired ones.

Flexible and context-aware discovery techniques thus become a key element to build pervasive learning infrastructures allowing a great personalization according to individual requirements, possibilities and context, also coping with the high differentiation of current mobile devices.

In spite of the growth in the diffusion of wireless-enabled handheld devices providing the necessary connectivity for complex applications, in general they are based on short range, low power technologies like Bluetooth (Bluetooth), which grant a limited interaction among hosts. Furthermore, as ubiquitous contexts are very volatile environments, some important issues are still open. Particularly, services or resources are often unavailable because the location of mobile providers can change unpredictably (Chakraborty et al., 2001). Hence, an advanced discovery of learning resources has to be flexible and decentralized, to overcome difficulties due to the host mobility.

We borrow languages and technologies from the Semantic Web vision and adapt them to pervasive contexts in order to produce a framework fully interoperable with fixed approaches as well as with accepted standards for learning contents modeling. In this paper, a coherent knowledge-based retrieval of mobile learning resources has been devised and implemented. Resources are advertised over a mobile ad-hoc environment as *learning objects* according to the LOM –Learning Object Metadata– standard (IEEE, 2002), supported by SCORM –Sharable Content Object Reference Model– specification (SCORM, 2004) for Learning Management Systems (LMS).

The learning content needs to be redesigned to meet the requirements of a mobile exploitation

(Keil-Slawik et al., 2005): in our approach learning resources have a formal characterization. Independently on the chosen syntax, learning modules are modeled as Learning Objects (LOs) according to LOM. But we propose to extend the standard to provide a semantic annotation unambiguously describing the learning object with respect to a specified ontology.

The context surrounding the learner is modeled by exploiting LIP –Learner Information Packaging– standard (LIP, 2001). Also in this case we extend LIP specification to deal with the semantic annotation of learner contextual information.

Given a learning need (user request), LOs are automatically retrieved following semantics of their descriptions. Furthermore they are ranked according to the degree of correspondence with the request. Both learning needs and resources have to be conveyed through annotations in OWL-DL (W3C, 2004). It is a formal representation language based on Description Logics (DLs) formalism (Baader et al., 2002), which allows interoperability with Semantic Web technologies and also enables a set of reasoning services. Abduction and contraction algorithms presented in (Di Noia et al., 2007) have been adapted for being performed by a mobile device.

A learner-centric perspective is adopted, providing *expertise on demand* solutions for self-training, *i.e.*, supporting a *pull* model for learning resource discovery and acquisition. Our approach is general and protocol-independent. Nevertheless it has been motivated in a cross-environment learning scenario where the semantic-enhanced Bluetooth/RFID discovery protocols presented in (Ruta et al., 2007b; Ruta et al., 2007a) are exploited as underlying interaction paradigms.

The paper is structured as follows. In the next section a motivating scenario should clarify the approach and the rationale behind it whereas in further section the proposed framework and algorithms are outlined. Finally, we comment on related work before conclusions.

MOTIVATING SCENARIO

Ubiquity, universality and efficiency are the main requirements for a knowledge-based framework aiming at supporting highly relevant and context-aware discovery and sharing of learning resources for self-training. In particular, our goal is to provide enough flexibility to support knowledge acquisition in informal and unstructured settings, in addition to more traditional and structured ones (*e.g.* real or virtual classrooms). This kind of use case can clearly show the benefits of adopting ubiquitous computing technologies in a multi-agent framework for goal-oriented knowledge acquisition and learning. The approach and algorithms are basically hardware and O.S. independent. Equipment features are taken into account when it comes to select best matching available learning resources.

Bluetooth technology is increasingly adopted in a variety of devices and appliances beyond desktop and mobile computers. This could allow exploitation of semantic-enhanced Bluetooth resource discovery protocol (Ruta et al., 2006) in many different contexts, in order to find learning modules matching with user's interests, needs and constraints. Pervasiveness is increased by embedding semantic-enhanced RFID technology into an environment (Ruta et al., 2007a). Objects can self-describe to nearby RFID-enabled mobile devices through their attached RFID tag, so becoming knowledge resources for helping the user to perform her intended task.

The proposed approach is described and motivated referring to a scenario outlining learning needs occurring to a woman, *Janet*, in typical daily activities.

In the morning, Janet is driving her newly purchased car to her workplace. She is still unfamiliar with advanced car controls. In particular, she is currently wondering how to store a station within the memory of the car radio system. She uses her Bluetooth-enabled mobile phone to discover such an information from learning modules supplied by her car's computer.

The car computer exposes the topics of the manual which can be discovered via the semantic-enhanced Bluetooth Service Discovery Protocol. Each topic is packaged as an atomic learning module, but dependencies and references between modules can be present. Each learning module is described by means of a semantic-based expression of its content and requirements for fruition. The mobile semantic matchmaker installed on the mobile phone could then perform a discovery process to find the learning resources best fitting the user's request and profile. Both are expressed in a reference ontology-based formalism in order to be matched with available LOs (whose semantic characterization follows the standards-compatible format extension outlined later on).

It is important to point out the differences between user request and profile. The request expresses the learning needs and goals of a user, whereas the profile describes her current context in terms of: background knowledge and training; time and place constraints; technological restrictions imposed by software/hardware features of the user device. Hence, the request varies with each knowledge discovery process.

The envisioned framework should support applications with both explicit and implicit user interaction paradigms. In the former case, a request is directly composed by the user and submitted to the embedded mobile matchmaking engine. In the latter case, the user implicitly triggers a support request by performing a particular interaction with elements of a smart pervasive computing environment.

The request is then built in a semi-automatic way, by interpreting the current user action and formalizing her intention into an information/knowledge need, while possibly leaving room for direct customization. On the contrary, the user profile changes with less frequency and generally in an automatic fashion, *e.g.* by updating the description of user location, characteristics of her device and the knowledge and experience she has gained.

User request and profile have their counterparts in the annotation of a learning module, in the form of description and requirements respectively. The description expresses the topics and contents of a learning resource in an unambiguous way, according to a reference ontology which models a broader discipline. On the other hand, requirements model necessary conditions for adequate fruition and comprehension of a learning module. They can concern (but are not limited to): (a) prerequisites on cultural or technical background knowledge; (b) time and location constraints for learning module fruition (*e.g.* a silent room is needed for an interactive pronunciation lesson); (c) constraints on hardware/software features for accessing a learning module (*e.g.* playing videos in a particular format with a certain minimum screen resolution). In a matchmaking session, both elements have to be taken into account. First of all, user profile must satisfy the prerequisites for fruition of a LO, otherwise knowledge acquisition cannot occur. Subsequently, the best matching descriptions with respect to user request are computed among available learning objects whose prerequisites are satisfied.

As a very small example, Janet's request can be stated as: user instructions on radio memory management for car sound system. At the same time, her profile can be modeled as: 5 minutes of available time for resource fruition, 240x320 pixel screen and support for Java ME and Flash Lite formats. Let us suppose that – among others – the following user manual topics are provided as learning objects:

A_1: user instructions on radio station memory management for Acme car sound system; length of activity is 2 minutes and format is Flash Lite.

A_2: user instructions on CD player for Acme car sound system; length of activity is 4 minutes and format is Flash Lite.

A_3: user instructions on air conditioning regulation: length of activity is 7 minutes and format is Flash Lite.

A_4: user instructions on radio station memory management; length of activity is 2 minutes and format is the old Flash one (note that it is incompatible with Flash Lite).

The detail level of descriptions reflects the "density" of learning resources for a given domain. In the previous example, single functionalities of station memory management (add, delete, modify) could be explained either within the same learning module or in separate ones. Furthermore, a car manufacturer could provide different sets of learning objects targeted to users and car electricians respectively. They would be annotated with respect to different reference ontologies. Hence, a requester could limit her search to the desired category through a preliminary ontology agreement with a provider of learning objects.

Learning objects and request are modeled in a similar way as in the previous use case, hence details are omitted for the sake of conciseness. They refer to an ontology modeling knowledge in the field of law. Differently from the previous case, the mobile device of the requester collects resources from multiple nodes. Mobile devices of co-workers are involved, as well as a Bluetooth zone server of the office, acting as a gateway toward learning material owned by the company. This use case is more similar to traditional e-learning approaches, which enable collaboration and resource sharing among learners, as well as access to a central repository of learning resources. Extensions of this use case may include semantic-based composition of learning objects to achieve an on-the-fly mobile courseware definition, given the background knowledge and the current learning needs of the user.

A semantically annotated learning object is directly associated to each painting via its RFID transponder. Janet's mobile phone comprises an RFID reader, which can access object descriptions. Let us suppose that the following works are currently in her radio range:

C_1: Renaissance oil painting with religious subject;

C_2: Renaissance oil self-portrait;

C_3: Renaissance tempera painting with mythological subject.

The mobile matchmaker embedded in the user's device matches her request with retrieved annotations and results are returned as a ranked list of learning objects. The user can select the ones she is interested in and access them. Otherwise, if she is not satisfied with results, she can start a Bluetooth-based discovery session by interrogating devices of other visitors in the same room. With respect to the previous use case, it can be noticed that no central repository of learning resources is present. A "virtual" knowledge base is built instead in a dynamic way, by accessing relevant content (*i.e.* referring to the same domain ontology as the request) provided through RFID. Moreover, a peer-to-peer user community is formed in an ad-hoc fashion in order to exchange learning resources. Such wireless ad-hoc network can be leveraged to satisfy further learning needs, beyond the original one. In the above example, after accessing the learning module about self-portrait C_2 (which is the best match for user request), Janet could search for further information about portrait as a genre across history. Even though that is not the subject of the exhibition, other visitors –who are supposedly interested in art as much as Janet– might possess and share learning material on such topic. User profiles and learning object requirements are involved into matchmaking, as already explained. This final use case shows the benefits of extending and integrating current smart identification technologies in an overall framework for knowledge discovery and mobile e-learning.

Figure 1. Basic architecture

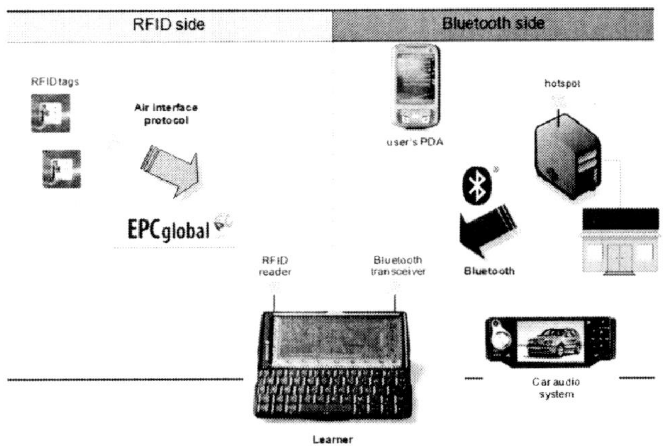

PROPOSED FRAMEWORK

In what follows we present a lightweight, hardware and O.S. independent framework to assist everywhere a learner equipped with a mobile device and wireless technologies like Bluetooth and RFID. The following subsections respectively outline the discovery architecture, the modeling approach to learning needs, learner profile and learning objects and finally the algorithms featuring the matchmaking process.

Prototype Architecture

The proposed approach and algorithms are fully reusable in several pervasive scenarios. In particular the wireless link toward a hotspot could be whatever. We have experienced and tested the framework on mobile devices equipped with Bluetooth and RFID connectivity. No specific requirements in terms of available memory and computational capabilities are mandatory. Bluetooth has been chosen as it is one of the most widespread wireless technologies. Common handheld devices usually integrate a Bluetooth connectivity because of its low cost and great diffusion.

Figure 1 shows the reference discovery architecture in terms of involved wireless technologies and protocols.

In a previous work (Ruta et al., 2007a), the EPCglobal standard for UHF tags and protocol was extended with semantic-based capabilities, while keeping full backward compatibility. An RFID tag could then store a rich description, expressed in DIG language (Bechhofer et al., 2003).

Tagged objects were then dipped into a mobile ad-hoc network, and could be dynamically discovered according to the degree of correspondence between their characteristics and a user request. As part of the solution, an effective compression algorithm was devised for semantically annotated object descriptions, in order to cope with the limited storage and transmission capabilities of RFID systems. To the best of our knowledge, this proposal represents the only framework devised specifically for pervasive RFID applications where item identification is not enough.

On the other hand EPCglobal RFID technology was also integrated at the application layer with a semantic-enhanced version of Bluetooth Service Discovery Protocol (SDP). It allows the management of both syntactic and semantic discovery of resources, by integrating a semantic layer within the OSI Bluetooth stack at service discovery level (Ruta et al., 2006). Hence, the standard was enriched by new functionalities which allow to maintain a backward compatibility (handheld device connectivity), but also to add the support to matchmaking of semantically annotated resources. To implement matchmaking and ontology support features, we introduced a Semantic Service Discovery functionality into the stack, slightly modifying the existing Bluetooth discovery protocol. In what follows, some basic features of both the Bluetooth and RFID side of the proposed architecture will be given.

We associated unused classes of 128 bit UUIDs in the original Bluetooth standard to mark each specific ontology calling this identifier OUUID (Ontology Universally Unique IDentifier). By means of OUUID matching, the context was identified and a preliminary selection of resource referring to the same request's ontology was performed. Within the environment, we assume each resource is semantically annotated and a description of the resource itself is available on the hotspot as a database record labeled with a unique 32-bit identifier. Each record contains general information about a single semantic enabled resource and it entirely consists of a list of resource attributes. In addition to the OUUID attribute, there are: *ResourceName* (a human–readable name for the resource), *ResourceDescription* (the resource description expressed using DIG syntax) and a variable number of *ResourceUtilityAttr_i* attributes, *i.e.*, numerical values used according to specific applications.

In general, they are associated to context-aware attributes of a resource; in the current implementation they are not exploited.

In (Ruta et al., 2006), by adding four SDP PDUs *SDP_OntologySearch* (request and response) and *SDP_SemanticServiceSearch* (request and response) to the original standard (exploiting not used PDU ID), together with the original SDP capabilities, further semantic enabled discovery functionalities had been introduced. The overall interaction was based on the original SDP in Bluetooth. No modifications were made to the original structure of transactions. There was just a different use of the SDP framework.

Similarly to the Bluetooth SDP enhancement, we add semantic based functionalities to an RFID application infrastructure. In our framework we refer to RFID transponders agreeing to the EPC standard for class I - second generation UHF tags. Memory of a EPC class I Generation-2 UHF RFID tag is divided in four logical banks: the *Reserved*, the *Electronic Product Code (EPC)*, the *Tag IDentification (TID)* and the *User* ones. EPCglobal Generation-2 UHF RFID air interface protocol is an Interrogator-Talks-First (ITF) protocol: tags only reply to reader commands.

An RFID reader can preselect a subset of the tag population currently in range, according to user-defined criteria, by means of a sequence of *Select* commands. With a *Select*, a bit string is sent to all tags in range. Each tag will compare it to a memory area specified by the reader, then it will set/reset one of its status flags according to the comparison result (match/no-match).

After this phase, the inventory loop begins. In each iteration the reader isolates one tag in range, reads its EPC code and has the opportunity to access its memory content.

Among the other available commands, only *Read* and *Write* ones are relevant for our purposes.

Read command allows to read from one of the four tag memory banks. *Write* command allows a reader to write a 16-bit word to one of the four tag memory banks.

In the proposed approach, we use two reserved bits in the EPC area within each tag memory. The first one –at 15_{hex} (10101_2) address– is exploited to indicate if the tag has a user memory (bit set) or not (bit reset). The second one –at 16_{hex} address– is set to mark semantic enabled tags. In this way, by means of a *Select* command, a reader can easily distinguish semantic based tags from current ones.

The EPC standard for UHF - class I tags imposes the content of TID memory up to $1F_{hex}$ bit is fixed. Optional information could be stored in the further additional TID memory. Generally these information are serial numbers or manufacturer data. Hence we use the TID memory area starting from 00100000_2 address. In that area we store the identifier of the ontology w.r.t. the description contained within the tag is expressed. Making the ontology support system proposed for the semantic based SDP in Bluetooth compliant with RFID systems, we set a bidirectional correspondence among OUUIDs stored in RFID transponders and those managed by Bluetooth devices. Hence we adopt a 128 bit structure for the RFID OUUIDs analogous to the one outlined before. Finally, in order to retrieve the OUUID value stored within a tag, a reader will exploit a *Read* command by adopting proper parameters.

The semantically annotated description of the good the tag is clung to is stored within the User memory bank. It is expressed in DIG formalism but, due to the strict amount of memory available, this annotation has to be compressed. For the sake of brevity, here we omit characteristics of the adopted encoding tool.

The extraction and the storing of a description carried out on a tag, can be performed by a reader by means of one or more *Read* or *Write* commands. Both commands are obviously compliant with the RFID air interface protocol.

Re-Conceptualizing Mobile LOs Structure

Learning approaches ask for a deep re-conceptualization in order to cope with pervasive environments (Keil-Slawik et al., 2005). In particular, learning content needs to be redesigned to meet the requirements of ubiquitous computing (Chan et al., 2004; Grasso & Roselli, 2005).

Even though mobile learning is often considered more suitable to informal learning (Sharples, 2007), we propose an integrated approach supporting the fruition of learning resources with both informal and formal content, in order to fit an everyday life scenario.

Independently on the format chosen for the learning modules, they have to be modeled according to LOM standard (IEEE, 2002) as Learning Objects (*i.e.* entities, digital or not, that may be used for learning, education or training). The fact of being conform to LOM standard also ensures interoperability with SCORM specification (SCORM, 2004) for LMSs.

Together with LOM, among basic references of the proposed approach, we have to take into account also the approach of Chan et al. (Chan et al., 2004), which proposed a standard for educational metadata for m-learning called MLM. It extends LOM rights category to also cope with validation and location restrictions needed for LOs bringing formal contents. In the same paper also the LIP standard (LIP, 2001) was extended to model learner and environment contextual settings.

We propose an extension of LOM and LIP standards for modeling mobile LOs content, learning context and learner features. Given the learner centric perspective we take, first of all there is the need to model all the significant contextual information about the learner. Crucial features to convey are reported in what follows:

- **Learning Need:** knowledge request originating the further process of learning resources retrieval.
- **Time:** interval the user prefers to spend in the learning activity she is asking for.
- **Equipment:** technological constraints of the learner, that is the technological features of the handheld device the learner holds.
- **Profile:** personal and cultural features of the learner.

Previous characteristics are highly depending on the context and they need to be analyzed every time a learning need arises and a learning process takes place. In particular, the learning need and the time availability have to be specified at each request, while the learner equipment and profile

Table 1. Reference LIP data structures

Data Structure	Description	Matchmaking information
Identification	Biographic and demographic data	Structural Elements of the *Profile*: Age, Language, Geographical Position
Goal	Learning, career and other objectives and aspirations	*Learning Need*: DL Description of the learning objective *Time*: Preferred Time availability for Learning
Qualifications, Certifications and Licenses	Qualifications, certifications and licenses granted by recognized authorities	Structural Elements of the *Profile*: learner formal qualifications
Activity	Any learning-related activity in any state of completion	/
Transcript	A record to provide an institution-based summary of academic achievement.	/
Interest	Hobbies and recreational activities	/
Competency	Skills, knowledge and abilities acquired in the cognitive, affective, and/or psychomotor domains	Cultural Elements of the *Profile*: DL description of the background knowledge
Affiliation	Membership of professional organizations	
Accessibility	General accessibility to the learner information as defined through language capabilities, disabilities, eligibilities and learning preferences including cognitive, physical and technological preferences	*Equipment*: learner technological constraints
Security key	Set of passwords and security keys assigned to the learner for transactions with learning information systems and services	/
Relationship	The set of relationships between the core components.	/

may be considered unchanged until the learner does not explicitly update them.

The retrieval process consists of finding resources satisfying the learning need as much as possible, among those learnable in the available time, through the mobile device at hand and compatible with the learner profile.

In the proposed approach, we model the learning request according to a subset of DLs and we exploit DLs also to model the cultural component of the learner profile as well as structural information, technological and time constraints, used to refine the discovery of learning resources in a given environment.

LIP standard makes the data structures in Table 1 available for storing learner information. In the third column we added significant information exploited in the semantic-based matchmaking.

In the learning resources discovery, we take into account only information stored in Identification, Goal, Competency and Accessibility data structures. We model the learning need as a DL description in a *Goal* data structure, which also embeds the time availability information. The learner profile is conveyed through *Identification*, *Competency* and *Accessibility* data structures: in particular, the technological preferences are stored in the *Accessibility* component whereas the cultural profile is a DL description within the *Competency* structure. Once the learner information is conveyed through LIP standard data structures, it can be automatically exploited as input of a retrieval process of learning resources described according to the LOM standard.

The LOM base schema describes LOs following the data element categories shown in Table 2. In the third column we added significant information exploited in the semantic-based matchmaking.

Table 2. Reference LOM categories

Category	Description	Matchmaking information
General	Learning object description as a whole	*Learning Content*: DL description of the LO content Structural Elements of the *Profile of the intended learner*: Language, Geographical Position
Lifecycle	History evolution and current state of the LO	/
Meta-metadata	Information about the metadata instance itself	/
Technical	Technical requirements and characteristics of the LO	*Equipment*: Technological Requirements for LO fruition *Time*: intended duration of the learning process
Educational	Educational and pedagogic characteristics of the LO	Structural Elements of the *Profile of the intended learner*: age range and formal qualifications Cultural Elements of the *Profile*: DL description of the required background Knowledge
Rights	Intellectual property rights and usage conditions for the LO	/
Relation	Relationship between a LO and other related Los	/
Annotation	Comments on the educational usage of the LO (information about when and by whom the comments were created)	/
Classification	Description of the LO w.r.t. a particular classification system	/

The *General* category is exploited both for modeling the learning resource content and important information like language and geographical coverage for the fruition according to DL formalism. The *Technical* category includes all the requirements for the fruition of the learning resource –in terms of software and hardware constraints– and the learning duration. The *Educational* category carries out important information like the age range and role of the user, together with a description of competences required for the fruition.

Matchmaking Algorithms

From now on we assume that learner request and profile as well as learning resources are annotated in a language whose semantics can be mapped to DL $\mathcal{ALN}(D)$, for instance (a subset of) OWL DL or the more compact XML-based DIG language. Formulas (concepts) in $\mathcal{ALN}(D)$ we use to represent user profile and learning request, are built according to the following rules where *CN* is a concept name, see Figure 2.

For what concerns the ontology $\mathcal{T}$ (Terminological Box in DL-words) we only allow relations between concept names in the forms, to represent respectively (1) subclass axioms; (2) equivalence axioms; (3) disjoint axioms, see Figure 3.

Furthermore, given a concept name *CN* we cannot have more than one equivalence axiom, with *CN* on the Left Hand Side (LHS) and if *CN* appears on the LHS of an equivalence axiom then it cannot appear on the LHS neither of a

Figure 2. Rules

$$C, D \rightarrow CN \mid \neg CN \mid \forall R.C \mid C \sqcap D \mid (\geq nR) \mid (\leq mR) \mid \geq_k g \mid \leq_k g$$

Figure 3. Concept names

$$CN_1 \sqsubseteq CN_2 \sqcap ... \sqcap CN_n \quad (1)$$

$$CN_1 \equiv CN_2 \sqcap ... \sqcap CN_n \quad (2)$$

$$CN_1 \sqsubseteq \neg CN_2 \sqcap ... \sqcap \neg CN_n \quad (3)$$

subclass axiom nor of a disjoint axiom. In order to avoid cycles within an ontology, we do not allow a concept name *CN* to appear, directly or indirectly, both on the LHS and on the right hand side of an axiom (Baader et al, 2002). Since the above conditions allow for representing concept taxonomies in a formal way, we will call such an ontology *formal-taxonomy*.

DL-based systems usually provide two basic reasoning services for $\mathcal{T}$, namely (a) Satisfiability and (b) Subsumption in order to check (a) if a formula C is consistent with respect to the ontology ($\mathcal{T} \not\models C \sqsubseteq \bot$) or (b) if a formula C is more specific or equivalent to a formula D ($\mathcal{T} \models C \sqsubseteq D$). Both Subsumption and Satisfiability are adequate in all those scenarios where an exact (yes/no) retrieval is required. For example, given a LO description and a Learning Need, respectively represented by LO and LN, we are able to determine whether they are compatible or not, i.e. whether LO models information which is not in conflict with the one modeled by LN or not. This task can be easily performed checking if $\mathcal{T} \not\models LO \sqcap LN \sqsubseteq \bot$ holds or not. On the other hand, using Subsumption we can verify if a LO satisfies a request LN. It is easy understandable that if the relation $\mathcal{T} \models LO \sqsubseteq LN$ holds, then LO results more specific than LN and contains at least all the requested features.

In (Colucci et al., 2003; Di Noia et al., 2003b) Concept Contraction and Concept Abduction, non-standard inference services for DLs, were introduced and defined for matchmaking scenarios. In the following subsections we briefly recall their definitions, explaining their rationale and the need for them in the proposed m-learning framework.

Concept Contraction and Concept Abduction

Starting with a LO description and a learning need LN, if their conjunction $LO \sqcap LN$ is unsatisfiable with respect to the ontology $\mathcal{T}$, i.e., they are not compatible with each other, our aim is to explain which part of the request LN is conflicting with the resource LO. If we retract conflicting requirements in LN, *G* (for Give up), we obtain a concept *K* (for Keep), representing a contracted version of the original request, such that $K \sqcap LO$ is satisfiable with respect to $\mathcal{T}$. In other words, *G* represents "why" $LN \sqcap LO$ are not compatible.

Definition 1. *Let $\mathcal{L}$ be a DL,* LO, LN *be two concepts in $\mathcal{L}$, and $\mathcal{T}$ be a set of axioms in $\mathcal{L}$, where both* LO *and* LN *are satisfiable in $\mathcal{T}$. A Concept Contraction Problem (CCP) is finding a pair of concepts $\langle G, K\rangle \in \mathcal{L} \times \mathcal{L}$ such that $\mathcal{T} \models LN \equiv G \sqcap K$, and $\mathcal{T} \not\models K \sqcap LO \sqsubseteq \bot$. We call K a contraction of* LN *according to* LO *and $\mathcal{T}$.*

We note that there is always the trivial solution $\langle G, K\rangle = \langle LN, \top\rangle$ to a CCP (with $\top$ we refer to the most generic concept in an ontology. We have instead used $\bot$ to denote the most specific concept, the unsatisfiable concept). This solution corresponds to the most drastic contraction, that gives up everything of LN. In our m-learning framework, it models the (infrequent) situation where, in front of some very interesting resource LO, incompatible with the requested one, a user just gives up completely his/her requirements LN in order to meet LO. On the other hand, when $LO \sqcap LN$ is satisfiable in $\mathcal{T}$, the "best" possible solution is $\langle \top, LN\rangle$, that is, give up nothing –if possible.

Since usually one wants to give up as few things

as possible, some minimality in the contraction must be defined (Gäardenfors, 1988; Colucci et al., 2003; Di Noia et al.,2004). Whenever the LO description and the request LN are compatible with each other, the partial specifications problem might still hold. That is, it could be the case that LO does not imply LN – though compatible with it.

Using DL syntax we write: $\mathcal{T} \not\models$ LO $\sqcap$ LN $\sqsubseteq \bot$ and $\mathcal{T} \not\models$ LO $\sqsubseteq$ LN. Then, it is necessary to assess what should be hypothesized (H) in LO in order to completely satisfy LN.

Definition 2. *Let $\mathcal{L}$ be a DL,* LO, LN, *be two concepts in $\mathcal{L}$, and $\mathcal{T}$ be a set of axioms in $\mathcal{L}$, where both* LO *and* LN *are satisfiable in $\mathcal{T}$. A Concept Abduction Problem (CAP) is finding a concept H $\in \mathcal{L}$ such that $\mathcal{T} \models$* LO $\sqcap$ *H* $\sqsubseteq$ LN, *and moreover LO $\sqcap$ H is satisfiable in $\mathcal{T}$. We call H a hypothesis about* LO *according to* LN *and $\mathcal{T}$.*

Note that if $\mathcal{T} \models$ LO $\sqsubseteq$ LN then we have $H = \top$ as a solution to the related CAP. Given LN and LO such that $\mathcal{T} \not\models$ LO $\sqsubseteq$ LN, H represents "why" the subsumption relation does not hold. H can be interpreted as what is requested in LN and not specified in LO. Hence, both Concept Abduction and Concept Contraction can be used for explanation of subsumption and satisfiability respectively.

Performing the Concept Contraction in $\mathcal{ALN}(\mathcal{D})$

An algorithm to solve CAPs for $\mathcal{ALN}$ has been proposed in (Di Noia et al., 2007) and it can be easily extended to deal with $\mathcal{ALN}(\mathcal{D})$. In this section we outline a novel algorithm to compute a possible solution to CCPs in $\mathcal{ALN}(\mathcal{D})$ given two concepts LO, LN and an ontology T built according to previous guidelines. Before computing solutions to a CCP, it is more convenient, from a computational perspective, to reduce both LO and LN to a common normal form. We use here a well know technique called *unfolding* (Baader et al, 2002) to syntactically transform two concept and preserve their formal semantics with respect to the ontology $\mathcal{T}$. Given a concept C the normalization process is performed applying recursively the following rewriting rules to each occurrence of the element appearing in the LHS of the rule. (See Figure 4)

Given a concept C and an ontology $\mathcal{T}$, we call *norm*(C, $\mathcal{T}$) the rewriting of C according to the above rules. (See Figure 5)

Logic-Based Matchmaking

In real u-learning scenarios, it is quite rare to find exactly the resource we are looking for. Often we have to reformulate a request in order to obtain satisfactory results in an approximate search. At this point a question arises: *what* should we change? Some suggestions would be useful. Both Concept Abduction and Concept Contraction can be used to suggest guidelines on what, given an offered resource LO, has to be revised and/or hypothesized to obtain a full match with the preference.

We now show how the previous inferences can help in an approximate semantic-based resource discovery, fully exploiting their structured description. Let us suppose to have a learning need LN, a resource LO and an ontology $\mathcal{T}$ such that $\mathcal{T} \square$ LN $\sqcap$ LO $\sqsubseteq \bot$, *i.e.* they are incompatible. In order to gain compatibility, a Concept Contraction is needed so that giving up G in LN, the remaining K could be satisfied by LO. Now, if $\mathcal{T} \not\models$ LO $\sqsubseteq K$, the solution H_K to the CAP $\langle \mathcal{L}, K,$ LO, $\mathcal{T} \rangle$ represents what is in K and is not specified in LO. (See Figure 6)

[lines 1-4]. Having a request LN and an offered service LO, if the conjunction of their descriptions is not satisfiable with respect to the ontology they refer to (*i.e.,* they are not compatible with each other for some concepts in their descriptions), first a contraction on LN is performed in order to regain compatibility **[line 2]** and then what has to be hypothesized in LO in order to completely satisfy LN (its contraction) is computed **[line 3]**. The returned values represent:

Figure 4.

$$
\begin{array}{ll}
\mathtt{CN}_1 \rightarrowtail \mathtt{CN}_1 \sqcap \mathtt{CN}_2 \sqcap \ldots \sqcap \mathtt{CN}_n & \mathtt{if}\ \mathtt{CN}_1 \sqsubseteq \mathtt{CN}_2 \sqcap \ldots \sqcap \mathtt{CN}_n \in \mathcal{T} \\
\mathtt{CN}_1 \rightarrowtail \mathtt{CN}_2 \sqcap \mathtt{CN}_3 \sqcap \ldots \sqcap \mathtt{CN}_n & \mathtt{if}\ \mathtt{CN}_1 \equiv \mathtt{CN}_2 \sqcap \ldots \sqcap \mathtt{CN}_n \in \mathcal{T} \\
\mathtt{CN}_1 \rightarrowtail \mathtt{CN}_1 \sqcap \neg\mathtt{CN}_2 \sqcap \ldots \sqcap \neg\mathtt{CN}_n & \mathtt{if}\ \mathtt{CN}_1 \sqsubseteq \neg\mathtt{CN}_2 \sqcap \ldots \sqcap \neg\mathtt{CN}_n \in \mathcal{T} \\
\mathtt{C} \sqcap \bot \rightarrowtail \bot & \\
(\geq \mathtt{nR}) \sqcap (\leq \mathtt{mR}) \rightarrowtail \bot & \mathtt{if}\ \mathtt{n} > \mathtt{m} \\
\mathtt{A} \sqcap \neg\mathtt{A} \rightarrowtail \bot & \\
(\geq \mathtt{nR}) \sqcap (\geq \mathtt{nR}) \rightarrowtail (\geq \mathtt{nR}) & \mathtt{if}\ \mathtt{n} > \mathtt{m} \\
(\leq \mathtt{nR}) \sqcap (\leq \mathtt{mR}) \rightarrowtail (\leq \mathtt{nR}) & \mathtt{if}\ \mathtt{n} < \mathtt{m} \\
\forall \mathtt{R.D}_1 \sqcap \forall \mathtt{R.D}_2 \rightarrowtail \forall \mathtt{R.}(\mathtt{D}_1 \sqcap \mathtt{D}_2) & \\
\forall \mathtt{R.}\bot \rightarrowtail \forall \mathtt{R.}\bot \sqcap (\leq \mathtt{0R}) &
\end{array}
$$

- $\langle G, K\rangle$: the first item is what has to be given up in the request –*G*– in order to continue the process, or, in other words, why LN is not compatible with LO. The second item is the contracted request *K* that is no more in conflict with the request.
- H_K: after the contraction of LN, the request is represented by *K*, *i.e.*, the portion of LN that is compatible with LO. H_K represents what has to be hypothesized in LO in order to completely satisfy *K*, or, in other words, why LO does not completely satisfy *K*.

[lines 5-7]. If the conjunction of LN's and LO's description is satisfiable with respect to the ontology they refer to, then no contraction is needed and only an abductive process is carried out. Notice that $H = abduce(\mathtt{LO}, \mathtt{LN}, \mathcal{T})$ **[lines 3,6]** determines a solution *H* for the CAP $\langle \mathcal{L}, \mathtt{LN}, \mathtt{LO}, \mathcal{T}\rangle$, while $\langle G, K\rangle = contract(\mathtt{LO}, \mathtt{LN}, \mathcal{T})$ **[line 2]** determines a solution $\langle G, K\rangle$ for the CCP $\langle \mathcal{L}, \mathtt{LN}, \mathtt{LO}, \mathcal{T}\rangle$.

As the obtained LO is an approximated match of LN, then evaluating how good is the approximation would be extremely useful. Given more than one resource, which is the best approximation? How can a numerical score be assigned, based on *K*, *H* and *G*, to the approximation in order to rank the learning resources? The algorithm *explain* does not depend on the particular DL adopted. Based on the minimality criteria proposed in (Colucci et al., 2003), the length *H* of the solution to a CAP for $\mathcal{ALN}(\mathcal{D})$ can be computed in a similar way as the one proposed in (Di Noia et al., 2003a). Hence, a relevance ranking score can be computed by a utility function defined as $U(G, K, H_K)$.

Dealing with User Preferences

In a matchmaking process, a user request can be often split into two separate parts: **strict** requirements and **preferences**. *Strict* requirements represent what, in the request, has to be strictly matched by the retrieved resource descriptions. *Preferences* can be seen as soft user requirements. In our scenario, the user will accept even a LO whose description does not represent exactly what she prefers. Usually, a weight is associated to each preference in order to represent its worth (absolute or relative to the other preferences).

Figure 5. Algorithm 1--Concept Contraction

```
Algorithm: Contract (ALN(D), LO, LN, T)
Input: ⟨LO, LN, T⟩ where T is a formal-taxonomy and ©LO, LN™ χ ALN(D)
Output: ⟨G, K⟩ with concept G, K χ ALN(D)
if LN = ζ then
    return ⟨⊤, ζ⟩;
else
    G := ⊤;
    K := ⊤ ⊓ LN;
    for each concept name A in K do
        for each concept name A' ∈ norm(A, T) do
            if there exists B in LO such that B = ¬ A' then
                G := G ⊓ A;
                remove A from K;
            end if
        end for
    end for
    for each concept (≥ x R) in K do
        if there exists (≤ y R) in LO and y < x then
            replace (≥ x R) with (≥ y R);
            G := G ⊓ (≥ x R);
        end if
        for each concept ∀ R.E in K do
            if there exists ∀ R.F in LO then
                ⟨G', K'⟩:= contract(ALN(D), E, F, T);
                G := G ⊓ ∀ R.G';
                replace ∀ R.E in K with ∀ R.K';
            end if
        end for
    end for
    for each concept (≤ x R) in K do
        if there exists (≥ y R) in LO and y > x then
            replace (≤ x R) with (≤ y R);
            G := G ⊓ (≤ x R);
        end if
    end for
    for each concept ≥ₓ g in K do
        if there exists ≤ᵧ g in LO and y < x then
            replace ≥ₓ g with ≥ᵧ g;
            G := G ⊓ ≥ₓ g;
        end if
    end for
    for each concept ≤ₓ g in K do
        if exists ≥ᵧ g in LO and y > x then
            replace ≤ₓ g with ≤ᵧ g;
            G := G ⊓ ≤ₓ g;
        end if
    end for
end if
return ⟨G, K⟩;
```

Figure 6. Algorithm 2--Results explanation algorithm

Algorithm: *explain* ($\mathcal{L}$, LO, LN, $\mathcal{T}$)

Input: LO, LN concepts in $\mathcal{L}$ such that $\mathcal{T} \models$ LO and $\mathcal{T} \models$ LN

Output: ©*G, K*™, *H, i.e.,* the part of LN that should be respectively retracted (*G*) and kept (*K*) and the part of LO that should be hypothesized (*H*) to find a full match between LO and LN.

```
if T ⊨ LN ⊓ LO ⊑ ⊥ then
    ⟨G, K⟩ := contract(LO, LN, T);
    H_K := abduce(LO, K, T);
    return ⟨G, K⟩, H_K;
else
    H := abduce(LO, LN, T);
RETURN ©T, LN™, H;
```

Hence, for a learning need LN we distinguish between a concept $\mathrm{LN_S}$ representing strict requirements and a set of weighted concepts LN, *v*⟩ where LN is a DL concept and *v* is a numerical value representing preference worth. It should be clear that a matchmaking process has not to be performed with respect to $\mathrm{LN_S}$. It represents what the user is not willing to risk on at all. He does not want to hypothesize nothing on it. An approximate solution would not be significant for $\mathrm{LN_S}$. Actually, performing a matchmaking process between preferences and a LO description makes more sense. After all, preferences represent what the user *would like* to be satisfied by LO. Hence, even though a preference is satisfied *with a certain degree* (not necessarily completely) the user will be satisfied *with a certain degree* as well.

Given an ontology $\mathcal{T}$, a LO description, a strict requirement $\mathrm{LN_S}$ and a set of preferences $\Pi = \{\langle \mathrm{LN}_i, v_i\rangle\}$ we compute a global ranking score using the following Algorithm 3. Here we retrieve, *i.e.* assign *score* greater than zero, only LOs whose description fully satisfies user strict requirements. Once we have a resource such that $\mathcal{T} \models \mathrm{LO} \sqsubseteq \mathrm{LN_S}$, then we compute how much it satisfies user preferences. For each preference we take into account both *U*(*explain* (LO, LN_i, $\mathcal{T}$, $\mathcal{L}$)) *i.e.,* the similarity degree computed using non-standard reasoning, and the value expressed by the user to represent preference worth. (See Figure 7)

Illustrative Example

A very small example can further clarify the approach and the rationale behind it. A request for car radio instructions with diagrams may be modeled as shown as Figure 8.

At the same time, user profile can be modeled as shown as Figure 9.

This expresses a maximum duration of 5 minutes for resource fruition, as well as restrictions on screen resolution and content format. Let us suppose that – among others – the following user manual topics are provided as learning objects, see Figure 10.

The list of results is arranged according to an overall match score from 0 to 100. It is computed by means of the formula:

$$s = 100\% \begin{cases} 1 - p & p > T \\ 0 & otherwise \end{cases}$$

where the semantic penalty function *p* is computed as:

Figure 7. Algorithm3--Preference retrieval

Algorithm: *preference retrieve* (LO, LN_S, $\mathcal{P}$, $\mathcal{T}$, $\mathcal{L}$)
Input: LO, LN concepts in $\mathcal{L}$ such that $\mathcal{T} \models$ LO and $\mathcal{T} \models$ LN
Output: score
1: score := 0 ;
2: **if** $\mathcal{T} \models \text{LO} \sqsubseteq \text{LN}_S$ **then**
3: **for each** ©D_i, v_i ™ χ $\mathcal{P}$ **do**
4: score := score + $v_i \cdot U$ (*explain* (LO, LN_i $\mathcal{T}$, $\mathcal{L}$));
5: **end for**
6: **end if**
7: **return** score;

Figure 8.

$$\text{UserInstructions} \sqcap \geq 1\ \text{HasDiagrams} \sqcap \exists\text{HasReferenceFunction} \sqcap \forall\ \text{hasReferenceFunction.CarRadio} \sqcap \exists\text{hasTopic} \sqcap \forall\ \text{hasTopic.Memory}$$

$$p = W \cdot contract + (1 - W) \cdot abduce$$

where *contract* is the penalty calculated by the contraction procedure between the local user's request and the learning object description, while *abduce* is the penalty value of the abduction procedure between the consistent part *K* of the request and description. The scoring mechanism is regulated by two user-adjustable parameters, the threshold value *T* and the weight *W*, both between 0 and 1. *T* influences the sensitivity of the discovery while *W* determines the relative weight of explicitly conflicting elements in the description of the learning object with respect to the demand.

After matchmaking the user is presented with the ranked list. In our example, with *W*=0.7 and *T*=0.6 outcomes are as reported in Figure 11. A_4 is immediately discarded because it is incompatible with the learner's technical requirements; it will not be shown among available LOs. On the contrary, A_3 will be shown among results, but its high penalty (due to distance in content from the request) makes it fall below the threshold and take a zero score. A_2 is close in content but not a full match (learner has to give up some of her requirements), whereas A_1 is a nearly exact match.

Figure 9.

$$\leq_{\text{minutes}} 5 \sqcap \leq_{\text{h\_display\_px}} 240 \sqcap \leq_{\text{v\_display\_px}} 320 \sqcap \forall\ \text{hasFormat}\ (\text{FlashLite} \sqcap \text{JavaME})$$

Figure 10.

A_1	*Radio station memory management, illustrated in Flash Lite format*
Description	UserInstructions ⊓ ∃hasReferenceFunction ⊓ ∀hasReferenceFunction.(CarRadio ⊓ ∃hasMaker ⊓ ∀hasMaker.Acme) ⊓ ∃hasTopic ⊓ ∀ hasTopic.Memory
Requirements	$\geq_{minutes}$ 2 ⊓ ∀hasFormat.FlashLite

A_2	*Car CD player, illustrated in Flash Lite format*
Description	UserInstructions ⊓ ∃hasReferenceFunction ⊓ ∀hasReferenceFunction.(CarRadio ⊓ ∃hasMaker ⊓ ∀hasMaker.Acme) ⊓ ∃hasTopic ⊓ ∀hasTopic.CdPlayer
Requirements	$\geq_{minutes}$ 4 ⊓ ∀hasFormat.FlashLite

A_3	*Car air conditioning regulation with diagrams, illustrated in Flash Lite format*
Description	UserInstructions ⊓ =3 hasDiagrams ⊓ ∃hasReferenceFunction ⊓ ∀hasReferenceFunction.AirConditioning ⊓ ∃hasTopic ⊓ ∀hasTopic.Regulation
Requirements	$\geq_{minutes}$ 7 ⊓ ∀hasFormat.FlashLite

A_4	*Radio station memory management, illustrated in old Flash format and incompatible with Flash Lite*
Description	UserInstructions ⊓ ∃hasReferenceFunction ⊓ ∀hasReferenceFunction.(CarRadio ⊓ ∃hasMaker ⊓ ∀hasMaker.Acme) ⊓ ∃hasTopic ⊓ ∀hasTopic.Memory
Requirements	$\geq_{minutes}$ 2 ⊓ ∀hasFormat.(Flash ⊓ ¬FlashLite)

RELATED WORK

In the literature there are several approaches trying to exploit mobile technologies to assist learners in the real world and in everyday life.

(Yang, 2006) proposed a ubiquitous learning environment which uses semantics to model both learner and resource profiles. It aims at identifying – in a real world context – the right contents, collaborators and services that can help or interest the occasional learner. In that approach every learner becomes a peer, so every user can barter resources, information and help with whoever has the same interest in the same moment and in the same place. This is an interesting P2P collaborative approach, but learners can retrieve or share only non-standard resource types. A major requirement of our proposal is to preserve compatibility with standard e-learning technologies as much as possible.

With the same intention, (Chan et al., 2004) presented Mobile Learning Metadata (MLM), an educational metadata for mobile learning system. The authors enhanced existing standards and

Figure 11. Outcomes of semantic matchmaking and utility function

Supply	**A_1**	**A_2**	**A_3**	**A_4**
Requirement Give Up (G_R)	⊤	⊤	⊤	∀hasFormat.(FlashLite ⊓ JavaME)
Preference Give Up (G_P)	⊤	∀hasTopic.Memory	∀hasReferenceFunction.CarRadio ⊓ ∀hasTopic.Memory	n.a.
Preference Keep (K)	UserInstructions ⊓ ∃hasReferenceFunction ⊓ ∀hasReferenceFunction.CarRadio ⊓ ∃hasTopic ⊓ ∀hasTopic.Memory	UserInstructions ⊓ ∃hasReferenceFunction ⊓ ∀hasReferenceFunction.CarRadio ⊓ ∃hasTopic	UserInstructions	n.a.
Preference Hypothesis (H)	≥1 hasDiagrams	≥1 hasDiagrams	⊤	n.a.
contract	0	0.3	0.6	n.a.
abduce	0.1	0.1	0	n.a.
p	0.03	0.25	0.42	n.a.
s	97%	75%	0	n.a.

specifications for modeling learning objects, in order to support mobile and informal learning. In particular, they amended and added fields for managing both the access rights and the history of the learner. With our approach, instead, we are able to manage Mobile Learning Objects (MLO) according to current standards also allowing to semantically annotate learning contents and learner profiles. This enables a principled matchmaking for ranking available resources.

In (Castillo & Ayala, 2008) a computational model for Mobile Learning Objects and an architecture for mobile learning environment with MLOs is outlined. They are a generalization of Mobile Interactive Learning Object (MILO) introduced by (Holzinger et al., 2005) to model knowledge for fruition in mobile environments. The authors stated that, in mobile applications, some characteristics take a particular relevance when designing learning objects: among others, the availability of an MLO, the interest of other learners (to promote collaboration), the needed experience level. Hence they proposed a three-part model: the first one adapts learning contents to learner's needs according to context and location; the second one is a collaboration model the learner can construct by annotating comments and sharing them within the learning community; the third one is a personalized model where interests and capabilities of the learner are considered. Upon this model, a multimedia-based application is designed and presented. An alternative approach to the definition of learning objects is given by the same authors in (Holzinger et al., 2006).

(Hwang et al., 2008) proposed criteria and strategies to establish a context-aware ubiquitous learning environment. Exploiting parameters referred to wired scenarios and related to personal and environmental data, they proposed twelve u-learning contextual models to assess learning performance of the students based on their real-world behavior. Basically, those strategies represent specific kinds of interaction between a learner and the system, based on different possible situations. The main innovation of that proposal is in the support level offered for learning activities, but the relevance of resource discovery seems to be underestimated.

In (El-Bishouty et al., 2007) a client-server application, useful to assist learner in real-world life is presented. RFID technology is adopted to detect learner position, while so-called Environmental Objects (EO) allow the system to map the physical space. When the user is in front of an EO, the system provides learning material related to that object, and suggests where are the nearest users, so that a message can be sent to them to ask for collaboration.

A further approach targeted at supporting language learning in real-life situations is presented in (Ogata, 2008). A ubiquitous learning environment is build to support vocabulary learning through RFID and language learning through a sensor network; GPS is exploited to detect the user location. Furthermore, a solution to record and reuse knowledge in real-world problems was proposed. A problem-solving procedure or action is linked by an RFID code within a tag. Keyword-based discovery is the main drawback of that approach; it limits matches to exact code-based correspondences. RFID technology is also used as a mere link to records in a data store.

(Chen et al., 2008) proposed a ubiquitous learning system, exploiting wireless technologies to aid in text composition in a context-aware fashion. The system assists students providing relevant information. It helps to observe and perceive the environment or to collaborate either with other students or with the teacher. A three-tier architecture is adopted to allow the learner to assess learning content, by using RFID for location detection. An experimental comparison between traditional methods and the proposed one, pointed out the latter resulted efficient and pleasant for students.

A framework has been proposed in (Motiwalla, 2007) for supporting distance learning through mobile devices. The impact of the approach on learning has been evaluated through student feedback analysis, both in terms of content learned and opinions about the learning methodology. Results show the leverage effect of the integration of mobile technologies with traditional education methods.

Learning support in a pervasive environment is also the goal of the GlobalEdu architecture, proposed in (Barbosa et al., 2005). It is targeted to the definition of an agent-based framework for pervasive learning rather than to its implementation through mobile devices. Similarly, a system based on agents is outlined in (Ling et. al, 2006) which provides personalized e-learning solutions. They have been deemed to have a positive effect on Life Long Learning, which is recognized as one of the key goals of e-learning technologies.

The LOM standard allows the composition of learning content in modules called Learning Objects defined in the standard itself. Research about m-learning has then addressed the topic of redesigning LOs with the aim of making their content usable through a mobile environment. In (Loidl, 2006) the features of a free and open virtual learning environment, WeLearn.Mobile, have been presented. An application allowing to present Content Packaging Specification on mobile devices is built upon the above framework.

In (von Hessling et al., 2004) a mobile environment is presented, where semantic services are matched against semantic user profiles. Here, if there is no intersection between user interests and service offers, authors conclude the user is not interested in the service. A complete and integrated solution for matching degree determination is not available.

CONCLUSION

We have proposed a ubiquitous learning approach, allowing to satisfy learning needs of a user whenever and wherever they arise. It can be useful not only to acquire knowledge, but also to solve cross-environment everyday problems.

A novel discovery framework specifically devised for mobile ad-hoc contexts without stable and fixed network infrastructures has been adapted to assist the user in retrieval and acquisition of knowledge she requires in her daily life (at home or at work, while travelling or shopping). Abduction and contraction algorithms presented in (Di Noia et al., 2007) have been revised to be widely exploited in client-server and/or peer-to-peer wireless scenarios. Thanks to a logic-based matchmaking procedure and by means of a semantic-based Bluetooth/RFID discovery protocols, users –equipped with an handheld device– can discover in the environment learning objects (modeled according to current e-learning standards) suitable for satisfying their needs.

REFERENCES

W3C Recommendation. *OWL Web Ontology Language. Retrieved from: http://*www.w3.org/TR/owl-features/, February 10, 2004.

Baader, F., Calvanese, D., Mc Guinness, D., Nardi, D., & Patel-Schneider, P. (2002): *The description logic handbook*. Cambridge University Press.

Barbosa, D., Geyer, C., & Barbosa, J. (2005). Globaledu: An architecture to support learning in a pervasive computing environment. *In Proceedings of the IFIP Workshop on Educational Technology.* Perth, Australia. Springer-Verlag.

Bechhofer, S., Möller, R., & Crowther, P. (2003). The DIG description logic interface. *In Proceedings of the 16th International Workshop on Description Logics (DL'03), volume 81 of CEUR Workshop Proceedings*. Rome, Italy.

Bluetooth. Core system architecture. *Retrieved from:*http://www.bluetooth.com/Bluetooth/Technology/Works/Core_System_Architecture.htm#10, July 26, 2007.

Castillo, S., & Ayala, G. (2008). ARMOLEO: An architecture for mobile learning objects. *In Proceedings of 18th International Conference on Electronics, Communications and Computers* (pp. 53-58). Cholula, Puebla, Mexico. IEEE Computer Society.

Chakraborty, D., Perich, F., Avancha, S., & Joshi, A. (2001). DReggie: Semantic service discovery for m-commerce applications. *In Proceedings of Workshop on Reliable and Secure Applications in Mobile Environment.*

Chan, T., Sharples, M., Vavoula, G., & Lonsdale, P. (2004). Educational metadata for mobile learning. *In Proceedings of the 2nd IEEE International Workshop on Wireless and Mobile Technologies in Education* (pp. 197-198). Washington, DC, USA. IEEE Computer Society.

Chen, T.-S., Chang, C.-S., Lin, H.-S., & Yu, H.-L. (2008). Context-aware writing in ubiquitous learning environments. *In Proceedings of Fifth IEEE International Conference on Wireless, Mobile, and Ubiquitous Technology in Education* (pp. 67-73). Beijing, China.

Colucci, S., Di Noia, T., Di Sciascio, E., Donini, F. M., & Mongiello, M. (2003). Concept abduction and contraction in description logics. *In Proceedings of the 16th International Workshop on Description Logics (DL'03).* Volume 81 of CEUR Workshop Proceedings. Rome, Italy.

Di Noia, T., Di Sciascio, E., & Donini, F. M. (2004). Extending Semantic-Based Matchmaking via Concept Abduction and Contraction. *In Proceedings of the 14th International Conference on Knowledge Engineering and Knowledge Management (EKAW2004)*. Whittlebury Hall, UK. Springer.

Di Noia, T., Di Sciascio, E., & Donini, F. M. (2007). Semantic matchmaking as non-monotonic reasoning: A description logic approach. [JAIR]. *Journal of Artificial Intelligence Research, 29*, 269–307.

Di Noia, T., Di Sciascio, E., Donini, F. M., & Mongiello, M. (2003a). A system for principled matchmaking in an electronic marketplace. *In Proceedings of International World Wide Web Conference (WWW '03)* (pp.321-330). Budapest, Hungary. ACM Press.

Di Noia, T., Di Sciascio, E., Donini, F. M., & Mongiello, M. (2003b). Abductive matchmaking using description logics. *In Proceedings of Eighteenth International Joint Conference on Artificial Intelligence (IJCAI-03)* (pp. 337-342). Acapulco, Mexico. MK.

El-Bishouty, M. M., Ogata, H., & Yano, Y. (2007). PERKAM: Personalized knowledge awareness map for computer supported ubiquitous learning. *Journal of Educational Technology & Society, 10*(3), 122–134.

Gäardenfors, P. (1988). *Knowledge in flux: Modeling the dynamics of epistemic states.* Bradford Books, MIT Press, Cambridge, MA.

Grasso, A., & Roselli, T. (2005). Guidelines for designing and developing contents for mobile learning. *In Proceedings of the 3rd IEEE International Workshop on Wireless and Mobile Technologies in Education (WMTE 2005)* (pp. 123–127). Tokushima, Japan. IEEE Computer Society.

Holzinger, A., Nischelwitzer, A., & Kickmeier-Rust, M. D. (2006). Pervasive e-education supports life long learning: Some examples of x-media learning objects (xlo). *In Proceedings of 10th IACEE World Conference on Continuing Engineering Education*. Vienna, Austria.

Holzinger, A., Nischelwitzer, A., & Meisenberger, M. (2005). Mobile phones as a challenge for m-learning: Examples for mobile interactive learning objects (milos). *In Proceedings of the Third IEEE International Conference on Pervasive Computing and Communications Workshops* (pp. 307–311). IEEE Computer Society.

Hwang, G.-J., Tsai, C.-C., & Yang, S. J. H. (2008). Criteria, strategies and research issues of context-aware ubiquitous learning. *Journal of Educational Technology & Society, 11*(2), 81–91.

IEEE (2002). IEEE Standard for Learning Object Metadata, std 1484.12.1-2002 edition.

Keil-Slawik, R., Hampel, T., & Esmann, B. (2005). Reconceptualizing learning environments: A framework for pervasive e-learning. *In Proceedings of the Third IEEE International Conference on Pervasive Computing and Communications Workshops* (pp. 322–326). IEEE Computer Society.

Ling, Z., Ying, Y., Qi, L., & Jia, Y. (2006). Research on personalized e-learning system by using mobile agents. *In Proceedings of the 1st International Symposium on Pervasive Computing and Applications* (pp. 311–315). IEEE Computer Society.

LIP (2001). IMS Learner Information Packaging Information Model Specification.

Loidl, S. (2006). Towards pervasive learning: WeLearn.Mobile. A CPS package viewer for handhelds. *Journal of Network and Computer Applications, 29*, 277–293. doi:10.1016/j.jnca.2005.03.002

Maurer, H. (1998). Web-based knowledge management. *Computer*, *31*(3), 122–123. doi:10.1109/2.660197

Motiwalla, L. F. (2007). Mobile learning: A framework and evaluation. *Computers & Education*, *49*(3), 581–596. doi:10.1016/j.compedu.2005.10.011

Ogata, H. (2008). Computer Supported Ubiquitous Learning: Augmenting Learning Experiences in the Real World. *In Proceedings Fifth IEEE International Conference on Wireless, Mobile, and Ubiquitous Technology in Education* (pp. 3-10). IEEE Computer Society.

Ruta, M., Di Noia, T., Di Sciascio, E., & Donini, F. M. (2006). Semantic-enhanced Bluetooth discovery protocol for m-commerce applications. *International Journal of Web and Grid Services*, *2*(4), 424–452. doi:10.1504/IJWGS.2006.011713

Ruta, M., Di Noia, T., Di Sciascio, E., & Donini, F. M. (2007b). Semantic based collaborative p2p in ubiquitous computing. *Web Intelligence and Agent Systems*, *5*(5), 375–391.

Ruta, M., Di Noia, T., Di Sciascio, E., Scioscia, F., & Piscitelli, G. (2007a). If objects could talk: A novel resource discovery approach for pervasive environments. *International Journal of Internet and Protocol Technology*, *2*(3-4), 199–217. doi:10.1504/IJIPT.2007.016221

SCORM. (2004) Advanced Distributed Systems (ADL) lab, Sharable Content Object Reference Model.

Sharples, M. (Ed.). (2007). *Big issues in mobile learning. Report of a workshop by the Kaleidoscope Network of Excellence Mobile Learning Initiative.* Nottingham, UK: University of Nottingham, Learning Sciences Research Institute.

von Hessling, A., Kleemann, T., & Sinner, A. (2004). Semantic user profiles and their applications in a mobile environment. *In Proceeding of Artificial Intelligence in Mobile Systems'04* (UbiComp'04).

Yang, S. J. H. (2006). Context aware ubiquitous learning environments for peer-to-peer collaborative learning. *Educational Technology & Society*, *9*(1), 188–201.

Chapter 3
Designing Effective Pedagogical Systems for Teaching and Learning with Mobile and Ubiquitous Devices

Wan Ng
La Trobe University, Australia

Howard Nicholas
La Trobe University, Australia

Seng Loke
La Trobe University, Australia

Torab Torabi
La Trobe University, Australia

ABSTRACT

The aim of this chapter is to explore issues in effective system design to bring about pedagogically sound learning with mobile devices, including the emerging generation of new devices. The authors review pedagogical models and theories applicable to mobile learning (or m-learning) and ubiquitous learning (or u-learning, also sometimes called pervasive learning, or p-learning), consider the technological support available, and describe scenarios and case studies that exemplify the achievements and challenges for each paradigm. They will also consider possible abstractions that relate ways in which learners can work within varied pedagogical model(s) to make use of relevant supporting technologies, e.g., the notions of "personal learning workflows" and "group learning workflows."

INTRODUCTION

The concept of 'ubiquitous computing' was first articulated by Mark Weiser to mean technologies that are being used unconsciously as they weave themselves into the fabric of our everyday lives (Weiser, 1991, 1996). In this regard, ubiquitous computing is 'calm technology' (Weiser and Brown, 1996). O'Malley & Fraser (2006) described it simi-

DOI: 10.4018/978-1-60566-703-4.ch003

larly as 'technology that is so embedded in the world that it disappears' (p12). When computing becomes ubiquitous, it has the capacity to support ubiquitous learning (*u-learning*, also sometimes called pervasive learning, or *p-learning*), that is, learning whenever and wherever it might take place. 'Calm' or invisible technology does not occupy the learners' attention all the time but is able to be moved seamlessly and effortlessly between the learners' central and peripheral attention. Hence, *u-learning* is not constrained by physical space, plans or timetables but is pervasive because it can occur anywhere at anytime and is supported rather than constrained by the technology that mediates it.

In this chapter, we discuss the concept of *u-learning* and its relationship to mobile learning (*m-learning*). We discuss mobile devices as potentially ubiquitous learning tools. We review theories and pedagogical models applicable to *m-learning* and *u-learning* and consider possible abstractions that relate ways in which learners can work within the pedagogical models to make use of relevant supporting technologies, for example, the notions of 'personal learning workflows' and 'group learning workflows'. We propose some scenarios and describe the systems architecture associated with them that would be required to bring about learning with mobile devices.

UBIQUITOUS AND MOBILE LEARNING

Ubiquitous learning is characterised by two dimensions: (1) it is not constrained by physical space, plans or timetables but is pervasive and occurs anywhere at anytime and (2) as a consequence of the distributed nature of the immediate access to a variety of sources of information or means of reflecting on experiences in interaction with others, ubiquitous learning is characterized by the transformation of understanding and the ability to question experiences and information. Informing this view and in line with the views of Schenker, Kratcoski, Lin, Swan and van 't Hooft (2007), we understand learning as 'the processing of encountered information [extended to include experiences, values or representations] that leads to changes in knowledge, skills, beliefs, abilities, and behaviours' (p172). The notion of 'ubiquitous' learning builds on this to emphasise how this processing can occur through a variety of modes and modalities unconstrained by time and location. Because of their capacity to situate the experiences and transforming reflections in both the immediate and more removed contexts, mobile technologies offer a powerful means to enable ubiquitous learning in being able to provide a portable, interactive learning environment capable of both multimedia functions and Internet access and supporting both self-directed, independent learning and interactivity with others. Consequently, mobile technology should be able to foster active and creative learning as it is capable of creating opportunities for students to collaborate with peers in project work (e.g., via phone to phone Bluetooth, or IR beaming) and to undertake independent research (via wireless networking) and be engaged in problem solving in real-life contexts. Its portability allows for context-based data collection and 'just-in-time' learning.

At a broad level, *u-learning* encompasses electronic learning (*e-learning*) usually associated with *m-learning*. *M-learning* is defined by different people differently. Keegan (2005) has a more technology-centred definition of *m-learning* stating that it is the ability of mobile devices (PDAs, mobile and smart phones) in providing education that constitutes *m-learning*. Considering *m-learning* from a more learner-centred approach, Georgiev et. al. (2004) define the term as the ability to learn everywhere, anytime without the need for permanent physical connections to cable networks. Vavoula and Sharples (2002) suggest that learning is closely linked to mobility and that there are three ways in which learning can be considered mobile: (1) space, that is, learning can happen anywhere – at home, school/workplace and at places of leisure (2) between areas of life, that is, learning may related to work, self-improvement or leisure and (3) time, that is, learning can happen anytime and any day. A broader definition on which this chapter is based is O'Malley, Vavoula, Glew, Taylor, Sharples and Lefrere's (2003):

Any sort of learning that happens when the learner is not at a fixed, predetermined location, or learning that happens when the learner takes advantage of the learning opportunities offered by mobile technologies. (p. 6)

A narrower view of the relationship between *m-learning* and *u-learning* is described by Dochev and Hristov (2006) citing Casey (2005) as the contribution of mobile technologies to *e-learning* resulting in ubiquitous access. One limiting consequence of this perspective is that it presents ubiquitous learning as one form of *e-learning* that is linked with a particular kind of communication technologies, mobile technologies. This view excludes consideration of the purposes that learners have in undertaking the learning. Considering *u-learning* from the perspective of the learners, there are specific requirements placed on technologies if they are to be used for ubiquitous learning. Ogata & Yano (2004), referencing the work of Chen, 2002 and Curtis et. al., 2002 identified these requirements as (1) **permanency,** where learners never lose their work unless it is deleted on purpose (2) **accessibility,** where learners are able to access their files, documents and data from anywhere (3) **immediacy,** where learners are able to obtain information immediately and (4) **interactivity,** where learners are able to interact with teachers, peers or experts through synchronous or asynchronous communication, enabling knowledge development and transformation to occur more quickly and readily. There is also a crucial requirement of ubiquitous learning to make explicit the mutual engagement of multiple learners through (5) **situating of instructional activities,** where learning is embedded in the learners' daily lives and (6) **adaptability**, where learners are able to get the right information at the right place in accessible ways. These characteristics are not unique to but are strongly enabled by the educational affordances that mobile technology such as the personal digital assistant (also called pocket PCs and Palm) can offer, although there are some constraints with mobile phones such as the inability to Word process or use Excel for graphing purposes.

MOBILE DEVICES AS UBIQUITOUS LEARNING TOOLS

In education, mobile devices offer the potential for *u-learning* through new ways of accessing information and thinking both individually and within networked communities, where collaborating with others supports developing new understandings and arguing for new solutions. As shown in Table 1, mobile devices are classified into four types (Dochev & Hristov, 2006): personal digital assistants (PDA), smart phones, mobile or cell phones, tablet PCs and Notebooks. PDAs have significant processor

power. They offer a range of computing capabilities such as processing with Word and Excel, PowerPoint editing and display, accessing the Internet via WiFi, sending files via infrared or Bluetooth, voice and video recording, image taking and supporting multimedia display using Flash and Media Player (Nicholas & Ng, in press). There is also a range of educational software available for PDAs that supports learning. In contrast, mobile phones are mainly communication devices for voice communication or sending/receiving text messages. The more powerful mobile phones are able to access the Internet. Smartphones are hybrids between PDAs and mobile phones with the combined capabilities of both. Screen sizes are smaller than those of PDAs but larger than mobile

Table 1. Mobile devices and features that support u-learning

Mobile device	Features supporting *u-learning*
Personal digital assistants (PDA)	Computing capabilities: Word and Excel processing, PowerPoint editing and display; Internet access via WiFi; sending files via infrared or Bluetooth; voice and video recording; image taking; displaying multimedia using Flash and Media Player; educational software that supports learning
Mobile or cell phones	Communication devices - voice or text messages; more powerful ones able to access the Internet
Smart phones (hybrids between PDAs and mobile phones)	Combined capabilities of both
Tablet PCs, laptops and notebooks	Larger portable devices than PDAs with larger processing capabilities and larger memory

phones. Tablet PCs, laptops and Notebooks are much larger portable devices than PDAs, mobile phones or smartphones and have larger processing capabilities and larger memory. The above mentioned technologies enable learning to be increasingly more pervasive with the potential that young people's thinking can be shaped by connectivity through these devices (Aleven, Stahl, Schworm, Fischer & Wallace, 2003; Hargreaves, 2003, BECTA, 2003). Consequently, the need to think in new ways and engage with others in that thinking is increasing all the time both to respond to the new potential and to increase the potential of the technologies to benefit the interactions between teachers and students.

LEARNING THEORIES SUPPORTING MOBILE AND UBIQUITOUS LEARNING

Mobile Learning Theories

Sharples, Taylor & Vavoula's (2007) theory of learning for a mobile age focuses on the communicative interactions between technology and people for the advancement of knowledge. They see learning as conversational processes (citing Pask, 1976) taking place across multiple contexts within systems where people and technology are in 'continual flux'. According to them, conversation requires the learner to describe him/herself and his/her actions, explore, extend, converse and externalize that description or understanding to subsequent activities. Technology can be involved as a conversational partner mediating conversations bridging formal and informal learning. Rochelle, Patton and Tatar (2007) view handheld computers as learning technologies that are representational tools for augmenting cognition and a mediational tool for social and cultural participation. They see learning with handhelds and within networked communities as the merging of cognitive augmentation and social mediation perspectives that link formal and informal learning, and the linking of representations mediated by mental operations and social practice. To this we would add the requirement that the learner using the handheld needs to be acknowledged as a 'partner' and collaborator in the conversation and learning rather than just a 'recipient' of information. In other words, the learner is a shaper and presenter of information and values.

Socio-Constructivist Learning Theory

In a technology-mediated learning environment, involving for example the use of mobile devices, the interactive, open and non-linear nature of learning require learners to be actively analyzing, evaluating and making decisions while manipulating the information at hand in order to construct new knowledge or solve a problem. They will constantly have to compare their own prior knowledge of a body of information with that presented in the learning environment and seek means of either re-confirming their prior knowledge or deconstructing and re-constructing new meanings. Given the view of ubiquitous learning as transforming understandings and relationships, theories that support learning with mobile devices (for example, Rochelle, Patton & Tatar, 2007; Sharples, Taylor & Vavoula, 2007) commonly embrace aspects of the socio-constructivist learning theories. Constructivism is a learning theory that has been highly influential in Western education over the last three decades and a vast amount of literature has been accumulated on it, particularly in the area of science education (Matthews 1993, 1998; Fensham, Gunstone & White, 1994; Solomon 1994; von Glasersfeld 1995; Phillips 1995, 2000). It is a theory of knowledge that offers explanations about how we have come to know what we know. The two notable theorists associated with socio-constructivism were Jean Piaget and Lev Vygotsky. They offer differing perspectives on constructivism. Piaget's (1955, 1972) constructivism is also known as personal constructivism and is based on his cognitive developmental theories, which propose that concept formation in the individual follows a clearly defined set of stages that must be experienced sequentially by that individual. In light of this, Piaget's constructivism is also called cognitive constructivism. The underlying principle in cognitive constructivism is that knowledge resides in individuals and that it cannot be given or transmitted complete to them by their teachers. Learners must construct their own knowledge in their minds and build upon the knowledge that is based on prior experiences. Learners learn only when they are actively engaged in the process, either at the operational level where learners are engaged in physical manipulations or at the cognitive level where they are mentally processing information or stimuli. Social constructivism (Vygotsky, 1962 & 1978) makes similar assertions to Piaget's cognitive constructivism about how learners learn in regard to knowledge being progressively built up and continually re-interpreted. However, Vygotsky places more emphasis on the social context of learning. In his theory, the learning process centrally involves interaction with other individuals where culture and society will influence the learning. A difference between cognitive and social constructivism is that in the former, the teacher plays a limited role, acting as a facilitator, whereas in the latter, the role of the teacher is crucial and involves mediating students' grasp of concepts by guiding and encouraging group or other analytic work. Socialization within Vygotsky's theory is not confined to teacher-student interactions. The interaction between students is also pivotal in aiding students to construct and build knowledge. These cognitive and social theories of Piaget and Vygotsky form the basis of socio-cognitive constructivism. Socio-cognitive constructivism as applied to many face-to-face and online learning situations, draws on and sees learning as a dynamic and social process. The social aspect of learning supported by technology such as mobile devices and online management systems (WebCT, Blackboard or Moodle) is supported theoretically by computer supported collaborative learning theory.

Computer Supported Collaborative Learning (CSCL) Theory

CSCL has been researched since the early nineties and Stahl, Koschmann & Suthers (2006) have provided a historical perspective of CSCL. Lipponen (2002) defines CSCL as how collaboration and technology are able to facilitate the sharing and distribution of knowledge and expertise among members in a community to enhance peer interactions and group work. Through directed discourse and negotiated meanings with others, CSCL supports collaborative learning for the construction of shared knowledge. Supporting socio-constructivist learning theories for CSCL are also Bakhtin's (1986) dialogical theories of learning.

Central to knowledge construction in CSCL is a socially-immersed learning environment where the students create a virtual community with peers and teacher mediating cognition enabling them to construct knowledge in an interactive and iterative process across the dimensions of conceptualisation and construction (Ng & Nicholas, 2007). The requirement for a social environment is based in the belief that there are pedagogical benefits in having access to discussions generated in learning dialogues, such as in developing critical reflections and constructing knowledge (Anderson, 2004; Lipman, 1991; McConnell, 2000). Fung (2004) and Stacey (1999) assert that discussions that negotiate meanings expose participants to different perspectives in collaborative learning environments. As a result of such discussions, students are able to develop critical thinking and judgment skills that value, support or oppose the different views.

Examples of Socio-Constructivist and CSCL Supported Learning Online

There are numerous empirical-based reports of socio-constructivist learning within CSCL frameworks. We illustrate here with a couple of examples, one of which is school-based and the other conducted at higher education.

Anderson and Witfelt (2003) reported on a project where 12-13 year-old students studying home economics worked in groups to solve problems in a MOO environment. A MOO is an Internet-based multi-user environment that allows multiple users to share a community of rooms and virtual spaces to explore specific concepts or engage in particular kinds of interactions and objects based on text, and to interact with each other (Holmevik & Haynes, 1998). Such an environment offers students the opportunity to undertake problem-solving tasks where they explore, investigate and create narratives based on their own ideas and experiences. A MOO-environment encourages collaborative teamwork and promotes constructivist learning as the students apply concepts learned in home economics, for example principles of cooking and concepts of energy and nutrition, into designing a 'storyline' within the MOO environment. An example of a 'storyline' reported was along the theme of throwing a party. The students took up employee roles in various sections of an imaginary ecological restaurant, such as the economic department, the production planning department and the kitchen. They designed a business plan, investigated the cost of production of different cooked products and physically experimented with various recipes to find out about the cheapest and best way of producing each recipe. Students applied skills in Science, Mathematics, English and Home Economics to develop this particular storyline. In working through such a group-oriented task, the MOO-based virtual environment enabled collaboration and promoted higher order thinking skills in students. In undertaking this task, students were actively constructing new knowledge and learning at an individual and a social level was occuring.

Another example of socio-constructivist learning within CSCL frameworks is reported by Ng (2008) on the ways that tertiary students worked within a web-based learning management system (WebCT) in small virtual teams to complete two semi-structured online tasks in order to develop conceptual understanding on the topic of ethics. Guidelines were provided for the tasks but learning was open and required the teams to work together, to research and share ideas and knowledge, to construct an essay, and to peer-review another team's essay. The open nature of the learning required students to actively construct understanding through research initially at an individual level and to collaborate later online with team members to come to a common understanding of the topic under study in order to produce the essay. Apart from posting messages online, communication technologies such as emailing and sending text messages were used between group members. The study showed that socio-constructivist learning

Figure 1.

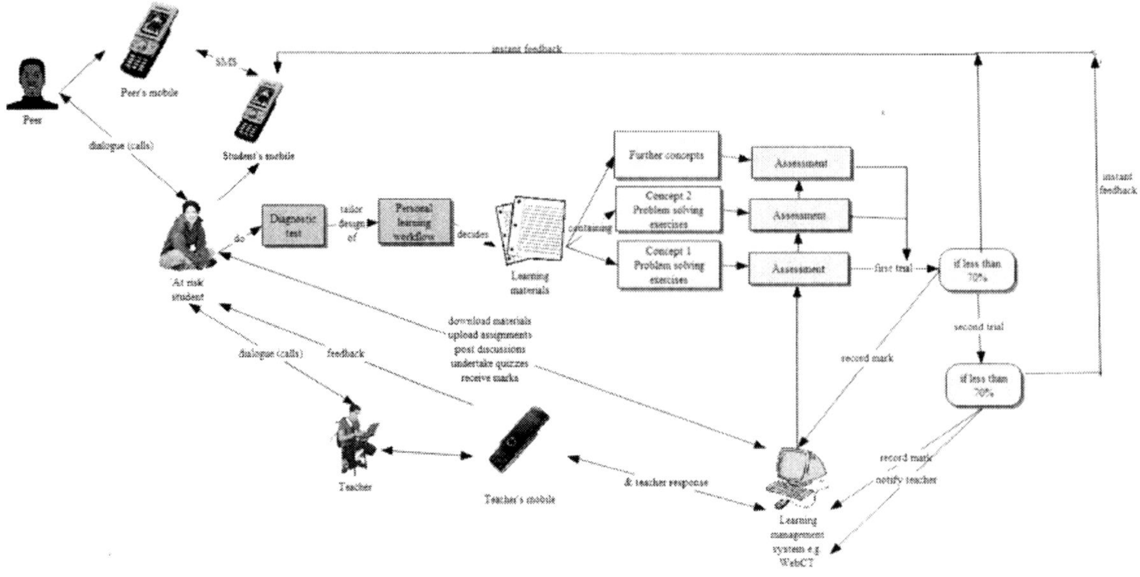

with web-based technologies with this group of students was effective.

In both the examples described above, the students were interacting with web-based technologies and sometimes with mobile technology, to complete set tasks within the guidelines provided by their teachers. In the next section, we describe scenarios that utilise similar technologies to promote constructivist learning but within the framework of personal (or group) learning workflows that have an automated component associated with them.

Scenarios for *U-Learning*

To combine the pedagogic and the systems perspectives (considered in a later section), we consider two scenarios of how *u-learning* principles can be implemented.

Scenario 1

This involves working with 11-13 year old students at risk in mathematics and science learning. These students could be from anywhere in Victoria, Australia. To assist the students to learn in a structured, personalised and ubiquitous manner, mobile phone technology and an online learning management system such as WebCT are used. The scenario, as depicted in Figure 1 has the following features:

1. **A networked community of people and devices (mobile and desktop).** The students learn collaboratively by talking directly in person to peers and/or instructor. Reminders, assessment marks or congratulatory remarks will be sent via SMS to students and instructor at different stages of the students' learning.
2. **Development of curriculum materials in mathematics and science for these 'at risk' students.** Each topic in mathematics or science will be divided into different concepts to be learned. Students will have personal learning workflows facilitated by mobile devices. Learning materials could be downloaded from WebCT in the form of notes, podcasts or videocasts to the student's mobile phone or to his/her home desktops.

The students learn in a personalised, self-directed manner as they actively construct understanding of a concept and discuss collaboratively with peers when problems are encountered. Assessment will be given after engaging with each concept, and instant feedback automatically provided by the system. If a students does not receive 80% in the assessment, he/she will be given another opportunity to undertake another assessment task related to the same concept. If he/she is unable to achieve 80% by the second trial, a notice will be given to the instructor letting him/her know which questions were wrong in the student's first and second trial. (S)he can then decide to allow the student to go on to the next concept or provide some online synchronous feedback (maybe with a small group of students with similar difficulties).

Scenario 2

A teacher education program is structured to develop (1) a personal and (2) a small-group philosophy of what makes a good science teacher. The broad framework for this scenario (see Figure 2) is where students (the to-be science teachers) keep a journal of beliefs that evolve as the semester progresses. The final task requires all students to present their personal philosophy of what makes a good science teacher, the format which is open – students could submit an essay, make an audio recording, construct a PowerPoint, webpage or concept map or cartoons etc to demonstrate how their thinking on the concept of a 'good science teacher' has evolved and been influenced. This task requires the students to look out for good teaching in different contexts and capture them e.g. at home, in a childcare centre, in museum and such places to help them shape their philosophy of good teaching. They are also encouraged to capture 'things' that could help them teach science better. Upon completion, the piece of work submitted to the central system is slotted into the group's space. A congratulatory message is sent to all group members to congratulate them for finishing the task and the lecturer is notified of the submission.

As a second component of the pre-service teachers' education, they will work in small groups of four to develop a philosophy of what makes a good science teacher. They work collaboratively to discuss beliefs and come to a consensus about what a good science teacher is. The group will be required to construct a concept map at the beginning of the semester identifying keywords, important attributes and factors that contribute to good science teaching. The concept map will be submitted to the central system that places the concept map in the 'space' for the group. The system will be adapted so that the lecturer is able to view the concept maps to have a sense of where each group of students is starting from. The system is able to remind the leader in each group at regular intervals to look at the concept map and see if he/she and other members of the group want to change (add or delete) keywords or linking phrases. At the end of the semester, each group will submit a final concept map. The system will need to be configured to detect the receipt of the final concept map from a group, to accept the concept map and place it in the space designated for the group and then to send both the initial and final concept maps to the group leader for dissemination to the group members.

A final requirement of the task is to write a reflective essay or produce a voice recording of beliefs that have evolved over the semester. Both the personal and small group activities are structured (while still maintaining flexibility) via *learning workflows* facilitated by mobile devices.

Figure 2.

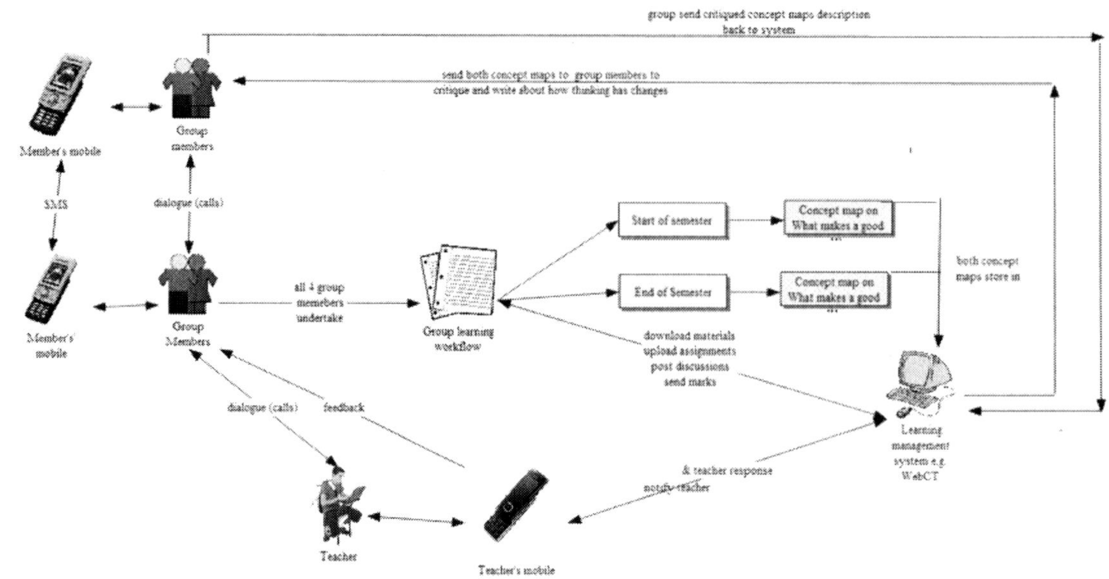

DESIGN ISSUES AND IMPLEMENTATION ABSTRACTIONS IN ADAPTING THE MOBILE PHONE FOR PEDAGOGICALLY SOUND EDUCATION

Having discussed pedagogical issues, we consider complementary issues from the systems perspective. The issues range from the challenge of a suitable choice of implementation abstractions, dealing with material that may be too dense (large) for the screen resolution of a mobile device (e.g., how does the server know when material has been accessed on a low resolution screen and keep track of what has been accessed?) to deployment issues such as who would have what kind of access to which resources. The issues also include the role that today's feature-packed mobile devices can play – from being convenient tools for data gathering (e.g., taking photos of plant specimens) to such devices as "social portals" to learning groups/communities. We also consider hardware and software capabilities, what type of modifications are needed for enhanced usage with learning e.g. downloading materials, providing feedback, sending messages, participating in discussions etc, i.e. technologies to enable the notion of learning workflows, personalized or group-based, as abstractions for system implementation.

Personal Learning Workflows

Scenarios 1 and 2 above can be realized with the concept of a workflow through which the student steps. The workflow can also be viewed as guide for the student as (s)he attempts to learn a particular module. A student may step through several of such workflows, each corresponding to a particular unit of knowledge. A workflow can be personalized for a student, or be adapted to the student, and, in the scenarios discussed above, the workflow is delivered to the student via a mobile device. The workflow consists of a set of tasks which the student undertakes. Figure 1 illustrates the process. The student first takes a diagnostic test whose results are then used by the system to personalize a workflow for the student. Note that a workflow has a particular learning goal (e.g., as in Scenario 1 or 2 above). The example in Figure 1 illustrates different concepts to be learnt, each corresponding to a workflow task that the student undertakes, and the corresponding assessment to be completed for each task. The student progresses through the assessment. On failure of an assessment, the instructor/teacher could be informed perhaps triggering a response from the instructor, and several measures taken, from retaking assessments, discussion with the instructor, or the instructor providing further information. The instructor is, hence, kept in the loop and informed of the student's progress (or non-progress) through the workflow. Setting up systems so that they will effectively address the issues involved in supporting interaction between learner, teacher and technology require a different kind of discourse so that the computer science and education can work together. For the combination of education and computer science to be effective, each must understand the requirements of the other. Each must also acknowledge the histories of the other, where educational objectives are currently framed by flexibility and computer science has a stronger emphasis on predictability in sequences of actions. Here we outline from a systems perspective, how the workflow needs to be thought about so that both predictability and flexibility can be incorporated.

From a programming perspective, we can define a personal learning workflow as an ordered set of triples:

{(p1,c1,a1),...,(pN,cN,aN)} combined using operators SEQUENCE (denoted by ";"), OR (denoted by "|"), and PARALLEL (denoted by "||"), where each triple (pi,ci,ai) represents a task within the workflow, p*i* is the person (an identifier) who is to perform the task, c*i* is the content of the task (e.g., materials to be learnt, an exercise or information for the student, a concept to be learnt), and a*i* is an assessment (viewed broadly here to mean a condition which must be fulfilled for the task to be said to be completed). The person to perform the tasks is typically the student. However, a workflow might also include the tasks of the instructor specified in the workflow itself. Where a triple denotes a task of an instructor, the "assessment" is merely a precondition of the task. The operators are what might be found in a typical workflow system.[1] For example, (p,c1,a1);(p,c2,a 2);...;(p,cN,aN) is a sequential personal learning workflow comprising N tasks all to be done by one person identified via "p", the tasks joined via the SEQUENCE operator, and indicates that the tasks should be completed in sequential order. Suppose such a sequential workflow is running, at any time, all or part of the workflow can be sent to the mobile device – a typical scenario is to effectively send one triple (or task) at a time to the user's mobile device, and on completion of this task, the next triple (or task) is then sent, etc. However, this need not be the case strictly, tasks can be prefetched or a chunk containing multiple tasks can be sent to the mobile device. A workflow for a person "p" such as

(p,c1,a1);((p,c2,a2)|(p,c3,a3));((p,c3,a3)||(p,c 4,a4));(p,c5,a5)

means do (p,c1,a1) first, and then there is a choice of tasks (p,c2,a2) or (p,c3,a3) to do (perhaps the choice of the student) and thereafter, (p,c3,a3)||(p,c4,a4) means that both these tasks should be done concurrently. Finally, (p,c5,a5) must be done.

Social (or Group) Learning Workflows

Figure 2 illustrates Scenario 2 where a group undertakes a group learning workflow. The personal learning workflows is easily generalized to a workflow where the tasks are ones that involve or encourage interaction among group members i.e., the workflow consists of triples from a set of the form {(g1,t1,c1,a1),...,(gN,t1,cN,aN)}, where g*i* denotes a group id mapped to a set of persons {p1,...,pk}, and t*i* denotes an interactional requirement on the task. We make such interactional requirements explicit to emphasize to students the need to interact – *ti* would be null if the group requires only one person for a task. A weak form of conforming to such requirements can be imposed by specifying that communications (via devices) that can be tracked unobtrusively must have specific features or will be specified to contain these features (e.g., cc-ing messages to the instructor) – of course, not all communications (e.g., verbal, face-to-face, etc) can (or should) be tracked.

Refinements

In executing a workflow, or when a student is "experiencing" a learning workflow, as noted in the scenarios above, the system may at pre-specified stages of a workflow remind the student of particular tasks yet to be completed or send encouragement messages to students, etc. A specification of such system messaging can be included in an extended definition (or specification) of a personal learning workflow via *workflow messaging rules* of the form (for a given task (pl,cl,al)):

- On completion of (pl,cl,al) send <encouraging message>
- On non-completion of (p1,c1,a1) by <date> send <reminder message>
- On assessment_failure of (p1,c1,a1) by <date> send <instructor message>

We deem such workflow rules to be separate from the learning workflow itself, serving as augmentations of an existing workflow. This separation of concerns we think convenient for designers of personal learning workflows.

Ubiquitous Learning and Context-Awareness

With the expected proliferation of positioning enabled devices (e.g., GPS phones or Wi-Fi positioning), we can employ *location-based learning*, where certain modules or concepts should be learned (the learning task triggered or initiated when at a particular location) or must be learned (the learning task cannot be done except at a particular location), specified via an annotation of personal learning workflow tasks. For example, given two triples (pi,ci,ai) and (pj,cj,aj) from a workflow {(p1,c1,a1),...,(pN,cN,aN)}, we can annotate as follows:

- location_trigger(pi,ci,ai) = at_school
- location_prerequisite(pj,cj,aj) = in_zoo

i.e., the task (pi,ci,ai) is triggered when the person (student) arrives at school, and the task (pj,cj,aj) must be completed while in the zoo. We need not stop at location but exploit other context information, including time, actvitiy,[2] and nearby persons and objects. Learning of a concept or a module can be triggered by the presence (or absence) or the close proximity of particular objects or persons. Consider a typical museum scenario whereby a person goes from one exhibit to another, and at each exhibit, a task of a personal learning workflow is triggered, and on larger geographical scale, a person walks through a historic city, and different learning tasks of a "city history" learning workflow are triggered and downloaded to the mobile device at different sites. In situ, situated and opportunistic learning (ubiquitously) can, hence, be supported, while maintaining a global goal as the tasks are connected via a workflow specification.

Moreover, device context (e.g., as can be specified using standardized vocabularies such as device apabilities and preferences profiling[3]) can be used to describe the deployment environment for triples (each task or triple encapsulated as a package of data and code), so that, tailoring of the data and code can be done to suit the device being used. Knowing the device context, such as the screen resolution and size as well as graphics capabilities, an adapted multimedia version or text version can be deployed (e.g., one task deployed via SMS for a low-end mobile phone and another media-rich task done via a laptop).

CONCLUSION AND THE FUTURE

The chapter has discussed learning in a mobile setting, highlighting the notion of the personal learning workflows as an implementation abstraction. We think that the metaphor of the personal learning workflow is useful in a variety of settings, for example in encouraging structured self-learning in situations where the instructor can be scarce (e.g., in rural settings), or where flexibility (while maintaining structure) in learning is required, or where device form factors are relatively small or device capabilities and connectivity may be limited (since the workflow metaphor encourages modularizing learning into connected tasks or units which can be "consumed" progressively), where modularity enables learning separate concepts in different contexts at different times and locations (the notion of *context-aware learning*), or where the learner self-constructs knowledge or a whole picture as gleaned from multiple tasks (the workflow helps the learner maintain the overall goal and activity of the learner while the learner can be concerned with one task at a time), where interruptions to learning can be tolerated (as decoupling of time and space between tasks can be tolerated), and where, in the future, multiple devices can be exploited to support learning (in which case, the workflow could involve some tasks done on a mobile device, some tasks done on a desktop, and some other tasks done using some new generation devices such as Microsoft's surface computer[4]).

In concluding, the personal and group learning workflows scenarios described in this chapter demonstrate the potential of *u-learning* resulting from ubiquitous access to information and human resources enabled by mobile technologies. The learning is both mobile (*m-learning*) where the learning is not constrained by space or time and pervasive (*p-learning*) where learning pervades all aspects of the students' lives and could take place anytime and in between other activities, for example, while eating or watching television. Students are able to seek assistance from their instructors or peers when the need arises, submit work completed anytime and be guided by the workflow structures to continue learning independently and in a personalised way.

We are working towards an implementation of the system discussed above.

REFERENCES

Aleven, V., Stahl, E., Schworm, S., Fischer, F., & Wallace, R. M. (2003). Help seeking and help design in interactive learning environments. *Review of Educational Research, 73*(3), 277–320. doi:10.3102/00346543073003277

Anderson, K., & Witfelt, C. (2003). Educational game design: bridging the gab between computer based learning and experimental learning environments. *Conference proceedings of the Sixth International Conference on Computer-based learning in Science, 1*, 402-413.

Anderson, T. (2004). Towards a theory of online learning. In T. Anderson & F. Elloumi (Eds.), *Theory and practice of online learning* (pp. 33-60). Canada: Athabasca University Press.

Bakhtin, M. (1986). The problem of speech genres (V. McGee, Trans.). In C. Emerson & M. Holquist (Eds.), *Speech genres and other late essays* (pp. 60-102). Austin: University of Texas Press.

BECTA (British Educational and Technology Agency). (2003). *Becta ICT Research Online*. Retrieved July 30, 2007, from http://www.becta.org.uk.

Casey, D. (2005, October 24-28). u-Learning = e-Learning + m-Learning. *Proceedings of e-Learn 2005: World Conference on e-Learning in Corporate, Government, Healthcare & Higher Education*, Vancouver, Canada. Retrieved June 8, 2008 from http://neptune.netcomp.monash.edu.au/staff/dcasey/papers/2005/E-Learn2005/paper6792.doc.

Chen, Y. S., Kao, T. C., Sheu, J. P., & Chiang, C. Y. (2002). A Mobile Scaffolding-Aid-Based Bird -Watching Learning System. In U. Hoppe, M. Milrad, & Kinshuk (Eds.), *Proceedings of IEEE International Workshop on Wireless and Mobile Technologies in Education (WMTE'02)* (pp.15-22). New York: IEEE Computer Society Press.

Curtis, M., Luchini, K., Bobrowsky, W., Quintana, C., & Soloway, E. (2002). Handheld Use in K-12: A Descriptive Account. In U. Hoppe, M. Milrad & Kinshuk (Eds.), *Proceedings of IEEE International Workshop on Wireless and Mobile Technologies in Education (WMTE'02)* (pp.23-30). New York: IEEE Computer Society Press.

Dochev, D., & Hristov, I. (2006). Mobile learning applications: ubiquitous characteristics and technological solutions. *Cybernetics and Information Technologies*, *6*(3), 63–74.

Fensham, P. J., Gunstone, R. F., & White, R. T. (1994). Science content and constructivist views of learning and teaching. In P. J. Fensham, R. F. Gunstone, & R. T. White (Eds.), *The content of science* (pp. 1-8). London: The Falmer Press.

Fung, Y. H. (2004). Collaborative online learning: interaction patterns and limiting factors. *Open Learning*, *19*(2), 135–149. doi:10.1080/0268051042000224743

Georgiev, T., Georgieva, E., & Smrikarov, A. (2004, June 17-18). M-learning: a new stage of e-learning, *Proceedings International Conference on Computer Systems and Technologies - CompSysTech'2004,* Rousse, Bulgaria. Retrieved June 8, 2008 from http://ecet.ecs.ru.acad.bg/cst04/Docs/sIV/428.pdf

Hargreaves, A. (2003). *Teaching in the Knowledge Society*. Maidenhead: Open University Press.

Holmevik, J. R., & Haynes, C. (1998). *High wired. On the design, use and theory of educational MOOs*. University of Michigan Press.

June 8, 2008 from http://www.mlearn.org.za/CD/papers/keegan1.pdf

Keegan, D. (2005, October). The incorporation of mobile learning into mainstream education and training. *MLearn 2005 conference*, Cape Town. Retrieved

Lipman, M. (1991). *Thinking in education.* Cambridge: Cambridge University Press.

Lipponen, L. (2002). *Exploring foundations for computer-supported collaborative learning*. Retrieved October 12, 2007 from http://newmedia.colorado.edu/cscl/31.html

Matthews, M. R. (1993). Constructivism and science education: Some epistemological problems. *Journal of Science Education and Technology*, *2*(1), 359–370. doi:10.1007/BF00694598

Matthews, M. R. (Ed.). (1998). *Constructivism in science education: A philosophical examination.* Dordrecht: Kluwer.

McConnell, D. (2000). *Implementing computer supported cooperative learning* (2nd ed). London: Kogan Page.

Ng, W. (2008). Virtual teamwork: students learning about ethics in an online environment. *Journal of Research in Science & Technological Education*, *26*(1), 13–29. doi:10.1080/02635140701847421

Ng, W., & Nicholas, H. (2007). Conceptualising the use of online technologies for gifted secondary students. *Roeper Review, 29*(3), 190–196.

Nicholas, H., & Ng, W. (in press). Ubiquitous learning and handhelds: An overview of theory and pedagogy. *Encylopedia for online and distance learning.*

O'Malley, C., & Fraser, D. S. (2006). *Literature review in learning with tangible technologies.* Bristol: Futurelab. Retrieved June, 6, 2008 from http://palmsource.palmgear.com/index.cfm?fuseaction=software.category&catid=209&PartnerREF=&siteid=7

O'Malley, C., Vavoula, G., Glew, J. P., Taylor, J., Sharples, M., & Lefrere, P. (2003). *MOBIlearn project: Report of WP 4 "Guidelines for Learning/Teaching/Tutoring in a MobileEnvironment.* Retrieved June 8, 2008 from http://www.mobilearn.org

Ogata, H., & Yano, Y. (2004). Context-aware support for computer supported ubiquitous learning. In J. Roschelle, T.W. Chan, Kinshuk, & S.Yang (Eds.), *Proceedings of the 2nd IEEE (WMTE2004)* (pp. 27-34). New York: IEEE Press.

Pask, G. (1976). *Conversation Theory: Applications in Education and Epistemology.* Amsterdam and New York: Elsevier.

Phillips, D. (Ed.). (2000). *Constructivism in education: Opinions and second opinions on controversial issues.* Chicago: University of Chicago Press.

Phillips, D. C. (1995). The good, the bad, and the ugly: The many faces of constructivism. *Educational Researcher, 24*(7), 5–12.

Piaget, J. (1955). *The construction of reality in the child.* London: Routledge & Keegan Paul.

Piaget, J. (1972). *Psychology and epistemology: Towards a theory of knowledge.* London: Penguin

Schenker, J., Kratcoski, A., Lin, Y., Swan, K., & van 't Hooft, M. (2007). In M. van 't Hooft & M. Swan (Eds.), Researching ubiquity: ways to capture it all. In *Ubiquitous computing in education: invisible technology, visible impact* (pp. 167-186), Mahwah, NJ: Lawrence Erlbaum Associates.

Sharples, M., Taylor, J., & Vavoula, G. N. (2007). A Theory of Learning for the Mobile Age. In R. Andrews & C. Haythornthwaite (Eds.), *The SAGE Handbook of E-learning Research* (pp. 221-247). London: Sage.

Solomon, J. (1994). The rise and fall of constructivism. *Studies in Science Education, 23,* 1–19. doi:10.1080/03057269408560027

Stacey, E. (1999). Collaborative learning in an online environment. *Journal of Distance Education, 14*(2), 14–33.

University Books Roschelle. J., Patton, C., & Tatar, D. (2007). Designing networked handheld devices to enhance school learning. In *Advances in Computers, 70,* 1-60.

Vavoula, G. N., & Sharples, M. (2002). KLeOS: A personal, mobile, knowledge and learning organisation system. In M. Mildrad, U. Hoppe, & Kinshuk (Eds.), *Proceedings of the IEEE International Workshop on Mobile and Wireless Technologies in Education (WMTE2002),* (pp.152-156), Vaxjo, Sweeden.

von Glasersfeld, E. (1995). *Radical constructivism: A way of knowing and learning.* London: Falmer Press.

Vygotsky, L. (1978). *Mind in society: The development of higher psychological processes.* Cambridge, M: Harvard University Press.

Vygotsky, L. S. (1962). *Thought and language.* Cambridge, MA: MIT Press.

Weiser, M. (1991). The computer for the twenty-first century. *Scientific American,* 265(3):94-104. September 1991.

Weiser, M. (1996, March). *Open house. ITP Review 2.0*. Retrieved March 15, 2008 from http://www.itp.tsoa.nyu.edu/~review.

Weiser, M., & Brown, J. S. (1996). *The coming of age of calm technology*. Retrieved March 15, 2008 from http://www.ubiq.com/hypertext/weiser/acmfuture2endnote.htm.

ENDNOTES

1 http://www.wfmc.org/standards/framework.htm

2 The notion of activity-based computing involves understanding the current activity of the user and aiding the user correspondingly. See http://www.activity-based-computing.org/

3 http://www.w3.org/TR/CCPP-struct-vocab

4 http://www.microsoft.com/surface

Chapter 4
Text Messaging to Improve Instructor Immediacy and its Role in Multiplatform E-Learning Systems

Paul Hayes
National College of Ireland, Ireland

Stephan Weibelzahl
National College of Ireland, Ireland

ABSTRACT

Text messaging has been exploited for supporting learning in a variety of educational settings. However, evidence for its effectiveness and impact is limited. This chapter demonstrates how the use of text messaging can contribute towards enhanced quality of learning. In particular the chapter focuses on the use of text messaging as a means of improving immediacy between instructors and students in third-level education. Immediacy is defined as behaviour which increases psychological closeness between communicators. The results of research in instructional communication suggest that improved immediacy leads to more positive student-instructor relationships engendering positive attitudes, increased interest and motivation by students as well as improved attendance, improved retention, improved student engagement and improved learning. This chapter outlines a theoretical basis for the effect of text messaging on instructor-student relationships, provides empirical evidence for the impact of text messaging on immediacy and discusses the integration of text messaging for improving immediacy in Multiplatform E-Learning Systems.

LEARNING EXPERIENCE OF STUDENTS

The quality of student learning is paramount to the success of any educational institution. There has been a plethora of initiatives recently in third level colleges aimed at improving the quality of the learning experience of students. Some of these have shown to be quite successful while others have not. The purpose of this chapter is to demonstrate how the use of mobile communication technologies can

DOI: 10.4018/978-1-60566-703-4.ch004

contribute towards enhanced quality of learning. In particular the chapter will focus on the use of text messaging as a means of improving communications between instructors and students in third-level education. Research has shown that the communication between instructors and students is one of the key factors in the quality of the learning experience of students (Christensen & Menzel, 1998; Christophel, 1990; Ellis, 2004).

The chapter will outline how mobile communication can be used to improve the learning experience of students, in particular third-level students, by improving their perception of instructor communication behaviour. Many benefits arise from improving the learning experience of students including improved attendance and retention. These benefits, which provided motivation for the research, will also be outlined. A theoretical basis for the effect of text messaging on instructor-student relationships will be provided, as will empirical evidence in the form of the findings of a year-long study into the use of text messaging and its effect on student perception of instructor communication behaviour. The chapter will discuss these findings and the integration of text messaging in Multiplatform Learning Management Systems. It will also give some recommendations for effective text messaging in educational settings.

BACKGROUND AND MOTIVATION

Text messaging has been exploited for supporting learning in a variety of ways and in different educational settings. New communication technologies such as mobile text messaging, known as SMS in many countries, provides a means of facilitating frequent and meaningful interaction amongst students and instructors. This interaction engenders feelings in students of being valued, leading to better attendance, student retention and deeper and more meaningful engagement in learning. Text messaging in particular is suitable for supporting out-of-class (OOC) communication between students and instructors since it has the property of being asynchronous, as with email, whereby both parties do not have to be using their devices at the same time in order to send or receive messages. It also has the important advantage of being ubiquitous as there are very few students and instructors these days who do not own at least one mobile device capable of sending and receiving text messages.

There have been numerous examples recently of where text messaging has been used to support education. An interesting research study by Griffith University in Australia relates the experience of a female instructor using OOC text messaging as a means of staying in touch with her students and how it can be used as a way of providing connection and community for first year students (Horstmanshof, 2004). Another study by Kingston University in the UK used OOC text messaging to provide a form of 'mobile scaffolding' at a fundamental level to support the needs of first-year students, and guide students towards independent self-management (Stone, 2004). SMS text messaging may also be used to encourage interactivity in the classroom. This results in a more active learning environment, facilitating the building of learning communities. It provides greater feedback for lecturers, and aids student motivation (Markett, Weber, Sanchez, & Tangney, 2006).

The use of mobile devices in education, also known as mobile learning, is nothing new. There are numerous areas in education where the functionalities of mobile devices have been used to support learning, including interaction and learning in collaborative groups, enquiry-based learning, constructivist and socio-constructivist learning activities, peer-to-peer communication and OOC communication between instructors and students (Hoppe, Joiner, Milrad & Sharples, 2003; Houser, Thornton & Kluge, 2002; Roschelle, 2003; Sharples, 2002). However the vast majority of cases where it has been used up to now have been attempts to enhance cognitive learning among students either individually or in groups. Most of the literature that has appeared regarding the use of mobile communication in education has also been chiefly concerned with enhancing cognitive learning. There has been very little mention of the huge potential of mobile communication technologies to enhance affective or psychomotor learning. In particular this chapter is concerned with enhancing affective learning, which has long been overlooked.

One of the key factors in affective learning is the quality of the communication between students and instructors. Mobile devices provide a means of improving the quality of this communication by the use of out-of-class mobile text messaging. A theoretical basis will be presented that provides the link between the use of text messaging as a means of communication between students and instructors, student perception of their quality of learning and improved affective learning. Empirical evidence will also be presented from a year long study of student perception to a text messaging service offered to them by their instructor.

Affective Learning

When we talk about student learning we are usually only referring to one type of learning behaviour, known as the cognitive domain. To understand the different types of learning that take place we have to refer to Bloom's Taxonomy (1956). Bloom's Taxonomy categorises the hierarchy of learning behaviours into three interrelated and overlapping learning domains; the cognitive (knowledge), affective (attitude) and psychomotor (skills).

Learning in the cognitive domain involves mental processes such as knowledge manipulation and the development of intellectual skills. These include the recall or recognition from memory of specific facts, pattern recognition and concepts that help in the development of intellectual abilities. There are six major categories of cognitive learning behaviours and these categories can be thought of as degrees of difficulties. That is, the first one must be mastered before the next one can take place. The six categories are: knowledge, comprehension, application, analysis, synthesis and evaluation (Bloom, 1956).

In contrast, learning in the affective domain includes the manner by which we deal with things emotionally, such as feelings, values, appreciation, enthusiasms, motivations and attitudes. There are six categories of affective learning behaviours: listening and awareness, responding and active participation, valuing, organisation and internalising values. Learning in the psychomotor domain includes physical and kinesthetic movement, coordination and mastery of activity. While all three domains are important, educational institutions usually place most emphasis on student cognitive learning at the expense of the other two.

The imbalance that exists between the cognitive and affective learning domains is further underlined in the following quote from the seminal publication "Affective Learning – A Manifesto":

The use of the computer as a model, metaphor, and modelling tool has tended to privilege the 'cognitive' over the 'affective' by engendering theories in which thinking and learning are viewed as information processing and affect is ignored or marginalised. In the last decade there has been an accelerated flow of findings in multiple disciplines supporting a view of affect as complexly intertwined with cognition in guiding rational behaviour, memory retrieval, decision-making, creativity, and more. It is time to redress the imbalance by developing theories and technologies in which affect and cognition are appropriately integrated with one another. (Picard et al, 2004, p.1)

As with all areas of research in education, exploration of mobile learning has also placed too much emphasis on cognitive learning at the expense of affective learning. There has been very little research done in the area of affective learning even though recent research has shown that it is critical to the quality of the learning experience of students and in producing lifelong learners (McCombs, 1991).

Motivation

When it comes to instructional communication, results of research suggest that instructor communication behaviour may have its strongest impact on student learning behaviour in the affective domain, out of these three domains (McCroskey, 1994). In fact, affective learning has been identified as the central causal mediator between instructor communication behaviours and cognitive learning amongst students (Andersen, 1981; Rodriguez, 1996). The communication between instructors and students is one of the key factors in the quality of the learning experience of students (Allen, Witt & Wheeless, 2006; Andersen, 1981; Gorham, 1988; Pogue & Ahyun, 2006).

Enhanced communication between instructors and students has been linked to positive student-instructor relationships engendering positive attitudes, increased interest and motivation by students (Christensen & Menzel, 1998; Christophel, 1990; Ellis 2004). A review of research literature in higher education reveals positive relationships between enhanced student perception of instructor communication behaviour and the following desirable qualities, many of which have now become critical to some institutions: improved attendance, improved retention, improved student engagement, improved cognitive learning, improved affective learning, improved classroom behaviour, as well as improved student satisfaction (Allen, Witt & Wheeless, 2006; Kearney, Plax & Wendt-Wasco, 1985; Rocca, 2004; Witt, Wheeless & Allen, 2004).

The motivation behind our research is to try to enhance the communication channel between students and instructor through the use of a text messaging service and thereby improve the quality of the learning experience of students. Our aim is to improve student perception of their instructors' communication behaviour and reap the benefits gained from this enhanced communication which are detailed above.

INSTRUCTOR IMMEDIACY

A significant body of research has found that positive open communication behaviours by instructors are central to the learning process. Positive open communication behaviours have been found to promote affective and cognitive learning in traditional instructional settings. This would suggest that improved instructor communication behaviour can lead to enhanced affective and cognitive learning and may also positively affect student perception of the quality of their learning experience.

The area of instructional communication is based on the assumption that verbal and nonverbal messages conveyed by instructors have the potential to significantly affect student learning outcomes (Witt 2000). When it comes to instructor communication behaviour one important construct is that of instructor immediacy. Immediacy is defined as behaviours, both verbal and nonverbal, that reduce physical or psychological distance between individuals (Andersen, 1979; Mehrabian, 1969, 1971, 1981). The results of a significant body of research conducted on instructor immediacy behaviours indicate that it can have a positive influence on student learning outcomes. For this reason instructor immediacy should be treated with great importance by any person or institution concerned with improving the quality of student learning (Witt, 2000).

Immediacy is based on the principle of approach-avoidance that "people approach what they like and avoid what they don't like" (Mehrabian, 1981, p. 22). Research on instructional communication suggests that nonverbal immediate behaviours by instructors such as physical proximity (Argyle & Dean, 1965; Mehrabian, 1971), direct eye contact (Argyle & Dean, 1965;

Kendon, 1967), smiling (Ekman & Friesen, 1975) and head nods (Mehrabian & Williams, 1969) can be used to express affinity with or liking for students (Witt, 2000). In traditional classrooms students generally perceive the immediate behaviour of their teachers as expressions of personal warmth and affinity toward the students (Ryans, 1964), which in turn enhances student affinity for the teacher, course, and subject matter (Andersen, 1979). Andersen's (1979) study was the first to document a significant relationship between student perceptions of teacher nonverbal immediacy and learning outcomes (Witt, 2000).

Immediacy research entered a new era with Gorham's (1988) investigation of verbal immediacy. Up to that time the actual content of messages had not been considered much in research (Witt, 2000). Research in verbal immediacy found that perception by students of instructors' verbal messages was always within a context that was influenced by the instructors' nonverbal immediacy behaviours (Witt, 2000). Therefore researchers concluded that, in relation to student affective learning, perceived "nonverbal behaviour of teachers served as mediators for teachers' verbal behaviours" (McCroskey & Richmond, 1992). Thus both verbal and nonverbal behaviours should be taken into account when evaluating the effects of instructor communication on student affective learning.

Gorham (1988) developed a more unified model of immediacy, integrating both verbal and nonverbal behaviours. According to her model, teachers should employ verbal strategies to "reduce psychological distance by recognizing individual students and their ideas and viewpoints, by incorporating student input into course and class design, by communicating availability and willingness to engage in one-to-one interactions, and by enhancing their 'humanness' via humour and self-disclosure" (Gorham, 1988, p. 52).

Research studies have shown a linear relationship between student reports of teacher immediacy behaviours and perceptions of state motivation, and of cognitive, affective and behavioural learning (Christensen & Menzel 1998; Pogue and Ahyun, 2006; Witt & Wheeless, 2001). This relationship has been shown to hold true for divergent classes (Kearney, Plax, & Wendt-Wasco, 1985) and also in multi-cultural studies (McCroskey et al., 1996)

EVIDENCE OF EFFECT OF TEXT MESSAGING ON IMMEDIACY

As shown in the previous section, there is no doubt that instructor immediacy has a significant bearing on student affective learning and hence their perception of their learning environment. In third-level education contact time between instructors and students is limited to usually only a few hours a week. It is usually the case that students have very little interaction with their instructors. These are some of the constraints that prompted us to ask the following types of questions: What if we could increase the availability of instructors to the students at any time in a way that would not impact too much on the mobility of instructors and their busy schedules? What if a student could send a query to an instructor from anywhere and outside normal class times? What if an instructor could choose when and where they would deal with and respond to the student's query?

We felt that this type of availability would greatly improve student perception of instructor immediacy and hence student affect which is the goal of this research project. This level of availability needed a system that was both asynchronous like email and also ubiquitous, so that it could be used anytime and from anywhere. Such availability was implemented by making a mobile text service available to students for OOC communication with their instructor.

We not only wanted to provide a theoretical basis for the improvement in student affect as a result of increased availability to their instructor through the use of a text messaging service, we also

Table 1. Categorisation of groups participating in the study

	Full-time	Part-time	Total
Treatment group	3	2	5
Control group	1	2	3
Total	4	4	8

wanted to demonstrate the effect empirically also. We designed a research study to do just that.

We set up a study to explore whether and to what degree OOC communication between instructors and students using text messaging improved student perception of instructor immediacy. Measuring immediacy effects in real world settings is a complex task as many other factors that may influence immediacy are also at play. Immediacy may depend on factors that are not under the instructor's control such as duration of exposure, subject domain, student status (part-time vs. full-time) and class size. The question was if the effect on the immediacy could be attributed to the text messaging and not other factors. For this reason we were very careful in our experimental design so that we could isolate other factors that may affect the immediacy.

Study Set-Up

In total 101 participants from eight different classes took part in the study, four classes of full-time students and four classes of part-time students (see Table 1). Five of the classes were offered a text messaging service by their instructor (treatment groups) while the remaining three were not (control groups). All classes covered technical domains such as databases and networking for computing students.

Each student who agreed to use the text messaging service was requested to fill in a consent form giving permission the instructor to send them text messages. Over 95% of students who were offered the text messaging service agreed to participate. Reasons mentioned by a few students for not participating included having been the victim of prank and hoax calls in the past and simply wishing to keep their mobile numbers private. Participants provided their mobile phone number and received a contact number from the instructor in return and were given guarantees of confidentiality concerning their numbers and their communication.

Throughout the college semester, the treatment group received a number of messages of different types, some for administration purposes (e.g., change of room), some based on the course content (e.g., multiple-choice question), some designed to encourage students to attend class and other miscellaneous messages. The messages were sent and received by a smartphone connected to the instructor's computer in order to ease the administration work needed to run such a service, including message-writing.

Measurement

In order to measure the impact of the text messaging service we assessed the level of students' immediacy at the end of the treatment period of 12 weeks anonymously, using two standardised scales. The immediacy levels of students in both the treatment groups and the control groups were measured to allow for comparison. The first scale used was the Generalised Immediacy Scale (GIS) (Andersen, 1979). The GIS measures a general or gestalt impression of an individuals overall level of immediacy. It comprises 9 items using a 7-point Likert-Scale. Accordingly, the score ranges between 9 and 63. Typically the GIS is used to measure students' perceptions of their

Table 2. Number of messages sent and received by instructor

	class A	class B	class C	class D	class E	total
teaching mode	part-time	full-time	full-time	part-time	full-time	
sent to individuals	22	43	46	22	23	**156**
sent to class	34	38	27	14	14	**127**
sent total	56	81	73	36	37	**283**
received	38	80	76	28	23	**245**
received/sent ratio	68%	99%	104%	78%	62%	**87%**

instructor's level of immediacy (Witt, Wheeless & Allen, 2004). GIS is a highly reliable scale with estimates ranging from .84 to .97 (Andersen, 1979; Kearney, Plax & Wendt-Wasco, 1985; Plax, Kearney, McCroskey & Richmond, 1986). The scale correlates highly with other measures of self-efficacy (Andersen, 1979) and students' affective learning (Kearney, Plax & Wendt-Wasco, 1985). The second scale used was the 10-item Revised Nonverbal Immediacy Behaviours (NIB) instrument (McCroskey, Fayer, Richmond, Sallinen & Barraclough, 1996). The scale has also been checked for reliability and validity with consistently high and positive results (e.g., Christophel, 1990; Gorham, 1988; Gorham & Zakahi, 1990). The NIB scale was chosen as we expected to see an effect related to non-verbal behaviour as opposed to verbal behaviour.

Participants were presented with a series of statements describing nonverbal behaviours of instructors. Participants were asked to indicate the frequency of their instructor's use of each behaviour. Having completed the two assessments, participants in the treatment group were also asked to fill out a comprehensive questionnaire about their attitudes and perceptions of the text messaging service and in what way, if any, it had impacted on them and their learning experience. A series of open questions explored their individual perception of the instructor and reasons given. Moreover, to get an even more detailed picture, students in some of the treatment classes were interviewed about their perceptions and attitudes by an independent person.

Results

Out of the 101 participating students, 63 received text messages on a regular basis from their instructor, while the remaining 38 did not. The text messaging service was well accepted by the participating students. The fact that none of the students in the treatment group dropped out, suggests that the messages were generally welcome and appreciated. In a free form comment at the end of the questionnaire several students expressed their appreciation of the service ("good service to students", "very good idea") and recommended further extension ("should be used with all lectures", "should be applied to all").

A total of 283 messages were sent out to students, 156 of them broadcast messages to the whole class and 127 messages to individual students usually answering individual questions (see Table 2). This shows that many students not only received messages but actively participated in the communication.

The treatment groups perceived the instructor as significantly more immediate than the control groups. This effect was observed with both the General Immediacy Scale, GIS (see Figure 1; $F=10.4$, $p=.007$), and the Nonverbal Immediacy Behaviour Scale, NIB (see Figure 2; $F=9.0$; $p=.004$), i.e. students who received messages perceived their instructor on average 6% - 10% more immediate than those who did not receive messages. As expected, the two scales correlate highly ($r = .40^{**}$).

We expected that student status and previous

Figure 1. Effect of text messaging service on general immediacy scale (GIS)

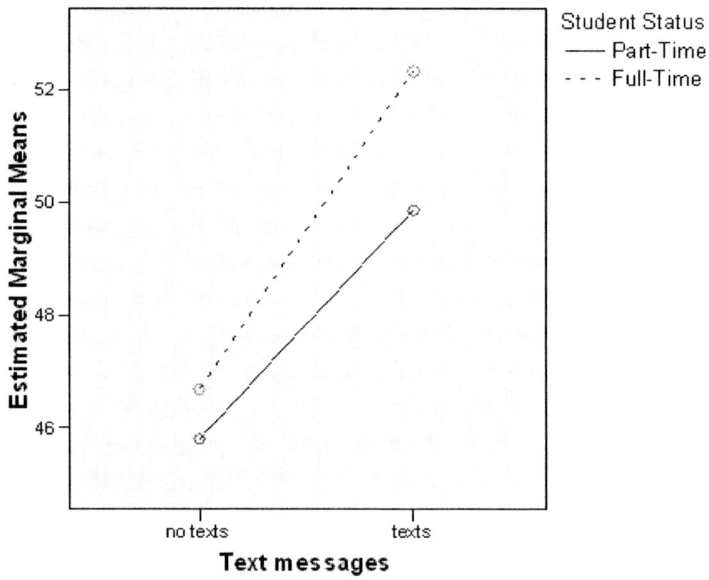

exposure to the instructor might have an effect on immediacy, too. In order to control for these two factors, we computed an ANCOVA with the number of semesters of previous exposure as covariate and student status as additional factor (see Table 3 and Table 4). The results demonstrate that neither status nor previous exposure can explain the observed differences between the groups. While part-time students score lower on average, there is no significant difference between the groups. An interaction effect was also not observed. The treatment effect on GIS is considered to be very large ($\eta 2= .473$), while the effect on NIB seems to be smaller in comparison ($\eta 2=.117$), but is still considered to be a large effect.

The questionnaire gave students the opportunity of stating their opinions on the text messaging service and its effects. The comments of the students are very revealing and they evaluated the service positively. They mentioned that the text messaging encouraged them to view their instructor as being more open and available to them i.e., more immediate. Students could see a variety of benefits of the service, including the availability of immediate feedback to questions and issues arising as well as the opportunity to stay in touch with the instructor ("made me feel closer", "easy to stay in touch").

When asked for their opinions about being in contact with the instructor, who was their lecturer, by text messaging students typically replied that it "shows a good example of commitment to students", it is "good because it helps to keep us up to date. More friendly and natural relation with your lecturer" and "if you have any questions about the class you can text your lecturer and get the answers you're looking for". Very few negative comments were received. Critique referred to costs ("it wastes my credit"; "cost for lecturer"), time of sending ("might be intrusive if messages are sent at in-appropriate times") and content ("sometimes it can be difficult what to say in the texts"). When asked about what they thought about the effect of the text messaging service on their education they typically replied, "it has made me feel closer to the lecturer, more comfortable therefore I am more comfortable asking questions in class or outside of class about my course. I feel I have

Figure 2. Effect of text messaging service on nonverbal immediacy behaviour scale (NIB)

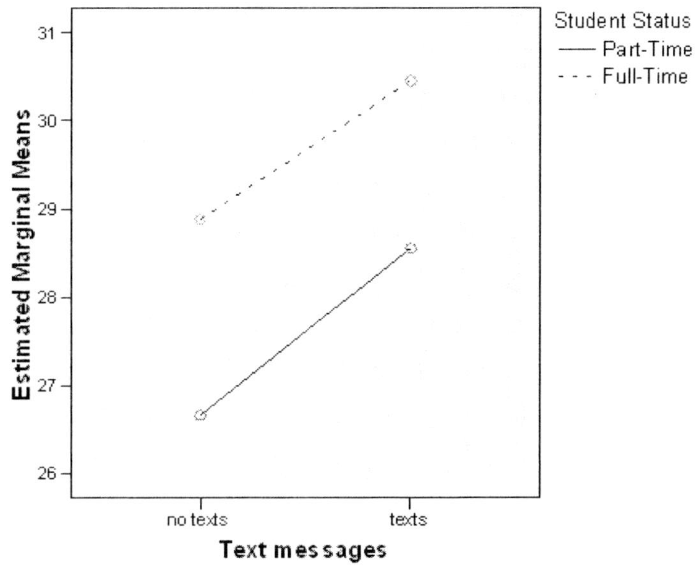

benefited greatly from this", "it is good because the lecturer asks you questions in texts and that refreshes your memory about the class" and it is beneficial "for the simple reason that my lecturer informs me on what's going on to give me time to prepare for the class". Again there were very few negative comments. When asked for their opinions about the effect of the text messaging on their class and their relationship with their instructor the comments were again very positive. When asked about the effect on their learning some students mentioned "encouragement (positivity)", "think it just makes you feel at ease coming to class" and "it has motivated me into attending lectures & seeing the lecturer as a person". This is further evidence of enhanced instructor immediacy and improved affective learning as a direct result of the text messaging.

Discussion of Study Results

We have seen from the results that OOC communication using text messaging has a positive effect on student perception of instructor immediacy, as evidenced by the increased scores on the GIS and NVIB scales and from feedback in the form of comments by individual students. The students who received the text messages perceived the in-

Table 3. Results of ANCOVA showing effect of duration of previous exposure to instructor (covariate), status of student (full-time vs. part-time) and text messaging on general immediacy scale (GIS)

Source	df	F	Sig.	Partial η^2
Intercept	1	278.6	.005	
Previous Exposure	1	.6	.433	
Status	1	15.5	.769	
Text Messaging	1	10.4	**.007**	.473
Status * Text Messaging	1	.1	.705	

Table 4. Results of ANCOVA showing effect of duration of previous exposure to instructor (covariate), status of student (full-time vs. part-time) and text messaging on nonverbal immediacy behaviour scale (NIB)

Source	df	F	Sig.	Partial η^2
Intercept	1	692.4	.000	
Previous Exposure	1	.1	.764	
Status[1]	1	-	-	
Text Messaging	1	9.0	**.004**	.117
Status * Text Messaging	1	.5	.863	

structor as closer and were therefore more likely to ask questions in class and engage in discussions with the instructor. Other studies have shown that immediacy enhances affective learning, leading to improved attendance, retention and student engagement as well as other desirable traits.

While immediacy is very important to the success of learning it is of course not the only factor. However its contribution to affective learning should not be underestimated. While every effort was made to ensure that there was no bias in the treatment of classes, double-blind treatment was not possible. The instructor was aware of the different treatments of groups but students were not. Every precaution was taken to ensure that bias was not introduced.

Of course such a service to students requires some additional effort on the part of instructors, however the effort from our experience is minimal and the benefits gained are well worth the extra effort, including better class attendance and engagement, better rapport with instructors and better student learning experience in general.

Costs of sending messages were hardly mentioned by students. On average each student sent very few messages in comparison to the total of messages sent by the instructor. Free text message contracts are becoming more and more common and some students had free text messaging with the instructor. These particular students were more likely to communicate using the text messaging service. Universities might introduce text messaging as part of student services in the same way as email services are considered a standard now.

Despite these overwhelmingly positive experiences, the study design imposes a number of limitations. First, the study involved only five different classes, all of them on technical subjects. We argue that text messaging is popular among students of all disciplines, however, some subject areas might not lend themselves to texting in the same way e.g., due to the restrictions of expressiveness of text (cf. mathematical or chemical formulas; special characters in foreign languages), and due to restriction in length of messages to 160 characters (cf. philosophical arguments).

Secondly, our classes are relatively small. Entering communication with classes of hundreds of students might be unfeasible or impossible. In that case messaging might be restricted to sending broadcast messages.

Thirdly, the study explored the impact on immediacy in detail, combining qualitative and quantitative methods. However, the consequences of increased immediacy on learning effect and other factors were not assessed. The close relationship between learning and immediacy has been demonstrated many times, but we cannot prove that students actually benefited.

TEXT MESSAGING IN LEARNING MANAGEMENT SYSTEMS

Mobile text messaging has some advantages for increasing perceived instructor immediacy and decreasing social distance between instructor and student over other means of on-line communication. This is due to its ubiquitous property as discussed earlier and also because of the fact that mobile text messaging is generally perceived by students as being personal in nature (Horstmanshof, 2004; Naismith, Lonsdale, Vavoula & Sharples, 2004).

Garrison, Anderson and Archer (1999) showed how critical inquiry could be supported in a text-based on-line environment and the importance of cognitive presence, social presence and teaching presence as elements of an educational experience. Mobile text messaging also supports critical inquiry in much the same manner, with the added advantage that students may send or receive messages at any time and do not have to be logged-in to a system. There is no reason why existing on-line text-based communication cannot be integrated with mobile text messaging. Such a communication system would support e-moderation requirements as well as provide students with support when they are not on-line.

Integration of Text Messaging

For this study the instructor used a laptop with a special software package that worked just like an email program, recording each message sent and received. While this worked fine for the purpose of this study, experience shows that for text messaging usage as described here to be used to improve immediacy on a larger scale the following requirements would need to be met. First of all, integration with the student record system would be required in order that instructors would not have to collect student phone numbers individually. In many cases these numbers are included in the student records anyway.

Secondly, students should complete a consent form when registering to allow all instructors to send text messages to them as part of the teaching. The consent form should outline students' rights and obligations as discussed below.

Thirdly, text messaging needs to be integrated with an existing Learning Management Systems (LMS) to reach its full potential. The LMS would allow instructors to address complete classes by a single click. It would also serve as a permanent record of all communication. Such integration would also allow easy reference to subjects and learning resources (e.g., URL included in message text). In the context of this study all these things had to be added manually.

Fourthly, a set of guidelines and rules need to be established that describes how the service is to be used and what students can expect. Enhancing immediacy through text messaging directly affects the student-instructor relationship.

Guidelines for Text Messaging

To avoid wrong expectations and misunderstandings the following aspects of text messaging should be stated explicitly from the outset of the service: firstly, the quality of service should be defined including maximum response times and office-hours. For example, some students noted that they do not want to receive messages in the evening. Sending messages during exam periods to individual students might be perceived as unfair advantage by their peers.

Secondly, the type and content of text messages should be defined. For example, information sent out to students should be redundant, i.e., messages should be posted as email or through other available channels as well and a permanent record should be kept. Students may be encouraged to ask questions when they arise rather than waiting for the next session or not asking them at all. Normally, answers to individual students would be available to the whole class to avoid any unfair advantage.

Thirdly, the ad-hoc use and often colloquial nature of communication via texts might potentially lead to misuse of the service, a phenomenon often observed in the early days of the introduction of email in organisations. Students should be made aware that text messages in this context are still part of the learning experience and thus need to comply with general communication rules. Our experience shows that there is a fine but significant line between high perceived instructor immediacy and close personal friendship. Students might misinterpret the higher availability and closer interaction with the instructor as a kind of peer relationship. They might then be surprised or disappointed when the instructor executes the necessary duties of his/her role such as disciplining students or allocation of marks.

IMMEDIACY THROUGH OTHER COMMUNICATION CHANNELS AND DEVICES

In this chapter we have shown how the use of mobile communication technology in the form of OOC text messaging can be used to increase instructor immediacy. Text messaging of course has the properties of being asynchronous and ubiquitous which make it very suitable for supporting OOC communications. There are a number of other channels of communication which can also be used to enhance instructor immediacy.

The Internet in particular offers a plethora of possible communication channels by which students could have contact with their instructors. We are not advocating the use of all these technologies but are simply pointing out that they provide possible channels of communication both now and in the future. The technologies include simple asynchronous email, instant messaging (IM) services such as MSN and Google Chat and also voice-over-IP (VOIP) services including Skype and Google Talk. Going even further in exploiting latest technologies, some universities already have a presence on the virtual reality sites such as Second Life.

Mobile technologies also offer the options of speech, text messaging, multimedia messaging, instant messaging, wireless networking and Bluetooth connections. Many mobile devices also have access to the Internet, making available services such as MSN and Skype.

Immediacy and other variables such as social distance and teaching distance (Garrison, Anderson & Archer, 1999) are being viewed with more and more importance in the field of e-learning and distance education. With the rapid adoption of the Internet into a mainstream communication medium, there has been a recognition of the importance of the dynamics of interpersonal communication in the online environment. There is significant overlap between discussion of traditional immediacy producing behaviours and discussions of online interpersonal communication (Woods & Baker, 2004).

While text messaging has the advantage that nearly everyone has a mobile device that supports it, be it a standard mobile phone, smartphone or PDA, it has not yet been fully incorporated into existing multiplatform e-learning systems such as LMSs. There have been some efforts made to implement a text messaging interface for Moodle and a few other types of LMS but it has not yet become a standard feature. As e-learning systems develop and become increasingly multiplatform more and more systems will have the capacity to communicate with registered students by sending and receiving text messages.

The other communication channels we have mentioned as a possible means of facilitating communication between instructors and students are increasingly available on different platforms. For example, a lot of mobile devices and laptops now support Bluetooth, regardless of system architecture or operating system. Any device with a web browser, large or small, can access MSN and social networking sites. Instant messaging is becoming more common not just on PCs and laptops but also on mobile devices.

CONCLUSION AND FUTURE PERSPECTIVES

In this chapter we provided evidence that text messaging between students and instructor can increase perceived immediacy and affective learning. Both quantitative measures of immediacy and qualitative feedback from students show that the instructor is perceived as closer, more positive and responsive when text messaging services are offered. Students expressed their wish to increase the use of text messaging in the future. Similar services might be offered through other channels (e.g., instant messaging, Bluetooth in the classroom) and on a variety of devices. Even virtual realities environments (such as Second Life) may be used to increase instructor immediacy.

However, experience in real world settings also makes it obvious that clear rules and guidelines for such services are required to set students' expectations. Integrating messaging with existing infrastructure such as student registration and learning management systems would make these services easily available to instructors and students on a large scale.

Future work will be needed to explore the scalability of the results reported in this chapter across organisations and subject domains. Text messaging might also be used in school settings with similar effect. More work is also required on the impact of text messaging on other learning factors. For example, we anticipate that both student motivation and learning affect would benefit.

In conclusion, we have shown that increasing instructor immediacy has great potential in positively affecting student behaviour and learning. Our results indicate that text messaging may be used for this purpose. Moreover, we argue that other channels and devices used for in-class and out-of-class communication may have similar potential.

REFERENCES

Allen, M., Witt, P. L., & Wheeless, L. R. (2006). The role of teacher immediacy as a motivational factor in student learning: Using meta-analysis to test a causal model. *Communication Education, 55*(1), 21–31.

Andersen, J. F. (1979). Teacher immediacy as a predictor of teaching effectiveness. In D. Nimmo (Ed.), *Communication Yearbook 3* (pp. 543-559). New Brunswick, NJ: Transaction Books.

Andersen, J. F. (1981). Three investigations exploring relationships between perceived teacher communication behaviors and student learning. *Communication Education, 30*(4), 377–392.

Argyle, M., & Dean, J. (1965). Eye-contact, distance and affiliation. *Sociometry, 28*(3), 289–304. doi:10.2307/2786027

Bloom, B. S. (1956). *Taxonomy of educational objectives: The classification of educational goals.* New York: McKay.

Christensen, L. J., & Menzel, K. E. (1998). The linear relationship between student reports of teacher Immediacy behaviors and perceptions of state motivation, and of cognitive, affective, and behavioral learning. *Communication Education, 47*(1), 82–90.

Christophel, D. M. (1990). The relationships among teacher immediacy behaviors, student Motivation, and learning. *Communication Education, 39*(4), 323–340.

Ekman, P., & Friesen, W. V. (1975). *Unmasking the face: a guide to recognizing emotions from facial clues.* Prentice-Hall, Englewood Cliffs, NJ.

Ellis, K. (2004). The impact of perceived teacher confirmation on receiver apprehension, motivation, and learning. *Communication Education, 53*(1), 1. doi:10.1080/0363452032000135742

Garrison, D. R., Anderson, T., & Archer, W. (1999). Critical inquiry in a text-based environment: Computer conferencing in higher education. *The Internet and Higher Education*, *2*(2-3), 87–105. doi:10.1016/S1096-7516(00)00016-6

Gorham, J. (1988). The relationship between verbal teacher immediacy behaviors and student learning. *Communication Education*, *37*(1), 40–53.

Gorham, J., & Zakahi, W. R. (1990). A comparison of teacher and student perceptions of immediacy and learning: Monitoring process and product. *Communication Education*, *39*(4), 354.

Hoppe, H., Joiner, R., Milrad, M., & Sharples, M. (2003). Guest editorial: Wireless and mobile technologies in education. *Journal of Computer Assisted Learning*, *19*(3), 255–259. doi:10.1046/j.0266-4909.2003.00027.x

Horstmanshof, L. (2004, 5-8 December). *Using SMS as a way of providing connection and community for first year students.* Paper presented at the Beyond the Comfort Zone: Proceedings of the 21st annual conference of the Australasian Society for Computers in Learning in Tertiary Education, Perth.

Houser, C., Thornton, P., & Kluge, D. (2002). Mobile learning: cell phones and PDAs for education. *International Conference Proceedings on Computers in Education* (pp. 1149-1150).

Kearney, P., Plax, T. G., & Wendt-Wasco, N. J. (1985). Teacher immediacy for affective learning in divergent college classes. *Communication Quarterly*, *33*(1), 61.

Kendon, A. (1967). Some functions of gaze-direction in social interaction. *Acta Psychologica*, *26*(1), 22–63. doi:10.1016/0001-6918(67)90005-4

Markett, C., Sanchez, I. A., Weber, S., & Tangney, B. (2006). Using short message service to encourage interactivity in the classroom. *Computers & Education*, *46*(3), 280–293. doi:10.1016/j.compedu.2005.11.014

McCombs, B. L. (1991). Motivation and lifelong learning. *Educational Psychologist*, *26*(2), 117–127. doi:10.1207/s15326985ep2602_4

McCroskey, J. C. (1994). Assessment of affect toward communication and affect toward instruction in communication. In M. Morreale & S. Brooks (Ed.), *1994 SCA Summer Conference Proceedings and Prepare Remarks* (pp. 55-65).

McCroskey, J. C., Fayer, J. M., Richmond, V. P., Sallinen, A., & Barraclough, R. A. (1996). A multicultural examination of the relationship between nonverbal immediacy and affective learning. *Communication Quarterly*, *44*(3), 297.

McCroskey, J. C., & Richmond, V. P. (1992). Increasing teacher influence through immediacy. *Power in the Classroom: Communication, Control, and Concern*, *45*, 200–211.

Mehrabian, A. (1969). Some referents and measures of nonverbal behaviour. *Behavior Research Methods and Instrumentation*, *1*(6), 205–207.

Mehrabian, A. (1971). *Silent Messages*. Belmont, CA: Wadsworth.

Mehrabian, A. (1981). *Silent Messages* (2nd ed.). Belmont, CA: Wadsworth.

Mehrabian, A., & Williams, M. (1969). Nonverval concomitants of perceived and intended persuasiveness. *Journal of Personality and Social Psychology*, *13*(1), 37–58. doi:10.1037/h0027993

Naismith, L., Lonsdale, P., Vavoula, G., & Sharples, M. (2004). *Literature Review in Mobile Technologies and Learning.* Bristol: NESTA FutureLab.

Picard, R. W., Strohecker, C., Papert, S., Bender, W., Blumberg, B., & Breazeal, C. (2004). Affective learning - A manifesto. *BT Technology Journal*, *22*(4), 253–269. doi:10.1023/B:BTTJ.0000047603.37042.33

Plax, T. G., Kearney, P., McCroskey, J. C., & Richmond, V. P. (1986). Power in the classroom VI: Verbal control strategies, nonverbal immediacy and affective learning. *Communication Education*, *35*(1), 43.

Pogue, L. L., & Ahyun, K. (2006). The effect of teacher nonverbal immediacy and credibility on student motivation and affective learning. *Communication Education*, *55*(3), 331–344.doi:doi:10.1080/03634520600748623

Rocca, K. A. (2004). College student attendance: impact of instructor immediacy and verbal aggression. *Communication Education*, *53*(2), 185–195. doi:doi:10.1080/0363452041000l682447

Rodriguez, J. I. (1996). Clarifying the relationship between teacher nonverbal immediacy and student cognitive learning: Affective learning as the central causal mediator. *Communication Education*, *45*(4), 293–305.

Roschelle, J. (2003). Keynote paper: Unlocking the learning value of wireless mobile devices. *Journal of Computer Assisted Learning*, *19*(3), 260–272. doi:10.1046/j.0266-4909.2003.00028.x

Ryans, D. G. (1964). Research on teacher behavior in the context of the teacher characteristics study. *Contemporary Research on Teacher Effectiveness*.

Sharples, M. (2002). Disruptive devices: mobile technology for conversational learning. *International Journal of Continuing Engineering Education and Lifelong Learning*, *12*(5/6), 504–520. doi:10.1504/IJCEELL.2002.002148

Stone, A. (2004). *Mobile scaffolding: An experiment in using SMS text messaging to support first year university students.* Paper presented at the IEEE International Conference on Advanced Learning Technologies, ICALT 2004, Joensuu, Finland.

Witt, P. L. (2000). *An experimental of teachers' verbal and nonverbal immediacy, student motivation, and cognitive learning in video instruction.* University of North Texas.

Witt, P. L., & Wheeless, L. R. (2001). An experimental study of teachers' verbal and nonverbal immediacy and students' affective and cognitive learning. *Communication Education*, *50*(4), 327–342.

Witt, P. L., Wheeless, L. R., & Allen, M. (2004). A meta-analytical review of the relationship between teacher immediacy and student learning. *Communication Monographs*, *71*(2), 184–207.doi:doi:10.1080/036452042000228054

Woods, R. H. Jr, & Baker, J. D. (2004). Interaction and immediacy in online learning. *International Review of Research in Open and Distance Learning*, *5*(2).

ENDNOTE

1 Due to lack of variance, degrees of freedom using Satterthwaite's method cannot be computed.

Chapter 5
The Role of Multi-Agent Social Networking Systems in Ubiquitous Education
Enhancing Peer-Supported Reflective Learning

Jonathan Bishop
Glamorgan Blended Learning LTD, UK

ABSTRACT

Knowledge it could be argued is constructed from the information actors pick up from the environments they are in. Assessing this knowledge can be problematic in ubiquitous e-learning systems, but a method of supporting the critical marking of e-learning exercises is the Circle of Friends social networking technology. Understanding the networks of practice in which these e-learning systems are part of requires a deeper understanding of information science frameworks. The Ecological Cognition Framework (ECF) provides a thorough understanding of how actors respond to and influence their environment. Forerunners to ecological cognition, such as activity theory have suggested that the computer is just a tool that mediates between the actor and the physical environment. Utilising the ECF it can be seen that for an e-learning system to be an effective teacher it needs to be able to create five effects in the actors that use it, with those being the belonging effect, the demonstration effect, the inspiration effect, the mobilisation effect, and the confirmation effect. In designing the system a developer would have to consider who the system is going to teach, what it is going to teach, why it is teaching, which techniques it is going to use to teach and finally whether it has been successful. This chapter proposes a multi-agent e-learning system called the Portable Assistant for Intelligently Guided Education (PAIGE), which is based around a 3D anthropomorphic avatar for educating actors ubiquitously. An investigation into the market for PAIGE was carried out. The data showed that those that thought their peers were the best form of support were less likely to spend more of their free time on homework. The chapter suggests that future research could investigate the usage of systems like PAIGE in educational settings and the effect they have on learning outcomes.

DOI: 10.4018/978-1-60566-703-4.ch005

INTRODUCTION

E-Learning is a term that describes electronically supported information systems that seek to impart knowledge. Or is it not as clear as that? Brown & Duguid (2002) ask whether there is something that the term knowledge catches that information does not and goes on to talk about networks of practice where actors within them have practice and knowledge in common. Mantovani (1996) suggests that knowledge is something that is constructed through social context and Suchman (1987) argues that plans, which can be seen as cognitive structures, are resources for action, suggesting that they are what affect the practice of an **actor**. These authors would seem to suggest that **knowledge** is what groups of actors construct through their individual interpretations of the information they share during their participation networks of practice.

Educational institutions that use electronic means to deliver their learning can be considered to be e-learning **networks of practice**. The use of e-learning systems in these networks opens up the possibility of increasing the flow of information and the use of knowledge, where the participants can support each other as critical friends. Such e-learning systems can make use of peer-assessment techniques, which can enhance self-directed learning and reflection (English, Brookes, Avery, Blazeby, & Ben-Shlomo, 2006). One social networking methodology that can achieve this is the Circle of Friends, which has the potential to act as a learner's list of approved peers, who could peer-assess their work critically and fairly. The **Circle of Friends** fits into a long history of using the Internet as an environment for developing relationships and increasing sociability (Weng, 2007). The first social networking service on the Web was Classmates.com, which launched in 1995 and used the Old School Tie social networking method, which is defined as a method for building networks of users using the schools and universities they graduated from. This was followed in 1997 with the launch of SixDegrees.com, which utilised the Web of Contacts model, which is defined as a technique for displaying social networks using social networking analysis that the user doesn't manage it. The advantage of the Circle of Friends, which was developed in 1999 as part of the Virtual Environments for Community and Commerce (VECC) Project (Bishop, 2002; Bishop, 2007a is that it allows the user to manage their network and decide who they want to be friends with. The 2001 implementation of the Circle of Friends as part of Llantrisant.com allowed users to classify their friends according to whether they trusted them or not, combining it with the Circle of Trust that was also developed in 1999. One of the potential problems of the Circle of Friends is that it might promote "friendship marking", which B. L. Mann (2006) defines as peer over-marking where there is a reluctance to provide critical comments and suggest that such problem may not be easily overcome by using online peer assessment. Understanding how people learn through peer-assessment and the role of technology such as the Circle of Friends in enhancing the learning outcomes of learners requires a deeper understanding of human behaviour within networks of practice.

BACKGROUND

Understanding networks of practice draws on various aspects of information science, including emerging fields such as post-cognitive psychology. It has been argued that there should be a framework for understanding actors based on ecological perceptual psychology (Kyttä, 2003). It is quite clear that any model to explain the behaviour of **actors** that ignores the possibly of direct perception, or one that ignores the role of the environment assuming that actors are wholly self-motivated and independent of their environment cannot fully explain the behaviour of actors in either physical or virtual environments. The **Ecological Cogni-**

Figure 1. The ecological cognition framework

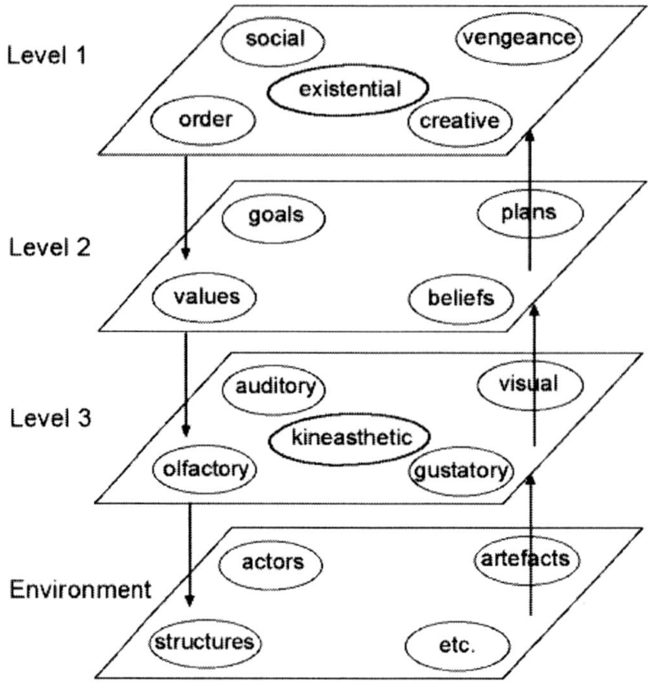

tion Framework proposed by (Bishop, 2007b provides a thorough understanding of how actors respond to and influence their environment. According to Rasmussen, Duncan, & Leplat (1987) the classification of human performance in skill, rule and knowledge-based situations behaviour is the role of the information observed from the environment. As can be identified in Figure 1, the ECF suggests that there are three levels that affect an actor's behaviour, connected through arrows that represent the process from the actor perceiving their environment through to making changes to it.

The six processes of the ECF according to Bishop (2007c are the Stimulus (from the Environment to Level 3), the Impetus (From Level 3 to Level 2), Intent (From Level 2 to Level 1), The Neuro-Response (from Level 1 to Level 2), The Judgement (from Level 2 to Level 3) and the Response (from Level 3 to the Environment). A stimulus is a detectable change in an actor's body's internal or external environment (Layman, 2004) and an actor will be influenced by a stimulus from those elements within the environment, such as structures, artefacts and actors, those actors being either themselves or others. The difference between how a stimulus is represented in the ECF and how classical psychologists thought of it is that in the ECF the stimulus is not connected to a so-called reflex, but creates the impetus for change in the actor. According to McClure (2003) a stimulus is anything that arouses activity either internally or externally. When an actor perceives a stimulus using their senses (Level 3 of the ECF) this results in an impetus, which is represented in the ECF by the arrow between Level 3 and Level 2. Bishop (2007c described five impetuses, which are affordances, resonances and cognizances, affectances and urgeances. Actors may experience many stimuli that initiate many impetuses, but nothing will come of these unless the actor creates the intent to exploit them, as represented by the arrow in the ECF between Level 2 and Level 1. Intent is created after an impetus interacts with

Table 1. Elements in the processes of the ECF

Process	Description
Stimulus	Connects the environment to Level 3. Can include Social (e.g.. what people hear), Gross (i.e. what people feel physically), Emotional (e.g. what people feel emotionally), Mental (i.e. what people sense through their thoughts) or Active (e.g. what people sense through sight)
Impetus	Connects Level 3 to Level 2. Can include Affordance (activated by Active), Resonance (activated by Social), Cognizance (activated by Mental), Affectance (activated by Emotional), Urgeance (activated by Gross)
Intent	Connects Level 2 to Level 1. The impetus will cause either a Doxa (activates a belief), Identification (activates a value), Bond (activates an interest) Enigma (activates a plan), or a Need (activates a goal)
Neuroresponse	Connects Level 1 to Level 2. The force resultant from the Intent causes a Neuroresponse, which can either be a Convention (creates a value), Relation (creates an interest), Episteme (creates a belief), Ambition (creates a goal), Desire (creates a plan).
Judgement	Connects Level 2 to Level 3. After a new cognition being created, the actor has to decide what do with it. The Judgement will be either Deference (cognition accepted without dissonance), Intemperance (cognition accepted after dissonance resolved), Retecance (cognition neither accepted or rejected and dissonance still present), Temperance (cognition rejected and dissonance resolved), Ignorance (cognition rejected without any dissonance)
Response	Connects Level 3 to the Environment. Acts on the Judgement, with the same elements as Stimuli, resulting in a change in the actor or their environment.

one of the cognitions at Level 2. Dependent on the intent, the actor next experiences a Neuroresponse, represented in the ECF as the arrow between Level 1 and Level 2. The Neuroresponse results in a change in the cognitions of an actor, which can result in them experiencing dissonance if the additional cognition is not consonant with their existing cognitions. In psycho-analytical theory (Freud, 1933) this stage, which is represented in the ECF by the arrow between Level 2 and Level 3, was thought to occur because of a conflict between the id and the ego, but in the context of the ECF it can be seen to occur because of a change in cognitions. The response to a stimulus was in classical psychology seen as a reflex, but as can be seen by the many stages of the ECF, where the response is represented as the arrow between Level 3 and the Environment, it occurs not simply as a result of a reflex, but as the result of the cognitive processes identified above. The response changes the environment, either intrinsically though modifying the actor, or extrinsically though them modifying their environment. This is summarised in Table 1.

Ecological Cognitive Learning Theory

The **Ecological Cognition Framework**, presented in Figure 1 provides the theoretical basis for understand how learning occurs in environments through the actors that take part in them. Figure 2 provides a visual representation of the **Ecological Cognitive Learning Theory** (ECLT) proposed by (Bishop, 2007d). The outer segments consist of the senses that form part of Level 3 of the ECF. The next segments inwards consist of the types of stimuli and response that are at between Level 3 and the Environment in the ECF. The five stages of learning can be seen in the arrows that makeup the outer circle. The five stages of developing learning materials can been seen in the inner circle, where 'who' the learning is targeted at is labelled Dependents. Yardsticks is the label for 'what' is being taught. Standpoints is the label 'why' it is being taught. Guides is the label for 'which' methods will be used. Understandings (making up the acronym DYSGU, the Welsh word for learn/teach) is the label for 'whether' it has been successful. The nature and focus of an

Figure 2. Ecological cognitive learning theory

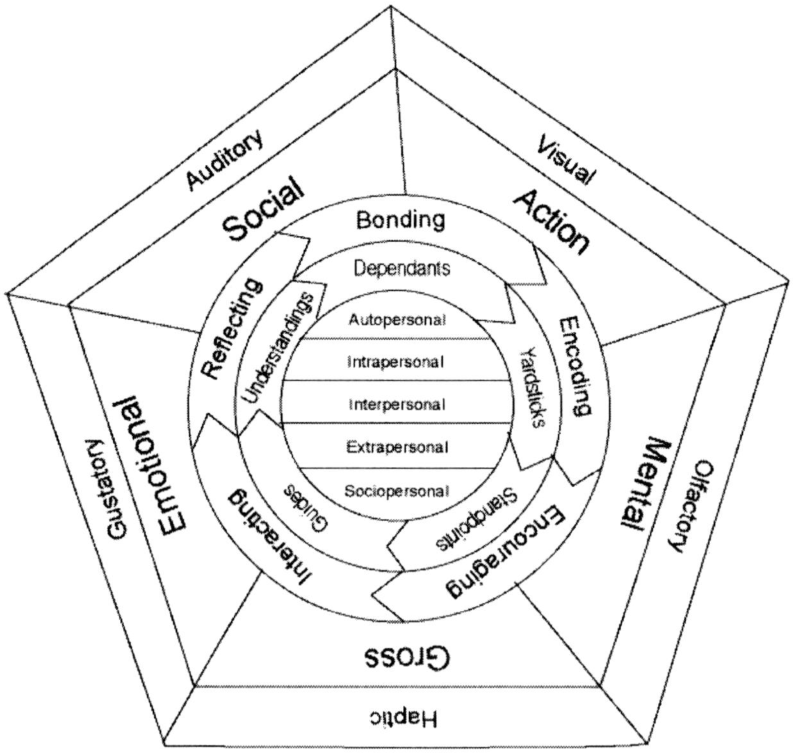

actor's cognitions are at Level 2 of the ECF and can be seen in the centre of the model.

According to Bishop (2007d) for an e-learning system to be an effective teacher it needs to be able to create the five effects in the actors that use it identified above, with those being the belonging effect, the demonstration effect, the inspiration effect, the mobilisation effect, and the confirmation effect. In designing the system a developer would have to consider who the system is going to teach, what it is going to teach, why it is teaching, which techniques it is going to use to teach and finally whether it has been successful.

Existing theories of learning such as **experiential learning** (Kolb, 1984) offer an incomplete understanding of the learning process because they miss out the important process of bonding (Bishop, 2007d). If Vygotsky (1930) is correct in his view that a actor can learn more with the aid of a more competent person, then the relationship between the learner and the educator needs to be cognitively developed through the learner experiencing the belonging effect and developing an interest in the educator. An actor can be made to experience the belonging effect when a source provokes an actor into developing an interest in the source. An interest can be provoked by a source through the *ingratiation process*, and in groups through the *bonding process*, whereby participants become more cohesive, and is achieved through specific types of '**ice-breaker**' (Jackson, 2001). An ice-breaker can be defined as a game or short activity that allows the educator and the learner to get to know each other at the start of a course, something which can also be used as a warm-up activity at the start of a session (Corder, 2002). Ceccucci & Tamarkin (2006) argue that ice-breakers can be suitable for electronic learning environments, particularly as they point out that these environments can be somewhat isolating to the learner. They argue that online ice-breakers can be used by educators to accomplish a variety of goals, such as introducing learners to one another, sharing experiences, benefiting from team learning, increasing participation, or encouraging learners to develop constructive online relationships throughout the course and allow students get to know one another on a personal basis. The ice-breaker could contribute to making a learner experience the belonging effect, in which they develop an interest in the e-learning system and the others on the same course.

An e-learning system or its participants that a learner has an interest in is well placed to help a learner develop a new skill by building on their knowledge through making them experience the demonstration effect. Rushby, 1979a, 1979b) argues that in order to guide learners effectively and to provide useful information to educators, an e-learning system must maintain records about the learner's performance and progress. Having a record of what a learner knows can be useful in targeting new information at them that is in keeping with what they believe and value, so as to create the demonstration effect in them. Through providing learning materials customised to the learner, they can be encouraged to build on what they already know and through the sub-conscious encoding process develop beliefs and values that are in their interests. For example an e-learning system could gather information on the learner's favourite television programmes. An e-learning system can create the belonging effect in the actors that use it, providing the foundation to create beliefs and values in the actor through creating the demonstration effect, but an actor will not be able to enact a demonstrated skill unless the e-learning system can create a goal in the learner through the inspiration effect, which goes beyond collecting data on the individual learner and providing customised learning materials. As LeWinter (2003) put it, "to teach, it is good / to motivate, it is better / but our goal, to inspire / can't be measured by a grade or a letter." An e-learning system should utilise a learner's values and inspire them to develop goals to utilise the knowledge they have gained through using the system. One of the most utilised processes in education that create the inspiration effect is the *encouragement process* as proposed by Dinkmeyer & Dreikurs (1963), which when applied by an educator can inspire a learner with the goals to do their best (Coop & White, 1974). A way to create the inspiration effect in actors is through persuasive dialogue, as recommended by Lickona (1983) for those persuading young learners to develop goals about their future. An e-learning system that has inspired a learner to create goals can utilise these to mobilise them to develop plans to utilise the knowledge they have gained, thus creating the mobilisation effect in the actor. One way an e-learning system can create the mobilisation effect in the learner is through the *interaction process*, which Allwright (1984) indicates changes the plans of educator and learner. Battistich, Solomon, & Delucchi (1993) suggested that the interaction process requires **collaborative learning** groups, though cites Webb (1982) who indicates that this process has not been fully investigated. An e-learning system could utilise the interaction process through allowing collaboration between the different users and through allowing the learner to interact with the system, which could suggest actions in order to create the mobilisation effect and allow the actor to develop skills. After a learner has experienced the mobilisation effect and made changes to their environment as a result, the next step for them is to experience the confirmation effect, through the reflection process. Much research has been done into the process of reflection, most notably by Kolb (1984) and Honey & Mumford (1986). These authors argue that the reflection process is essential after a learning experience, such as that created by the mobilisation effect. Other authors such as Vygotsky (1930) refer to this process as internalisation, whereby an external operation becomes internalised. One effective means of enabling learners to go through the reflection process to experience the confirmation effect in an e-learning system is the use of weblogs, as proposed by Bishop (2004).

TOWARDS AN MOBILE E-LEARNING SYSTEM USING SOCIAL NETWORKING AND AGENT-BASED TECHNOLOGIES

Forerunners to **ecological cognition**, such as activity theory have suggested that the computer is just a tool that mediates between the actor and the physical environment (Engeström, 1993; Kaptelinin, 1996; Nardi, 1996b). Despite its limitations activity theory has been applied to human computer interaction in general (Nardi, 1996a) and e-learning specifically (Uden, 2007). While activity theory has its theoretical limitations, which are addressed by ecological cognition, some of the conclusions drawn by Vygotsky (1930) can be explained by ecological cognition and are useful in developing e-learning systems. Vygostky's Zone of Proximal Development (ZPD) suggests there is a distance between the actual development level of a learner as determined by what they can achieve independently and the level of potential development as determined by what they can achieve with the support of more capable peers. Whilst some of those utilising Vygotsky's work have argued that the computer is no more than a mediating artefact, others have argued that a computer interface provides mediating artefacts (e.g. Bishop, 2005).

The Portable Assistant for Intelligently Guided Education

The author proposes a multi-agent e-learning system called the Portable Assistant for Intelligently Guided Education (PAIGE), which is based around a 3D anthropomorphic **avatar** for educating actors ubiquitously and was developed using Bishop (2007a's adaptation of the star lifecycle and the PASS approach proposed by Bishop (2004), which indicated that e-learning systems should be persuasive, adaptive, sociable and sustainable. According to Foroughi & Rieger (2005), in the area of e-learning, avatars have mainly been implemented in custom made applications for the purpose of on-the-job training, where the training is conveyed through role play scenarios with instructions with the main interactive avatar normally representing the learner and depending on the training task, while other avatars play other people involved in the learning scenario. Whilst the avatar in PAIGE is used for demonstrations, it does not represent the learner, but interacts with them.

Content Issues

The Content of PAIGE supports the learning programmes as part of Glamorgan Blended Learning's Emotivate Project. GBL teaches its course in three segments that take account of the Ecological Cognition Learning Theory presented above, which are 'Learn', where the belonging and demonstration effects take place, 'Create', where the inspiration and mobilisation effects take place and 'Communicate', where the confirmation effect occurs followed by the belonging effect once more. The learning outcomes of the Basic Football Skills learning programme include for learners to identify the appropriate playing area and the necessary equipment, describe and demonstrate the main rules of the game, set up small game situations to test the skills learned, perform simple skills on the football pitch and evaluate learning and performance.

Technology Issues

The mobile platform is well suited to using agents as its robust and fault-tolerant architecture means that it is able to overcome network latency problems that can occur with network-intensive agents (Lange & Oshima, 1999).

Figure 3. PAIGE – A ubiquitous multi-agent social networking system

Persuasive Architecture

The Persuasive architecture of PAGE includes multiple agents, which provide information to users through either prompts or responding to queries and persuade them to carry out particular actions or approach educational activities in a particular way.

Advice Agent

The Advice Agent assists learners with the Bonding Stage of the Ecological Cognition Learning Theory identified above, which creates the belonging effect. It has long been argued that computer systems should know their audience and present information in a language familiar to them and it has also long been recommended that a system be evaluated on whether it 'speaks the user's language' and it has been argued that developers should, 'use wording and language that can be understood entirely by the target audience' (Ozok & Salvendy, 2004).

Reddy & Lewis (2002) suggested that intelligent computer systems can detect and respond to stimuli in an environment, followed by Bishop (2003) investigating the use of an audio-visual stimuli detecting mobile phone for translating social stimuli for users with social impairments, finding that those with autistic spectrum disorders found the e-learning system useful for translating other's complex phrases into more understandable ones. Bishop's system, called the Portable Affect Recognition Learning Environment (PARLE), utilised affect recognition algorithms to translate communicated messages for those who had difficultly understanding them. The reliability of affect recognition algorithms to interpret the emotions held in the structures of faces and patterns of speech are fundamental in order for the mobile educational agent to be effective. There has been a significant amount of research into facial affect recognition (FAR) and speech affect recognition (SAR), investigating both the technical and practical aspects of implementing emotion recognition agents (ERAs) on desktop platforms. Facial expressions are generated by contractions of facial muscles, which results in temporally deformed facial features such as eye lids, eye brows, nose, lips and skin texture, often revealed by wrinkles and bulges. Fasel & Luettin (2003) found that some FAR agents are restricted due to the robustness of these systems and the constraints posed by recording conditions. A study by Petrushin (2000) investigated the potential of a SAR agent for distinguishing between 'calm' and 'agitated' emotional states of callers to a telephone support centre, delivering 77% accuracy and being flexible enough to be built into e-learning systems with the potential to assist individuals in developing emotional skills.

Pedagogical Agent

A pedagogical agent is a form of computational support, which enriches the social context in a social learning environment either by providing virtual participants to enhance the member multiplicity of communities or by supporting facilities to foster communication among real participants (Chou, Chan, & Lin, 2003). PAIGE extends this

Figure 4. Processes of the PAIGE System

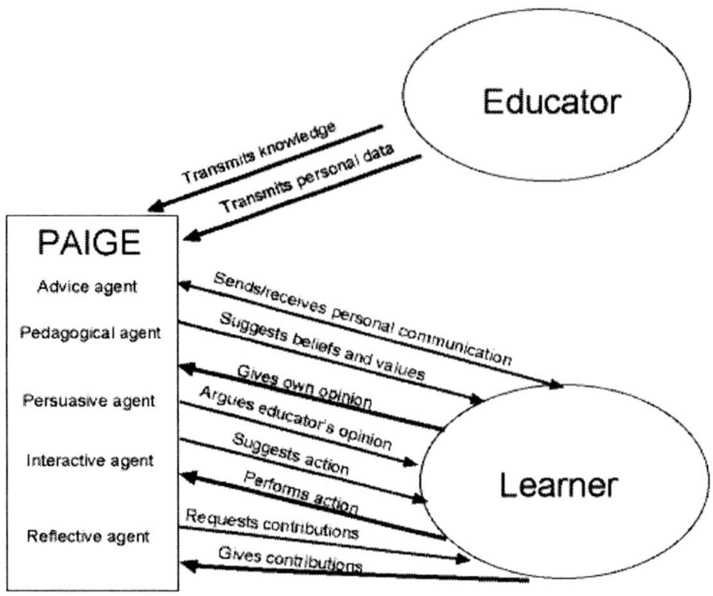

through assisting the learner with the Encoding stage of learning identified in Ecological Cognition Learning Theory, helping them to develop beliefs and values though creating the demonstration effect. The main difference between **Animated Pedagogical Agents** and other agents is that APAs use **avatars** to exhibit life-like characteristics and emotions in order to improve the attitude of users towards the VLE they are using and enhance their learning experience (Lester, Converse, Stone, Kahler, & Barlow, 1997). The persuasive functions of the APA include presenting information to learners to encourage particular actions and receiving information to personalise the system and make it more persuasive. Using the principles of argumentation theory, the APA encourages learners to think about the content on the current screen they are viewing in a particular way and carry out specific actions in order to beet the learning outcomes of the lesson. When applied to virtual environments, argumentation theory suggests that as users pursue their own goals, a system should be designed so that it uses negotiation techniques to achieve cooperation with the user in order to change their attitudes or behaviour (Kraus, Sycara, & Evenchik, 1998). This is implemented into the system through requiring the user to interact with the APA at specific occasions (e.g. when a screen loads or closes) in order to persuade them to carry out particular actions, such as encouraging those who are sharing the same machine to share the keyboard and mouse, or persuade them to provide information, such as their interests and hobbies in order for the content to be adapted. In addition to the persuasive functions, the APA uses the suggestive technology that formed part of the PARLE System (Bishop, 2003) to provide learners with additional information on the mean of words and phrases that are stored in the database. This function is activated through either the user clicking on a highlighted word or through them conducting a query, such as 'What is an agent?'. In order to provide the learner with the correct definition, the system is context aware to the degree that it recognises the subject and topic of the activity the user is currently carrying out and matches it to the appropriate word stored in the database. For example, if the user was studying ICT the definition of the word, 'agent' would be 'a program that works automatically on routine tasks specified by a user…', whereas if they were studying business it would be 'somebody representing somebody else in business…'.

Persuasive Agent

The persuasive agent assists learners at the Encouraging stage of learning identified by Ecological Cognitive Learning Theory, helping the learner develop goals appropriate to the learning activity though creating the inspiration effect. Sklar, Parsons, & Davies (2004) investigated a persuasive agent that utilised argumentation techniques, which is extended by PAIGE through encouraging learners to think through a problem through presenting them with a problem and a false solution, which may lead them to object to the solution that an agent could follow up with asking them to justify their reaction, as depicted in Figure 3.

Interaction Agent

The interaction agent assists a learner in developing plans to realise their goals utilising the Interaction stage of learning identified by Ecological Cognitive Learning Theory, and thus create the mobilisation effect. Decker (1998) proposed an agent-based system that helped individuals manage their plans and schedule them accordingly. PAIGE extends this by prompting the learner to carry out a specific task and received feedback from them on the result of their actions.

Reflection Agent

The reflection agent helps a learner with the Reflecting stage of Ecological Cognitive Learning Theory through support of their peers, encouraging them to reflect on the learning activity they have been involved in so they validate their interests, beliefs, values, goals and plans thus creating the confirmation effect and completing the learning process. Recognising the benefits of learners reflecting on their learning, Goodman, Hitzeman, Linton, & Ross (2003) proposed a system that allowed a system to work both with a **collaborative learning** system and with an electronic **avatar** as a peer, which can co-exist with human collaborators and interact as a partner to promote effective collaborative learning and problem solving through asking questions that get learners to think.

Adaptive Architecture

The Adaptive Architecture of PAIGE is based on adjusting the values of specific variables that control the text, graphics and functions of the system. The variables are set either locally via session variables by the user or though the override database table that is used by educators. This has the advantage of providing the user with a more personalised and engaging interface. Educators also have the option to embed parameters into learning material to personalise it with textual artefacts that have been defined by the individual learner. This is achieved through placing the parameter into a specific part of the text and surrounding it with parentheses. For example, if the learner's favourite actor was Tom Cruise, the text, "Write about a movie starring {User_Char_FavActor} that you enjoyed" would be converted into "Write about a movie starring Tom Cruise that you enjoyed", which should create a positive attitude towards the activity as it is about something the learner is interested in. Furthermore, as the interests of secondary school learners are likely to change frequently, educational material will always appear current and relevant if the learner is encouraged to update their profile. In addition PAIGE includes Adaptive Learning Levels (ALLs), which is based on the Zone of Proximal Development (ZPD) that was developed by Vygotsky (1930), which indicates that a learner will be able to perform at a higher level when under the supervision of an educator or more competent peer that when they are learner by themselves. This is achieved through the educator assigning an individual learner levels for each subject, one based on the level they think they are currently able to achieve under examination conditions, and the other what they are able to achieve with the support of the educator or a more competent peer. The wording of questions and statements is then adjusted based on the differing complexity of words in Bloom's Taxonomy. This enables learners to be required to approach a task from different perspectives depending on whether they have the support of an educator or not. For example a learner at NC Level 5 would be required to focus of the application of artefacts, such as text or graphics, and one at Level 6 would be required to analyse the problem more.

Sociability Architecture

The Sociability Architecture is based on the principle that learning is a social process through which information is shared and that collaboration amongst educators and learners is desirable for meeting learning outcomes. Educators are likely to need to communicate with both classroom and distance learners when they are conducting a lesson, providing support for distance learning, providing feedback on assignments and other activities, as well as responding to queries of learners when received by email or feedback forms. Learners are likely to need to communicate with each other when they are taking part in classroom activities, collaborating on group assignments, as well as when they want to discuss a topic outside planned activities. A chat facility is available in the system to enable learners to communicate with each other from device to device and allow distance learners to take part in classroom activities. A message board facility will enable learners to discuss topics set by the education as well as their own unrelated topics that would promote use of the VLE. The Weblog facility would enable learners to publish their reflections on classroom activities for others to see as recommended by Bishop (2004, 2007d)

Sustainability Architecture

To achieve the level of pervasiveness required to make the system sustainable, it has been designed to separate content from structural mark-up of nodes by storing it in a database. Through being separated from the mark-up, the artefacts that make up PAIGE, including text, graphics and downloadable files (e.g. Powerpoint, Word) are accessible by any node in the e-learning system, allowing them to be shared between educators and learners. The Shared Artefacts are also assigned a metadata descriptor, which is based on the e-Government Interoperability Framework (e-GIF). The e-GIF metadata is unable to describe the context of an artefact, meaning its use is limited to basic forms of filtering and recommendation. However, through using a standardised classification system, it makes it simpler for educators and learners to locate resources on any platform (Littlejohn, 2003).

EVALUATING THE MARKET FOR THE PAIGE SYSTEM

PAIGE with its persuasive, adaptable, sociable and sustainable architecture may fit into a specific market of young learners where portable devices are common. To explore this data was acquired from paired interviews and observations in schools and households with a national, in-home, Computer Assisted Personal Interviewing (CAPI) face-to-face survey of 1,511 school attendees in the UK aged 9-19 years old, together with a self-completion questionnaire to 906 parents of the 9-17 years olds, between 2003 and 2005. Some of the most widely used methods for researching online includes interviewing and observations (C. Mann & Stewart, 2000). Interviewing, it is argued, is the most widely applied technique for conducting systematic social inquiry in academic, clinical, business, political and media life and observational techniques are also widely used in conventional qualitative research, frequently used in conjunction with some form of interviewing (*ibid.*). Data was collected on young learner's use of the Internet for social purposes (It_1), including viewing friends' profiles, and for educational purposes (It_2), including homework, their use of mobile Internet devices and the support they get from friends and family.

Results

Young learners were classified according to whether they used their Mobile for accessing the Internet, with 205 young learners using WAP and 770 not using WAP. The participant's level of interest (It) in WAP as a technology was calculated on a scale from 1 (low interest) to 10 (high interest) a scale that Argyle (1992) suggested could measure interest. Interest can be predictive of behaviour when combined with belief as those individuals that have a strong association between their interest and their beliefs are more likely to show consistent behaviour (Ratner & Fitzsimons, 2002). The Dependency (Dy) of the young learners' interests on each other were calculated using the following formula assuming the level at which the cognitions were joindered (j) was 1. In this context, joinder is a value between 0 and 1 that reflects a combination of factors including the relatedness and cohesion of two cognitions.

$$Dy=Mean(It_1, It_2)*j$$

To explain the meaning of Dy an example will be used from Pavlov (1927). In his classic experiment Pavlov demonstrates that a canine actor will learn to salivate at the sound of a metronome in the same way it would at the sight of food. Before learning the canine actor's interest in the metronome is likely to be between 1 and 5 and its interest in the food between 5 and 10 and the level at which the cognitions were joindered between 0 and 0.5, making Dy between 0 and 3.75. Once the canine actor had learned to associate the metronome with food it's interest in the metronome would likely be similar to the food at between 5 and 10 and the level at which the cognitions were joindered between 0.5 and 1, making Dy between 2.5 and 10.

A summary of the results is presented in Table 2. The data was analysed using an independent samples t-test, showing that there was a significant difference between Dependency scores of the WAP and no WAP groups towards using the Internet for social purposes ($t=-40.21$, $p<0.001$) and educational purposes ($t=-82.01$, $p=<0.001$). The data indicates that there is a much stronger Dependency on using the Internet for social and educational purposes among those using WAP than those not using WAP. This suggests that those young learners using their mobile phone for accessing the Internet are dedicated to using it for social and educational purpose and also suggests that there is a segment of young learners that would benefit from utilising e-learning solutions in their educational and activities.

Table 3 shows a difference between the young people who saw their peers as the best source of support for homework and those who thought media such as television and books were the best help ($t=6.643$, $p<0.001$). The data shows that those that thought their peers were the best form of support were less likely to spend more of their free time on homework. This may be because they get greater support from peers while in school, though also suggests there is a market for a ubiquitous e-learning system that makes better use of a learner's peers in supporting them in their education.

CONCLUSION

Knowledge it could be argued is constructed from the information actors pick up from the environments they are in. Assessing this knowledge can be problematic in ubiquitous e-learning systems, but a method of supporting the critical marking of e-learning exercises is the Circle of Friends social networking technology. Understanding the networks of practice in which these e-learning systems are part of requires a deeper understanding of information science frameworks. The Ecological Cognition Framework (ECF) provides

Table 2. Mean (M) Dependencies (Dy) of young learners using WAP and those not doing so

Purpose	WAP Mean Dy	No WAP Mean Dy
Using the Internet for Social Purposes (It_1)	6.51	1.55
Using the Internet for Educational Purposes (It_2)	5.61	1.11

a thorough understanding of how actors respond to and influence their environment. Forerunners to ecological cognition, such as activity theory have suggested that the computer is just a tool that mediates between the actor and the physical environment. While activity theory has its theoretical limitations, which are addressed by ecological cognition, some of the conclusions drawn by Vygotsky (1930) can be explained by ecological cognition and are useful in developing e-learning systems. Utilising the ECF it can be seen that for an e-learning system to be an effective teacher it needs to be able to create the five effects in the actors that use it, with those being the belonging effect, the demonstration effect, the inspiration effect, the mobilisation effect, and the confirmation effect. In designing the system a developer would have to consider who the system is going to teach, what it is going to teach, why it is teaching, which techniques it is going to use to teach and finally whether it has been successful.

This chapter has proposed a multi-agent e-learning system called the Portable Assistant for Intelligently Guided Education (PAIGE), which is based around a 3D anthropomorphic avatar for educating actors ubiquitously. Whilst the avatar in PAIGE would be used for demonstrations, it would not represent the learner, but interact with them.

Table 3. Mean (M) hours spent on homework by the two groups

	N	M Hours	SD
Friends not best source of help	299	1.3961	1.12663
Friends best source of help	152	0.7168	.79119

An investigation into the market for PAIGE was carried out. Data was gathered in paired interviews and observations in schools and households with a national, in-home, Computer Assisted Personal Interviewing (CAPI) face-to-face survey of 1,511 school attendees aged 9-19 years old, together with a self-completion questionnaire to 906 parents of the 9-17 years olds. The data showed that those that thought their peers were the best form of support were less likely to spend more of their free time on homework. This may be because they get greater support from peers while in school, though also suggests there is a market for a ubiquitous e-learning system that makes better use of a learner's peers in supporting them in their education.

Limitations and Directions for Future Research

This study has looked at the educational theories and technologies that could respectively support and transform the learning process. The educational theory in question, Ecological Cognition Learning Theory was developed utilising a qualitative research method, and may need to be investigated through an empirical quantitative method to allow it to be verified. The study has shown that multi-agent social networking systems can be appropriate for ubiquitous education within networks of practice. Future research could investigate the usage of such systems in an educational setting and the effect it has on learning outcomes.

ACKNOWLEDGMENT

The author would like to thank the staff and students at Bryn Celynnog Comprehensive School and Ysgol Gyfun Rhydfelin for their support in the development of the virtual learning environment as well as all the anonymous reviewers who provided comment and feedback during the construction of this paper. In addition the author would like to thank S. Livingstone and M. Bober of the Department of Media and Communications at the London School of Economics and Political Science for collecting the data used in this study, which was sponsored by a grant from the Economic and Social Research Council. Glamorgan Blended Learning Ltd is a Knowledge Transfer Initiative, supported by the University of Glamorgan through the GTi Business Network of which it is a member.

REFERENCES

Allwright, R. L. (1984). The importance of interaction in classroom language learning. *Applied Linguistics*, *5*(2), 156–171. doi:10.1093/applin/5.2.156

Argyle, M. (1992). *The social psychology of everyday life*. London: Routledge.

Battistich, V., Solomon, D., & Delucchi, K. (1993). Interaction processes and student outcomes in cooperative learning groups. *The Elementary School Journal*, *94*(1), 19. doi:10.1086/461748

Bishop, J. (2002). *Development and evaluation of a virtual community.* Unpublished BSc(Hons), University of Glamorgan, Pontypridd.

Bishop, J. (2003). The internet for educating individuals with social impairments. *Journal of Computer Assisted Learning*, *19*(4), 546–556. doi:10.1046/j.0266-4909.2003.00057.x

Bishop, J. (2004). *The potential of persuasive technology for educating heterogeneous user groups.* Unpublished MSc, Univeristy of Glamorgan, Pontypridd.

Bishop, J. (2007a). Evaluation-centred design of E-learning communities: A case study and review. In V. Grout, D. Oram, & R. Picking (Eds.), *Proceedings of the second international conference on internet technologies and applications* (ITA07 ed.) (pp. 1-9). Wrexham: University of Wales Press.

Bishop, J. (2007b). Increasing participation in online communities: A framework for human–computer interaction. *Computers in Human Behavior*, *23*(4), 1881–1893. doi:10.1016/j.chb.2005.11.004

Bishop, J. (2007c). Ecological cognition: A new dynamic for human-computer interaction. In B. Wallace, A. Ross, J. Davies, & T. Anderson (Eds.), *The mind, the body and the world: Psychology after cognitivism* (pp. 327-345). Exeter: Imprint Academic.

Bishop, J. (2007d). *The psychology of how christ created faith and social change: Implications for the design of E-learning systems.* University of Winchester.

Brown, J. S., & Duguid, P. (2002). *The social life of information*. Cambridge, MA: Harvard Business School Press.

Ceccucci, W., & Tamarkin, D. (2006). *Ice breakers for the traditional or online classroom*. Palm Springs, CA.

Chou, C. Y., Chan, T. W., & Lin, C. J. (2003). Redefining the learning companion: The past, present, and future of educational agents. *Computers & Education*, *40*(3), 255–269. doi:10.1016/S0360-1315(02)00130-6

Coop, R. H., & White, K. (1974). *Psychological concepts in the classroom*. London: Harper & Row.

Corder, N. (2002). *Learning to teach adults: An introduction*. London, UK: Routledge Falmer.

Decker, K. (1998). Coordinating human and computer agents. *Coordination Technology for Collaborative Applications-Organizations, Processes, and Agents. LNCS, 1364*, 77–98.

Dinkmeyer, D., & Dreikurs, R. (1963). *Encouraging children to learn: The encouragement process*. Englewood Cliffs, NJ: Prentice-Hall.

Engeström, Y. (1993). Developmental studies of work as a test bench of activity theory: The case of primary care medical practice. In J. Lave, & S. Chaiklin (Eds.), *Understanding practice: Perspectives on activity and context* (pp. 63-104). Cambridge: Cambridge University Press.

English, R., Brookes, S. T., Avery, K., Blazeby, J. M., & Ben-Shlomo, Y. (2006). The effectiveness and reliability of peer-marking in first-year medical students. *Medical Education, 40*(10), 965–972. doi:10.1111/j.1365-2929.2006.02565.x

Fasel, B., & Luettin, J. (2003). Automatic facial expression analysis: A survey. *Pattern Recognition, 36*(1), 259–275. doi:10.1016/S0031-3203(02)00052-3

Foroughi, R., & Rieger, T. (2005). A virtual character for face to face interaction with children in E-learning. In D. Shepherd (Ed.), *Creative engagements* (pp. 107-111). Oxford: Inter-Disciplinary Press.

Freud, S. (1933). *New introductory lectures on psycho-analysis*. New York: W.W. Norton & Company, Inc.

Goodman, B., Hitzeman, J., Linton, F., & Ross, H. (2003). Towards intelligent agents for collaborative learning: Recognizing the role of dialogue participants. *Proceedings of the International Conference on AI in Education,* Honey, P., & Mumford, A. (1986). *Manual of learning styles*. Maidenhead, UK: Peter Honey Publications.

Jackson, P. Z. (2001). *The inspirational trainer: Making your training flexible, spontaneous & creative*. Kogan Page.

Kaptelinin, V. (1996). Activity theory: Implications for human computer interaction. In B. Nardi (Ed.), *Context and consciousness: Activity theory and human-computer interaction* (pp. 103-116). Cambridge, MA: The MIT Press.

Kolb, D. A. (1984). *Experiential learning: Experience as the source of learning and development*. London: Prentice Hall.

Kraus, S., Sycara, K., & Evenchik, A. (1998). Reaching agreements through argumentation: A logical model and implementation. *Artificial Intelligence, 104*(1-2), 1–69. doi:10.1016/S0004-3702(98)00078-2

Kyttä, M. (2003). *Children in outdoor contexts: Affordances and independent mobility in the assessment of environmental child friendliness*. (Doctor of Philosophy, Helsinki University of Technology).

Lange, D. B., & Oshima, M. (1999). Seven good reasons for mobile agents. *Communications of the ACM, 42*(3), 88–89. doi:10.1145/295685.298136

Lester, J. C., Converse, S. A., Stone, B. A., Kahler, S. E., & Barlow, S. T. (1997). Animated pedagogical agents and problem-solving effectiveness: A large-scale empirical evaluation, *2330*.

LeWinter, B. Y. (2003). *Inspire*. In A. Mack-Kirschner (Ed.), *Powerful classroom stories from accomplished teachers* (pp. 137-139). Thousand Oaks, CA: Corwin Press.

Lickona, T. (1983). *Raising good children: Helping your child through the stages of moral development*. London: Bantam Books.

Littlejohn, A. (2003). Issues in reusing online resources. In A. Littlejohn (Ed.), *Reusing online resources: A sustainable approach to E-learning* (pp. 1-6)

Mann, B. L. (2006). *Selected styles in web-based educational research*. New York: Information Science Pub.

Mann, C., & Stewart, F. (2000). *Internet communication and qualitative research: A handbook for research online*. London: Sage Publications.

Mantovani, G. (1996). *New communication environments: From everyday to virtual*. London: Taylor & Francis.

McClure, J. (2003). *How to find your dream job and make it a reality: Solutions for a meaningful and rewarding career*. Oxford: Trafford Publishing.

Nardi, B. A. (1996a). *Context and consciousnesses: Activity theory and human-computer interaction*. Cambridge, Ma: The MIT Press.

Nardi, B. A. (1996b). Studying context: A comparison of activity theory, situated action models and distributed cognition. In B. A. Nardi (Ed.), *Context and consciousnesses: Activity theory and human-computer interaction*. (pp. 35-51). Cambridge, MA: MIT Press.

Ozok, A. A., & Salvendy, G. (2004). Twenty guidelines for the design of web-based interfaces with consistent language. *Computers in Human Behavior*, *20*(2), 149–161. doi:10.1016/j.chb.2003.10.012

Pavlov, I. P. (1927). *Conditioned reflexes*. London: Routledge and Kegan Paul.

Petrushin, V. A. (2000). Emotion recognition agents in real world. *AAAI Fall Symposium on Socially Intelligent Agents: Human in the Loop*, Rasmussen, J., Duncan, K., & Leplat, J. (1987). *New technology and human error* John Wiley & Sons Inc.

Ratner, R. K., & Fitzsimons, G. J. (2002). When will consumers act on their attitudes?: New directions in attitude-behavior consistency: Special session summary. *Advances in Consumer Research. Association for Consumer Research (U. S.)*, *29*(1), 139–142.

Reddy, M., & Lewis, S. F. (2002). Building aunt hillary: Creating artificial minds with 'Neural nests'. *Proceedings of the 3rd International Workshop on Ant Colony Optimization, Ants 2002*, Brussels, Belgium.

Rushby, N. J. (1979a). *Computers in the teaching process*. New York: John Wiley & Sons, Inc.

Rushby, N. J. (1979b). *An introduction to educational computing*. Towbridge: Redwood Burn Limited.

Sklar, E., Parsons, S., & Davies, M. (2004). When is it okay to lie? A simple model of contraditcion in agent-based dialogues. *Proceedings of the First Workshop on Argumentation in Muliagent Systems*, Suchman, L. A. (1987). *Plans and situated actions: The problem of human-machine communication* Cambridge University Press.

Uden, L. (2007). Activity theory for designing mobile learning. *International Journal of Mobile Learning and Organisation*, *1*(1), 81–102. doi:10.1504/IJMLO.2007.011190

Vygotsky, L. S. (1930). *Mind in society*. Cambridge, MA: Webb, N. M. (1982). Student interaction and learning in small groups. *Review of Educational Research*, *52*(1982), 421-445.

Weng, M. (2007). *A multimedia social networking community for mobile devices.* Unpublished Interactive Telecommunications Program, Tisch School of the Arts / New York University.

KEY TERMS AND DEFINITIONS

Actor: An individual or entity that develops goals, plans, values, beliefs and interests and who engages with other actors in networks of practice in order to confirm or fulfil these.

Collaborative Learning: Learning that occurs within information systems that form part of networks of practice, where the participants have mutual goals.

Ecological Cognition: An evolving post-cognitivist paradigm for human-computer interaction within information science.

Ecological Cognition Framework: The founding model of the paradigm developed from a grounded theory literature review that can facilitate the explanation of phenomenon in networks of practice.

Ecological Cognitive Learning Theory: An understanding of learning derived from analysis and application of the Ecological Cognition Framework to learning problems.

E-Learning System: An electronic system such as a piece of software or other sequence of data that provides information to an actor or group of actors with the expectation that this will transfer knowledge from its originator.

Experiential Learning: Learning that occurs through an actor's participation in an event within a functional system where the outcome is uncertain and where the participant's goals, plans, values, beliefs and interests do not become clear to them until after the event has completed.

Information System: A unit that functions to fulfil the goals of the actors within in and through which information is shared and stored. It can contain e-learning and other systems to facilitate its purpose.

Network of Practice: A distributed group of actors connected by a shared purpose with shared knowledge. These networks cut across various information systems and can take the form of virtual communities.

Virtual Community: A platform, usually electronic, where the participants collaborate to achieve individual goals for a mutual purpose where they establish a presence through taking part in membership rituals.

Section 2
Design and Integration

Chapter 6
A Method for Generating Multiplatform User Interfaces for E-Learning Environments

Juan Manuel González Calleros
Université catholique de Louvain, Belgium

Josefina Guerrero García
Université catholique de Louvain, Belgium

Jaime Muñoz Arteaga
Universidad Autónoma de Aguascalientes, México

Jean Vanderdonckt
Université catholique de Louvain, Belgium

Francisco Javier Martínez Ruiz
Université catholique de Louvain, Belgium

ABSTRACT

In this paper the authors present a structured method for automatically generating User Interfaces for e-learning environments. The method starts with a definition of the learning scenario where the different goals, jobs (professor-student/trainer-learner), and tasks are described and stored in a template. After, the description is mapped to FlowiXML, a learning process authoring tool, where graphically trainers or content designers draw the overall process. A learning process is viewed as a workflow and modeled using Petri net notation. From each step in the process model more details are added using user task models; user's activity interacting with a user interface is stored in such diagrams. Then, a transformational method for developing user interfaces of interactive information systems is used that starts from a task model and a domain model to progressively derive a final user interface. This method consists of three steps: deriving one or many abstract user interfaces from the task model, deriving one or many concrete user interfaces from each abstract interface, and producing the code of the final

DOI: 10.4018/978-1-60566-703-4.ch006

user interfaces corresponding to each concrete interface. The models and the transformations of these models are all expressed in UsiXML (User Interface eXtensible Markup Language) and maintained in a model repository that can be accessed by the suite of tools. Developing user interfaces in this way facilitates its automated generation over multiple computing platforms while maintaining portability and consistency between the multiple versions. Our approach is illustrated on an open Learning environment using a case study.

INTRODUCTION

Knowledge acquisition in e-learning environments requires both, individualization of content and social interaction based on learning objects. The learning process links users to domain-specific information sources in collaboration spaces designed for knowledge transfer and knowledge generation. E-learning Communities describe social settings where knowledge can be exchanged effectively. Communities are composed of people who share common interests or needs following a set of rules or policies using computer technologies (Preece, 2000). In recent years, there has been a vast interest in how groups of people work together, and in how collaboration and cooperation might be supported. E-learning communities are formed and exploited by a variety of social and professional groups interacting via the Internet. An E-learning Community is a network of individuals who share a domain of interest about which they communicate online. The participants share the environment, resources, experiences, problems and solutions, tools, and methodologies.

Becoming efficient and stimulating for better and effective learning process using available technology requires a strategy to follow. The strategy must consider at least: design of learning content, design of different ways to present content (textual, graphical or mixed) considering different devices (PDA, mobile, laptop), and defining collaboration during learning process. In this chapter our objective is to describe a method to generate collaborative scenarios with learning objects that can be manipulated in different platforms. We refer to collaborative scenarios as a description of the foreseen interaction between instructors, learners and the system; this information is gathered in a template using MACOBA (Margain, Muñoz, & Álvarez, 2008). Each scenario is viewed as a workflow (depiction of tasks during which documents or information is passed for one participant to another according to a set of rules) that is recursively decomposed into tasks, that could be associated to a learning object (LO); then each task gives rise to a task model whose connection with others models allows the design of user interfaces (UIs) using a transformational approach (Vanderdonckt, 2005). Normally, interaction and communication among e-learning community members take place through a technological interface.

In the reminder of the chapter, we provided a brief background about e-learning environments and the generation of user interfaces, next we present the models involved and the method to generate collaborative multiplatform scenarios exemplifying with a case study. The chapter is wrapped up by summarizing our work, addressing future trends, and deriving conclusions.

BACKGROUND

E-Learning Environments

There is a plethora of computer-assisted e-learning environments/tools (Wainwright, Osterman, Finnerman, & Hill, 2007). The platform of choice

Table 1. Comparing different learning tools

	License	Language	UI Design	Collaborative facilities	Scalability
Claroline	O	PHP	+	+	++
Dokeos	F, O	PHP	+	+	++
Moodle	F, O	PHP	+	+	++
Sakai	F, O	Java	+	+	++
Blackboard	C	Java	+	+	++
Sloodle	F, O	PHP	+	+	+
ECOOL	F, O	Neutral	++	++	++

for most of the learning environments is the web browser (Bär, Häussge, & Rößling, 2007). Cesarini, Monga, & Tedesco (2004) contribute to have scalable and open architectures among existing solutions. Also, common elements are: tools for creating course material, assessment as well as collaborative tools (forums, emails and chats). These tools achieve the goal imposed by the first requirement because with them we could deliver interaction during and after the lecture, i.e. synchronous and asynchronous learning modes (Li, Lau, Shih, & Li, 2008).

Also, users and jobs are considered in most of the systems but collaboration is neither controlled nor modeled (Wainwright et al.., 2007). Some researchers are exploring the power of workflow modeling in ELE e.g. carrying on the e-learning process with a workflow management engine (Cesarini et al.., 2004) where the system is aware of the different interests of learners. These approaches are near to our work. However, our proposal expands the concept with the integration of User Interface (UI) definition and the modeling of Learning Objects (LOs).

Moodle an ELE to support lectures (Bär et al.., 2007; Dougiamas & Taylor, 2003; Wainwright et al.., 2007) is constrained to a single platform. It is possible to work on different platforms but the system itself is not capable of deliver such changes, so there is a need to program each added platform from scratch. Other possibility to overpass the problem of the extensions is using a Service Oriented Architecture (SOA) as in (Costagliola et al.., 2006) where some functionality is leaved outside the ELE and make it available though web services.

In ELE the process of creation of the UI follows the following approach: there is a general theme and platform, and results are always the same, this means no context adaptation mechanism is included. Major changes could be delivered by plug-ins (also to cover the scalability requirement) but this implies a separated design process and recoding. For instance, Yingling (2006) proposes a migration of Moodle to the mobile domain. The process involves manual changes in order to fit the new platform.

User Interface Development and context adaptation to the platform is considered in ECOOL (Martinez Ruiz, Gonzalez Calleros, Guerrero García, Muñoz Arteaga, & Vanderdonckt, 2008). Since ECOOL posses a meta-definition of the UI in order to provide portability and ease in the updating and migration processes. Besides that ECOOL includes workflow features that are missing or not integral part of other environments. Workflow includes users and jobs managements, working place definition and task specification. From this rich description User Interfaces are derived.

For the sake of simplicity, Table 1 presents only a few of the possible ELE existing in the market. The comparison symbols mean: + supported, ++

Table 2. Comparison of collaborative learning environment design methods

Criteria/Work	(Jonassen & Rohrer-Murphy, 2006)	(McDonald et al., 2005)	(Constance et al., 2002)	(Germán Sánchez et al., 2007)	Current work
Formal specification technique	*Activity theory*	*Conceptual framework*	*PBL*	*State machine*	*Workflow*
Environment	*NonA*	*C-Flow*	*STEP*	*Cated*	*Ecool*
Personalization	+	+	++	-	++
Multiple User interface.	--	+	+	--	++
Reverse engineering	--	+	+	--	++
Customization	+	+	++	+	++
Learning objects	+	++	++	++	++
Multimedia content	+	++	++	+	++
Traceability of collaborative learning	-	--	-	+	++

strongly supported, - not supported. The parameters used to compare the ELE correspond partially to main requirements identified. We use the following characteristics identified in (Bär et al.., 2007) to compare them: facilities to interact during and after the lecture (Collaborative facilities); an open architecture which should include the possibility of allow extensions (scalability), including that the system should be able to manage a single course or a whole organization. Support for designing UIs and the type of license, i.e., **C**ommercial, **F**ree software or **O**pen Source. Also the programming language used for adding content.

User Interface Development Methods for E-Learning Environments

In general, a method for designing and developing a collaborative e-learning environment could use a formal specification technique in order to modeling the evolution of learning process and also it could offer a collaborative learning environment where users can apply and execute some models. In the literature exists several methods to compare with our approach, we observe in Table 2 that there is not a lot methods to take into account all these criteria. Specially, two capital features in the analyzed methods: Customizable interfaces and Personalization. Most of the methods are unable to offer an ELE with UI capabilities similar to ECOOL (Martinez Ruiz et al., 2008).

COMPONENTS OF THE METHOD

There is a global consensus about the components of a User Interface (UI) development method (Vanderdonckt, 2005) which are:

- **Models:** A series of models pertaining to various facets of the UI such as: task, domain, user, presentation, dialog. These models will be defined in the next section and located on a reference framework.
- **Language:** In order to specify different aspects and related models, a specification language is needed that allows designers and developers to exchange, communicate, and share fragments of specifications and that enables tools to operate on these specifications. These models are uniformly and univocally expressed according to a single Specification Language, described in the language model section. The USer

Figure 1. Workflow learning process model

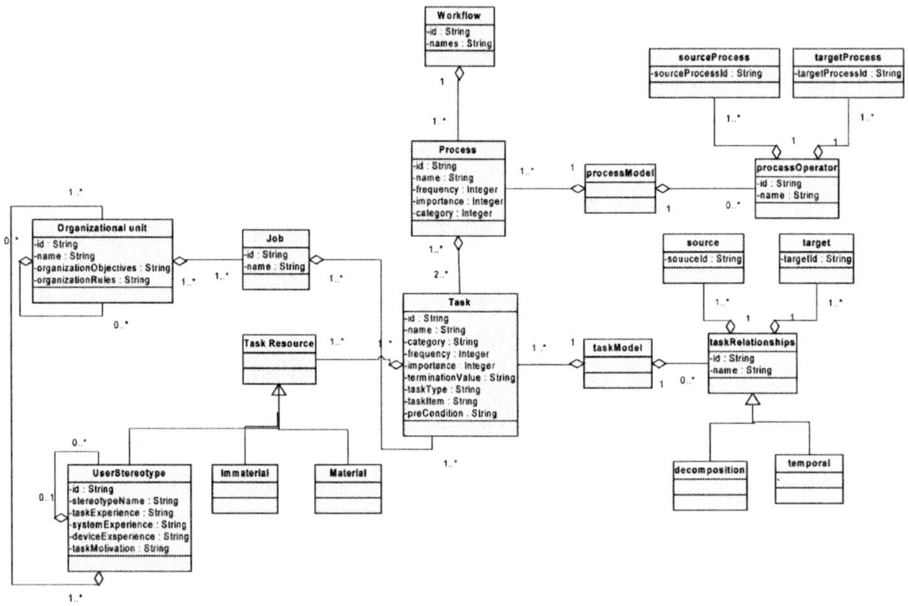

eXtensible Markup Language (UsiXML) (Limbourg & Vanderdonckt, 2004), a User Interface Description Language (UIDL). UsiXML has been selected because of its availability, capacity of expansion.

- **Approach:** An approach refers to the research line or paradigm to be followed by the method. The proposed method Section defines this approach by combination of models and model transformations so as to be compliant with the Model-Driven Architecture (MDA), to support Model-Driven Development (MDD).
- **Tools:** A suite of software engineering tools that supports the designer and the developer during the development life cycle according to the method.

In the following subsections the models and other components implied are introduced.

Learning Process Model

Learning process is described as a workflow, i.e. a progressive series of tasks (Guerrero, Vanderdonckt, González, & Winckler 2008) where the environment, the participants involved and the jobs that each plays, the resources used, and the tasks to be executed are defined. The workflow is recursively decomposed into processes which are in turn decomposed into tasks. Each task gives rise to a task model whose structure, ordering, and connection with other models allows the automated generation of their corresponding UI, using a transformational approach. Inside an e-learning community could be a teaching process for the trainers, a learning process for the learners, and an organizational unit for all participants. All these components interact to each other to form an overall learning workflow. In order to have a graphical representation, we can use Petri Nets (van der Aalst & van Hee, 2002) for the specification of processes, and CTT (Paternò, 1999) for the representation of tasks. Figure 1 reproduces a simplified version of the workflow model used, a complete version of it can be found in (Guerrero et al., 2008).

In recent years, there has been a vast interest in how groups of people work together, and in how collaboration and cooperation might be supported. At the same time technology to support working groups is rapidly growing in use, some very important trends are: multiple computing platforms, multiple channels, multiple modalities, multiple environments, and multiple users. Groupware is an application that covers some aspects of the current needs, such as collaboration, communication, coordination, and information sharing. Computer-Supported Cooperative Work (CSCW) is considered a generic term that combines the understanding of the ways in which people work in groups with the available technology of software, hardware, services, and related techniques. Collaborative work is hereby referred to a context of use where several users are worked with coordination, have communication among them, share information in a workspace to achieve a goal. We cover at some level this collaboration of users through the workflow model where jobs of interaction are specified.

Learning Object Model

A learning object (LO) is defined as a self-standing, reusable, discrete piece of content broken down into smaller chunks that can be reused in any environment in order to meet an instructional objective. The way LOs are conveyed includes: web pages, PDF documents, video and/or audio content, animations, virtual reality, to mention a few. LOs have been developed to support virtual learning using technology and pedagogical support. These products can be used under any condition or circumstance where the training or the distribution of the knowledge is required; classroom lessons, staff training in the industry, self-learning process, among others.

The structure of a LO showing in Figure 2 could be specified with: a name, a context, authors name, date, brief description, participants involved, pre and post conditions, and normal and/or alternative workflow. The LO model is a simplified version of SCORM standard. A LO can be part or be composed of other LOs. Also, it can be associated to exercises and/or assessments. The LO is part of a task and will be used in a task, this information is relevant when further we explain how a LO is mapped to a UI from a task model.

User Interface Model

Model-based user interface design process starts with a task model that is evolved through an incremental approach to the final UI (Cuppens, Raymaekers & Coninx, 2006). In this way, we selected UsiXML models to design UIs (Vanderdonckt, 2005). UsiXML relies on the Cameleon Reference Framework (Calvary, Coutaz, Thevenin, Limbourg, Bouillon, & Vanderdonckt, 2003). The simplified version, reproduced in Figure 3, structures four development steps: 1) *Task & Concepts* (T&C): describe the various user's tasks to be carried out and the domain-oriented concepts as they are required by these tasks to be performed. 2) *Abstract UI* (AUI): defines abstract containers and individual components, two forms of Abstract Interaction Objects by grouping subtasks according to various criteria, a navigation scheme between the containers and selects abstract individual component for each concept so that they are independent of any modality. An AUI is considered as an abstraction of a Concrete User Interface with respect to interaction modality. At this level, the UI mainly consists of input/output definitions, along with actions that need to be performed on this information. 3) *Concrete UI* (CUI): concretizes an abstract UI for a given context of use into Concrete Interaction Objects (CIOs) so as to define widgets layout and interface navigation. It abstracts a final UI into a UI definition that is independent of any computing platform. Although a CUI makes explicit the Look & Feel of a final UI, it is still a mock-up that runs only within a particular environment. A CUI can also be considered as a reification of an AUI at the upper level and an abstraction of the final UI with respect to the platform. 4) *Final UI* (FUI): is the operational UI i.e. any UI running on a particular

Figure 2. Learning object model

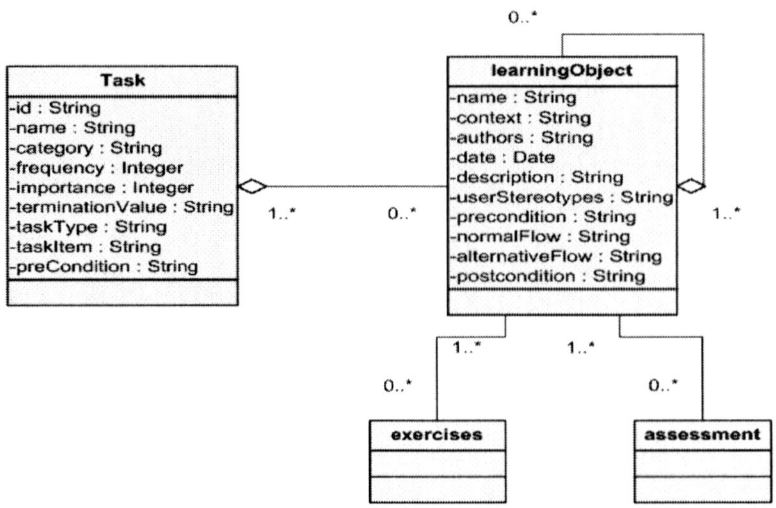

computing platform either by interpretation or by execution.

To support conceptual modeling of UIs and to describe UIs at various levels of abstractions, the following models have been involved:

- **domainModel:** is a description of the classes of objects manipulated by a user while interacting with a system.
- **mappingModel:** is a model containing a series of related mappings between models or elements of models.
- **auiModel:** is the model describing the UI at the abstract level as previously defined.
- **cuiModel:** is the model describing the UI at the concrete level as previously defined.
- **transformationModel:** Graph Transformation (GT) techniques were chosen to formalize explicit transformations between any pair of models, except from the FUI level.
- **contextModel:** is a model describing the three aspects of a context of use in which a end user is carrying out an interactive task with a specific computing platform in a given surrounding environment. Consequently, a context model consists of a user model, a platform model, and an environment model. This model plays and important job for multiplatform development and UI adaptation.
- **uiModel:** is the topmost superclass containing common features shared by all component models of a UI.

Language

Designing UIs to display Learning Objects requires a User Interface Description Language (UIDL), which consists of a high-level computer language for describing characteristics of interest of a UI with respect to the rest of an interactive application; it helps define user interfaces linguistically with a general trend to do so in an XML-complaint way. UsiXML, which stands for

USer Interface eXtensible Markup Language (Limbourg & Vanderdonckt, 2004; Vanderdonckt, 2005), has been selected as the UIDL because of its capabilities, such as: extensiveness, availability, central storage of models, and its transformational approach. Also, the graceful degradation of UI (Florins, Montero, Vanderdonckt, & Michotte, 2006) is a feature relevant to this project in particular when adapting the UIs to mobile devices.

Figure 3. UsiXML and Cameleon reference framework

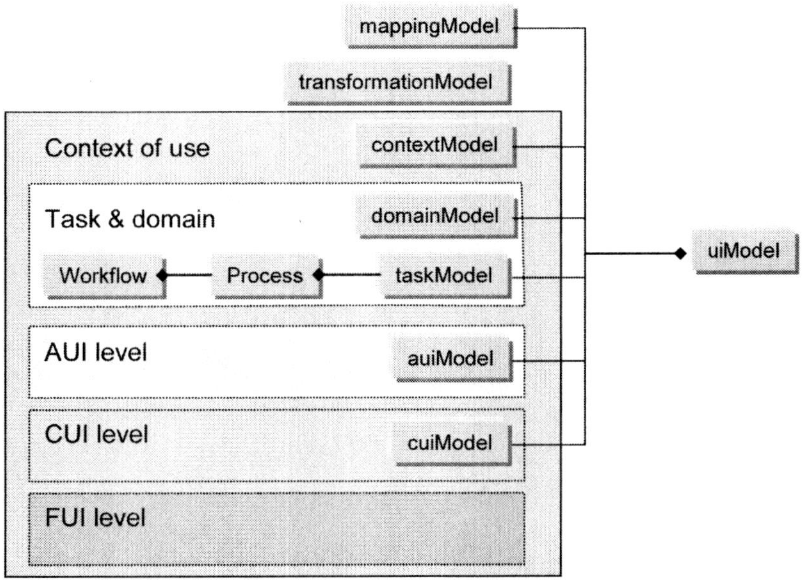

UsiXML supports platform independence: a UI can be described in a way that remains autonomous with respect to the various existing platforms, such as: mobile phone, Pocket PC, Tablet PC, kiosk, laptop, desktop, and wall screen.

For LO we use a simplified version of the Sharable Content Object Reference Model (SCORM), which offers a set of specifications for defining e-learning material. The standards define the communication scheme and how to create packages in the form of zip files. It is important to point out that SCORM is not attached to any instructional style, pedagogical methods of appearance (this is delegated to the ELE). The SCORM API includes a set of methods that can be implemented in any ELE. Additionally, this standard is scalable. Their core model is composed of three aspects: the content (which describes the content), the meta-data and packing. These models are capable of dividing the content into units called Sharable Content Objects (SCOs) and assets. These last are the format for representing media, text, images, sound, and web pages among others (the resources needed to show information to the student). The use of, partial or extended, standards such as SCORM is part of all ELEs, for instance Hard SCORM (Li et al., 2008).

Approach

MDA has been applied to many kinds of business and academic problems and integrated with a wide array of other common computing technologies, including the area of UIs. In MDA, a systematic method is recommended to drive the development life cycle to guarantee some form of quality of the resulting software system (Vanderdonckt, 2005). The transformational development of UI finds its root motivations in the concept of heterogeneousness. In this case the heterogeneousness concerns the variety of contexts of use (referred as a triple <user, computing platform, physical environment>) for which a UI is designed. This heterogeneousness stresses the need for abstractions able to factor out details relevant to specific contexts. From these abstractions, it is possible to obtain context specific representations by progressive refinements. The advantage of accessing to such representations is to be able to reason on one single model and obtain many different UIs.

To cope with the ever increasing diversity of markup languages, programming languages, tool kits, and interface development environments, UsiXML proposes a conceptual modeling of user interfaces (Vanderdonckt, 2005). The conceptual framework was created for specifying, designing, and developing user interfaces at a level of abstraction that is higher than the level where code is merely manipulated. For this purpose, a complete environment has been created based on conceptual modeling of user interfaces of information systems structured around three axes: the models that characterize a user interface from the end user's viewpoint and the specification language that allows designers to specify such interfaces, the method for developing interfaces in forward, reverse, and lateral engineering based on these models, and a suite of tools that support designers in applying the method based on the models. This environment is compatible with the MDA recommendations in the sense that all models adhere to the principle of separation of concerns and are based on model transformation between the MDA levels. The models and the transformations of these models are all expressed in UsiXML (User Interface eXtensible Markup Language) and maintained in a model repository that can be accessed by the suite of tools. Thanks to this environment, it is possible to quickly develop and deploy a wide array of user interfaces for different computing platforms, for different interaction modalities, for different markup and programming languages, and for various contexts of use.

Model-Driven Development of User Interfaces for E-Learning Environment

Considering that scenarios are a description of the foreseen interaction between instructors, learners and the system, they capture the context in which they occur, the process and associated data, and the order of executed tasks; our proposed method starts by defining them in a formal definition using a workflow, i.e., a series of tasks, to specify learning process. In this step we can indicate the resources involved, the organizational unit where they work, how and in which order tasks will be executed (using task model); after, we associate the LO which are stored in a repository and were defined using MACOBA methodology (Margain et al.,2008), finally the whole set of UIs for the ELE is generated. The multiplatform feature is then also available by relying on UsiXML existing tools. In the following subsections these steps are introduced. (see Figure 4)

Specification of Learning Process

An e-learning environment could be viewed as a social space where multiple users produce and share information. Each user has a job (the total collection of tasks, duties, and responsibilities assigned to one or more users) and has a hierarchical position inside the social space. Between the principal users are: teachers, learners, domain experts, and managers. Inside of learning spaces users interact in order to learn, so, users could work as groups in cooperative, collaborative or competitive way. For instance: Dominique Stuart (user), a teacher (job) in computer science faculty (organizational unit), has 10 learners (users/job) which need to work in the resolution of simple equations (task). In order to facilitate the task and integrate all learners, she decides that learners will work in pairs (groups) and the couple that resolves the equations in less time (competition) wins an extra point.

Relying on (Guerrero, Vanderdonckt, González, & Winckler, 2008) learning process is described.

The scenario is described in a complete textual statement describing the learning process to be addressed. This is assumed to be capture using MACOBA template (see Table 4). This textual scenario is aimed to identify elements that are relevant for building a first version of models that will be further exploited in a model-driven engineering method of UIs.

Figure 4. Proposed method

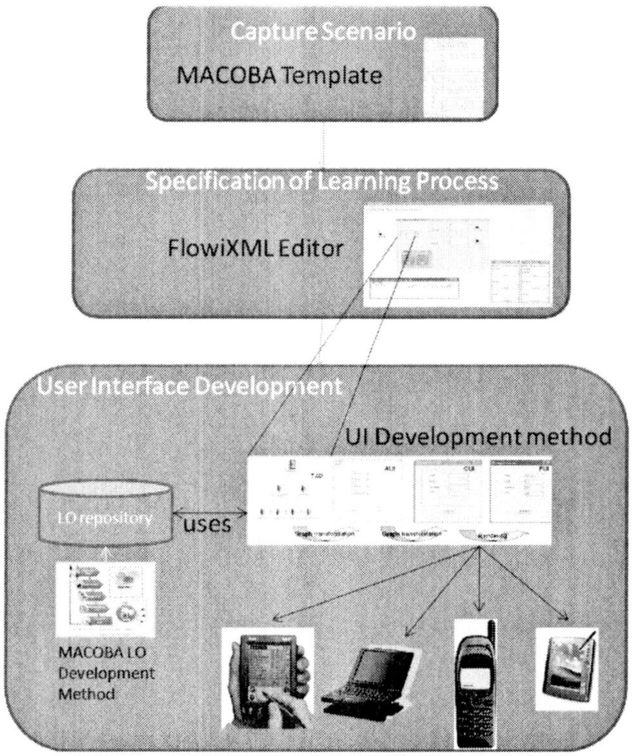

Figure 5. Designing UIs with UsiXML

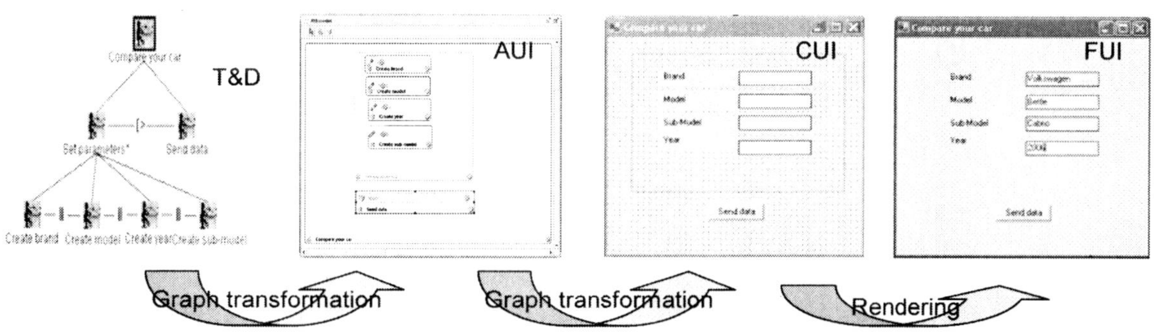

1. A workflow model is developed using Petri Nets to represent graphically the series of tasks that are involved in the collaborative scenario.
2. After the identification of the task, tasks models are designing using CTT.
3. Defining resources and their jobs in the organization it is an important aspect that we consider. It is not just about identifying different jobs (trainer, learner) and resources involved in the process but also, attributes of the job (specification, family, grade,

Table 3. Main laws of classical mechanics course

Table of Content
The first law
The second law
The third law
The Universal gravitation law
The free fall

privileges) and resource (level of experience, hierarchy level) that can lead to further define who is capable to perform a determined task. After defining tasks, jobs, and resources, tasks are allocated or offered to resources in different ways: direct allocation, delegation, history-based allocation, among others.

User Interfaces Development Method

As we mention before, user interface design processes starts with a task model that is evolved through an incremental approach to the final UI. Figure 6 shows the four levels that are involved in the design of a UI using UsiXML, i.e. the task model, the abstract UI, the concrete UI, and the final UI.

For instance, a task that is no further divided is represented with an Abstract Individual Component (AIC); for each AIC a facet (input, output, control, navigation) and transformational rules are

Table 4. MACOBA template to capture textual scenarios of collaborative learning

Pattern Name		**Collaborative learning**
Context		Classical mechanics, Physic and Mathematic
Authors:		Jaime Muñoz Arteaga and Lourdes Margain
Date:		10/08/2008
Description: This use case describes the collaborative learning with learners they work in groups of 2 or 3 learners who collaborate and they could interact through collaborative tools. This use case covers the curricula and allows access in the case of use; learners are located in different geographic places.		
Actors: Student Teacher	No. Actor 1 2	Jobs: Student Teacher
Preconditions: The Students are connected to internet from different places and collaborate between colleagues throughout the communication service of the learning environment.		
Normal Flow		
Jobs	**Case**	**Description**
Student	Initial exploration (I)	Browsing through the screens which will be explained through text, images, audio and video. These elements are defined as a learning object called "bodies in free fall Newtonian law".
Student	Auto-evaluation (II)	Students answer the quiz and get feedback from learning object. The quiz also gives the number of successes achieved.
Student, Teacher	Forum Discussions (III)	Teacher listen the Student's opinions and answer their doubts about bodies in free fall law.
Student	Develop report (IV)	Develop a report about the knowledge learned.
Teacher	Rate Results (V)	To analyze the strengths and weaknesses obtained from the Student and provide suitably qualified.
Teacher, Student	Feedback (VI)	Teacher gives feedback to the Student's strengths and weaknesses notify them by mail
Alternative Flow: None		
Pos conditions: The collaborative forum has been enabled.		

Figure 6. Rendering of the e-learning environment Icampus on different platforms

applied. The next step consists in code generation from CUI to obtain a FUI. For the complete definition of the method, we refer to (Limbourg & Vanderdonckt, 2004).

From a task model specification it is possible to derive as many UIs as devices have been specified in UsiXML. Figure 6 illustrates the rendering of the E-Learning Environment of icampus (http://www.icampus.ucl.ac.be/) on Mobile devices and a PDA. UsiXML approach covers the multi device and multiplatform support. The process is as follows

The MACOBA methodology (Margain et al., 2008) offers a framework to integrate the future of the collaborative teaching-leaning process based on patterns, in particular for designers. To construct collaborative learning objects with constructed by patterns MACOBA offers four basic levels and one final evaluation (Figure 7):

- **Requirement level:** teachers increase their experience with the planning process. The patterns in this level are guidelines for teachers (instructional designers).
- **Analysis level:** The analysis level implements UML as an innovative way of reusing the use cases patterns and sequence diagrams patterns for collaborative learning.
- **Design and develop level:** the technological designer personalizes the frames patterns and selects communication services as wiki, forum, and chat.
- **Implantation level:** responds to the learning scenario organization. In this level the elements are the activities, jobs, sequences. This level is bases on IMS-LD Model incorporating structure and collaboration activities.
- **The evaluation level:** considers a whole process review about the correct patterns application because is necessary to be sure about the application of collaborative process.

We rely on the existing set of LO resulting of MACOBA approach that can be found at http://ingsw.ccbas.uaa.mx/repo/galeria/. It is out of scope of this work to define LO we assume that the existing repository fulfills our needs. Even more, any new LO can be added following the MACOBA methodology.

TOOL SUPPORT

We designed the ECOOL architecture (Martínez et al., 2008) to support the method. It is composed mainly of two parts (Figure 8): server side and client side The first part includes an ELE composed of seven modules: the LO management module (in charge of LO modification), the assessment engine (in charge of recovering evaluations in order to present them to the user), the translation engine for transforming our definition from/

Figure 7. MACOBA methodology

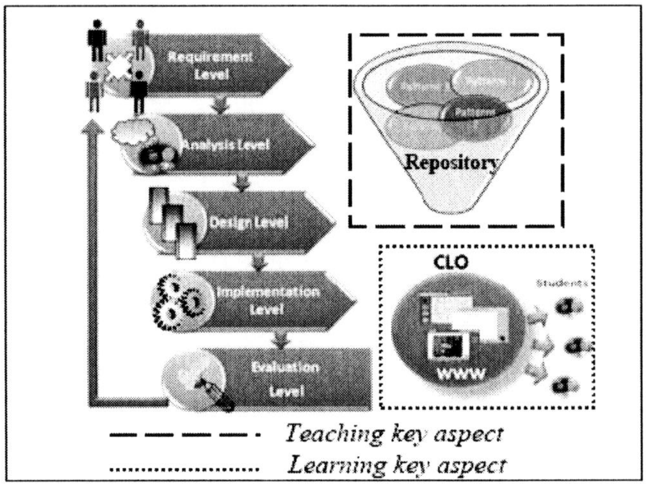

Figure 8. ECOOL architecture

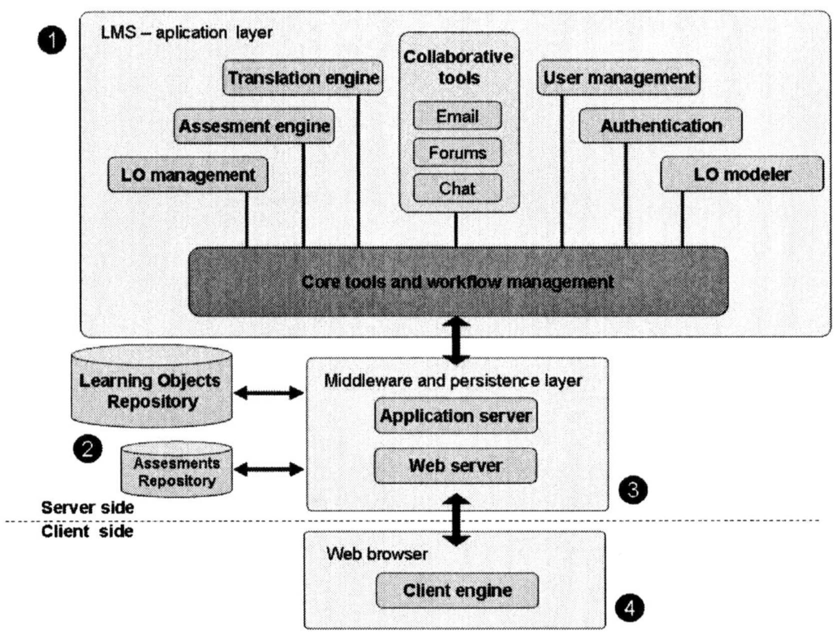

to known standards as well as including rules to produce a LO from the definition of a task to some specific UI, collaborative tools module (Email, forum and chat), the security module (in charge of authentication issues) and finally a designer tool. All these modules are in communication and under the authority of a major component: the workflow management tool. Indeed, this component will deliver the other part of the collaborative process, the modeling and control of the flow of all the activities in the E-Learning environment.

Also, this part deals with the data base control,

Figure 9. FlowiXML learning process editor

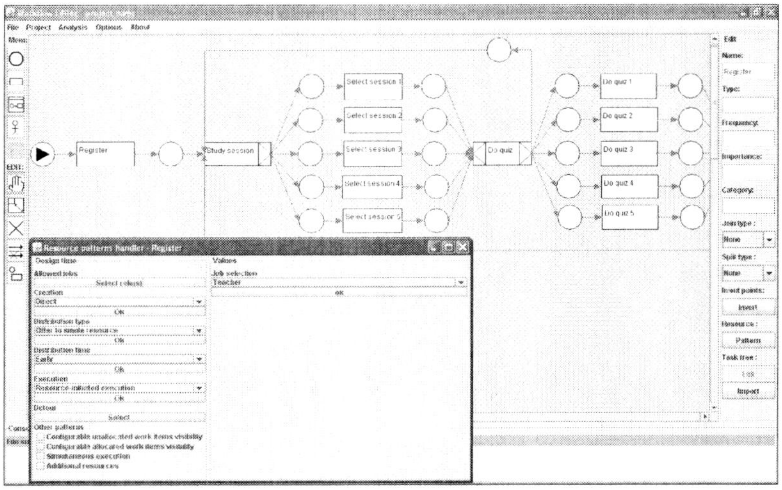

Figure 10. A multi-target user interface

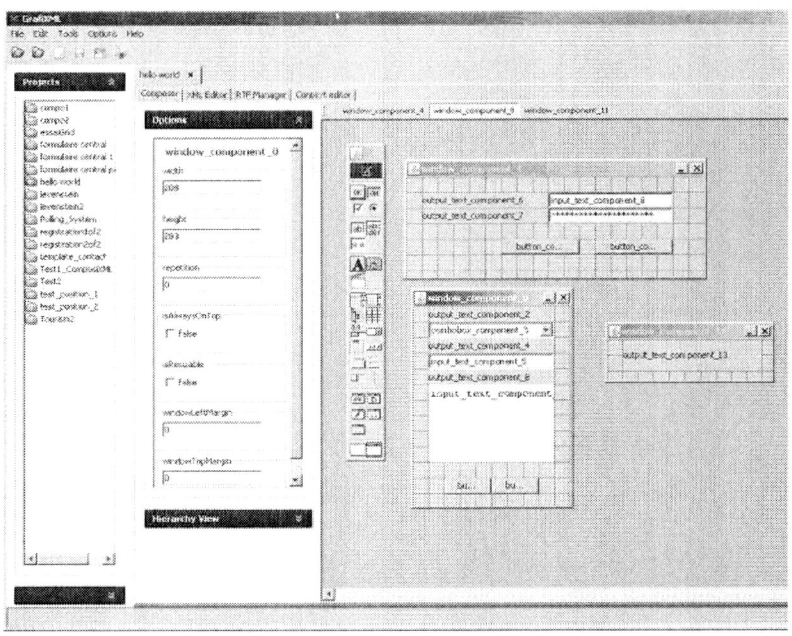

the management, and the application server (Figure 8 section 3). This section corresponds to any typical middleware architecture. The only novelty is the strict separation between the LOs and the assessment repositories (Figure 8 section 2). The second part, i.e. client side, is where after the first admission to the system, it would be mandatory to download a client engine/plug-ins to deal with the virtual renders and extra capabilities provided by the 3D world (González Calleros, Vanderdonckt, &

Arteaga, 2007), all models supported in UsiXML can be a dynamic way to present LOs.

In order to support the development of UIs from a workflow model to a task model, a workflow editor has been developed, Figure 9 (Guerrero et al., 2008). This editor allows modeling the general workflow defining processes and tasks models, defining organizational units, jobs and resources involved, allocation of tasks to resources, and to manage the flow of tasks.

Finally, if a GUI should run in many different contexts of use, then alternative GUI designs should be specified and added for each new context of use, thus leading to specifying a multi-target UI. GrafiXML (Michotte & Vanderdonckt 2008) an original user interface builder in that it enables designers and developers to design several UIs simultaneously for multiple contexts of use, i.e. for many users, platforms, and environments. For this purpose, it maintains coordination between three representations: an internal representation consisting of specifications in USer Interface eXtensible Markup Language (UsiXML), an external representation consisting of the interface preview, and a conceptual representation consisting of a user interface model. GrafiXML is an intelligent UI builder in that it maintains model consistency between these representations through a set of mappings based on an UI ontology. Thanks to this mechanism, GrafiXML provides a unique set of features for supporting designing interfaces for multiple targets. (see Figure 10)

CASE STUDY

In order to illustrate our approach this section presents a case study for the collaborative learning about laws of bodies in motion. We consider a course to study the laws of classical mechanics (see Table 3).

Each one of the topics of laws of bodies in motion will be encapsulated as a LO and the order of presentation to learners is defined through a workflow. Here the workflow is used to specify the collaborative learning about laws of bodies in motion in order to generate the graphical UI for our learning environment. With the workflow notations is possible specify different mechanisms to establish for example what learning objects are required to the scenario and/or what is the sequences of presentation of these objects in function of learners task.

Capture of the Scenario

For this course the trainer could propose different activities in group in order to teach this law to the

Figure 11. Use case of collaborative learning

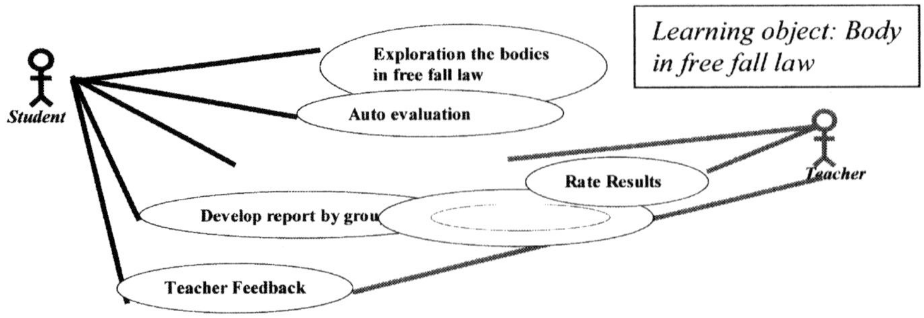

Figure 12. The course of mechanics modeled using petri nets in order to describe its workflow

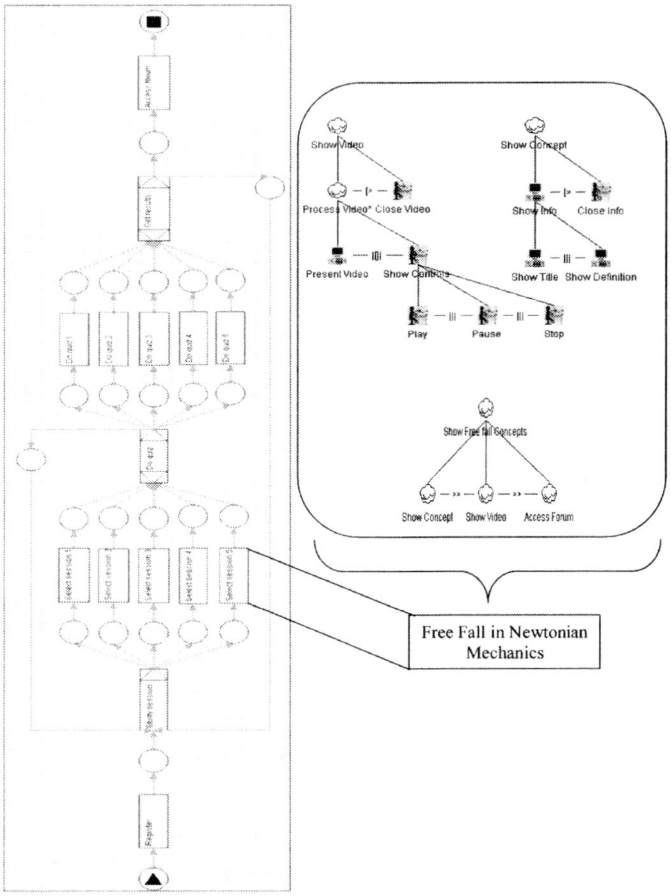

learners in a collaborative manner. These activities are specified with educative analyze pattern proposed in the MACOBA methodology (Margain et al., 2008), as follow in the Figure 11 and Table 4. According to the use case of Figure 11, every learner starts exploring the learning object of bodies in free fall law studying its concepts and interacting with different simulations for this law. An auto evaluation could be taken for learners and be ready to discuss between them and with their teacher throughout a forum web service.

Specification of Learning Process

The pattern for this use case describes the collaborative learning of bodies in free fall law. This pattern specifies that there are users which have the job of learners and other one as teacher. Also we have included a collaborative task that requires a group of three learners who will collaborate in a communication service (forum) see Figure 12 section a. Next, the designer has three options: Define new LOs (to fulfill the learning goal), Modify existing ones and reuse the ones stored in the repository. For this example, the first possibility is selected.

In Figure 12 we could see a list of the LOs proposed by the teacher to cover the laws of bodies in motion. For the sake of simplicity only one is selected: bodies in free fall Newtonian law. The objective is to present the concept of free fall to learners using a textual description followed by a

Figure 13. Table depicting the description of the design of a LO for introduce a concept

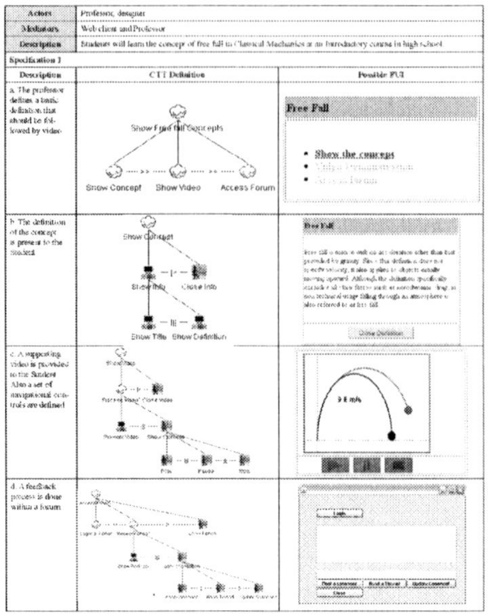

video presentation. So, we model the workflow, indicating which participants will be involved, and specifying the different tasks that are necessary to execute. The tasks will be grouped in task trees which will be translated through the successive process of translation into a FUI as in Figure 5. Here the resulting AUIs and CUIs of the example are shown in order to clarify the process and understand the translation process. For the sake of simplicity, we are only presenting some branches of the Petri Net. In Figure 13 and Figure 14 the development process is described.

According to the learning pattern in Table 4, certain tasks are proposed, they are numbered from one to six using roman numerals. Then, we could see expressed them in a more accurate way. The process is straightforward: for each case defined by the pattern, the professor and the designer have to introduce the needed tasks. For instance, the first case is the exploration of the available material as text, videos, and Web pages; and this process is dosed in a number of sessions that is specified by the professor. Next, learners are invited to evaluate themselves through quizzes and this is done in order to fulfill the second case of the pattern. The following step is discuss between other learners and the professor for resolving doubts and discuss hard topics or problems with the learned concepts, as we have seen until here, a third set of states are defined in our Petri net to model the interaction with the professor in a forum. Finally, the rest of cases are not included since they follow the same principle.

User Interfaces Development

The next step is model each one of the nodes of our workflow in order to define the neutral definition of the UI as in Figure 12. Since this process is the main goal of the method, it was presented before. Our method reminds silence in most of the decisions of the way to create the workflow. This is

Figure 14. Table depicting the possible translations of the tasks tree definitions

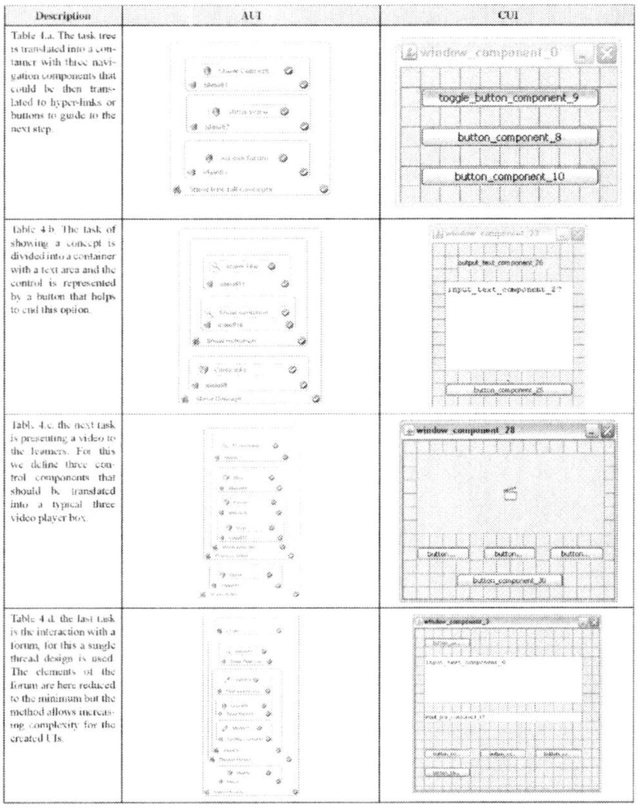

intentionally to avoid any problem of portability or assumption of a specific platform.

These task trees should be translated through the successive process of translation into a FUI as in Figure 5. Here the resulting AUIs and CUIs of the example are shown in order to clarify the process and understand the translation process. Again for the sake of simplicity, we are only presenting some branches of the Petri Net. In Figure 13 and Figure 14 the development process is described.

In this section we illustrate the method for developing Multiplatform User Interfaces for E- learning environments based on scenarios description captured in MACOBA template. The feasibility of this approach has been tested and validated.

FUTURE TRENDS

Until recently, e-learning environments, as well as their supporting UIs, have been condemned to stay pre-defined and fixed along most dimensions. In the near future, we expect that these constraints will be relaxed progressively so as to give rise to a new series of open questions including, but not limited to:

- **Platform variation:** As a consequence of the previous variation or independently of that, the worker may want to use another computing platform at run-time because it is more convenient for her. Therefore, the supporting UIs should support the same task on different computing platforms and accommodate their variations.

- **Ad-hoc adaptation of the UI:** When referring to change in the platform there is a need to define what information is relevant to be displayed. As the problem is not just about rendering on a small screen but also to keep the UI with the LO usable.
- **The migration of UIs:** Related to the previous tasks while referring to collaboration among any worker involved in the process there might be scenarios in which splitting the UI in several pieces is necessary but not just splitting but also migrate it to different devices. For instance, imagine a set of learners working on an assessment related to the lecture of the day they all could take a subset of questions migrate them to their mobiles and then put them back and present the UI with the whole assessment.
- **More Dynamic Content:** The use of virtual reality has been explored and successfully used for lectures. However, to present true virtual content depicting LOs has not been integrated in any ELE. For instance, many virtual reality models exist of the Human Body; this might be useful in an anatomy course.
- **Run-time adaptation to the context:** While our method serves to design UI that will be later run in different platforms, it would be interesting to investigate how to adapt to the context in run-time without using a predefined UI but more an automatic transformation for any platform.

CONCLUSION

An alternative method for the automatic generation of Collaborative Multiplatform Scenarios with Interactive LOs has been introduced. The overall learning process is seen as a workflow. This approach introduces a flow control that allows tackling at the same time, the problem of divergence in individual learning and the definition of the learning process in terms of collaboration agents and processes i.e. giving the system the capacity of managing collaboration between learners/trainers, as well as, defining in an explicit way the network and control points of collaboration. Other innovation is the modeling of the interaction of the LOs which includes the introduction of a meta-description (in UsiXML) that is going to aid in the process of generation of multiple UIs to be spread over multiple platforms. Instead of create a new environment we are going to work with one of the most known ones: Claroline (Docq, Lebrun, Smidts, 2007). The process of deliver a modify version is on its way. The next step is to gather all above concepts into an ELE framework. The idea of building learning spaces is that users interact intensively through them and share information in order to learn, then the users could work as groups in cooperative or collaborative way. The set of tools that are used in the method were described shortly because they have been widely described somewhere else. Finally a case study has been presented to illustrate our method.

RESOURCES

All resources related to UsiXML, FlowiXML can be found at: http://www.usixml.org. On this web page software can be downloaded, along with its user's manual, and case studies with examples.

ACKNOWLEDGMENT

We gratefully acknowledge the support of the CONACYT program (www.conacyt.mx) supported by the Mexican government and the SIMILAR network of excellence (http://www.similar.cc), the European research task force creating human-machine interfaces similar to human-human communication of the European Sixth Framework Programme (FP6-2002-IST1-507609).

REFERENCES

Bär, H., Häussge, G., & Rößling, G. (2007). An integrated system for interaction support in lectures. *In Proceedings of the 12th Annual SIGCSE Conference on innovation and Technology in Computer Science Education ITiCSE '07* (pp. 281- 285). New York: ACM.

Birchfield, D., Savenye, W., & Thornburg, H. (2008). *AMEEd Edulink*. Retrieved July 10, 2008. from http://ame4.hc.asu.edu/edulink

Blackboard. (2008). Retrieved July 10, 2008, from http://www.blackboard.com/us/index.bb

Bondendorf, E., Schertler, M., & Cohen, E. (2006) Producing reusable web based multimedia presentations. In Alex Koohang (Eds.), *Interdisciplinary Journal of knowledge and Learning Objects, 2*, 127-142.

Calvary, G., Coutaz, J., Thevenin, D., Limbourg, Q., Bouillon, L., & Vanderdonckt, J. (2003). A unifying reference framework for multi-target user interfaces. *Interacting with Computers, 15*(3), 289–308. doi:10.1016/S0953-5438(03)00010-9

Cesarini, M., Monga, M., & Tedesco, R. (2004, March 14-17). Carrying on the e-learning process with a workflow management engine. In *Proceedings of the 2004 ACM Symposium on Applied Computing* (pp. 940-945), Nicosia, Cyprus. SAC '04. New York: ACM.

Claroline (2008). Retrieved July 10, 2008, from http://www.claroline.net

Costagliola, G., Ferrucci, F., & Fuccella, V. (2006, July 11-14). SCORM run-time environment as a service. In *Proceedings of the 6th international Conference on Web Engineering* (ICWE '06), Palo Alto, CA, USA, (vol. 263, pp. 103-110). New York: ACM.

Cuppens, E., Raymaekers, C., & Coninx, C. (2006). A model-based design process for interactive virtual environments. In S.W. Gilroy & M.D. Harrison (Eds.), *Interactive Systems. (LNCS, 3941*, pp. 225-236.) Springer, Heidelberg.

Dillenbourg, P. (2000). Virtual learning environments. In *Proceedings of Workshop on Virtual Learning Environments during EUN Conf. 2000, Learning in the new millennium: Building new education strategies for schools*.

Docq, F., Lebrun, M., & Smidts, D. (2007). Claroline, une plate-forme d'enseignement/apprentissage sur Internet: pour propulser la pédagogie active et l'innovation? In M. Frenay, B. Raucent, & P. Wouters (Eds.), *Actes du 4ième colloque "Questions de pédagogies dans l'enseignement supérieur* (pp. 99–109). Presses universitaires de Louvain, Louvain-la-Neuve.

Dokeos, (2008). Retrieved July 10, 2008, from http://www.dokeos.com/

Dougiamas, M., & Taylor, P. C. (2003, June 23-28). Moodle: Using learning communities to create an open source course management system. In *Proceedings of the World Conference on Educational Multimedia Hypermedia & Telecommunication (ED-MEDIA'2003)*, Honolulu (pp. 171–178). Charlottesville: AACE Press.

Florins, M., Montero, F., Vanderdonckt, J., & Michotte, B. (2006, May 23-26). Splitting Rules for Graceful Degradation of User Interfaces. In *Proceedings of the 8th International Working Conference on Advanced Visual Interfaces (AVI'2006)* Venezia (pp. 59-66). New York: ACM Press.

Germán Sánchez, V., Téllez Mora, F., & Morales Gamboa, R. (2007). Collaborative environment for the composition of learning scenarios. In Gelbukh & Zechinelli (Eds.), *IEEE Editorial. ENC07 Conference,* Mexico.

González Calleros, J. M., Vanderdonckt, J., & Arteaga, J. M. (2007, June 6-8). A method for developing 3D user interfaces of information systems. In *Proceedings of the 6th International Conference on Computer-Aided Design of User Interfaces (CA-DUI'2006, 5*(7), pp. 85-100). Springer-Verlag, Berlin. Guerrero, J., Vanderdonckt, J., & Gonzalez, J.M. (2008). FlowiXML: A step towards designing workflow management systems. *Journal of Web Engineering, 4*(2), 163–182.

Guerrero, J., Vanderdonckt, J., González, J. M., & Winckler, M. (2008). Modeling user interfaces to workflow information systems. Paper presented in *4th International Conference on Autonomic and Autonomous Systems* (ICAS'2008). Springer-Verlag.

Jonassen, D. J., & Rohrer-Murphy, L. (1999). Activity theory as a framework for designing constructivist learning environments. *Educational Technology Research and Development, 47*(1), 61–79. doi:10.1007/BF02299477

Li, Q., Lau, R. W., Shih, T. K., & Li, F. W. (2008). Technology supports for distributed and collaborative learning over the internet. ACM Tranactions on Inter. Technology, 8(2).

Limbourg, Q., & Vanderdonckt, J. (2004). UsiXML: A user interface description language supporting multiple levels of independence. In M. Matera & S. Comai (Eds.), *Engineering Advanced Web Applications* (pp. 325-338). Paramus: Rinton Press.

MacDonald, C.J., Stodel, E.J., Thompson, T-L., Muirhead, W., Hinton, C., Carson, B., & Banit, E. (2005). Addressing the eLearning contradiction: A collaborative approach for developing a conceptual framework learning object. Interdisciplinary Journal of knowledge and Learning Objects, 2, 79-98.

Margain, L., Muñoz, J., & Álvarez, F.J. (2008). A *Methodology for Design Collaborative Learning Objects.* I*CALT 2008*, Santander, Spain.

Martínez Ruiz, F. J., González Calleros, J. M., Guerrero García, J., Muñoz Arteaga, J., & Vanderdonckt, J. (2008, June 9-11). ECOOL: Genaration of Collaborative Multiplatform Scenarios with Interactive Learning Objects. In M. Lozano & J.A. Gallaud (Eds.), *Servicio de Publicaciones de la Universidad de Castilla - La Mancha* (pp.361-370), Albacete.

Michotte, B., & Vanderdonckt, J. (2008, March 16-21). GrafiXML, A Multi-Target User Interface Builder based on UsiXML. In *Proceedings of 4th International Conference on Autonomic and Autonomous Systems (ICAS'2008)*, Los Alamitos: IEEE Computer Society Press

Paternò, F. (1999). Model-based design and evaluation of interactive applications. *Applied Computing.* Springer: Heidelberg.

Preece, J. (2000). Online communities: Designing usability and supporting socialability. John Wiley & Sons, Inc.

Sloodle. (2008). Retrieved May 10, 2008, from http://www.sloodle.org van der Aalst, W., & van Hee, K. (2002): *Workflow Management: Models, Methods, and Systems.* Cambridge: The MIT Press.

Steinkuelhr, C. A., Derry, S. J., Woods, D. K., & Hmelo-silver, C. E. (2002). The STEP environment for distributed problem based learning on the World Wide Web. In *Proceedings of Computer Support for Collaborative Learning (CSCL 2002).*

Vanderdonckt, J. (2005, June 13-17), A MDA-compliant environment for developing user interfaces of information systems. *Proc. of 17th Conf. on Advanced Information Systems Engineering (CAiSE'05)*, Porto (LNCS, 3520, 16-31). Berlin: Springer-Verlag.

Wainwright, K., Osterman, M., Finnerman, C., & Hill, B. (2007, October 7-10). Traversing the LMS terrain. In *Proceedings of the 35th Annual ACM Conference on User Services SIGUCCS'2007* (pp. 355–359), Orlando. New York: ACM Press

Yingling, M. (2006). Mobile Moodle. *Journal of Computing Sciences in Colleges, 21*(6), 280–281.

Chapter 7
Cross Platform M-Learning for the Classroom of Tomorrow

Daniel C. Doolan
Robert Gordon University, Scotland

Tracey J. Mehigan
University College Cork, Ireland

Sabin Tabirca
University College Cork, Ireland

Ian Pitt
University College Cork, Ireland

ABSTRACT

Mobile devices are becoming more and more commonplace across all walks of life from the workplace to leisure activities and even the classroom. Many schools shun the use of devices such as mobile phones in the classroom environment, but this will have to change as they become a more integral part of our daily lives. The ever increasing capabilities of these devices allow for opening up on new application domains. The ubiquitous use of mobile technology in the classroom may provide new and interesting ways for students to interact with subject matter. This chapter discusses the use of cross platform Bluetooth enabled mobile devices within the classroom setting to allow students to interact with subject matter in a new and interactive way using the ICT resources that are ever present in our daily lives.

INTRODUCTION

The revolutionary Irish writer Thomas Osborne Davis born in Mallow the crossroads of Munster, Ireland is renowned for the saying "Educate that you may be free". At the time of his birth in 1814 the primary tools of the classroom were books and writing tablets. Little has changed in almost 200 years as students still heavily rely on written text in print format, and copy books to answer solutions to questions proposed by the teachers. The majority of schools on the island of Ireland are equipped with at least one computer laboratory capable of facilitating about twenty or so students at any given time, both at primary and secondary school level. These facilities are however extremely under utilised, a class may expect to see the lab for as little as one

DOI: 10.4018/978-1-60566-703-4.ch007

hour per week, and many classes may go through school without ever seeing the lab.

This is where mobile technology can be used to reinvent the way in which classroom based learning is carried out. This can prove advantageous for both the teacher and student, and provide new and interactive ways to learn that can help to engage and entertain the students. In early 2008 the mobile phone market penetration rate for Ireland stood at well over 111% and in November 2007 it was announced that world wide mobile phone subscriptions surpassed 50% of the world's population. One can say that every school going student is an expert in the usage of a mobile phone, from the sending of text messages to the use of more advanced features such as video messaging. They therefore require little or no training in how to use a new application designed for the mobile platform. Given this, the mobile platform is the perfect mechanism to engage and interest students in topics that many may otherwise find boring and uninteresting.

To allow the teacher to interact and deliver content with the student's mobile, one may readily think of using SMS messages to transmit questions and answers to and from the student's phones. This however is hindered by the fact that each transmission would incur a financial cost. Similar financial costs are also incurred if one was to use a mechanism such as WAP or simple html pages to deliver content and collect feedback. One viable alternative is the use of Bluetooth technology. There are presently over two billion Bluetooth enabled devices in existence, with this being added to every day. A considerable proportion of all mobile phones now sold come with Bluetooth as standard, often conforming to Bluetooth 1.2, but more and more are we seeing low to moderately priced phones shipping with Bluetooth 2.0 as standard.

Bluetooth provides an excellent transmission mechanism as there are no financial costs incurred for inter-device communication. Many teachers may have access to devices such as laptop or PDA's which they can use to drive the cross-platform, mobile teaching/learning student experience. A perfect example of this is in the subject area of mathematics where by the teacher could send out a problem to be solved to the student. The students in turn would have to solve the problem and transmit the solutions back to the teacher's device. Such a mechanism can provide the teacher with accurate feedback as to how well each individual student is performing, as well as the class as a whole, with just a single glance of their master terminal.

Bluetooth application programming generally requires substantial development time, primarily due to the searching for the devices, creation of connections and maintenance of those connections. The Mobile Message Passing Interface (MMPI) provides the developer with a simple to use mechanism that abstracts them from all of the underlying Bluetooth application development necessary for even the simplest of programs. It allows for rapid application development with minimal code development. It is therefore the perfect development architecture to employ for the efficient development of a Bluetooth enabled application. The multi-platform Student / Teacher mathematics tool presented in this chapter makes use of this library to provide the end user with a simple and engaging application that allows for classroom based learning to move the traditional booked based paradigm into 21st century technology based pedagogy.

BACKGROUND

To facilitate the effective development of Bluetooth based applications it is necessary to provide the developer with some form of framework that abstracts them from lower lying Bluetooth specific code development. This allows for Bluetooth applications to be developed in a more efficient manner, simplifying the development process and reducing potential errors. A number of such frameworks are in existence providing the developer with varying degrees of abstraction from the Bluetooth Client / Server development process.

The BlueCube project (Chang et al., 2003) was a pilot study in creating a parallel computation environment using Bluetooth. It was based on the concept of linking the devices together in a Hypercube fashion. The system was divided into three distinct phases: Ring Construction, Scatternet Construction and BlueCube Construction. The main focus of the work was on the creation of a scatternet environment that allowed for inter-device communication, however there was no discussion regarding how one could actively develop a parallel program using such a system.

BlueGrid (BlueGrid 2008) is a middleware system that allows for parallel computing over J2ME using a Bluetooth infrastructure for communication. As the J2ME platform does not support dynamic bytecode loading a distributed scripting language called GScript was provided that is based on a JavaScript like open source scripting language called FScript. Blocks of parallel code are designated in between the *parallel* and *endparallel* constructs. The architecture of the system maintains the star topology network; therefore all the Client devices are connected to a common node (Broker). It therefore allows for the total number of two-way connections to be N-1, but as such a Client device is unable to directly communicate with another Client.

DynaMP (Shepherd et al., 2004) is a dynamic message passing architecture focused on Scatternets that uses on-demand routing based on the Ad-hoc On Demand Distance Vector (AODV) routing algorithm. The target devices for the system are the TINI (Tiny Internet Interface) microcontroller based Single Board Computer (SBC) that uses Java as its native environment and the iPAQ PocketPC running a Linux operating system and the Kaffe JVM. The topic of dynamic class loading is discussed in some detail, but no mention is given regarding what forms of communication can be achieved, or what the structure of a program using the system would look like.

The Bluetooth Multiplayer Games Framework (BlueMGF, 2007) is a project available from sourceforge that provides a framework to create both multiplayer games and applications running over a Bluetooth network. A two player tank combat game is included within the package as an example of the frameworks usage. The framework consists of a set of nine Java classes within the package panic.bluemgf. To create a Bluetooth application using the package one must extend the BlueMIDlet class. This performs the initialisation of the system and displays the game mode configuration form. The configuration form allows the user to choose between the options: Stand Alone, Server, Client and Configuration. The configuration option allows for the user to select the level of logging feedback to report (using log4j2me on sourceforge) and the type of connection protocol to use (L2CAP or SPP).

If the user chooses the option of "Stand Alone" then the role of the framework is concluded as no inter-device communications are required. Choosing the "Server" or "Client" options will instigate use of the framework for the carrying out of communications in a Client / Server, star network topographic manner. When creating a MIDlet to use the framework one needs to call two methods in the constructor setNumberOfPlayers(. . .) passing a parameter for the number of players to partake in the game, and setServiceUUIDAsString(. . .). The second method requires the developer to specify the Universally Unique Identifier (UUID) for the game.

The last element to implement is the handleEvent(. . .) method which is in turn called by the framework. Two parameters are passed to it from the framework, the first being a string representing if the application has been started as a stand alone application, a Client or a Server. The second parameter returns a Vector of Connections if the application has been started as a Server, or just a single instance of a Connection object in the case of a Client. The disadvantage of this is that when a developer starts to code for a multiplayer game they must deal with Connection objects, data streams and communications error handling. Another disadvantage is due to the star network configuration because of the fact that if a message was intended to be passed between Clients it would have to be routed through the Server, requiring both extra communications and code development time. The framework provides resources only for the communications element, and provides no aid to speed up the development of the game elements itself such as the management of sprites, and various layers.

Peer2Me (Wang et al., 2007) is a rapid application framework for mobile peer-to-peer applications. The framework is based on a layered architecture, and comprises several distinctive units. The Node is a logical representation of a peer (mobile device) running the framework. A Group is a collection of nodes that know of the existence of each other. A Service is a description and identifier of an application using the framework. A Network is an abstraction of the network layer representing the communications medium. A Message is an entity that can be exchanged between nodes. The Session keeps track of known nodes, groups and network media. The Framework is the core entity between the application and the rest of the system. The final unit is the Application in which a Peer2Me application is implemented as a MIDlet that makes use of the framework for communications.

Any message that is sent can be comprised of three main types: a file, a primitive data type or a serialisable object. As J2ME provides no serialisable interfaces, such functionality had to be included in the framework to allow for serialisation and deserailisation. To send a message a Message object must first be instantiated. The data that is to be sent is then added to the message by calling the addElement(. . .) method.

The method takes two parameters: the data and the type. The type is represented as a string, for example the type "info" is used for data in the form of a string of text. One can add several elements to a message before sending. The message is transmitted by calling the sendMessage(. . .) method of the framework object. Correspondingly there is a messageReceived(. . .) method that reads the type information and then reads the data that relates to the type, for example a call to message.getString(. . .) is executed in the case of a string data type.

An alternative approach to achieving a level of mobile parallel computation lies within the realm of Remote Method Invocation (RMI) (Wei et al., 2002). The framework consists of two sets of software infrastructures. A set of layers called JavaBT was created for handling the Bluetooth protocol stack. The Java RMI layer was then implemented on top of the JavaBT system. The RMI implementation runs directly on top of the L2CAP layer instead of the traditional TCP/IP communication associated with RMI. The system was tested across desktop computer systems with Bluetooth dongles.

MOBILE MESSAGE PASSING

The majority of Bluetooth applications use the point-to-point network structure. The most widely used example is hand free kits for the mobile phone, in which the user can use a Bluetooth enabled headset in conjunction with the phone. Many Bluetooth games also use this structure, allowing just two people (devices) to partake. The Point to Multi-Point configuration is a more complex network structure in which the Master device must maintain a list of all clients connected to it. This type of network has a Star topology in which all outlying nodes are connected to one central node, and none of the outlying nodes can communicate directly with each other.

The MMPI system (Doolan et al., 2006) allows for the creation of a fully interconnected mesh network structure allowing every device within the Piconet to communicate with every other device independently. Should one wish to build larger networks then a Scatternet topology can be employed at the cost of the removal of direct Client to Client interconnections. This architecture requires a message routing system to facilitate packets destined for outlying nodes to be routed through the network from the source to the destination through the available network links.

The overall structure of the MMPI library consists of three main classes. The MMPI class is the main class that is used to carry out the primary message passing functions. The two remaining classes BTClient and BTServer are required for creating the underlying Bluetooth connections. The MMPI class will instantiate only one of the Bluetooth classes depending on a parameter value that is sent to the constructor. With the channels of communication established the MMPI class is capable of sending or receiving messages simply by accessing an element in an array of either DataInputStreams or DataOutputStreams. With the MMPI object instantiated, the size and rank can be established by calling the getSize() and getRank() methods of the MMPI object. By knowing these values an application using the MMPI system can divide up the task appropriately in preparation for the communication of data to all nodes in the system.

In MMPI Point-to-Point communication is carried out using the send(. . .), recv(. . .) methods, in the same manner as MPI itself. The sending and receiving of data is carried out by specifying the device number to which data is to be sent, and the device number from where data will be received from. In MPI programming a Send on one node must match up with a Receive on the destination node. Hence if some data is sent from device 0 to device 1, device 1 must have a corresponding recv(. . .) method call with device 0 specified as the sending device. Data is sent in the form of an Array of data which is in turn passed to the MMPI methods in the form of a standard Java Object. The parameters required by these methods include: the input or output buffers, the starting position of the array, the number of elements to send or receive, the data type and the identifier of the node with which data will be sent to or received from. The library also provides the application developer with a number of global communication methods allowing for operations such as broadcast, scatter, gather and reduce.

To write any MMPI application a few standard lines of code are required. Between the creation of the MMPI object and the closing down of the world (with the finalize() method) parallel computation may be carried out. In many programs the first methods to be called are getSize() and getRank() so the node is informed about its environment. This is followed by the parallel code the user wishes to execute. The library may be used in myriad of mobile computing application domains where inter-device communication is required. This can include such application areas as parallel computing, multiplayer games, distributed mLearning, and parallel graphics.

THE USE OF ICT IN THE CLASSROOM

Benefits Arising from Cross Platform Pedagogical Applications

Within the last number of years, there has been a shift in emphasis toward the incorporation of mobile devices into the conventional classroom environment for learning purposes. To date, significant amounts have been invested in digital classroom initiatives (Prensky, 2005). Due to advances in wireless technologies, there is now a great opportunity through 'Blended Learning' techniques (Singh, 2003), to connect students with their teacher and colleagues for specific learning activities. This connectivity can be facilitated through the incorporation of Bluetooth enabled mobile phones or PDAs and notebook computers, when combined with specific learning applications. These devices are a compelling choice for classroom situations as they can be used frequently for integral purposes (Roschelle, 2003). Handheld devices, when used in this context, provide enhanced social interaction and an increased motivation to learn (Zurita & Nussbaum, 2005).

Because wireless devices are small, mobile and readily available in the student's pocket, the incorporation of this type of device into daily classroom proceedings provides an opportunity to facilitate modern learning in the modern school. Today's mobile devices combine high processing power and capabilities with new and interactive wireless technologies such as Bluetooth. This capability can be harnessed to create fun and interactive learning sessions to reengage a new generation of learners.

Mobile devices offer schools an economic and accessible option for the provision of wireless "one to one" and collaborative computing environments for students (Lei et al., 2007). The mobile device is an existing resource, with most students today, in possession of at least one device. In most cases, this device is a mobile phone. The incorporation of these devices into the classroom ensures that each child has access to a machine. This incorporation provides an equitable system of education putting an end to the looking over your classmates shoulder scenario that has existed in school computer labs for some time now. Bringing mobile phones and PDAs into the classroom can also provide daily interaction with technology, thus removing the need for the expensive, fixed location computer labs, which restrict teaching capacity and access to learning via technology (Roschelle, 2003).

Today's learner has different requirements then the classroom inhabitants a decade ago. These children have developed alongside technology and in most cases 'need' to interact with technology in order to maintain an interest in their environment. Schools that do not incorporate technology as part the regular classroom routine, run the risk of 'loosing' their students altogether. These 'Digital Natives' (Prensky, 2001) learn in a different way to their teachers. Because of this, teachers today are left frustrated as they find their task of teaching very much a one way street (Russell & Pitt, 2004). Therefore, both today's learners and educators must be facilitated through newly devised methods of teaching. Schools who ask students to 'switch off' their phones when they go to class are immediately focusing the student's attention away from the classroom's events and onto the device in their pocket. Teenagers, as they are very adept at using their mobile device for socialising and for downloading games and music, should be encouraged to use it also as a learning tool. Students, in most cases are agreeable to the use of mobile phones in educational settings (Lubega et al., 2004).

Many students today find it difficult, under traditional roles and teaching methods, to participate in class. Students can find many reasons not to join in classroom discussion, but the main reason for non participation can be a simple lack of interest in the class material due to the way it is presented. This can be a particular problem in the teaching of mathematics. This subject can be particularly difficult to learn, but when coupled with a culture of unwillingness to participate, the result is that the learning process is greatly hindered; the student becomes confused and is quickly left behind the other students. By including interactive applications, through the incorporation of Bluetooth enabled devices, the student can become engaged in the learning process while at the same time, the teacher is empowered. The teacher can now assess the students' abilities thus given the means to solve problems where they first arise. The student, through classmate anonymity, is given the ability to participate without the potential embarrassment of failure.

Therefore, multi-platform learning can maximise the benefits of conventional classroom experiences to enhance the overall learning process. An enthusiastic learning environment can be created to overcome student inhibitions by shifting away from traditional teaching methods. This new generation of learners holds high expectations from technology. These expectations, when combined with both existing and developing theories of learning, enhance classroom learning as a whole. Until now, mobile devices have been viewed as a distraction to learning with teachers viewing them as a detriment to the challenging task that is teaching. However, because today's students feel so attached to their mobile devices (Al-khamayseh, 2007), the combination of advanced wireless technology with face to face, teacher led didactic instruction, represents an essential part of emerging and engaging learning processes for the future.

PEDAGOGICAL RATIONALE BEHIND CROSS PLATFORM PEDAGOGICAL APPLICATIONS

Blended learning can be described as a method used to facilitate different learning styles through different delivery methods for teaching and learning models. A key element of blended learning is the creation of communication between the involved parties. The applications outlined in this chapter attempt to incorporate the learning style of modern children, who are members of a new digital world, through the use of technology in the blended classroom. Most modern children can find it difficult to learn via traditional teaching methods due to their exposure to and dependence on digital technologies. Relying on traditional methods in absence of technology could lead to the total exclusion of one or many students within the everyday classroom session. The student can become bored and unmotivated by the content of a traditional lesson. This can be further heightened by a developing lack of interest where the student is facing a problem within the area that has been overlooked unknowingly by the teacher.

Learning styles refer to the many ways in which people learn, and in this case blended learning has been employed to create activities for the classroom through the use of technology for the inclusion of a new type of learner. The applications attempt to reach new learning styles that in many cases can not be reached by traditional means. This new generation of learners has developed their own learning style based on the modern technologies of today where they have a need to interact with technology. These applications facilitate communication between the instructor and the students and also allow for peer interaction. The instruction in this case is represented by traditional methods where the student is taught the methodologies of algebra in the normal classroom environment. This is then blended with technology through mobile type devices to generate motivation among the students based on a quiz like session to follow the traditional instruction session.

The students are connected directly to their teacher via a device such as a mobile phone, PDA, or laptop. The cross platform facility provides convenience over single platform deployment, thus allowing a student with a PDA and a student with a laptop to communicate simultaneously with their teacher who is perhaps using a mobile phone. Therefore there is no restriction placed on the student or teacher in relation to the device category that they have available to them during the session.

The applications provide a competitive learning environment for the classroom where students are motivated to learn on an equal level with their peers through digital technologies. Group sessions can be established to achieve a competitive environment, where students collaborate together on the solution for an equation, or individuals can be given the opportunity to display their own level and capabilities directly to the teacher. The student can benefit from the anonymity provided by the connection of their individual device to the teacher's device. This anonymity gives the teacher the ability to discover problems pertaining to individual students. The teacher can use the information provided to them on their device to address existing problems at an earlier stage.

This allows the whole class group to move on to the next topic at an earlier stage with everyone in the class having achieved a level of necessary competence within an area topic. Individuals who are in difficulty with a topic can be helped prior to moving on to the next topic thus removing the potential for isolation within the classroom. This eliminates the need for the student to put forward to the teacher the fact that a problem exist. Most students in such a situation would be reluctant to approach a teacher due to shyness or embarrassment within their peer group.

CROSS PLATFORM MLEARNING APPLICATION IN PRACTICE

Mathematics can be a very difficult subject to learn. Some students can have difficulties in expressing their problems within a topic due to their ineptness to communicate and interact in a classroom environment. Their unwillingness to participate can leave the student confused or uninterested. The teacher can be left unsure of the individual students' ability to complete the tasks prescribed to the class. By developing interactive and engaging applications using Bluetooth we can reduce this problem by empowering both the teacher and the student within the classroom. The student can be given the ability to interact in an interesting environment, anonymous from their classmates, while the teacher can be given the means to assess students and solve problems that exist at an individual student level.

To highlight the potential for this type of learning scenario, an example application was developed focusing on the Algebraic Linear Simultaneous Equations. This equation area was selected because within the subject area of algebra, it can represent an area of particular difficulty for students.

The application is in essence a Bluetooth game designed for deployment within a collaborative classroom environment. Cross platform devices are employed over a piconet to facilitate the deployment of the system. It is envisaged that the application could create a fun and interactive method for both teaching and learning algebraic equations in the classroom. The images provided in Figure 1 show the system in action, running on a teacher's mobile phone in conjunction with a student laptop.

Application Structure

The structure of the application employs the MMPI library and is designed for a Client / Server type architecture. J2ME is used to develop a mobile phone specific MIDlet offering GUI input forms which allow the incorporation of both the teacher and students' mobile phones into the classroom mathematics teaching session. Java Swing is also employed to develop a desktop version of the application to provide GUI input forms that facilitate the incorporation of both laptop computers and PDAs. The application allows for the client side, represented by the student's mobile device, to employ the equation information received from their server side teacher's mobile device, to calculate a solution for the relevant equation. The operations for the applications are outlined in Algorithm 1.

The application is designed for use with different devices such as mobile phones, PDAs and Laptops to provide a simple means of learning for secondary school students through cross platform communication. To begin the application session, using either a laptop, PDA or a mobile phone, the teacher starts the communication on the chosen Master (Server) device. He or she then proceeds to manually or randomly input values for the equation. The values are transmitted to the Slave (Client) devices where they are output to the device screen as an equation.

When each connected student receives the transmitted information in the form of an equa-

Figure 1. Example of system running on teacher's phone and student's laptop

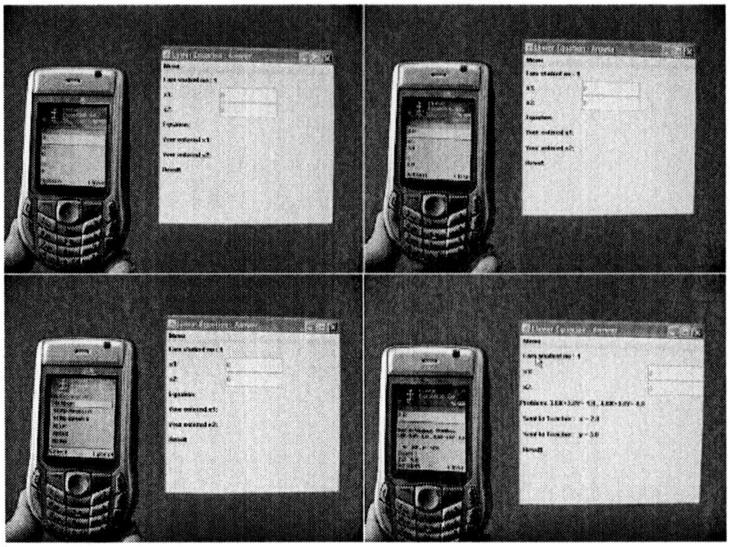

Figure 2. The server side (teacher) interface

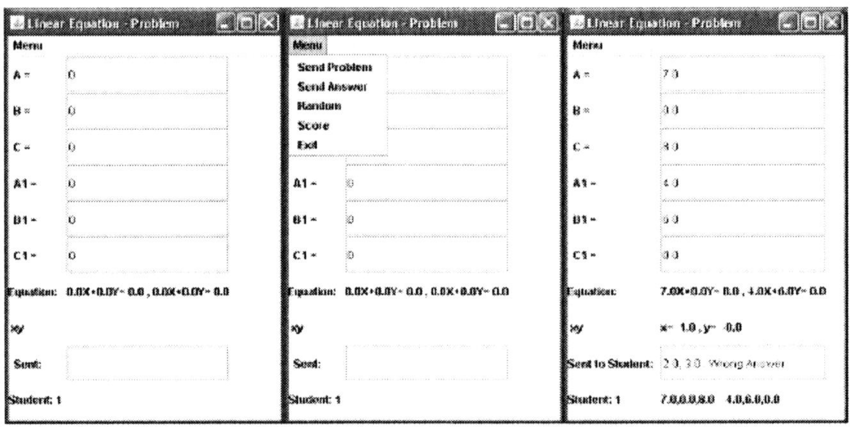

tion output to their screen, they are required to employ taught methods to find the solution to the problem. The student individually transmits their resulting values back to the teacher's Master device where it is automatically assessed. The result of the solved problem is then transmitted back to the individual student's node device.

The Graphical User Interface

The teacher, when employing the system, has the option of using either a mobile phone, PDA or laptop device to run the server side aspect of the classroom session. The GUI is similar in each instance with differences generally only relating to the programming systems used to develop the interface. The Server Side GUI can be seen in Figure 2.

The student also has a choice, they can either employ a mobile phone which uses a MIDlet, or they can easily incorporate their PDA or laptop running a Java Swing Application (see Figure 3). As a result, any combination of devices can be employed for a learning session.

The Teacher Running the Server Side System Through their Mobile Phone

Where the teacher wishes to run the system on a mobile Phone, a Server side input form is created using J2ME for specific use on the Master device. This enables the teacher, through the use of textFields, to enter values for the equation (see Figure 5). In the case of the Linear Simultaneous Equation Game, six textFields will be displayed for value entry.

The teacher can either enter the values manually using the device keys or select the Random Command from the MIDlet menu (see Figure 4). The equation is then output to the Master screen in the form of text through the use of a stringItem. Dynamic textFields are created to log all the student devices connected on the client side indicating the student's individual device number. The number of student textFields therefore correlates with the number of client nodes connected with the server for the particular session.

When the student sends their answer values to the teacher, these textFields are set with the student answer values. In turn, when the device has assessed the values, these dynamic textFields are set with both the individual's answer and their result which are logged by the application. This enables the teacher to assess where problems exist within the classroom group.

The Student Running the Client Side System Through their Mobile Phone

On the Student's Mobile Phone the Client Side GUI is also developed using J2ME. The GUI offers the student a simplistic user input form which is composed of two textFields supported by a send command available as a menuItem (Figure 6). These TextFields facilitate the student in entering their answer which should be comprised of two values. The values they provide are then displayed as text to their node screen through the use of stringItems. Once their answer has been sent to their teacher, this text indicates the answer which has been sent.

Text is sent to their teacher only when the "Send Command" has been selected from the menu. The student is then required to wait for the teacher's device to respond with their result. When the individual's result has been transmitted to and

Figure 3. The client side (student) interface

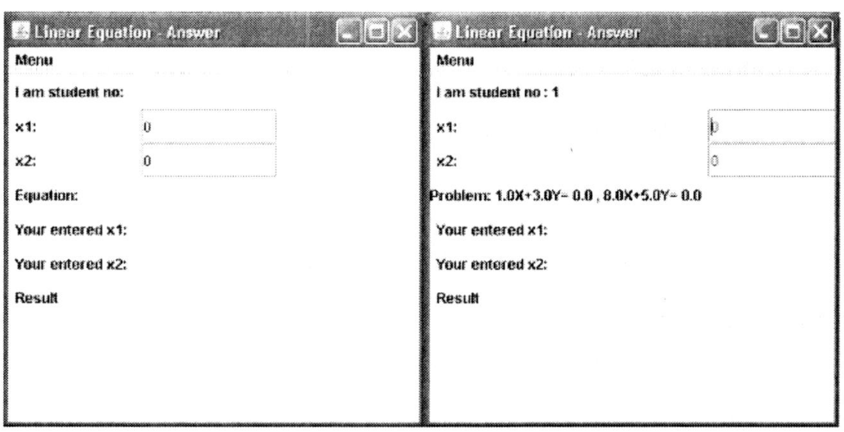

Figure 4. Menu screen as displayed on the teacher's mobile phone

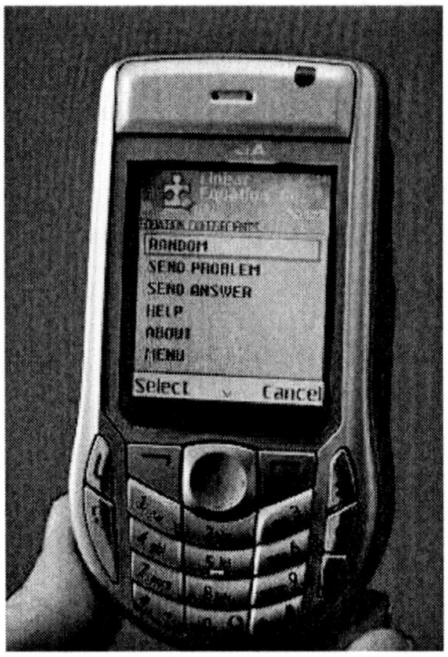

received by the device, it is output to the student's screen using a stringItem. The display remains on screen for a number of minutes before the screen is cleared allowing the student to receive a new equation from their teacher's device.

The Teacher Running the System Through their Laptop Computer or PDA

Where the teacher requires the use of their laptop computer or their PDA, a server side input form is available as part of a Java Swing Application. The Input form is designed in very much the same manner as the one provided by the J2ME MIDlet, the only difference lying with the programming style required for a Swing application.

The Swing input form, contained within the JFrame, is comprised JTextFields which allow the teacher to enter relevant values. Again, the teacher has the option of either entering the values manually or through a Random command. The Random Command is accessed from the application menu.

Once the values have been entered, the equation generated by the application is output to the device application screen using a JLabel to display the text. Dynamic JTextFields are also provided for in the application version of the system to log each student device connected on the client side indicating the student's individual number. The answer values received from the students are set into the JTextFields. The application assesses the values and sets the JTextFields with the individual

Figure 5. Example of system running on teacher's mobile phone and student laptop

Figure 6. The student interface as displayed on a mobile phone

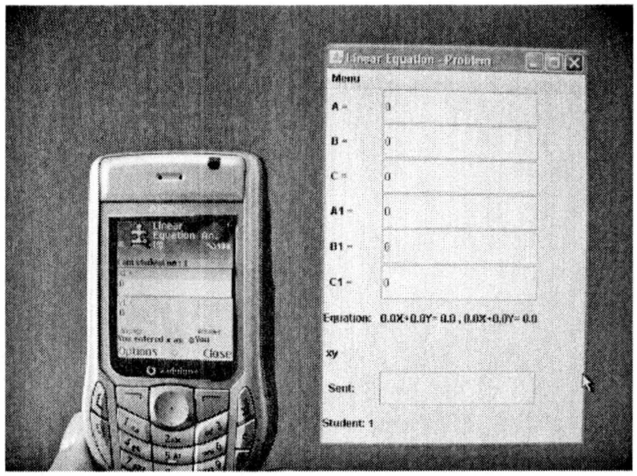

students' answer and result. These are also logged by the application in this scenario to inform the teacher of any existing problems. An example of the server side application running on a PDA can be seen in Figure 7 & 8.

The Student Running the System Through their Laptop or PDA

Here again the student is provided with a GUI input form comprising of Swing components. The simplistic user form is designed with two JTextFields which allow the student to enter their values. Once the values for the solution to the received equation are entered into the form via the device keys, the 'Send Command' from the menu is used to transmit the answer to the teacher. This is illustrated in Figure 9 & 10.

The student then waits for the teacher's device to respond with their individual result. When this is received by the device, it is output to the student's screen using a JLabel to display the text.

TESTING AND EVALUATION

The applications developed through NetBeans IDE ran successfully on combinations of PDA and Mobile Phone devices, including the Dell Axim X51v PDA, Nokia 6620 mobile phone and windows based desktop / laptop environments. Samples of the applications under execution may be seen in figures 5, through 9.

The only limitations imposed were represented by the size of the PDA and Phone device screens.

Figure 7. Teacher interface displayed as on a PDA

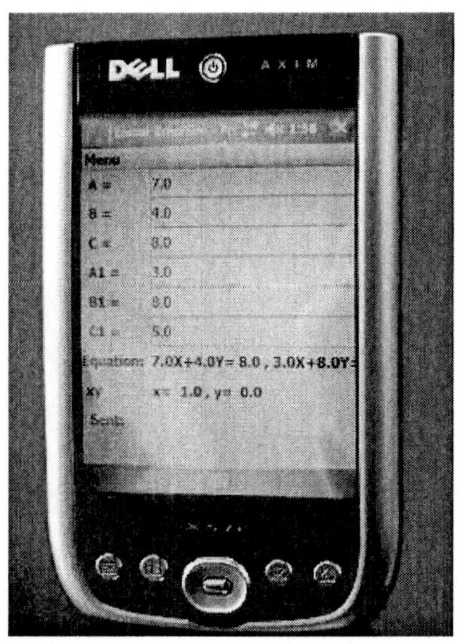

Figure 8. Example of system running on the teacher's PDA

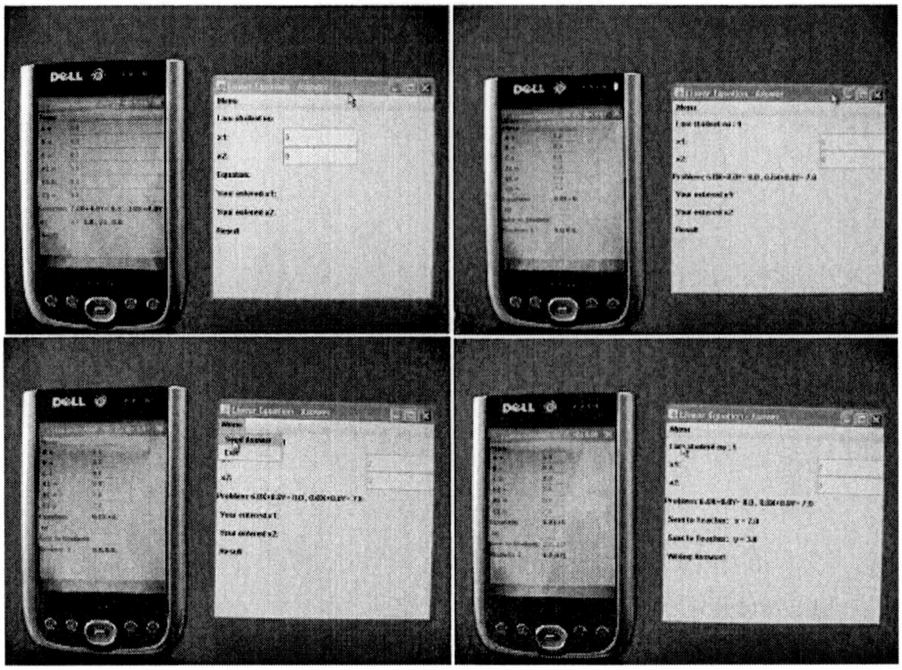

This limitation was apparent only where these devices were employed by the Teacher. This limitation was evident only in the creation of a need for scrolling as the entire user interface could not be displayed on screen simultaneously. This limitation did not pertain to the student user interface when using these devices. Therefore the student's experience of the application was not hindered by the limitation.

An efficient message passing function was offered by the MMPI Library to enable communication between all nodes, even where multiple platforms categories were used. The message passing was in no way hindered by the normal limitations posed by mobile devices.

A usability study questionnaire was completed by twenty-five second year students and their teacher at a local secondary school who employed the applications as part of their routine mathematics class. The employment of the system looked to include all three device categories in all potential combinations. This test was the first experience that the selected teacher and his students had with mobile blended collaborative learning. The study revealed positive results and feedback as can be seen in Table 1 which outlines some of the usability questions on the questionnaire.

The findings show that 88% of the students who completed the questionnaire enjoyed and benefited from the learning experience provided by the applications. The majority of students who tested the applications found them to be both challenging and user friendly. A high number of students, 84%, found the experience enabled them to participate in the class where they normally would not do so, thus improving the quality of their overall classroom experience. 90% saw that the applications created both a challenge and a competitive atmosphere within the classroom, thus increasing their motivation to successfully solve the equations put forward by the applications. 76% of these students suggested that they would like to see similar applications incorporated into their classroom program in the future. The teacher

Figure 9. Example of system running on student's laptop / teacher's phone

Figure 10. The student's interface as displayed on laptop or PC

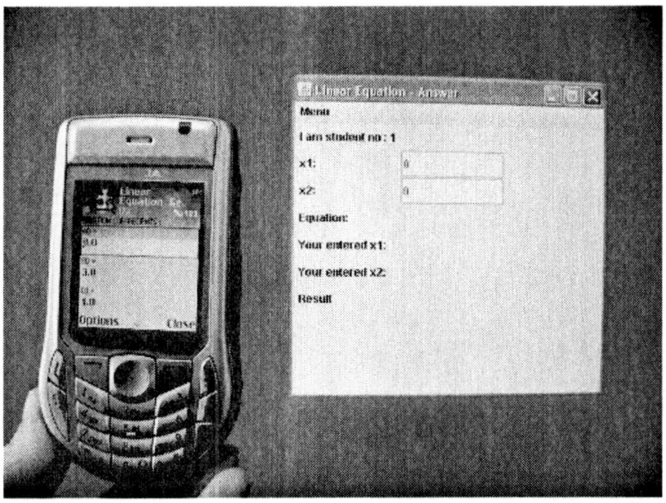

saw an increase in learning results through the use of the applications via the results recorded for each student on the master device. The increase was particularly significant for a number of student's whom usually would achieve at a lower level. The teacher also found an increased level of concentration and collaboration within the classroom session.

Therefore it has been demonstrated that such cross platform learning applications, can provide benefits to secondary level students when incorporated into a classroom environment. The overall learning process is enhanced by these modern teaching methods allowing the student to learn through inclusion. Learning applications, when combined with mobile technology, such as Bluetooth, can be employed to remove student inhibitions and to facilitate the creation of a more enthusiastic learning environment within mainstream education. As such technologies continue to advance their future for education is greatly significant.

CONCLUSION

This chapter has reviewed a number of the Bluetooth frameworks that are available for simplifying multiuser application development. An example has been presented in which the MMPI library is

Table 1. Usability study results

QUESTION	ANSWER	PERCENTAGE
Did you enjoy the overall mLearning experience offered by the applications?	YES NO	88% 12%
Did the applications bring an element of fun to the learning experience?	YES NO	75% 25%
Did you find the applications were beneficial to learning in the classroom environment?	YES NO	87% 13%
Did you find the applications to be user Friendly?	YES NO	78% 22%
Did you find the applications challenged your mathematical ability?	YES NO	90% 10%
Did the applications increase your level of class participation and the quality of your classroom experience?	YES NO	84% 16%
Would you like if this type of application was incorporated into your classroom routine in the future?	YES NO	76% 24%
Was your motivation increased by the application?	YES NO	77% 23%
Did the applications increase your level of competition against your classmates during the classroom session?	YES NO	91% 9%
Was your potential learning score increased by the incorporation of the applications?	YES NO	67% 33%

employed to facilitate the rapid development of classroom based teaching tool. The learning tool focuses on the subject area of mathematics but can readily be applied to any topic a teacher may require. The architecture provides both students and teachers with a new means on interacting within the classroom environment that can both stimulate education and entertain using the cross platform mobile technology that surrounds us all today. Teaching methodologies have changed little in the past few hundred years, relying on antiquated tools such as the blackboard and chalk. Today in the early years of the 21st century we now have the means and the technology to provide teaching resources suitable for mobile world that we now live in.

ACKNOWLEDGMENT

Funding for this project was provided by the "Irish Research Council for Science, Engineering and Technology funded by the National Development Plan in collaboration with the Digital Hub".

REFERENCES

Al-khamayseh, S., Zmijewska, A., Lawrence, E., & Culjak, G. (2007). Mobile learning systems for digital natives. In *Proceedings of the Sixth IASTED International Conference on Web Based Education* (pp. 252–257). Chamonix, France.

BlueGrid. (2008). *Grid computing for J2ME Bluetooth devices.* Retrieved July 20, 2008, from https://bluegrid.dev.java.net

BlueMGF. (2007). *Bluetooth multiplayer games framework.* Retrieved July 20, 2008, from http://sourceforge.net/projects/bluemgf

Chang, C. T., Chang, C. Y., & Sheu, J. P. (2003). BlueCube: Constructing a hypercube parallel computing and communication environment over Bluetooth radio system. In *International Conference on Parallel Processing* (pp. 447–454).

Doolan, D. C., Tabirca, S., & Yang, L. T. (2006). Mobile parallel computing. In *5th International Symposium on Parallel and Distributed Computing* (pp. 161-167).

Lei, J., Conway, P. F., & Zhao, Y. (2007). *The digital pencil: One-to-one computing for children.* New York: Lawrence Erlbaum Associates.

Lubega, J., McCrindle, R., Williams, S., Armitage, U., & Clements, I. (2004). Uses of mobile phones in higher education. Cantoni & McLaughlin (Eds.), In *Proceedings of EDMEDIA 2004* (pp. 3951-3956). Lugano, Switzerland.

Prensky, M. (2001). Digital natives, Digital immigrants. *On the Horizon, 9*(5). MCB University Press.

Prensky, M. (2005, September/October). Engage me or enrage me: What today's learners demand. EDUCAUSE Review, 40(5), 60–64.

Roschelle, J. (2003). Unlocking the learning value of wireless mobile devices. Journal of Computer Assisted Learning, 19(3), 260–272. doi:10.1046/j.0266-4909.2003.00028.xdoi:10.1046/j.0266-4909.2003.00028.x

Russell, G., & Pitt, I. (2004). Visions of a wireless future in educational technology. In *Proceedings of Informing Science and Information Technology Education Joint Conference* (pp. 747-752). Rockhampton QLD, Australia.

Shepherd, R., Story, J., & Mansoor, S. (2004). Parallel computation in mobile systems using Bluetooth scatternets and Java. In *Proceedings of the International Conference on Parallel and Distributed Computing and Networks*. Retrieved July 20, 2008, from http://www.informatics.bangor.ac.uk/~robshep/publications/DynaMP.pdf

Singh, H. (2003). Building effective blended learning programs. Educational Technology, 43(6), 51–54.

Wang, A. I., Bjornsgard, T., & Saxlund, K. (2007). Peer2Me – Rapid application framework for mobile peer-to-peer applications. In *International Symposium on Collaborative Technologies and Systems*. Retrieved July 20, 2008, from http://www.idi.ntnu.no/grupper/su/publ/alfw/peer2me-cts2007.pdf

Wei, P. C., Chen, C. H., & Chen, C. W. (2002). Support and optimisation of Java RMI over a Bluetooth environment. In *Concurrency and computation: Practice and experience* (pp 1-21).

Zurita, G., Nussbaum, M., & Salinas, R. (2005). Dynamic grouping in collaborative learning supported by wireless handhelds. Educational Technology & Society, 8(3), 149–161.

Chapter 8
Plastic Interfaces for Ubiquitous Learning

José Rouillard
Laboratoire LIFL - Université de Lille 1, France

ABSTRACT

This chapter presents research around pervasive and ubiquitous computing, particularly oriented in the field of human learning. We are studying several solutions to deliver content over a heterogeneous networks and devices. Converting and transmitting documents across electronic networks is not sufficient. We have to deal with contents and containers simultaneously. Related work in interface adaptation and plasticity (the capacity of a user interface to withstand variations of both the system physical characteristics and the environment while preserving usability) is presented and some examples of context-aware adaptation are exposed. We present an adaptive pervasive learning environment, based on contextual QR Codes, where information is presented to learner at the appropriate time and place, and according to a particular task. This learning environment is called PerZoovasive, where learning activities take place in a zoo and are meant to enhance classroom activities.

INTRODUCTION

In the coming years, learning through heterogeneous telecommunication networks will probably become the rule and not the exception. Studies show that information which circulated in the world is progressively being stored in a numerical form. In theory, we should take advantage of improving access to this information, since it is immediately available and consumable. But in fact, with the multiplicity of possible connections to Internet and over heterogeneous networks, is it not always the case.

The new kinds of networks such as WIFI or Bluetooth offer new perspectives of research for the ICT-based education and hypermedia community. The goal is to satisfy all types of users by proposing data-processing solutions available on almost all the products or peripherals available on the market. The rise of a great number of hardware such as mobile phones, PDA or WebTV, having

DOI: 10.4018/978-1-60566-703-4.ch008

variable capabilities, leads to a reflection on independent interfaces specification, in order to avoid specific developments. The ubiquitous role of the computer makes each day more unsuitable the screen-keyboard-mouse model posed on a corner of a desk (Beaudouin-Lafon, 2000).

The large success and rise of the Internet network is mainly due to the technical standards used and the adoption of languages such as HTML, WML (WML), or VoiceXML (VoiceXML) . But, we observe some incompatibilities in spite of the standards promulgated. Indeed, on the one hand, various types of media such as texts, graphics, sounds or video can easily be used and transmitted through networks, but on the other hand, the fact that machines are not necessary from the same vendors, or do not support the same operating system, leads to situations where information processing systems and/or databases are incompatible with other data coded in particular formats. Therefore this consumes additional costs and time for each end-user, like students and teachers, whatever his platform, can obtain a product or a satisfying service. The need for easier access to information; whether at the office, home, in the train, etc., is felt all the more with the arrival of new materials, and the success of the pocket computers as well as mobile telephones. One wishes to lean towards the transformation of end-user's interface for "anyone and anywhere" (Lopez & Szekely, 2001). With the multiplicity of the means of connecting to Internet, it is necessary to conceive generic interfaces and mechanisms of transformation to obtain concrete interfaces for each platform. That's why the W3C launched an activity in the Device Independence field.

Adaptable interfaces in multi-device e-learning environment or m-learning are playing a very important role in improving the accessibility of these applications, and are leading to their increased acceptance by the users. After the switch from Learning to E-Learning, we are now facing another switch towards M-Learning. Thus, we are entering an era of pervasive computing with the challenge of providing services available anytime and anywhere. In this context, data management is obviously the heart of concerns in what is now called pervasive or ubiquitous computing.

Consequently, recognizing that mobile computing is one of the most rapidly growing areas in the software market, some researches explore the role of adaptation in ubiquitous learning and particularly in the area of mobile computing. Mobile computing has a very strong potential due to the extremely high market penetration of mobile and Smartphone. The significant development of wireless networks and mobile devices, such as phones and laptops, PDA, sensors, or Smartphone, that we know since the past fifteen years leads to profound changes in applications and services offered to users. The terminals available on the market today are more and more powerful. Their autonomy is sometimes limited, but they provide equipment increasingly rich, with multiple connections, GPS receivers, etc.

These systems operate in a dynamic environment particularly because of frequent disconnections or user mobility. They must be able to respond dynamically to changes in these different settings. They therefore must be sensitive to the context in order to be able to adapt dynamically and so provide an important quality of service to users. If only a few applications accessible to the general public have now emerged, some should be available soon in areas such as transport, health, commerce or education.

This document presents some aspects of this scientific problem and is structured as follows: Section two explains the background and motivation of our work. Section three gives an overview about interface adaptation. Section four addresses the plasticity of the user interface, context-awareness and adaptations. Section five exposes in details the problem of adapting to the platform, the user, the task and the environment. Future trends and ideas for further work are presented in section six before the conclusion of the chapter.

BACKGROUND AND MOTIVATIONS

Since several years, we are seeing the miniaturization of electronic devices and their integration into everyday life objects. For example, mobile phones are almost all equipped with a good quality camera, diverse connections to networks such as WiFi or GPRS, "free hand" feature, etc. With some kinds of personal digital assistants (PDA) that use GPS, users can be helped and vocally guided to follow a specific route. This trend that consists on systematically digitalizing resources enabling access to data needed anywhere, anytime is sometimes called, in the literature, ubiquitous. However, there is a wide variety of terms used to describe this paradigm that is opposed to the more conventional desktop metaphor, with one computer per person. This is known as ambient intelligence, ubiquitous and pervasive computing. This refers to the increasing use of widespread processors that exchange small and spontaneous communications with each other and with sensors. Thanks to their much smaller size, these sensors will be integrated into everyday objects, until it become almost invisible to users.

Indeed, as Mark Weiser explains: "*The most profound technologies are those that disappear. They weave themselves into the fabric of everyday life until they are indistinguishable from it.*" (Weiser, 1991). He gives here a first definition of pervasive computing. The pervasive computing refers to access the same service through various channels of communication, such as a desktop computer, a PDA or even a phone to use voice, phone keypad (DTMF) or SMS, depending on the needs and constraints of the user. Adam Greenfield, meanwhile, uses the word "Everyware." This word, formed from "everywhere" and "hard/software" is a neologism to encompass the terms of ubiquitous computing, pervasive computing, ambient computing and tangible media. He explains in more details his thought: "*When I talk about surfacing information that has always been latent in our lives, I mean putting precise numerical values on one present location, on what task we might happen to be currently engaged in, and in whose company.*" (Greenfield, 2006). He therefore suggested that information as diverse as the tone of our voices, our caloric intake, or the composition of our urine can be used, sometimes even without our knowledge.

The question of adaptation in computer sciences is often studied from three different points of view. The first one, coming from the Artificial Intelligence domain, tries to increase the performances by providing an adapted system to the user. Thus, some help messages could be adapted (Browne, et al., 1990); some adaptive hypermedia documents could be generated (Brusilovsky, 2001) etc.

With the second point of view, close to the software engineering domain, the focus is put on the adaptation according to available resources changing. The system is supposed to be adapted in an autonomously and optimally way. The principle of aspect-oriented programming (AOP) (Kiczales et al., 1997), Model Driven Architecture (MDA) and Model Driven Engineering (MDE) are often used. The basic principle of MDA (Gokhale et al., 2002) is developing models of "Platform Independent Model" (PIM) and transforming them into "Platform Specific Model" (PSM) for the concrete implementation of the system. It is not required that all the code is generated automatically, but the overall architecture of the system at least must be obtained as well.

The third point of view, referring to Human-Computer Interaction (HCI) domain, is mainly dedicated to the user interface. The idea is to strictly split the design and the execution levels. The interface is specified only once, at a high level, and transformed according to different interaction contexts (Thevenin & Coutaz, 1999). Unfortunately, none of these approaches allow adapting to an unknown context. This way, we agree with the criticism of (Balme, 2008): "*The contexts of interaction targets should be known by beforehand, which is contrary to the principles of ubiquitous computing where unexpected and opportunism prevail.*"

Another interesting and complex aspect of the adaptation is the semantic facet. Actually, the intention of the author must be respected during the different levels of adaptation. For example, a teacher proposes an interactive e-service composed by a video with some explanations: if a learner accesses to this document through a small device, which is not able to play the video, the intention of the author is not respected, because for him, the video is not separable from the rest of the documents, and a simple picture, for example is not sufficient to replace this video.

We will present in the rest of this chapter some researches around the notion of adaptation, context awareness, and plastic interfaces for ubiquitous learning. We will see how to provide support for the learner, particularly in mobile pervasive situations, where information is presented to learner in appropriate time and place, and according to a specific task.

INTERFACE ADAPTATION AND RELATED WORK

The notion of adaptive interfaces has many facets. Some general works and taxonomies are available (Malinowski et al., 1992), (Brusilovsky, 1996), but concepts change from time to time according to authors, study domains and points of view. As (Paramythis, 2004) said: *The field of adaptive systems is infamous for its lack of standards, or even commonly accepted approaches in this respect.*

A first perspective on adaptation is provided by looking at the temporal sequences involved in adaptation processes. Different tasks that occur in adaptation processes can be grouped in stages. From a system-centered point of view, the main stages cited by (Totterdell & Rautenbach, 1990) are the following: **variation**, **selection,** and **testing**. From a user-centered point of view, adaptation stages are known as **initiative** (it is the decision of one of the agents to suggest an adaptation), **proposal** (alternatives proposed for adaptation), **decision** (one of the alternatives is chosen) and **execution** (Dietrich, et al., 1993).

In order to achieve a specific goal, a designer creates a system according to a personal point of view in a context C. A user may have a different point of view from the designer's, if s/he is in a context C'. Hence, an adaptation of the system is necessary to switch from C to C'.

According to (Trigg et al., 1987) there are four ways in which a technical system can exhibit adaptability:

- A system is **flexible** if it provides generic objects and behaviours that can be interpreted and used differently by different users for different tasks.
- A system is **parameterised** if it offers a range of alternative behaviours for users to choose among.
- A system is **integrable** if it can be interfaced to and integrated with other facilities within its environment as well as connected to remote facilities.
- A system is capable of being **tailored** if it allows users to change the system itself, by building accelerators, specializing behaviour, or adding functionality.

For (Oppermann, 1994), *application systems are not designed for a particular user and a particular task. They are designed and distributed for a class of users and a set of tasks*. According to this author, a way to increase user freedom without increasing the complexity of the system is to provide systems that can the tailored. The two terms commonly found in the literature are adaptable and adaptive systems. A system is called **adaptable** if it provides the user with tools that make it possible to change the system characteristics. A system is called **adaptive** if it is able to change its own characteristics automatically according to the user's needs.

Figure 1 . Adaptation concepts (limited map)

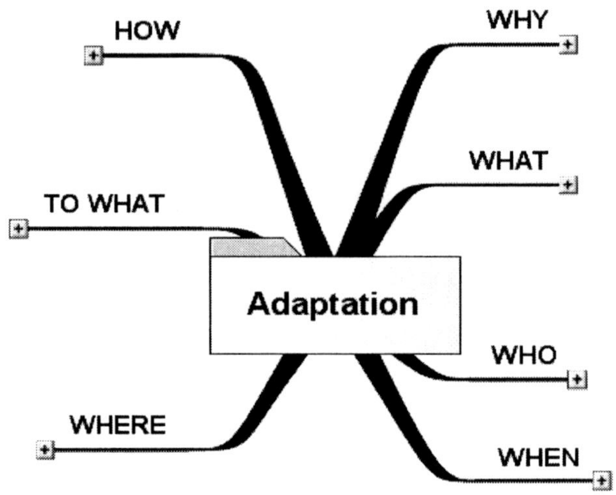

For other researchers, **adaptability** can be considered as a prerequisite for achieving **adaptivity** and vice versa. However, in the literature, adaptation and adaptivity are often used synonymously (Khalil, 2001). Sometimes, the terms "configurable" or "customisable" are used to refer to this kind of adaptation (Weibelzahl, 2002).

We have represented some of the main dimensions of adaptation on a map. Each branch presented in Figure 1 is detailed in the following paragraphs. The map is not exhaustive but it helps clarify the main dimensions of adaptation.

Why

The first question about adaptation is "why adapt?" Basically, the main goal of adaptation is to speed up and simplify usage by presenting to the user what s/he wants to see and thus make complex systems more usable. Presenting easy, efficient, and effective interfaces is the main goal of adaptation (Malinowski et al., 1992). *To reach these goals it is necessary to have a user interface that is suitable for heterogeneous user groups and considers increasing experience of a user* (Dietrich, et al., 1993).

The role of adaptivity in the digital document field is to minimize the user's effort devoted to the exploration of the capacities of the system in order to optimize the effort necessary for the resolution of a problem. As we know, it is difficult to write software that will fit millions of users perfectly, but it is possible to develop systems able to adjust the interface according to the user's skills, knowledge and preferences, for instance. In order to achieve this adaptation, we will see that underlying models of at least user and task are essential, as well as separating the user interface from the application (Fischer, 2000). Adaptive Systems are used in many domains to solve different tasks. The following list of functions, given by (Weibelzahl, 2002) and adapted from (Jameson, 2001) gives some example applications: help the user find information, tailor information to the user, recommend products, for example for e-commerce, help with routine and repetitive tasks, give appropriate help, support learning, conduct a dialog, and support collaboration….

What

Another important interrogation is to consider what part of the application is adapted. Sometimes just the help or tutorial is adapted. In other examples, the dialogue between the system and the user is modified by the adaptation. The different dimensions of adaptation can be illustrated with some examples of adaptation to the delivered content, the navigation, the presentation, the functionalities, etc. Generally, only the presentation interface is adapted. In other cases, both the presentation and some functions of the system are adapted. This last point is called **malleability** by some authors (Morch, 1995), (Morch & Mehandjiev, 2000) and is close to the notion of **flexibility** proposed by (Scapin & Bastien, 1997).

Who

In adaptive interfaces, both user and system can be in charge. Who controls the adaptation and who is really doing adaptation? According to (Browne et al., 1990), when the system controls adaptation, we talk about **personalization** and more generally about **adaptivity**. When the user controls adaptation, we talk about **customization** (choosing among several options) and more generally about **adaptability**. Naturally, it could be a mix of the two, when adaptation is controlled by both the system and the user. Personalization and customization aim at answering the needs and particular characteristics of each user (Cingil, et al., 2000). Indeed, these words come from the e-commerce field and insist on an individual dimension of the adaptation. According to (Rosenberg, 2001), personalization is specific to the end user and based on implied interest during the current and previous sessions.

Some authors point out **personalization**, **customization** and **adaptation** are synonymous in the works of (Mobasher et al., 2000), (Kappel et al., 2000) or (Rossi et al., 2001).

When

Adaptation may take place at very different times. **Design time** adaptation is often opposed to **runtime** adaptation. Some authors also differentiate static adaptation (before and between sessions) and dynamic adaptation (Dietrich, et al., 1993). During a session, adaptation can be done continuously, at predefined junctures, before or after predefined functions, in special situations or on user's request.

In (Stephanidis et al., 1998) terminology, "adaptive" denotes adaptations that occur at runtime and that may be produced both by the system and by the user, and "adaptable" denotes adaptations before runtime, e.g., when the system is first installed (Kobsa et al., 2001).

Where

Adaptation can be internal or external. **Internal**, also called "closed", indicates that adaptation mechanisms are embedded in the system itself while **external**, also called "open", assumes that adaptation is done outside of the system, for instance with a Web service (Oreizy, et al., 2004). Open adaptation mechanisms seem to be preferred in projects related to pervasive/ambient technologies because they allow the possibility to discover new services at runtime, and to do adaptation on the fly, in the service oriented architecture approach philosophy. We will return to that point in section 4.

To What

What is the target of the adaptation? In other words, what does the system adapt to? Indeed, it can be adapt to the **user**, to the **platform**, to the **environment** or a mix of them. Users can be seen as typical user, individual user, subcategories of users such as groups, categories, etc. Adaptation can also take place according to user's roles, rights, skills, abilities, preferences, handicaps, culture.... Adaptation to the platform implies software and/or hardware adaptation. It means that the same application can be executed on e.g., different operating systems or physical devices, or that the application can adapt to the computing resources available. Adaptation to the environment requires means of sensing physical variables, such as location, ambient light, or temperature.

When interfaces are able to adapt to the usage context, some authors talk about **plasticity** (Thevenin & Coutaz, 1999). The term **plasticity** is inspired from materials that expand and contract under natural constraints without breaking, thus preserving continuous usage.

How

The answer to the question "how to adapt applications?" is treated in the literature from two main points of view: strategies and methods. Four basic strategies of adaptation are proposed by (Cockton, 1987): enabling, switching, reconfiguring, and editing. **Enabling** is adaptation by activation/deactivation of components and features in process control design systems. **Switching** is adaptation by selecting one of several different user interfaces, preconfigured user interface components, like dialog configuration, or user interface settings, like colors, font or size. Adaptation by **reconfiguring** is the modification of a user interface using predefined components. **Editing** is adaptation without any restrictions on the basis of the dialog model. Other features were added to this list: adjusting, transforming, altering/merging and exchanging/combining (Balint, 1995). Concerning the methods, there are four main ways identified in order to achieve an adaptation. It can be done by **translation** by **reverse engineering or migration**, by **markup languages** and by **model-based approach**.

The two first methods, translation and reverse engineering, are considered as bottom-up adaptations, because they use preexistent software or system and try to adapt them according to a new context, different from the original one. The two last methods, markup languages and model-based approach, are considered as top-down adaptations. They start from specifications and try to create interfaces adapted to a particular context. Obviously, a mix is possible when adaptation is made thanks to the two approaches presented above used conjointly.

We have presented different dimensions of adaptation. As we have seen, there is not yet a commonly adopted definition of adaptation. For us in this paper, adaptation is the ability for a system to reconfigure itself or otherwise perform actions in reaction to changes in the context of use, while preserving its usability.

The first works around adaptation where principally based on the four questions: What, When, Why and How to adapt? (Karagiannidis et al., 1996). Now, more criteria are used and design spaces for context-sensitive user interfaces are provided (Vanderdonckt et al., 2005), nevertheless it is difficult to choose a toolkit or architecture according to the needs of development, e.g. adaptation targets, models needed, distribution and dynamicity envisaged.

PLASTICITY OF THE USER INTERFACE AND CONTEXT-AWARE ADAPTATIONS

Developers have to design and realise interfaces having in mind many constraints such as production optimization, conception cost reductions, portability, scalability, reusability, quick prototyping, and easy maintenance etc. Currently, a few abstract representation languages have been developed in order to achieve this goal. The availability of many types of computers and devices has become a fundamental challenge for designers of interactive software systems. So, to reach N information through M peripheral is equal, for the developer, to write N * M programs, as we can see on Figure 2.

However, from the user point of view, the service offered should always be the same one, for example to reach an E-learning service. It does not matter that the user reaches this service by a telephone Wap, Palm Pilot, a PC connected

*Figure 2. The N * M Problem*

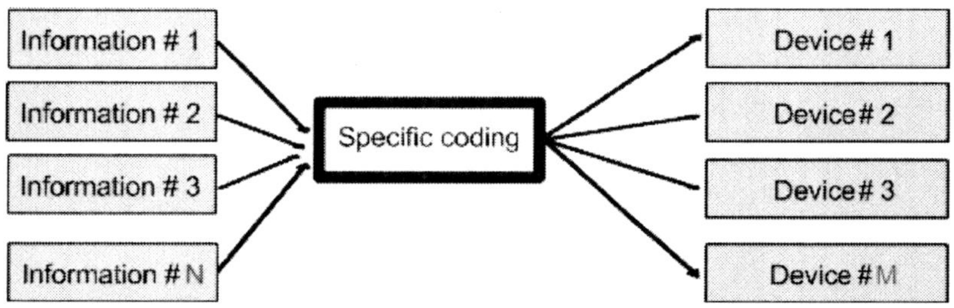

to Internet, or any other means; its goal remains the same, and only the way to interact with the collaborative system is different. Instead of coding N * M applications, researchers try to offer one model for many interfaces (Paternò & Santoro, 2002); see, for example, the Cameleon[1] project for more details. Those kinds of languages allowed describing interface on an abstract way. This description is used to generate adapted interfaces. This factor is called plasticity of the user interface (Thevenin, 2001). Plasticity is the ability of a user interface to be re used on multiple platforms that have different capabilities. The personalization may depend on many factors such as the client device, the user profile with roles, access rights, skills, abilities, handicaps, etc., the location, the access history, and so on.

For the World Wide Web, HTML combines data and presentation into one document. Oppositely, XML (Extensible Markup Language) provides separation between data and its presentation format. Languages like UIML (User Interface Markup Language) are used to describe and generate such kind of interfaces. UIML is an XML language used for defining the actual interface elements. This means the buttons, menus, lists and other controls that allow a program to function in a graphical interface like Windows or Motif.

It is important to identify the different elements which will take a part in the personalization. The three primary axes to investigate are the target level, the conception level, and the runtime level. Concerning the target level, there are three sub axes that can be followed: the user model, the device model, and the environment model. Concerning the conception level, tools of production can be qualified by two aspects: the specification of the models and the transformation of those models. The specification should describe the abstraction level (model or Meta model) and the scope of the specification. This scope could target one or many models among the concepts, the task, the user, the device, the environment, the evolution, and the transition. In addition, transition models are used to cover the transition between concepts to tasks. This in turn, will provide the final abstract interfaces, followed by the concrete interfaces and final interfaces. Concerning the runtime level, is it important to notice the capabilities of the interface to recognize context shifts, context capture, and identification of a particular context during the execution such as noise, light or network's break. At this level, the software must be able to recalculate a part or the entire interface, in order to present a user interface adapted to this new context.

ADAPTING TO WHAT (SLIGHT RETURN)

Among all the dimensions of adaptation presented on the map in Figure 1, one of the most important questions is: "Adapting to what?" We will illustrate some aspects of possible answers to that question, across different models such as user, platform, task or environment, used separately or concomitantly. Within the framework of the plasticity of the interfaces, we conceived and developed the Plastic ML (Rouillard, 2003). This language allows developers to describe input and output elements, on a high level, without referring to the peripheral which will be used at the runtime. Thanks to XSLT transformations, the Plastic ML documents are automatically translated to markup languages: HTML for the Web, WML for the WAP, and VoiceXML for the phone. It is possible to declare inputs, outputs, choice, secret elements, on an abstract level, and thanks to XSL files, the interface will be displayed on the appropriate device.

Adapting to the Platform

Let's envisage a scenario where a developer wants to prepare a generic document, for instance, poetry, and would like to distribute this document among different devices. Plastic ML was used for our prototypes. Hence, a document based on XML is created, containing different tags: a title, a body, an author name, and a few other emphasized elements. Using XSLT transformation, it is possible to obtain an HTML document, which is viewed on a single scrollable page, all the poetry, with a picture of the author, and the other emphasized elements, in bold, as we can see on the left of the Figure 3.

The same transformation, based on another XSL document is possible in order to obtain a WML document. Each part of the poetry is sent to a suitable card, and the user can navigate from one card to another. No picture is created for this device, even if a WBMP transformation is possible, and emphasized elements are presented in bold font; we also tried underline, italic and other presentations that were available.

The right of the Figure 3 represents the result of the transformation for a cellular phone using WAP protocol. Contrary to the web version which comes with the picture of the author, here the presentation is simpler. The user can only see one strophe by card. The document presentation is guided by the device properties; a small screen for example. XSL transformation allows choosing the appropriate tag that match a specific device. For example, the tag <emphasis> of the original XML source document was transformed in <B>Helen</B> in HTML and <U>Helen</U> in WML.

Naturally, the presentation of the same document on a mobile phone will be entirely aural. We use the VoiceXML language to generate the document. Emphasized elements are presented with the <emp> tag of the VoiceXML language, which allows changing the sound of the TTS (Text to Speech) synthesis. It allows to make "voice effect", lower or louder voice for instance, to express important words. Thus, speech synthesis plays a major role for a correct restitution of the text and so for a good understanding by the student. The user can also navigate through the document with the DTMF (Dual-Tone Multi- Frequency) buttons or by pronouncing oral commands such as "next" or "previous," in a multimodal way.

Adapting to the User

One tag particularly used in Plastic ML documents is the <block> tag. Indeed, according to that block and the associated role, it is possible to show or hide certain parts of the original document. This feature is used in order to adapt documents according to the different models. For instance, a first transformation is made according to the role, and then a second adaptation is done according to the device used.

It is also possible to declare that the roles are not hierarchic for a determined block. If the structure is flat that means that roles are not dependants from each other. For example, in Figure 4, a French user will have the role number 1 and an English user will have the role number 2. They will see the "same" document, but it will be adapted to them, according to their chosen language. The system respects the semantic cutting and decomposition proposed by the author of the Plastic ML document. For example, if it's possible, all the information

Figure 3 . Web browser and WAP views of the same poetry

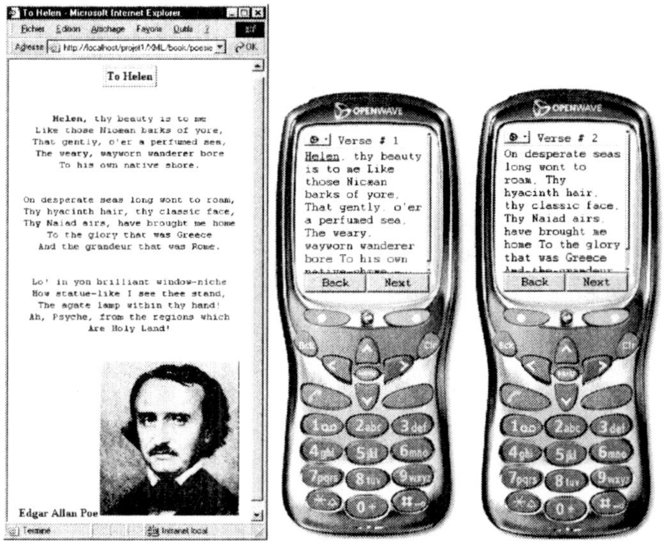

Figure 4. Plastic ML document integrating information blocks coupled with non-hierarchical roles

```
<?xml version="1.0" encoding="iso-8859-1"?>
<plasticML version="2.0">
<interface title="Adaptative Hypermedia, document fragmentation and interface plasticity">
<block role="1" align="center">
<output><important2>Hypermedia Adaptatif, document fragmentaire et interfaces plastiques</important2>
</output>
</block>
<block role="2" align="center">
<output>
<important2>Adaptative Hypermedia, document fragmentation and interface plasticity</important2>
</output>
</block>
<block role="ALL">
<output><important1>Introduction</important1></output>
</block>
<block role="1" align="left">
<output>
Cette partie du document est en français. Le rôle que nous utilisons ici est défini de manière non hiérarchique.
Cela signifie que selon le paramètre utilisé, il n'y aura qu'une partie du document qui sera présentée à l'utilisateur
: celle qui correspond à son rôle déclaré.
</output>
</block>
<block role="2" align="left">
<output>This part of the document is in English. The role that we use here is defined in a no-hierarchical
manner. It means that according to the used parameter, there will only be a part of the document that will be
presented to the user: the one that corresponds to his/her declared role.</output>
</block></interface></plasticML>
```

contained in a block will be sent to a card for a Wap phone device.

The role "ALL" will be used to present information, whatever the role. In our example, we must be careful to use words understandable in the two target languages, such as "introduction", "conclusion", etc.

Figure 5 shows the result obtained in a Web page after processing of Plastic ML in HTML, with the role number 1. The text of the document is then displayed in a French-language version. More exactly, only fragments of the document source, whose blocks are marked as "role = 1" are used to dynamically rebuild the target document.

Figure 5. The document obtained for the role 1 presents only the French part

Hypermedia Adaptatif, document fragmentaire et interfaces plastiques

Introduction

Cette partie du document est en français. Le rôle que nous utilisons ici est défini de manière non hiérarchique. Cela signifie que selon le paramètre utilisé, il n'y aura qu'une partie du document qui sera présentée à l'utilisateur : celle qui correspond à son rôle déclaré.

Figure 6. The document obtained for the role 2 presents only the English part

Adaptative Hypermedia, document fragmentation and interface plasticity

Introduction

This part of the document is in english. The role that we use here is defined in a no-hierarchical manner. It means that according to the used parameter, there will only be a part of the document that will be presented to the user: the one that corresponds to his/her declared role.

Figure 6 shows the result obtained in a Web page after processing of Plastic ML in HTML, but with the role number 2 instead of number 1. Hence, the text is written in English.

Obviously, adaptation could be made according to more than one factor at the same time.

Adapting to the Task

Adapting to the task is not an easy problem, because tasks described, envisaged at the design time and tasks realized, detected at the runtime, are not always systematically the same. Carboni (2004) explains this point of view: "*Someone having a breakfast in his kitchen would check his mailbox without turning on the computer but rather by means of his interactive TV. So tasks models should be tailored according to the context of use, e.g. place, time, and device available*" (Carboni, 2004). Sometimes, the task requires the user to act with their hands. Hence, it's more difficult for the user to interact with the interface without disturbing their activity. Figure 7 shows two examples where the hands of the user are required to execute the task; cooking or repairing a car, for instance. The interface is adapted to the user's language, but the aural channel is also used in order to take into account the fact that the user can not use the stylus during the entire session. Other parameters for the adaptation of the system can be used, such as the number of people available for a task. The instruction given to set up a furniture kit will not be the same for one, two, three, or more persons.

Adapting to the Environment

As we previously explain, the context of use for a significant adaptation can be taken among the subsequent: user's profile, current task, device used, location, time, and environment of the interaction. A context can be considered as

Figure 7. System adapted to the task, such as repairing a car or cooking

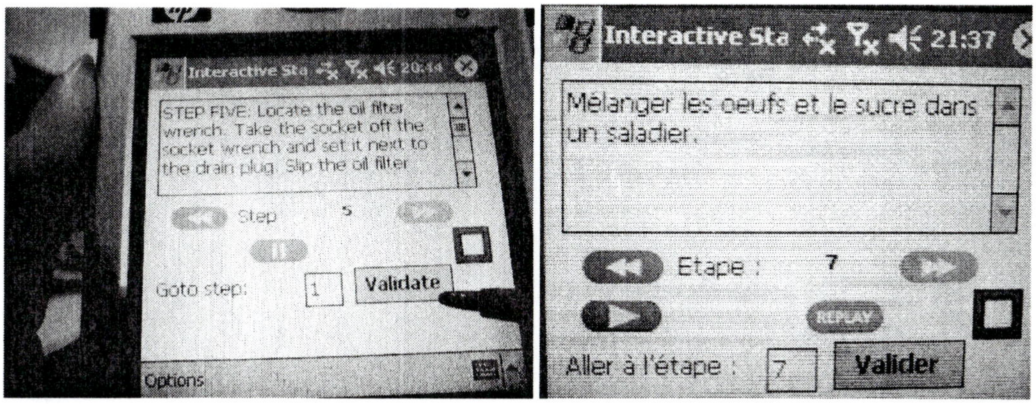

shared or individual. Individual context includes information relevant to the interaction between the learner and the M-learning applications. For the learning domain, shared context includes information relevant to collaborative group work or learners sharing common interests. Individual context can be viewed as specifically tailored to each learner. Shared context is more relative to the collaborative work.

The environment can be subdivided into many parts (physical, people, resources…). Smart systems are capable of detecting the environmental particular data influencing the task of the user. If the user is moving, the environment is too noisy or if there isn't enough light to achieve a task, for example, the system will propose adapted solutions, in the respect of the usability rules.

Adaptation in Ubiquitous Learning

We are involved in researches on new interactive systems for ubiquitous e-Learning within the P-LearNet project (P-LearNet) which is supported by the ANR (Agence Nationale de la Recherche Française - National French Research Agency). P-LearNet means Pervasive Learning Networks. It's an exploratory project on adaptative services and usages for human learning in the context of pervasive communications. The main goal of this project is to explore the potential of ubiquitous and pervasive communications, over heterogeneous networks, for a large and important field of application, e.g. human learning. To achieve this, we take into account the maximum available amount of information including places, times, organisational and technological contexts, individual and/or collective learning processes, etc.

We will now present some results of our work in adaptive pervasive learning environment, based on contextual QR Codes. This is where information is presented to learner at the appropriate time and place according to a particular task. (Specht & Zimmermann, 2006) already showed the interest of a contextualization in the learning experiences. Five fundamental categories for context information were identified in (Zimmermann, et al., 2007) as follows: individuality context, time, location, activity, and relations context.

The notion of contextual QR Codes was proposed in previous recent work (Rouillard, 2008). It can be defined as the following: it's the result of a fusion between a public part of information, encoded in a 2D barcode named QR Code, and a private part of information, the context, provided by the device that scanned the code. Figure 8 shows the public and private parts of a contextual QR

Code. The private part can be one or more of the following user's profile, current task, device, location, time, and environment of the interaction. The mobile device decodes the QR Code and merges it with private data obtained during the interaction. Next, the XML (Extensible Markup Language) resulting file is sent to a web service created in our laboratory that computes the code and returns personalized messages. Some private information can be stored in the profile owner's of the phone, the class level for example, and some others are given directly by the user when the interaction takes place. It's the case for the language or the task chosen, for example. A previous study of people engaged in a location-based experience at the London zoo was reported by O'Hara and colleagues. In this experience, location-based content was collected and triggered using mobile camera phones to read 2D barcodes on signs located at particular animal enclosures around the zoo. "*Each sign had an enticing caption and a data matrix code (approx. 7x7 cm) which encoded the file locations for the relevant media files.*" (O'Hara et al., 2007). By capturing a 2D barcode, participants extracted the file's URIs from the codes and added the corresponding preloaded content files (audio video and text) into their user's collection. The fundamental distinction between that approach and our system is that the London zoo system always provides the same content to the user, while the PerZoovasive system provides tailored information according to a particular context. Now let us present the system functionalities through the following scenario.

PEDAGOGICAL SCENARIO

A French elementary school decides to organize a visit to the zoo. This visit allows a follow-up of

Figure 8. Public and private parts of a contextual QR Code

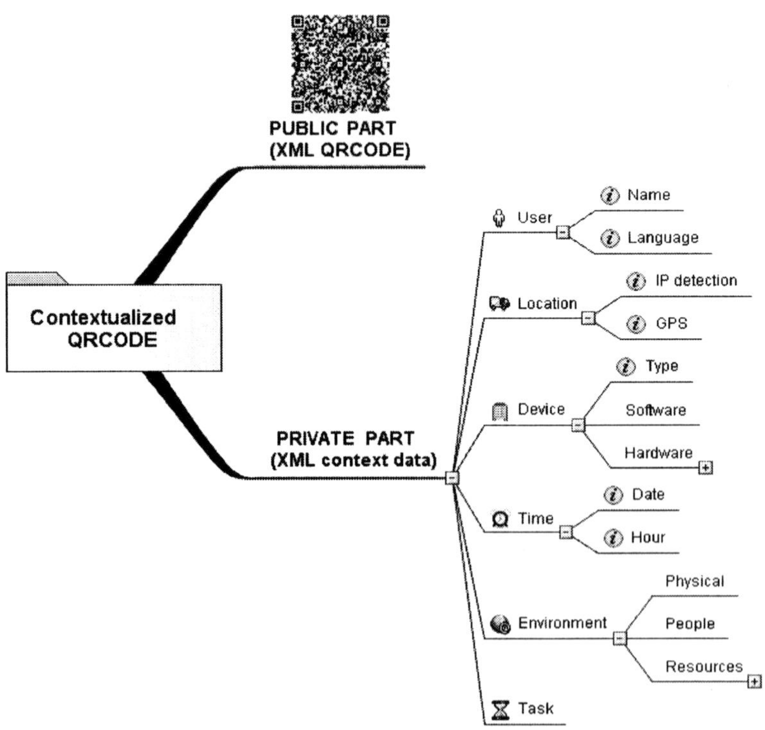

the pupil's educational curriculum. The subject is the different levels of pupils and number of teachers accompanying them. The teachers usually do not have an-indepth knowledge of every subject. Question that could arise are: Do we have to display, disseminate, and post the same information to all pupils? How do we make sure each group of students gets the appropriate information? The accompanying teachers usually have a cellular phone. We propose to use the phone to provide adapted information. When the group gets to the zoo, each teacher's cellular phone connects to a dedicated server where applicable software is access for set up. For instance, Mrs Martin is preparing a visit to a zoo for her CE1[2] class. She thinks it is a good idea to establish a link between the lessons that she gave in her classroom during the morningabout the animals and their environments in the real world (see Freinet's pedagogical method). She starts her visit with her group and stops in front of the Gibbons' cage. She takes a photo of the QR code attached to the cage. This image is automatically sent to the server. Instantly, a text in French is adapted to the pupils' level and a description of the Gibbon is then displayed on the teacher's phone. The pupils then can listen to a synthesized voice deliver the information on Mrs. Martin's mobile phone. At the same time, another class from an English school stops in front of the same cage. Mr. Ford, the teacher of this class (a different level than the French one) will perform the same operation as Mrs. Martin with his mobile phone. The text presented is adapted to the level, the language, and the task of these pupils. The two groups continue their visit and the same thing happens again in front of each subsequent cage.

TECHNICAL SCENARIO

Figure 9 presents a mobile application, called PerZoovasive, developed in our laboratory. It was written in the language of C# and runs, thanks to Tasman[3] library, on a smartphone HTC TyTN II, also known as "Kaiser", supporting Windows Mobile 6. The name of the user and the level of the class are automatically detected in the mobile registry (Control Panel\Owner). The user clicks on the upper left radio button if the task is a lesson or on the upper right radio button if the task is a Quiz knowledge control.

In addition to the plain text, it is also possible to choose the flag corresponding to the appropriate language, and to select the TTS option in order to obtain a Text-To-Speech response. The trace option is a debugging tool that displays the XML code data sent and received. The camera manager is invoked by clicking on the "Capture QRCode" button. Each cage number is coded with a particular QR Code. Then, according to the selected representation, the decoded information is presented to the user. The application becomes extremely context-aware by using a combination of many parameters. The web service that receives the contextual QR Code opens a file which has the following canonical form: Cage_Level_Language_Task.txt. For example, the file named

Figure 9. Example of generated text (cage=123, level=CM2, language=French, task=lesson)

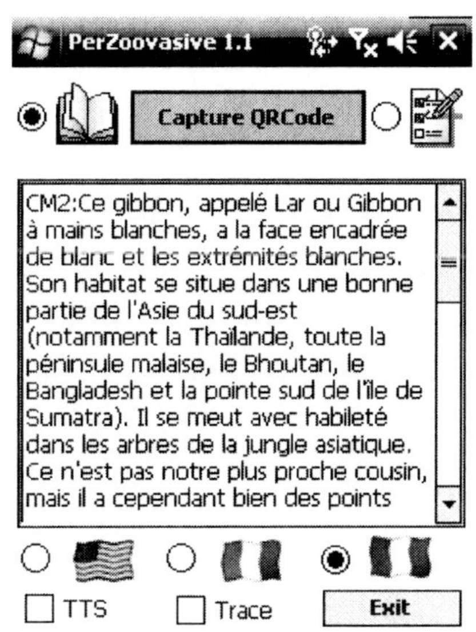

"8245_CM1_FR_Quiz.txt" is related to a Quiz in French, for the CM1 class (level 3) and for the cage number 8245.

In Figure 9, since the teacher did not select the TTS option, she can read the data provided by the system on her mobile screen and give the information to her pupils as she sees it.

Figure 10 shows a French-speaking teacher in a zoo, using the PerZoovasive application with a 3G connection (provided by Orange France) scanning a contextual QR Code in order to obtain information about gibbons (left) and turtles (right). After a few seconds, she can read her mobile screen and give relevant information about these animals to her pupils.

We can see the same teacher asking information about the turtle of another cage. She could retrieve general information such as specie, origin, speed or food, about this kind of animal, but also personal data about that particular turtle like its name, age, birthday, etc. Figure 11 shows the PerZoovasive application used by an English-speaking user. As she selected the Quiz task and the TTS option, she obtained on a multimodal way, that uses screen and speech, some questions about the panther, living in the cage number 8245, adapted to the level of her class.

FUTURE TRENDS

The idea behind ubiquitous ICTs is the convergence, or even fusion of real-world objects with information systems. We are changing from "e-everything" to "u-everything". Physical objects equipped with barcodes or RFID chips, for instance, will be reachable in a numeric world. Multimodal and multi-channel adaptation will be the kernel of future ubiquitous, plastic and intelligence systems. Our future work is oriented to the investigation of the traces/logs and original strategies to capture environmental information. This will lead to documents being capable of modification on the fly, but also will help us to better understand the behaviour of the entire system driven by multiple and sometimes contradictory policies. Technologies and solutions like IPv6, SOA (including web services approaches) are already used in order to deliver the appropriate information to the appropriate person at the appropriate place and time. But the challenge will be, for the next generation of systems, to adapt themselves smoothly to situations not yet encountered. This will be possible only with a real dialogue between the systems and the users. Hence, the future ubiquitous systems will have to show their seams, instead of trying to hide them, to facilitate their usage.

Figure 10. A teacher is scanning a Contextual QR Code to retrieve information about gibbons and turtles

Figure 11. Example of generated quiz (cage=8245, level=CM1, language=English, task=quiz)

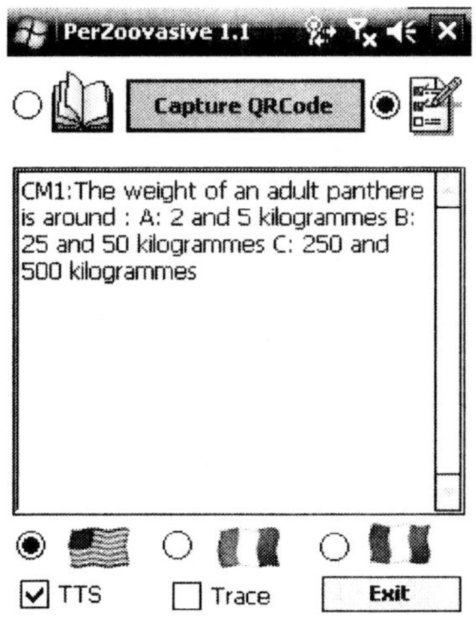

CONCLUSION

We have shown in this chapter that interface adaptation is a challenge to implement, accomplish, and manage. In view of the fact that for years, many studies and works in various domains have been attempting to expose the why, what, when, where and how to create an adaptable interface. With the emergence of ubiquitous computing, the difficulties are still growing. It's not only a matter of transformation or conversion from a format to another. It also includes for the most part many semantic issues. Research oriented around pervasive and ubiquitous computing, especially in the field of human learning, works to create a model precisely for the user, the task, the device, which additionally includes some parts of the environment in which the interaction will take place. Moreover, relevant adaptations have to "understand" the meanings of the manipulated components. Now it is technically possible to change an element by another or to transform an aspect from an abstract level to a concrete one, yet meanings and intentions are crucial for relevant adaptations. Furthermore, that fact that people are mobile and want to use multimodal and multi-channel systems directs us to take into account real issues encountered by users *in situ*. The preliminary results of our P-LearNet project shows that we have to deal with contents and containers simultaneously. Context-aware adaptations must be made, both in the respect of the usability of the systems and in the respect of the author intentions. We are confident that education will be increasingly performed across mobile devices, anytime and anywhere. But, the challenge is to offer intelligent and adapted tools, for cohabitation between smart people and smart environment.

ACKNOWLEDGMENT

We are grateful to ANR P-LearNet project for providing support for this research and to Tasman, for special tools provided. The author gratefully acknowledges M. Kent Washington for his help in improving this paper.

REFERENCES

Balint, L. (1995). Adaptive interfaces for human-computer interaction: a colourful spectrum of present and future options. *Systems, Man and Cybernetics, Intelligent Systems for the 21st Century, IEEE International Conference, 1*(22), 292 - 297.

Balme, L. (2008). *Interfaces homme-machine plastiques: Une approche par composants dynamiques*. Unpublished doctoral dissertation, University Joseph Fourier, Grenoble, France.

Beaudouin-Lafon, M. (2000). Ceci n'est pas un ordinateur. Perspectives sur l'Interaction Homme-Machine. *Technique et Science Informatique, 19*(1-2-3), 69-74.

Browne, D., Totterdell, P., & Norman, M. (1990). *Adaptive User Interfaces*. Computer and People Series. Academic Press, ISBN 0-12-137755-5.

Brusilovsky, P. (1996). Methods and techniques of adaptive hypermedia. *User Modeling and User-Adapted Interaction, 6*(2-3), 87–129. doi:10.1007/BF00143964

Brusilovsky, P. (2001). Adaptive hypermedia. In Alfred Kobsa (Ed.), *User Modeling and User Adapted Interaction, Ten Year Anniversary Issue, 11*(1/2), 87-110.

Carboni, D. (2004). *Services in Pervasive Computing Environments: from Design to Delivery*, Unpublished doctoral dissertation, CRS4 – Center for Advanced Studies, Research and Development in Sardinia, Italy.

Cingil, I., Dogac, A., & Azgin, A. (2000). A broader approach to personalization. *Communications of the ACM, 43*(8), 136–141. doi:10.1145/345124.345168

Cockton, G. (1987). Some Critical Remarks on Abstractions for Adaptable Dialogue Managers. In D. Diaper & R. Winder (Eds.), People *and Computers III. Proceedings of the Third Browne Conference of the British Computer Society, Human Computer Interaction Specialist Group* (pp. 325343). University of Exeter.

Dieterich, H., Malinowski, U., Kühme, T., & Schneider-Hufschmidt, M. (1993). State of the Art in Adaptive User Interfaces. In M. Schneider-Hufschmidt, T. Kühme, & U. Mallinowski (Eds.), *Adaptive User Interfaces, Principles and Practice* (pp. 13-48). Included in series Human Factors in Information Technology, 10.

Fischer, G. (2000). User Modeling in Human Computer Interaction [UMUAI]. *User Modeling and User-Adapted Interaction, 11*, 65–86. doi:10.1023/A:1011145532042

Gokhale, A., Natarjan, B., Schmidt, D. C., Nechypurenko, A., Wang, N., Gray, J., et al. (2002). CoSMIC: An MDA Generative Tool for Distributed Real-time and Embdedded Component Middleware and Applications. *Proceedings of the OOPSLA 2002 Workshop on Generative Techniques in the Context of Model Driven Architecture*, Seattle, WA.

Greenfield, A. (2006). Everyware: The Dawning Age of Ubiquitous Computing. *New Riders Publishing., ISBN-13*, 978–0321384010.

Jameson, A. (2001). *Systems That Adapt to Their Users: An Integrative Perspective*. Unpublished doctoral habilitation, Saarbrücken: Saarland University.

Kappel, G., Retschitzegger, W., & Schwinger, W. (2000). Modeling Customizable Web Applications – A Requirement's Perspective. In *Proceedings of the International Conference on Digital Libraries, (ICDL 2000),* Kyoto, Japan.

Karagiannidis, C., Koumpis, A., & Stephanidis, C. (1996). Deciding `What', `When', `Why', and `How' to Adapt in Intelligent Multimedia Presentation systems. ECAI'96. *Workshop on Towards a standard reference model for intelligent multimedia presentation systems*, Budapest, Hungary.

Khalil, C. I. J. (2001). *The Design of a Multimedia Adaptive Interface for Process Control Using a Multi-Agent Approach*. Unpublished doctoral dissertation, Computer Science, Loughborough University, UK.

Kiczales, G., Lamping, J., Mendhekar, A., Maeda, C., Lopes, C., Loingtier, J.-C., & Irwin, J. (1997). Aspect-Oriented Programming, Proceedings of the *European Conference on Object-Oriented Programming*, 1241, 220–242.

Kobsa, A., Koenemann, J., & Pohl, W. (2001). Personalized Hypermedia Presentation Techniques for Improving Online Customer Relationships. *The Knowledge Engineering Review, 16*(2), 111–155. doi:10.1017/S0269888901000108

Lopez, J., & Szekely, P. (2001). Automatic web page adaptation. *CHI2001-Workshop,* Transforming the UI for anyone anywhere, Seattle.

Malinowski, U., Kuhme, T., Dietrich, H., & Schneider-Hufschmidt, M. (1992). A taxonomy of adaptive user interfaces. In Monk, Diaper, Harrison (Eds.), *People and Computers VII*, (pp. 391-414). Cambridge University Press.

Mobasher, B., Cooley, R., & Srivastava, J. (2000). Automatic Personnalization Based on Web Usage Mining. *Communications of the ACM, 43*(8), 142–151. doi:10.1145/345124.345169

Morch, A. (1995). Three levels of end-user tailoring: customization, integration, and extansion. *Third Decennial Aarhus Conference* (pp. 41-51), Aarhus, Denmark.

Morch, A., & Mehandjiev, N. D. (2000). Tailoring as Collaboration: The Mediating Role of Multiple Representation and Application Units. *Computer Supported Cooperative Work (CSCW'00)*, (pp. 75-100), Philadelphia, Pennsylvania, USA.

O'Hara, K., Kindberg, T., Glancy, M., Baptista, L., Sukumaran, B., Kahana, G., & Rowbotham, J. (2007). Collecting and Sharing Location-based Content on Mobile Phones in a Zoo Visitor Experience. *Computer Supported Cooperative Work (CSCW), 16*(1-2,34), 11-44. Springer

Oppermann, R. (1994). Adaptively supported adaptability [Duluth, MN, USA: Academic Press Inc.]. *International Journal of Human-Computer Studies, 40*(3), 455–472. doi:10.1006/ijhc.1994.1021

Oreizy, P., Gorlick, M.M., & Taylor, R.N., Heimbigner, D., Johnson, G., Medvidovic, N., et al. (1999). An Architecture-Based Approach to Self-Adaptive Software. *IEEE Intelligent Systems, 14*(3), 54–62. doi:10.1109/5254.769885

P-LearNet project. Pervasive Learning Network, Retrieved November 17, 2008, from http://p-learnet.univ-lille1.fr/

Paramythis, A. (2004). Towards Self-Regulating Adaptive Systems. In S. Weibelzahl & N. Henze (Eds.), Proceedings of the Annual *Workshop of the SIG Adaptivity and User Modeling in Interactive Systems of the German Informatics Society* (ABIS04), (pp. 57-63), Berlin.

Paternò, F., & Santoro, C. (2002). One Model, Many Interfaces. *International conference Computer-Aided Design of User Interfaces, CADUI'02*, Valenciennes, France.

Rosenberg, M. (2001). *The personalization story, ITworld.com.* Retrieved November 17, 2008, from http://www.itworld.com/Man/2676/ITW010511rosenberg/

Rossi, G., Schwabe, D., & Guimarães, R. (2001). Designing Personalized Web Applications. *10th International World Wide Web Conference (WWW10)*, Hong Kong.

Rouillard, J. (2003). HCII 2003. *Plastic ML and its toolkit*, Heraklion, Crete, Greece.

Rouillard, J. (2008). Contextual QR Codes. *Third International Multi-Conference on Computing in the Global Information Technology, ICCGI 2008*, Athens, Greece.

Scapin, D. L., & Bastien, J. M. C. (1997). Ergonomic criteria for evaluating the ergonomic quality of interactive systems. *Behaviour & Information Technology, 6*(4-5), 220–231.

Specht, M., & Zimmermann, A. (2006). An Architecture for Contextualized Learning Experiences. Proceedings of the *Sixth IEEE International Conference on Advanced Learning Technologies (ICALT)*, (pp. 169-173). ISBN:0-7695-2632-2.

Stephanidis, C., Paramythis, A., Sfyrakis, M., Stergiou, A., Maou, N., Leventis, A., et al. (1998). Adaptable and adaptive user interfaces for disabled users in AVANTI Project. In S. Triglia, A. Mullery, M. Campolargo, H. Vanderstraeten, & M. Mampaey, (Eds.), *Intelligence in Services and Networks: Technology for Ubiquitous Telecom Services* (pp. 153–166). Springer.

Thevenin, D. (2001), *Adaptation en Interaction Homme-Machine: Cas de la plasticité*, Unpublished doctoral dissertation, Université Joseph Fourier, Grenoble, France.

Thevenin, D., & Coutaz, J. (1999). Plasticity of user-interfaces: framework and research agenda. In *Proceedings of 7th IFIP conference on human-computer interaction, INTERACT 1999*, (pp. 110-117), Edinburgh, Scotland.

Totterdell, P., & Rautenbach, P. (1990). Adaptation as a Problem of Design. In: Browne, D., Totterdell, P., Norman, M., (Eds.): *Adaptive User Interfaces*. Academic say, Press, London, (pp. 5984).

Trigg, R., Moran, T., & Halasz, F. (1987). Adaptability and Tailorability in NoteCards. In Bullinger & Shackel (Eds.), *Proc. INTERACT '87*, North-Holland.

Vanderdonckt, J., Grolaux, D., & Van Roy, P. Limbourg, Macq, Q., & Michel, B. (2005). A Design Space For Context-Sensitive User Interfaces. *International Conference on Intelligent and Adaptive Systems and Software Engineering (IASSE'2005)*, (pp. 207-214). Toronto, Canada.

Voice, X. M. L. Forum. Retrieved November 17, 2008, from http://www.voicexml.org

Weibelzahl, S. (2002). *Evaluation of Adaptive Systems*. Graduate Programme Human and Machine Intelligence, Kunzenweg 21, University of Education Freiburg, Germany.

Weiser, M. (1991). The Computer for the 21st Century. *Scientific American, 265*(3), 94–104.

WML. Wireless Markup Language and WAP: Wireless Application. Retrieved November 17, 2008, from http://www.wapforum.org

Zimmermann, A., Lorenz, A., & Oppermann, R. (2007). An Operational Definition of Context. In B. Kokinov et al. (Eds.), *Proceedings of Context 07*, LNAI 4635, pp. 558–571, Springer-Verlag Berlin Heidelberg.

ENDNOTES

1. See Cameleon project (http://giove.cnuce.cnr.it/cameleon.html)
2. Grade 2 in French elementary school (CP, CE1, CE2, CM1, CM2).
3. Tasman barcode recognition for developers (http://www.tasman.co.uk/index.html)

Chapter 9
Co-Design and Co-Deployment Methodologies for Innovative m-Learning Systems

David Millard
University of Southampton, UK

Yvonne Howard
University of Southampton, UK

Lester Gilbert
University of Southampton, UK

Gary Wills
University of Southampton, UK

ABSTRACT

Building innovative m-learning systems can be challenging, because innovative technology is tied to innovative practice, and thus the design process needs to consider the social and professional context in which a technology is to be deployed. In this chapter the authors describe a methodology for co-design in m-learning, which includes stakeholders from the domain in the technology design team. Through a case study of a project to support nurses on placement, they show that co-design should be accompanied by co-deployment in order to manage the reception and eventual acceptance of new technology in a particular environment. They present both our co-design and co-deployment methodologies, and describe the techniques that are applicable at each stage.

INTRODUCTION

In the last decade we have seen many types of mobile (m-) learning tools, from simple systems that allow access to existing content and functionality on-the-move (Flynn *et al.*, 2000, Collins, 2005) to more targeted applications, that take advantage of the mobility or locality of the applications users, for example, to provide location-based information (Abowd *et al.*, 1997, de Crom, 2005), or to support fieldtrips (Kravcik *et al.*, 2004, Weal *et al,* 2006). However, these tools often replicate existing learning activities, rather than changing the nature of the activity itself.

DOI: 10.4018/978-1-60566-703-4.ch009

This is understandable as changing practise also requires a change in philosophy or of culture, and is far more difficult to achieve even in small problem areas. It requires an appreciation of the problem space that is usually beyond the technology experts, and an understanding of the technology that is usually beyond the domain experts. This creates a tension in the design process as no single expert has the necessary knowledge or skills.

Design methodologies are therefore required to drive the creation of novel tools, ensuring that they are both useful and sustainable in practice. Many methodologies and models for the design of information and e-learning systems take a layered approach, separating design issues to allow independence (Wills *et al.*, 2003): mapping the domain (in terms of its structure, content, work flow, *etc)*, analysing the associations and relations in that domain, and presenting the information to appropriate users.

Our co-design approach has the objective of ensuring that an m-learning system is both feasible and useful by explicitly including the expertise of the people in the intended domain, and as such it is similar to the socio-cognitive engineering approach proposed by Sharples *et al.* (2002), which seeks to develop a theory-based framework of the user's underlying cognitive and social processes. Our co-design methodology is more lightweight, and does not attempt to build a theoretical model of users' practice, but instead works closely with users as part of the design team in an agile way—something that is essential for tools that innovate practice. Our methodology involves domain experts as 'first-class' members of the design team, using a number of techniques that brings the design team together and helps it to converge on a joint understanding of the conceptual space, and focusing on tools to tackle real problems in the domain. We have done this by integrating techniques found in HCI (personas, scenarios, and storyboarding) with agile software development techniques (iterative and incremental delivery) and lightweight software engineering (use case, activity, and iteration diagrams).

We have also discovered that the engagement with the eventual user group must continue into the deployment phase of prototyping, using co-deployment methodologies that emphasise a continuing conversation between a range of stakeholders in the user community and the design team.

In the rest of this chapter we present our co-design and co-deployment methodologies, using a case study to demonstrate how some of the techniques work, and showing the consequences of underestimating the importance of the deployment phase. Our hope is that these methodologies will help other m-learning developers to design new technology and applications that create genuine innovations in the domains in which they are deployed.

BACKGROUND: DESIGN METHODOLOGIES

Our work on co-design and co-deployment builds on traditional software engineering practices for m-learning, and especially on participatory design and agile development. Co-design and co-deployment are focused on enabling innovation in a domain through the use of technology, and as such can also benefit from the experiences of management methods for institutional change.

Traditional Software Engineering

Mobile learning development has often followed traditional software engineering methodologies where teams of developers envision, implement, and deploy systems. In so doing, mobile learning systems are not immune from the software engineering paradox described by Lehman (1980) satisfaction declines unless steps are taken to constantly improve systems. This effect can be detected even before a system has been completed, so that when a system has been completed to an agreed specification, on deployment, users feel dissatisfied with it. Lehmann suggests that the activity of envisioning a system creates technology transfer from the technology specialist to the domain experts, increasing the domain's awareness of opportunities for change and its expectations for the deployed system beyond the originally agreed goals.

Efforts to mitigate these effects have included agile development with its principles of domain expert involvement in the design team throughout the design phase, and participatory design methods where users are first class members of the design team.

In mobile learning, traditional software engineering methodologies have been challenged to adapt to the specific needs of designing for the mobile learning experience. Parsons *et al.* (2006) examined m-learning in particular, and identified four design concerns:

- **Generic mobile environment issues:** Such as communication support and the device interface
- **Learning context:** Including the roles of users, the collaborations and activities
- **Learning experiences:** Structuring the learning, e.g. through cinematic or game metaphors
- **Learning objectives:** The desired goal, for example improved skills or social abilities

Participatory Design

Participatory design is a long established practice in Human-Computer-Interaction (Grudin and Pruitt, 2002). It challenges the traditional view that after the requirements gathering/ elicitation stage, end users are not required and that they should let the 'experts' get on with the job and design the system. Participatory design brings the end users into the design team as equal participant (members) of the design process. For instance, participatory design has been used in the design of mobile systems (Svanaes and Seland, 2004); and Massimi *et al.* (2007) have used participatory design to design mobile phones for the elderly.

Other approaches have also been developed that include users in the later stages of design. In their work on m-learning, Sharples *et al.* (2002) developed a socio-cognitive engineering approach, a theory-based framework of the user's underlying cognitive and social processes. Socio-cognitive engineering is wary of user-centred design as users are not always able to articulate their own working pattern and methods; instead the method seeks to develop a theory-based framework of users' underlying cognitive and social processes. In practical terms this requires two studies: an investigation into how user activities are performed in their normal contexts, and a theoretical study of the underlying cognitive and social processes.

Agile Development

Agile methods are a number of software development methods that were proposed in the mid 1990s as a reaction to traditional approaches. An agile method could be defined as an adaptive process run by talented and creative people and controlled with iterative and incremental development (Abbas *et al*,. 2008). Although agile methods were initially described as development methodologies, the term *agile* represents an attitude, a philosophy, and a way of thinking that was presented through the principles and practices in the agile manifesto (Highsmith *et al.,* 2001). This way of thinking can be applied to many other aspects of software creation including design and modelling. Agile techniques share common principles such as (Larman, 2004):

- Delivering working software usually within a short timescale
- Close communication
- Simplicity
- Preferring programming over documenting

- Customer involvement
- Encouraging rapid and flexible response to change.

We advocate agile methodologies for co-design and co-deployment, as these are the most transparent to non-technical team members, and also can react quickly to changes in context or circumstance.

Change Management

Our approach to co-design is intended to encourage innovation and change of practice. Introducing technology usually changes current practice, at a minimum changing the standard operating procedures. However, introduction of technology also has a wider effect than just the immediate application, department, or division (Yusop *et al.*, 2005). Hence the change has to be managed carefully. This is especially true in the case of co-design, which encourages the team to not only think about implementing a solution, but to re-examine underlying assumptions and practices, encouraging innovation and change in practice.

The process of change needs to be managed with care to ensure that all stakeholder are positively engaged, especially those who have the power to implement the change (primary stakeholders), and those who have influence over opinion within the organization. Hence it is essential to carry out a full stakeholder analysis. As with any change management, when it comes to implementing the change it is important to identify champions in each of the stakeholder groups, coupled with clear and regular communication.

CO-DESIGN METHODOLOGY

An overview of the five stages in our co-design process is shown in **(**Error! Reference source not found.**). Each stage is supported through workshops and design meetings, attended by both the technical and domain experts in the design team. While there is a natural flow from stage one through to stage five, the design methodology (Figure 1) is a natural cycle involving feedback from later to earlier stages, and design teams will typically undertake several iterations of the last three stages. Each stage, and the techniques used in that stage, is explained in the following sub-sections.

For clarity we have not shown the usual project quality reviews and documentation that takes place between each stage. Similarly with the development of code and in-line with good practice, unit tests and integration tests were written before the code was produced.

Scoping - Primary Stakeholder Scoping (Establishing Domain Partner)

Before co-design as such can begin, there is a stage where the co-design itself is defined and planned. The first priority is to establish the domain and technology partnerships in the venture from which the management team will be drawn.

The objectives of the planning team are to ensure that co-design principles are observed and that the design outcomes are achieved. It defines the scope of activity and identifies co-design goals. It also identifies, selects, and recruits co-design participants, identifies, articulates and shares common goals and purposes with co design participants, and defines and carries out the co-design plan to fit overall delivery iterations.

Stakeholder Analysis. One of the first activities of the planning team is to select and recruit the co-designers using stakeholder analysis to ensure that the domain within scope is fully explored. There are a number of stakeholder analysis models, but the model we have chosen (Dix *et al.*, 1993) divides stakeholders into four categories:

- **Primary stakeholders:** People who use the system directly

Figure 1. Overview of agile co-design methodology

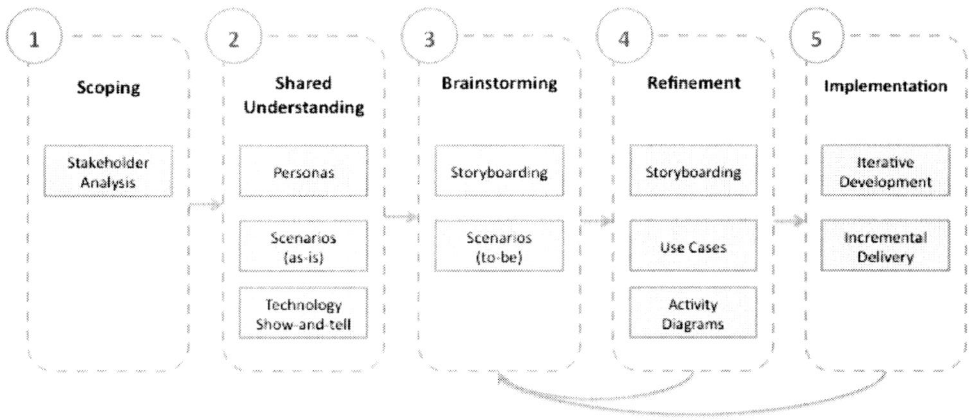

- **Secondary stakeholders:** People who do not use the system directly but receive output from it or provide input to it (indirect users)
- **Tertiary stakeholders:** People who do not interact with the system either directly or indirectly but who are affected by its success (or failure)
- **Facilitating stakeholders:** People who are involved with the design, development and maintenance of the system.

Having identified and categorized the domain stakeholders in scope, each stakeholder's concerns and characteristics are defined and their influence and impact in the domain is assessed. When these aspects are fully understood, then the co design team is more likely to achieve its goals.

The planning team selects the most important stakeholders from each category, and recruits representatives to join the co-design participants. The planning team should aim to recruit a group of between six to eight co–designers with a focus on those primary, secondary, and facilitating stakeholders who are closely identified with the needs of the primary users in the domain scope. The planning team's expertise and experience is crucial in identifying potential co-design participants who have the enthusiasm and commitment to achieve the co-design goals.

Shared Understanding (of Problem Domain)

Once the stakeholders have been identified, and have agreed to come into the team, the next priority is to create a *Shared Understanding* of the design space. Typically, this means that technicians learn about the values of the application domain, and that domain experts are introduced to the technology, its scope, potential, and limitations. The intention is not to bring everyone in the team to the same level of expertise (this is not possible in sophisticated domains or with sophisticated technologies) but to enable an informed conversation to occur in the other stages.

Personas and scenarios are a lightweight method for capturing and recording the requirements of a system from an end user's viewpoint (Cooper and Reimann, 2003). A persona describes an end user in some detail; their background, job function, and situation in the organization. Scenarios are textual descriptions of how a persona interacts with the system and other personas when using a system. The scenarios are independent of any technology and they may represent either current practice (as-is) or improved practice (to-be). In this stage they will typically reflect the domain 'as-is'.

Technology Show-and-Tell. It is important that all members of the design team are familiar with both the current technology being used in the domain, and with new technologies that could create new opportunities. We have found that a show-and-tell event is a useful way of bringing developers and users together (often for the first time), breaking down social barriers, and creating a common vocabulary for the design team to move forward. Sometimes developers will have to visit key locations to understand the context of technology use, but usually a workshop can be held that is focused around a number of key props. Props can include pen and paper based systems, as well as electronic and computing devices.

Brainstorming

The next stage involves *Brainstorming* ideas for new applications and tools. It relies on the shared understanding established in stage two, and builds a number of initial design artefacts based on the common vocabulary, in particular new scenarios and storyboards of potential applications.

Scenarios (to-be) and Story Boarding. A second set of 'to-be' scenarios can be written to capture ideas created by the group. These are textual descriptions that can draw on the same personas used in the earlier stage. The 'to-be' scenarios describe how the personas might interact with potential applications to fulfil existing needs (or, for innovating developments, new needs). Using the scenarios, story boards can be created to represent the user interface design (UI) of a given tool. This is a standard technique used in HCI development and is very effective when used in a participatory (or co-design) process. Both the end users and developers (HCI experts) are engaged with designing the UI. During this process, the scenarios can be clarified and modified if required.

Refinement

The next stage is one of *Refinement*, where the informal ideas captured in stage three are converted into a more concrete set of software requirements and specifications. We take an agile approach, and use lightweight documentation as a means to drive development, rather than as a passive record of activity. We have found three formalisms particularly useful: ontological modelling, use cases, and activity diagrams.

Ontological Modelling. Identifying key resources and mapping the relationships between them is a significant part of any co-design process involving conceptual spaces. Often domain experts will not be aware of what types of structure exist within the conceptual space of their domain (for example, is it a taxonomy or just a hierarchy, do richer relationships exist, if so what are the constraints?). Key information structures may have evolved rather chaotically, and modelling them may be a useful point of reflection for domain experts. Ontological modelling is expressive, makes no assumptions about the underlying information models of the domain, and can be easily communicated to domain experts in the form of entity-relationship diagrams.

UML Use Cases and Activity Diagrams. Use Cases are an excellent high-level (and implementation independent) starting point for describing functionality in the context of a given system and user. We use standard UML 2.0 use cases, consisting of a use case diagram, with success scenarios for each case. A brief narrative description is held alongside the diagram as a whole, as well as for each individual use case. From an agile point of view they are effective because they are relatively informal, yet help to define and capture a problem space in detail that can be understood by the whole team, including the end users.

Implementation

The fifth and final stage is *Implementation*, where the design artefacts created in stages three and four are used to drive the creation of the application or tools. We use agile software engineering practices based around iterative development and incremental deployment, and assume that other activities key to the software engineering process (such as architecture design, testing and evaluation) are consistent with these principles.

Iterative Development and Incremental Deployment. Agile methods are a number of software development methods which were proposed in the mid 1990s as a reaction to the limitations of traditional software development methodologies. Although these methods vary in practice, they share common principles, such as delivering working software frequently within a short timescale, close communication within the team and with the customer, simplicity, and programming over documenting (Larman, 2004). We have applied the spirit of these principles to the earlier stages of co-design, but they are especially important in the final implementation stage. Development should be focused in a number of small iterations, with the design team (including the users) reviewing the progress of each iteration in a design meeting. The development is shadowed by incremental deployment of the application, which first engages users with a simple (but stable) application from early iterations, and then gradually introduces new functionality. In some cases this will require the team to revisit earlier stages, to reassess assumptions and revise requirements.

CASE STUDY: MPLAT

In this section we present a case study of using this agile co-design methodology to help create an m-learning tool for nursing students on placement. The Mobile Placement Learning and Assessment Toolkit (mPLAT) project aimed to provide a mobile learning toolkit to support practice based learning, mentoring, and assessment to these nursing students. Sloan and Delahoussaye (2003) have shown that nurses benefit from mobile access to information, and the mPLAT system applies this idea to student nurses. Our belief going into the co-design process was that practice-based learning and the mentoring process would be improved with tools that connected the student *in situ* with the competency model against which they were being assessed, and were required to learn.

Motivation

The following scenario illustrates the problems and the need for such a toolkit:

Pre-registration nursing students spend 50% of their 3-year programme in clinical practice undertaking a series of placements in different areas of the healthcare system. Mentors support students for the duration of their placement. Mentors assess the students' competence in practice against a set of learning outcomes detailed in the practice assessment booklet or practice portfolio. These are summative assessments which students are required to pass in order to register as a nurse at the end of their programme. Students are expected to complete a preliminary, an interim and a final interview with their mentor. The interim interview is crucial as it is at this point that the student who is failing to progress is likely to be identified and action plans can be put into place. This good practice feature of induction, interim, and final assessment is common to most educational situations where students experience work-based learning situations.

Issues around ensuring that students are fit for practice at the point of registration were brought home recently following a report by Duffy (2004) which found that mentors were reluctant to fail students due to a number of factors, including lack of confidence, concerns over personal consequences (for student and self), and leaving it too late to implement formal procedures (the preliminary and interim interviews missed or undertaken so late that action plans to assist a student who is not progressing cannot be implemented).

The Co-Design Process

The co-design process was initiated with a number of workshops. We invited domain experts and stakeholders to join the project team. The main purpose of the first session was to acquaint the co-design team with our individual expert areas, and to allow us to exchange knowledge so that we built a shared understanding of the domain and technologies in preparation for the following two co-design sessions. The second co-design workshop focused on a brainstorming exercise to facilitate thinking about small novel applications that could help solve some of the problems identified in the first workshop. The third co-design workshop was a refinement process, where we selected three candidate applications and explored their requirements in more detail. It became clear at an early stage that the main concern of the nursing team was in trying to connect student's practice with the conceptual competency framework used by the School of Nursing. As a result we identified a number of potential tools based around this concern:

1. **Profile Placement Tool** would provide guidance for mapping the domain and competencies to the experiences (opportunities) offered in the placement area (e.g. care delivery in a medical ward).
2. **Learning Contract Builder** would draw on the student's and mentor's experience and the placement profile to create an action plan concerning what the student will achieve, how they will achieve it, what evidence is required, and which learning resources are appropriate.

Personas and Scenarios

Personas and scenarios were actually written by the end users, with a little guidance. In addition to the different institutions, there were also the requirements from the professional body (Nursing and Midwifery Council) and the British National Health Service (NHS) to capture, so we developed a number of personas and scenarios for each role. Student surveys have identified the character of a good and poor mentor (Gray and Smith, 2000). A summary of three of our personas for the mentors were:

- The 'gold-standard' mentor: one who facilitates learning appropriate to the student's level of ability and makes an appropriate assessment of the ability.
- The 'toxic' mentor: The term 'toxic mentor' was initially coined by Darling (1984), but discussions with mentors and students currently in practice indicate that this is still relevant. Darling described a gallery of toxic mentors as Avoiders, Dumpers and Blockers.
- The 'hero' mentor: the mentor who refuses additional help with a student despite potential difficulties such as personality clashes, professional value issues, etc. It appears that these mentors believe that they can turn anyone into a nurse and that if they don't it is their fault and not the student's.

These were accompanied by scenarios describing the situations in detail and the problems mentors faced while trying to carryout this function and still perform their duties on the ward. These personas and scenarios were developed from a combination of a literature review and interviews with current mentors and student advisors. Those listed above aided the first set of co-design workshops and helped the technical members of the design team become familiar with the issues faced by the end users.

Figure 2. Initial competency model identified through ontological modelling

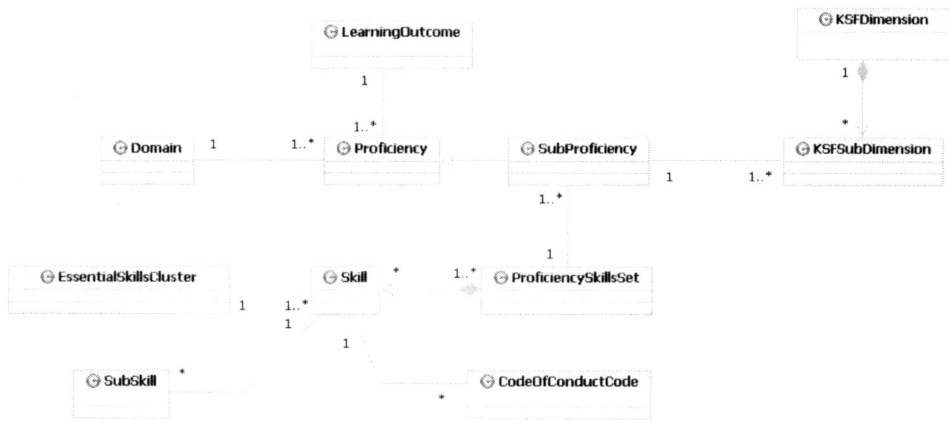

Storyboarding and Ontological Modelling

From the co-design session, personas, scenarios, and the competency profile of the student user, we were able to create storyboards for the tools quite quickly. We first identified the key features from the scenario and sketched out the initial ideas. In parallel with the storyboarding we were developing our use case and activity diagrams. Our preliminary storyboard presented the basic ideas regarding the user interface, for example using tabs to allow users to recognize the available functionality.

We also explored the competency model that lay behind the portfolio (a simplified overview of this is shown in Figure 2). Through the modelling process we discovered that certain terms used were ambiguous or overlapping, which seems to have occurred as a result of combining several other competency models from professional bodies, the UK government, and the University's School of Nursing.

Use cases and Activity Diagrams

The development of the use case and activity diagrams was again a co-design exercise but on a small scale; just one or two members of the Nursing team joined in the activity. Figure 3 shows the use case diagram for the system. This was supported by normal use and alternative use scenarios for each use case.

When supporting students to assess their competency for a task it was necessary to decide how much 'scaffolding' support was required. Should we build in a very structured approach, directing them in the way they should go, or a looser approach and let them find out themselves? We decided to adopt something between the two extremes, with a leaning toward the more structured approach. The reasons for this were:

- Part of the learning process involved students becoming reflective practitioners; being too prescriptive would not give students the opportunities they needed to self-assess properly.
- There was value to a degree of structure, provided students were allowed to reflect on their learning. Giving them some structure would aid them to start their reflection.

As this was a new tool, and the entire system of practice assessment had changed, it was felt that a structured approach would make it easier for users to understand and use. A less structured approach was the goal, but was not realistic for an initial implementation. Rather than just providing a tool that mimics the multiple views of the current paper-based competency system, we wanted to provide functionality that supported the preparation for the initial meeting wherein the Action Plan is co-created by the student and mentor.

Implementation

The self assessments, placement profiles, and action plan applications were developed in .NET for the Windows Mobile platform. We developed from the start with the mobile platform in mind, i.e. the less-featured platform first. We developed separate projects under a Visual Studio 'solution' which represent different areas of programming focus, for example, database functions, data objects, GUI, and file I/O. Because of the limited storage of mobile devices, our code was combined into one project when deployed to a mobile device.

One of the essential uses of the tool was for students to be able to understand the competence model and how it was applied to their work placement. The Learning Contract Tool gave an overview of the competency model, which included a graphical view of the competency network and the student's progress within it, as shown in Figure 4.

Figure 3. Use case for the system

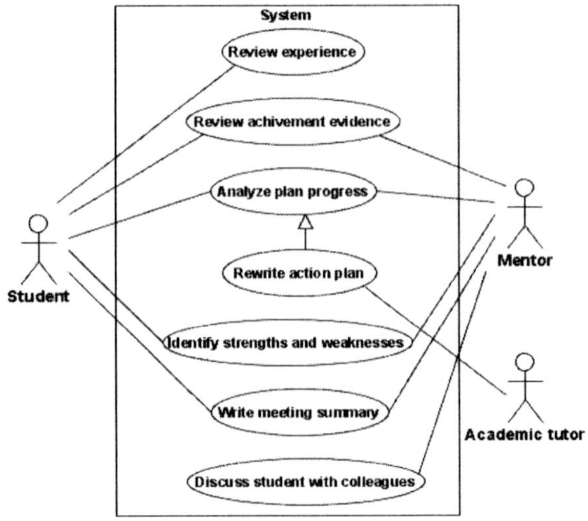

EXPERIENCE OF DEPLOYMENT

In mPLAT we used the co-design methodology but did not at first extend the principles into the deployment phase. Instead we planned and carried out a careful, traditional deployment. Nursing students and their mentors undertook a six-week placement in a clinical ward setting supported by the mPLAT tools. The deployment was supported by a domain expert who had been a member of the co-design team, shared the vision of project, and understood both the set of tools and the users. Student nurses volunteered to join the trial. We planned two-stage training for them, firstly in using the mobile device and its native tools (calendar, phone, email etc.) and then in the mPLAT tools.

Figure 4. Mobile application showing an action plan (left) and competency network (right)

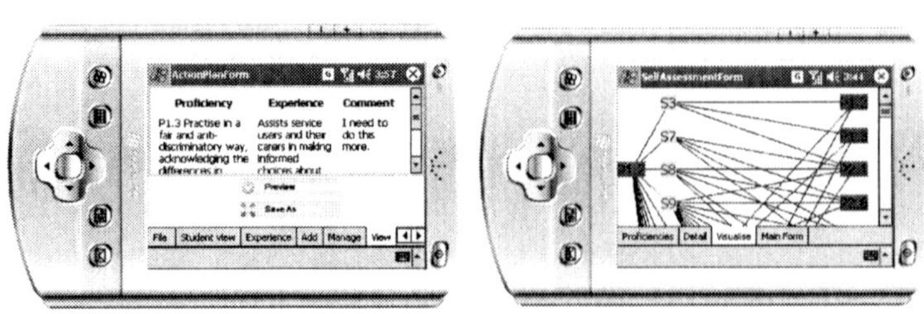

We set up a helpdesk and website that provided guidance and training materials for the users to browse or download. We informed the deployment locations (clinical wards in hospitals) that some of their placement students were taking part in a mobile learning trial that would involve the use of mobile devices in ward settings.

All of this preparation should have ensured a successful trial that would have enabled us to evaluate the goals of the project: Had student's holistic understanding of nursing proficiencies and skills improved? Were mentors more empowered to assess students? Were students more able to remain connected to their learning environment?

Unfortunately, the trial did not go according to our plans. Many of the students gave up using the mobile tools within the first few weeks of deployment, and by the end of the trial there were few users.

Using focus group meetings with the student nurses to find out what went wrong, we heard reports of ward staff demanding that the mobile devices were 'put away' immediately, and that they were not be used in a ward environment. Some student nurses were accused of 'texting their friends instead of working'. Some students found the mobile device itself technically challenging, and with the generally negative atmosphere did not feel motivated to overcome the difficulties. This happened even though domain partners were sponsors of the innovation and even though the deployment areas were informed about the trial.

Unfortunately we had overlooked some important truths: the technology and tools that we were deploying were more than beneficial innovations in placement practice; they represented change to a community with an identifiable and strong culture of norms, practices, and processes. Our student nurses, who were spearheading this innovative practice, were the least powerful members of the community we were deploying into. (It should be noted that these tools were new for nurses but similar devices were already established for other professionals working in the same environment.)

In our focus groups, we were able to find examples where determined students made good and effective use of the tools in their practice: looking up web resources to support their studies, and making journal entries in the placement context to help them write up their portfolios later.

This sobering experience helped us to recognize that co-design for creating useful, innovative tools is not enough to ensure innovation in practice; it requires a method that we are calling *co-deployment* which recognizes the difficulties of deploying tools which may challenge long-held practice, creates initiatives to mitigate them, and brings all stakeholders in the domain community to work together to accept beneficial innovation.

In the second mPLAT trial, we have been developing a co-deployment methodology that attempts to meet some of these challenges. Already we have improvements in domain community understanding of the scope for innovation and a lowering of some of the barriers to change. Co-deployment has brought together a team with a shared understanding of the benefits of innovation and a commitment to make the deployment successful through working with the community stakeholders. Through conversation and co-operation, barriers to change are being uncovered and strategies to overcome them developed.

CO-DEPLOYMENT METHODOLOGY

Co-design is focused on producing innovative applications, which leverage new technologies to create beneficial changes in practice. However in the deployment phase these changes are likely to be resisted, regardless of their value, as people working in a domain are already invested in their existing processes and methods, and may have suspicions about new technology and the motivations behind its introduction.

Co-deployment is the process of involving the domain community in the gradual deployment, evaluation and revision of the m-learning application. The goals are to create community awareness,

Figure 5. Stages of co-deployment

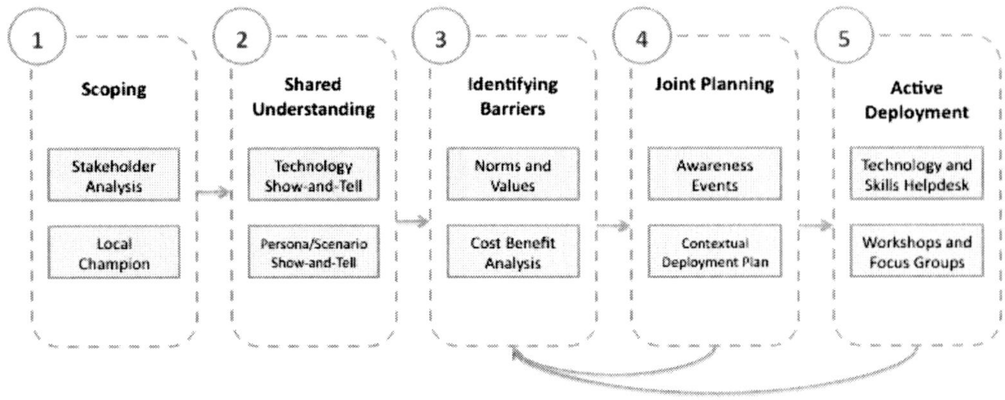

engage with existing groups, build skills, and enable a conversation between the development team and the domain stakeholders. It is based on the notion that any m-learning deployment has secondary and tertiary stakeholders, who must also be managed if innovation is to occur within a particular setting. Co-deployment has the same general shape as co-design, but is more focused on a particular context. It should start and run alongside the Refinement and Implementation stages of co-design.

Figure 5 shows the stages of co-deployment, which are further explained in the following subsections.

SCOPING

The planning team has similar responsibilities in the co-deployment phase as in the co-design phase of establishing the co-deployment team and guiding its activities to a successful outcome.

Stakeholder Analysis. We use stakeholder analysis to identify potential participants for the co-deployment sessions. However, here the focus shifts to the specific deployment context which may be different than the general co-design context. The stakeholder analysis is extended to identify co-deployment team members for use in planning for active deployment. The set of tertiary stakeholders is expanded to include people in the deployment community whose concerns and attitudes may influence deployment success. Once again the concerns, characteristics, and spheres of influence of each stakeholder are assessed together with another aspect: their power within the deployment community.

Representatives from each category (but not necessarily from the expanded set of tertiary stakeholders) are invited to join the co-deployment team, again using the experience of the planning team to primarily select individuals who will have the enthusiasm and commitment to achieve a successful deployment. Since one of the goals of using co-design is to innovate practice using new technologies, there are likely to be groups of individuals to whom such changes initially seem threatening. The planning team should seek out influential representatives of tertiary stakeholders to join the co-deployment team, members who understand and may even share these concerns and who are also influential representatives of their stakeholder group. The other tertiary stakeholders are later addressed in stage 5, active deployment.

Local Champion. At the same time as establishing the co-deployment team from the stakeholder analysis, the planning team should recruit a project support person or 'champion' who will not only support users and the deployment community during deployment, but will champion the goals of the deployment with stakeholders. The support person is a first class member of the co-deployment team.

Shared Understanding

The early sessions of co-deployment are very similar to those of co-design. Once appropriate representatives from the stakeholder groups have agreed to join the team, the first priority is to build a shared understanding of the deployment space. The technologists learn about the practices, processes, and behavioural norms of the deployment domain culture, and also the practical limitations of schedule and timing. The domain stakeholders learn about the technology, its potential for innovation in practice, and its limitations.

Co-deployment uses similar techniques to create shared understanding of the deployment context:

Technology Show-and-Tell. We have found that a show and tell event helps to create a shared understanding and helps the domain community members to quickly feel familiar with the technologies and tools being deployed. The event demonstrates the tools, applications, and technology that will be deployed, showing potential and scope but also limitations. Deployment community team members describe the environment and its physical and cultural characteristics as the context for deployment. The event helps to break down social barriers within the team and create a shared vocabulary.

Persona and scenario show and tell. In a similar way to co-design, personas and scenarios are a useful way of exploring technology in a deployment context, creating narratives of use with recognizable personas. Personas can be derived from both co-design and co-deployment stakeholder analyses describing their motivations and behaviour in the deployment context. Scenarios should be firmly situated in the deployment context.

Identifying Barriers

There are going to be barriers when introducing technology into an organization. Stage three of co-deployment involves identifying these barriers and knowing which are the ones that can be worked with, although possibly limiting the application, and which are the ones that should be challenged.

Within a given working domain or context, there is generally a difference between the designers norms and values and those of the users of the technology. This is in addition to the recognized barriers to the introduction of technology: senior management commitment, buy-in from users, use of technology that has been fully tested, good communication, *etc.* (Brown *et al,*. 2007). Resistance to change can come from many quarters when introducing mobile technology. The specific issues are:

- **Intrusion:** Mobile devices can be seen as an intrusion into current work practice.
- **Privacy:** Many of these devices can capture audio and video, and if used unprofessionally may compromise a client's privacy.
- **Digital divide:** We now have the situation of digital natives and digital immigrants (Prensky, 2001), that can threaten the hierarchies and social norms in the work place.

This stage is concerned with identifying *Barriers to Change*, and the *Barriers to Challenge*. We use two different techniques to achieve this:

Norms and Values (Identifying Barriers to Change). The values of a community are often related to their culture and belief system (so for example, in nursing, respect for the person and caring for their needs are very strong values). The norms are closely related to values, concerned with the ways the community normally works (professionalism), and the implicit structures that exist within the community (Kling, 1996). Value

Exchange recognizes that people will do things if it is of value to them or they get something of value in return. For instance, people will take the time to learn a new system for many reasons not directly related to money (Gordijn and Akkermans, 2001): they may want to make the task easier, want to be seen to be in the 'know' or up to date, believe it will help the patient for whom they are working for, etc. We use semi-open interviews and small forums to try and identify norms and values.

Cost Benefit Analysis (Identifying Barriers to Challenge). CBA is a group of techniques that assign a monetary value to activities or artefacts in order to ascertain if a project is financially worth starting or continuing with, indicating when and at what rate the return on investment will be realised. We seldom do a full and formal CBA, but the process can be useful in identifying which of the norms may be worth challenging, and which are linked to values that may not be apparent to developers from a different working culture.

Joint Planning

The co-deployment team, having established a shared understanding of the opportunities afforded by the new technologies and also of the difficulties of deploying the technology in practice, need to establish a context for the deployment.

Awareness Events. In this activity, the concerns of secondary and tertiary stakeholders are addressed by creating awareness through workshops, presentations, demonstrations, and interviews. The deployment community is given the opportunity to voice their concerns and the co-deployment team can demonstrate where and how those concerns have been addressed, and also dispel misunderstandings. The co-deployment team needs to be open about the deployment and demonstrate their willingness to both listen and respond to legitimate concerns, be robust in dispelling misunderstandings, but do so in a manner sensitive to long held beliefs.

The co-deployment team should take special care to create and promote positive relationships with any powerful tertiary stakeholders whose consent is required for successful deployment. Their positive influence can be beneficial in generating acceptance from other tertiary stakeholders.

Contextual Deployment Plan. Whilst preparing the deployment, the co-deployment team develops a full plan for deployment in the domain context. Scenarios created in the earlier stages help the team focus on how users and stakeholders can be prepared for and supported in deployment. The plan should be informed by:

- Awareness of the context created when working with the domain community
- Analysis of the 'as-is' skills of the users compared with the skills required to use the new tools
- Physical deployment of tools and equipment, including, for instance, the mobile device itself; application and tool set up; and in a mobile context, provision of communications access including how it will be paid for.

For example: the users of the new tools may need training on how to use the physical device, the native tools available, and the applications provided by the technologists. They may need training on the use of the device in the deployment context, including guidelines for professional behaviour, and how to get the most out of the available tools. Secondary and tertiary stakeholders may need information on how the mobile devices might be used in their context and guidelines to support effective use. The preparation and dissemination of information, demonstrations, and presentations should all be included in the co-deployment plan.

Active Deployment

The conversational interaction started in co-design and extended to the deployment community, continues throughout deployment.

Technology and Skills Helpdesk. Primary users are supported with: personal contact with a project support person, a technology helpdesk, drop-in technology 'surgeries', a web site with browsable and downloadable content, and on-going skills development with a support person.

Workshops and Focus Groups. Secondary and tertiary users are supported with: a web site, personal contact with a project support person, workshop sessions on deployment in practice, and guidance on supporting users in the deployment context. The successes, concerns, and issues are fed back to the technologists and management team through evaluations, interviews, and focus groups.

CONCLUSION

We have developed an agile co-design methodology that includes end-users (domain experts) in the design team, in order to help create m-learning applications that innovate practise in a particular domain (as well as using innovative technology).

Our methodology brings together techniques found in HCI (personas, scenarios, and storyboarding), agile software techniques (iterative development and incremental delivery) and lightweight software engineering practice (use cases, simple ontological modelling, and activity diagrams). The methodology is an agile approach that gives us a lightweight method of capturing and recording requirements and feeding these though the design cycle in such a way that they are integral (as opposed to tangential) to the software development process.

Through a case study of creating an m-learning tool for nurses on placement we have shown how the co-design methodology can be effective in identifying subtle requirements, and can result in appropriate application development. The case study also shows that there is a need to take the principles of co-design through to co-deployment, in order to identify and manage the challenges of introducing new technology into an environment with its own particular values and established ways of working.

The co-deployment methodology we have developed in response includes key stakeholders in the deployment team in order to raise awareness of the deployment, and enable the team to identify and address concerns that arise.

We believe that using the co-design methodology can drive the creation of more sophisticated tools that don't just replicate existing practice in a digital form, while co-deployment can help support and foster innovative new practise in the domain based around these tools. Our experiences demonstrate how agile methods throughout the design and deployment can help ease the process of creating new m-learning tools and allow a flexible and adaptable approach that is more sensitive to the eventual end-users.

ACKNOWLEDGMENT

The authors would like to thank the extended mPLAT team, in particular our colleagues Hugh Davis and Mary Gobbi at Southampton, and Balbir Barn at Thames Valley. The mPLAT project is funded by JISC.

REFERENCES

Abbas, N., Gravell, A., & Wills, G. (2008, June 10-14). Historical Roots of Agile Methods: Where did "Agile Thinking" Come from? In *Agile processes and eXtreme programming in Software Engineering*, Limerick, Ireland.

Abowd, G. D., Atkeson, C. G., Hong, J., Long, S., Kooper, R., & Pinkerton, M. (1997). Cyberguide: A mobile context-aware tour guide. *Wireless Networks*, *3*(5), 421–433. doi:10.1023/A:1019194325861

Brown, S. A., Chervany, N. L., & Reinicke, B. A. (2007, September). What matters when introducing new information technology. *Communications of the ACM, 50*(9), 91–96. doi:10.1145/1284621.1284625

Collins, T. G. (2005). English Class on the Air: Mobile Language Learning with Cell Phones. In *Fifth IEEE International Conference on Advanced Learning Technologies (ICALT'05),* pp. 402-403.

Cooper, A., & Reimann, R. (2003). *About Face 2.0: The Essentials of Interaction Design*. John Wiley & Sons.

Darling, L. (1984, October). What do nurses want in a mentor? *The Journal of Nursing Administration, 14*(10), 42–44.

de Crom, N. (2005, October). The "ME"-Learning Experience: PDA Technology and eLearning in Ecotourism at TUT. *Fourth World Conference on Mobile Learning (mLearn).*

Dix, A., & Finlay, J. Abowd, G., & Beale, R. (1993). *Human-Computer Interaction*. Prentice-Hall

Duffy, A. (2004). *Failing students - a qualitative study of factors that influence the decisions regarding assessment of students, article for Nursing and Midwifery Council.* http://www.nmc-uk.org/aArticle.aspx?ArticleID=1675 (last accessed Dec, 2007)

Flynn, M., Pendlebury, D., Jones, C., Eldridge, M., & Lamming, M. (2000). The satchel system architecture: Mobile access to documents and services. *ACM Journal of Mobile Networks and Applications*, *5*(4), 243–258. doi:10.1023/A:1019172931873

Gordijn, J., & Akkermans, J. M. (2001). A Conceptual Value Modeling Approach for e-Business Development. In *Proceedings of the First International Conference on Knowledge Capture, Workshop Knowledge in e-Business (K-CAP 2001)*, (pp. 29-36) http://www.cs.vu.nl/~gordijn/er2001.pdf

Gray, M., & Smith, L. (2000). The qualities of an effective mentor from the student nurse's perspective: Findings from a longitudinal qualitative study. *Journal of Advanced Nursing*, 32(6), 1542–1549. doi:10.1046/j.1365-2648.2000.01606.x

Grudin, J., & Pruitt, J. (2002) Personas, Participatory Design and Product Development: An Infrastructure for Engagement. *Proceedings Participatory Design Conference (PDC*) 2002, 144-161.

Highsmith, J. Beck k., Cockburn A., & Jeffries, R. (2001). *Agile Manifesto.* from www.agilemanifesto.org (last accessed July 2008).

Kling, R. (1996) *Computerization and Controversy: Value Conflicts and Social Choices*., San Diego, CA: Academic Press.

Kravcik, M., Kaibel, A., Specht, M., & Terrenghi, L. (2004). Mobile collector for field trips. *Educational Technology & Society*, *7*(2), 25–33.

Larman, C. (2004). *Agile and Iterative Development: A manager's guide*. Pearson Education.

Lehman, M. M. (1980). Programs, life cycles and laws of program evolution. *Proceedings of the IEEE*, *68*, 1060–1076. doi:10.1109/PROC.1980.11805

Massimi, M., Baecker, R. M., & Wu, M. (2007, October 15-17). Using participatory activities with seniors to critique, build, and evaluate mobile phones. In *Proceedings of the 9th international ACM SIGACCESS Conference on Computers and Accessibility* (pp. 155-162), Tempe, Arizona, USA. Assets '07. New York: ACM

Parsons, D., Ryu, H., & Cranshaw, M. (2006). A Study of Design Requirements for Mobile Learning Environments. In *Proceedings of the Sixth IEEE International Conference on Advanced Learning Technologies* (ICALT'06) (pp. 96-100).

Prensky, M. (2001). Digital natives, digital immigrants. *Horizon*, *9*(5), 1–2. doi:10.1108/10748120110424816

Sharples, M., Jeffery, N., du Boulay, J. B. H., Teather, D., Teather, B., & du Boulay, G. H. (2002, January 16). Socio-cognitive engineering: A methodology for the design of human-centred technology. *European Journal of Operational Research, 136*(2), 310-323. Retrieved from http://dx.doi.org/10.1016/S0377-2217(01)00118-7

Sloan, H. L., & Delahoussaye, C. P. (2003). Clinical application of the Omaha system with the Nightingale Tracker: a community health nursing student home visit program. *Nurse Educator, 28*(1), 15–17. doi:10.1097/00006223-200301000-00008

Svanaes, D., & Seland, G. (2004, April 24-29). Putting the users center stage: role playing and low-fi prototyping enable end users to design mobile systems. In *Proceedings of the SIGCHI Conference on Human Factors in Computing Systems* CHI '04 (pp. 479-486), Vienna, Austria. New York: ACM.

Weal, M. J., Cruickshank, D. G., Michaelides, D. T., Millard, D. E., De Roure, D. C., Halloran, J., et al. (2006). A persistent infrastructure for augmented field trips. In *Proceedings of World Conference on Educational Multimedia, Hypermedia and Telecommunications, Ed-Media 06* (in press), Orlando, Florida, USA.

Wills, G., Woukeu, A., Carr, L., & Hall, H. (2003). The need for semantic in web design. *Semantic Web workshop, Fourteenth Conference on Hypertext and Hypermedia Nottingham,* UK

Yusop, N., Lowe, D., & Zowghi, D. (2005). Impacts of Web systems on their domain. *Journal of Web Engineering, 4*(4).

Chapter 10
Design and Implementation of Multiplatform Mobile-Learning Environment as an Extension of SCORM 2004 Specifications

Kiyoshi Nakabayashi
National Institute of Multimedia Education, Japan

ABSTRACT

A learner-adaptive self-learning environment has been developed in which both mobile phones and personal computers can be used as client terminals. The learner-adaptive function has been implemented using SCORM 2004 specifications. The specifications were extended to enable offline learning using mobile phones. Because the application-programming environment of mobile phones varies from carrier to carrier, a common content format was specified for the learning content and content-execution mechanisms were developed for each carrier's environment to maximize content-platform interoperability. The latest learning results achieved by using mobile phones were synchronized with the latest ones on the server-side sequencing engine so that the learner-adaptive function was available from personal computers as well. The system can provide adaptive courses such that the results of a pre-test taken on mobile phones can modify the lecture content on personal computers, fitting them to each learner's level of knowledge and understanding. The functionality and usability of the system was evaluated through two trial experiments, the first of which involved adult learners and the second with small children and their parents.

DOI: 10.4018/978-1-60566-703-4.ch010

INTRODUCTION

Mobile learning is becoming increasingly popular due to the rapid growth in the use of personal mobile devices and wireless networks (Leuhn & Chan, 2003). There are three main types of mobile learning:

- Distribution of learning materials, such as test questions, to mobile terminals (Thornton & Houser, 2004),
- Utilization of mobile devices for mentoring and scaffolding purposes (Stone, 2004; Wang et al., 2003), and
- Collaborative learning in a wireless environment (Cortez et al., 2004; Ogata & Yano, 2004).

Although personal digital assistants (PDA) are available (Shih et al., 2005) that have functionality as rich as that of personal computers, mobile phones are more popular as easy-to-use mobile terminals equipped with both voice-communication and Internet functions (Thornton & Houser, 2004).

This chapter discusses a self-learning environment in which mobile phones and personal computers are used to complement each other (Nakabayashi et al., 2007a; Nakabayashi et al., 2007b). The design goals of the system were:

1. To provide a standard-based mobile-learning infrastructure independent of device characteristics (often differing from mobile phone to mobile phone or from carrier to carrier) exploiting existing e-learning standards (Fallon & Brown, 2003; Nakabayashi, 2004),
2. To enable offline learning using mobile phones,
3. To implement learner-adaptive functionality with which learning materials and the learner's status are shared from mobile phones and personal computers and his/her status is reflected in the next learning activity from both environments, and
4. To exploit the advanced facilities of mobile phones such as cameras or contactless smart-card readers.

To achieve these goals, the system we developed uses a Sharable Content Object Reference Model (SCORM 2004, Advanced Distributed Learning Initiative, 2006) compliant learning management system (LMS) (Nakabayashi et al., 2006), content browsers, and a SCORM 2004 sequencing engine on mobile phones; there are protocol transformation servers between the LMS and the mobile phones. The content browser on the mobile phones is capable of displaying downloaded content offline. The SCORM 2004 sequencing engine on the mobile phones enables learner-adaptive functionality during offline learning. The learner's learning results on the browser are later sent to a protocol transformation server, which modifies the data format so that it is compliant with SCORM 2004 learner-tracking information. It is then forwarded to the SCORM 2004 compliant LMS. The LMS manages the tracking information on both mobile phones and personal computers. Based on the tracking information, the next learning activity is selected adaptively by the SCORM 2004's sequencing functionality. The learning-material format has partly been extended from the SCORM 2004 specifications to support mobile learning.

We bore several educational settings in mind in designing the system. The main educational setting was conventional self-learning where learners take a pre-test or post-test to check their knowledge on a certain subject using mobile phones while commuting on trains or buses. They then later use a PC-based environment in their offices or homes to strengthen their knowledge with the new content possibly tailored based on their previous test results. Another educational setting is that utilizing input devices for mobile phones to provide an "authentic" or "situated" learning opportunity where learners are asked to explore objects in the real world under certain circumstances and to input information about these to the system. The system can then provide them with various feedbacks.

The rest of this chapter is organized as follows. The next section discusses the background of this system, and then SCORM 2004 specifications are introduced in the following section. Design policies and system implementation are then discussed in the next two sections. The results of trial experiments are presented after that, and then future trends are discussed followed by concluding remarks.

BACKGROUND

As described in the introduction, this chapter discusses an adaptive self-learning environment in which mobile phones and personal computers are used. "Adaptation" in e-learning or mobile learning generally has two meanings where the first means learner-adaptation and the second means context-adaptation.

Learner-adaptation is a technique where the system adapts its behavior taking into account learners' preferences or status of understanding. Learner-adaptation techniques usually consist of so-called learner models that are representations reflecting the status of learners' knowledge or understanding (Fletcher, 1975; Wenger, 1987). The earliest and most basic learner model was called the overlay model (Carr & Goldstein, 1977; Wenger, 1987) where the system represented learners' knowledge as a collection of concepts. The estimated value of a learner's level of understanding is assigned to each concept. The system estimates the learner's level of understanding about a certain concept by providing some test questions or interactive simulations, then it tries to improve his/her level of understanding about the concept by providing hints, explanations, and/or remedial materials related to the concept. The overlay model, which originated in the early days of Computer Assisted Instruction (CAI) research, is still widely employed in various e-learning systems that have recently been developed (Brusilovsky, 2003; De Bra & Ruiter, 2001; Sosnovsky et al., 2007). This is because the model is simple, general, and easy-to-use independently of the subject domain, while other sophisticated learner models such as the bug model (Brown & Burton, 1978) can be highly dependent on the subject domain and thus difficult to use.

SCORM 2004, which was adopted in the system we developed, also provides learner-adaptation capabilities inherited from the overlay model. As will be described in the next section, SCORM 2004 provides learner-adaptation behavior taking into account "tracking information" assigned to separate learning activities and learning objectives. This tracking information corresponds to the level the learner understands the concepts in the overlay model.

Another "adaptation" in e-learning, especially in mobile learning, is context-adaptation. Context-adaptation means that the system adapts its behavior taking into account device capabilities, network capabilities, or the environment the learner is in. For example, since mobile devices usually have smaller screens than desktop PCs, the displayed images should be adjusted to fit their screen sizes. If the wireless network has a poor-quality connection, the system should send text instead of rich video, or even turn into the offline mode. If the learner moves from the dark indoors to the bright outdoors, the image needs to be adjusted for contrast. There have been numerous studies in this research area spanning the simple adaptation of specific media types such as images and video (Mukherjee et al., 2005; Vetro & Timmerer, 2005; Xie et al., 2006) to much more sophisticated adaptation techniques for general-content types and diverse learner situations or "contexts" (Lemlouma & Layaida, 2004; Schilit et al., 1994; Yang, 2006).

In terms of context-adaptation, the proposed

system in this chapter employs a rather simple yet practical approach. It allows the content designer to prepare different kinds of content corresponding to a variety of terminal devices in advance, then in actual learning the system selects the content matching the learner's device. It also provides the capability for offline learning using a mobile device that allows the learner to keep studying in environments with poor-quality or unavailable wireless networks.

The other aspect of mobile learning related to the proposed system is the "situatedness" (Lave & Wenger, 1991) or "authenticity" (Brown et al., 1989) of learning supported by mobile devices. In contrast to e-learning in the office or classroom using desktop PCs, mobile learning enables learners to learn while moving around in the real world and watching and/or touching real objects. In such learning environments, the learners have more opportunities to actively "engage" with the subject being learned based on their own interests or motivation, in contrast to classroom settings where learners tend to passively accept instructions by the teacher. Examples of research emphasizing this aspect of mobile learning include those dealing with outdoor observation (Chen et al., 2002), language learning in a real situation (Ogata and Yano 2004), and enhanced experiences in museums (Hsi & Fait 2005; Yatani et al., 2004).

The proposed system was designed with the intention of supporting such "situated" or "authentic" learning, similar to the studies previously mentioned. The system is especially capable of utilizing the advanced facilities of mobile phones such as cameras or contactless smart-card readers. As this allows learners to input information about "real-world objects" they are observing, the system can support certain areas of learning to enhance their learning experience.

In summary, the proposed system exploits SCORM 2004, which is not only a standard specification that is becoming increasingly popular in the e-learning community in corporate and educational sectors but also a learner-adaptive framework based on the overlay learner model, which is well established and widely used. The system also incorporates a practical context-adaptation mechanism capable of dealing with diverse terminal devices and poor-quality wireless networks. The system is also intended to facilitate learning in authentic situations by exploiting the advanced facilities of mobile phones as a means for learners to input real-world information from the environment. This bridges the system-centered learning environment provided by SCORM 2004 with the learner-centered learning environment where learners actively engage with subjects they are learning.

SCORM 2004

SCORM Overview

SCORM is a collection of standards and specifications for Web-based Training (WBT) content. The SCORM 1.2 published in 2001 has been widely adopted by numerous LMSs and authoring tools. However, in the field of Computer Assisted Instruction (CAI) and Intelligent Tutoring Systems (ITS), a great deal of research has been done on learner-adaptive mechanisms that make learning content sufficiently intelligent to dynamically change its behavior taking into account a learner's knowledge or his/her level of understanding (Wenger 1987, Murray et al. 2003). There are implementations of WBT systems featuring such learner adaptation functions (Nakabayashi et al., 1995). However, little effort has been expended to introduce these learner-adaptive functions into SCORM specifications. This is due to the fact that it is too difficult to design a single standard specification capable of dealing with diverse learner-adaptive technologies proposed in various CAI/ITS systems. Thus, SCORM specifications contain quite a few functions based on learner-adaptive technologies. This makes it almost impossible for content designers to develop sophisticated learner-adaptive content using these specifica-

Figure 1. Structure of SCORM content

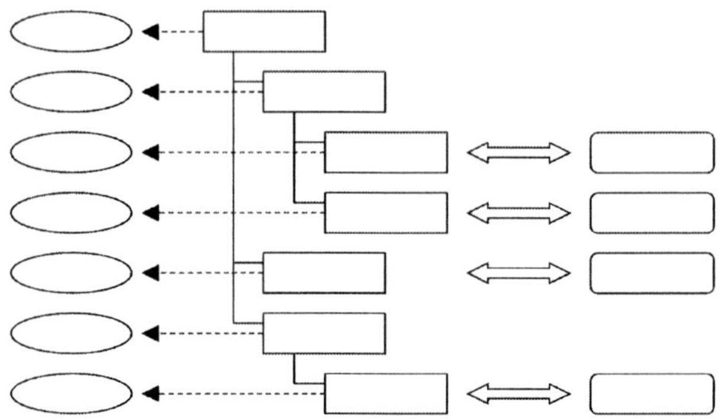

tions. This means that they have to reinvent their resulting proprietary learner-adaptive mechanisms that prevent them from taking advantage of interoperability and reusability, which are the most important benefits of standardization.

To overcome these problems, ADL published the SCORM 2004 specifications (Advanced Distributed Learning Initiative 2006). To provide sufficient learner-adaptive functions, SCORM 2004 defines a sequencing specification with which a system dynamically selects content pages that are most suited to the status of a learner's progress. Of course, it is not possible to cover all the potential learner-adaptive functions with this single sequencing specification, but now it is possible to design a variety of learner-adaptive content that is interoperable and reusable. Several research and development programs have appeared to exploit the benefits of SCORM 2004 (Kazi, 2004: Yang et al., 2004).

SCORM specifications deal with hierarchically structured content that runs on Web-based server/client environments. Figure 1 outlines the structure of SCORM content. The nodes in the hierarchical structure are called "activities". Leaf activities are associated with the multimedia learning resources presented to a learner. To run the courseware, the LMS reads the course-structure definition, selects one of the activities from this definition, and then sends the learning resources associated with the selected activity to the WWW browser on the client. The learning resources presented to the learner are WWW content consisting of HTML, JavaScript, Java Applets, and a variety of plug-ins. There are two types of learning resources: Sharable Content Objects (SCOs) and Assets. An SCO is a learning resource capable of communicating with the LMS, while an Asset has no communication capabilities. Using communication capabilities, an SCO retrieves useful information such as a learner's name, ID, and previous learning status from the LMS as well as transmits information reflecting learning results such as elapsed time, answers and scores to exercises, and updated learning status to the LMS.

SCORM1.2 includes:

- A content-aggregation specification consisting of a data model of the hierarchical course structure and its binding to XML, and
- A run-time environment (RTE) specification for SCO/LMS communication including a JavaScript Application Program Interface (API) and data model.

In addition to these two specifications above, SCORM 2004 introduces another two new ones:

- A sequencing specification. This allows the content designer to describe the learner-adaptive behavior of the content. The designer can describe the sequencing rules that control the behavior of the content taking into account the learner's status.
- A navigation specification. This allows an SCO to provide a navigation GUI representing commands to learners to navigate through the hierarchical course.

The following sections briefly describe the sequencing and navigation specifications.

Sequencing Specification

The SCORM 2004 sequencing specification allows a content designer to describe sequencing behavior attached to the course structure. On execution, the sequencing engine in the LMS reads the course structure together with the sequencing rules. The sequencing engine then repeats the sequencing process, which consists of receiving a request from the learner, interpreting sequencing rules, updating status information to reflect the status of his/her progress, and determining the next activity to deliver learning resources.

The sequencing specification consists of four basic elements:

- A "course structure" is the hierarchical structure of the activities. An activity is associated with one or more "objectives". The content designer can define global objectives shared with multiple activities.
- "Tracking information" represents the status of the learner's progress. Tracking information is associated with activities and objectives. This information is updated by the sequencing engine by propagating the status from the SCO toward the root of the activity tree.
- A "sequencing request" is a command to the sequencing engine such as "continue", "previous", "retry", and "exit". A sequencing request is generated either from the learner's input of a navigation command or post condition rules described below.
- A "sequencing rule" is the content designer's description of sequencing behavior. Sequencing rules are further categorized into three types. The first category is to define some restriction on sequencing behavior. This category includes a "control mode", a "limit condition", and a "precondition rule". The second category is called a "post condition rule", which produces a certain sequencing request when the rule condition is met. The third category is the "rollup rule", which describes how a parent activity's tracking information is updated using its child activities.

Learner-adaptive content can be implemented based on these specifications (Nakabayashi et al., 2006).

Navigation Specification

In SCORM 1.2, an SCO cannot issue navigation command such as "next" or "previous", and these commands are only issued from a special command frame in the browser. Since the GUI design of the command frame depends on the LMS, the content designer cannot modify it to be consistent with SCO design. Also, navigation commands cannot be issued from an SCO preventing simulation or exercise content from issuing such commands flexibly at the timing relevant to the operation. To avoid these problems, SCORM 2004 introduces specification which defines types of navigation commands and a mechanism to issue these commands from an SCO. This mechanism is specified as an extension of the RTE specification for SCO/LMS communication. There are additional data-model definitions for navigation commands transmitted from an SCO to the LMS.

Design Policies

The four design goals presented in the introduction are described here in detail. These are:

1. To provide a standard-based mobile-learning infrastructure independent of device characteristics. There is currently no content interoperability between mobile phones, since the environments for developing and executing application programs are often different from mobile phone to mobile phone or from carrier to carrier. To overcome this problem, the system design aims to establish a mobile-learning infrastructure exploiting existing e-learning standards.
2. To enable offline learning using mobile phones. It is highly probable that learning on commuter trains will become one of the scenarios most demanded for mobile learning. System design is intended to support continuous learning in such environments without stable quality in wireless communication.
3. To implement learner-adaptive functionality in which learning materials and a learner's status are shared between mobile phones and personal computers with his/her status is reflected in the selection of the next learning activity from both environments. This feature enables learning scenarios such as the results of pre-test using mobile phones which are reflected in the learning content that follows on personal computers, or remedial learning with mobile phones based on learning results using personal computers. This is intended to implement a self-learning environment that complements the portability of mobile phones and the rich multimedia environments of personal computers.
4. To exploit the advanced facilities of mobile phones such as cameras or contactless smart-card readers. System design is intended to implement functionality that allows input information from such devices to be reflected in the sequencing of learning content. With this functionality, it might be possible to provide learning activities that are aware of the context within which the learner is situated.

To meet these design goals, the system was designed according to the following design policies:

- To extend SCORM 2004 to mobile environments, and
- To develop a content-execution mechanism for mobile phones.

These policies are discussed in the following subsections.

EXTENSION OF SCORM 2004 TO MOBILE ENVIRONMENTS

The SCORM 2004 specifications explained in the previous section were adopted to meet the first design goal, which was "to provide a standard-based mobile-learning infrastructure" and the third design goal, which was "to implement learner-adaptive functionality in which learning materials and results from both mobile phones and personal computers are shared", since SCORM 2004 offers standard specifications for Web-based training (WBT) content capable of learner-adaptive functionality.

Learner-adaptive functionality on personal computers is provided by introducing SCORM 2004 specifications. However, there are several issues to be resolved for mobile phones to implement and make this adaptive functionality available. This is because mobile phones have several significant technical limitations that personal computers or PDAs do not. One of these limitations is their inability to run JavaScript (ECMAScript), which the SCORM RTE specification relies on for communication between the LMS and SCOs.

Another limitation is that it is difficult for mobile phones to render various rich media content available for personal computers due to their small screens and lack of plug-in software.

Taking these limitations into account, we derived the following design policies:

- Manifest files, which describe content on course structure and sequencing rules for learner adaptation, are shared for learning accessed from both mobile phones and personal computers. Since the SCORM 2004 sequencing engine on the LMS deals with shared manifest files, a learner's status is also automatically shared. This makes it possible to implement learner-adaptive functionality on both mobile phones and personal computers.
- The RTE specification for LMS-SCO communication will be extended to deal with the limitations of mobile phones.
- Two types of SCOs and Assets, one for mobile phones and one for personal computers, are prepared. During learning, suitable types of content are selected by checking the type of terminal device.

Content-Execution Mechanism on Mobile Phones

To meet the second design goal, which is "to enable offline learning using mobile phones", a certain mechanism should be implemented on mobile phones that is capable of dealing with a set of meaningful learning activities such as tests that ask questions, accept answers, score them and give feedback, or a chunk of adaptive-learning material that provides a pre-test followed by remedial learning and a post test. These executions should be done on mobile phones without communicating with the LMS.

Two techniques are usually possible to fetch content from networks and execute this on mobile phones. The first is to use a native (built-in) mobile phone browser. The second is to implement learning content as a downloaded application program and run it on mobile phones. The first does not meet the second design goal since built-in browsers are usually not equipped with a programming language for scoring and/or feedback functions on test content without communicating with the LMS. The second has the potential for developing rich content by fully exploiting the functionality of mobile phones. However, the developed content will strongly depend on a particular mobile phone's programming environment. Since this environment differs from carrier to carrier, it significantly reduces the interoperability of the content. This technique is thus not acceptable to achieve the first design goal concerning interoperability.

Taking these situations into consideration, we selected a third technique in which a general-purpose mechanism for executing content is installed on mobile phones. As outlined in Fig. 2, content that is compliant to a specific format is downloaded and executed on mobile phones. Although it is necessary to implement multiple mechanisms for executing content, each of which runs in the programming environments of different carriers, a standardized content format that is independent of the carriers' programming environments can be introduced. This solution thus meets both the first and the second design goals. In the actual implementation, we decided to use a sub-tree of SCORM 2004 content as a standardized format for the content to be downloaded. This naturally met the requirements discussed above. It is also possible to design the content-execution mechanism so that it is capable of managing several devices in mobile phones such as cameras or contactless smart-card readers. This satisfies the fourth design goal, which is to exploit such devices to input information that reflects the sequencing of learning content.

Figure 2. Content sharing by introducing content-execution mechanism

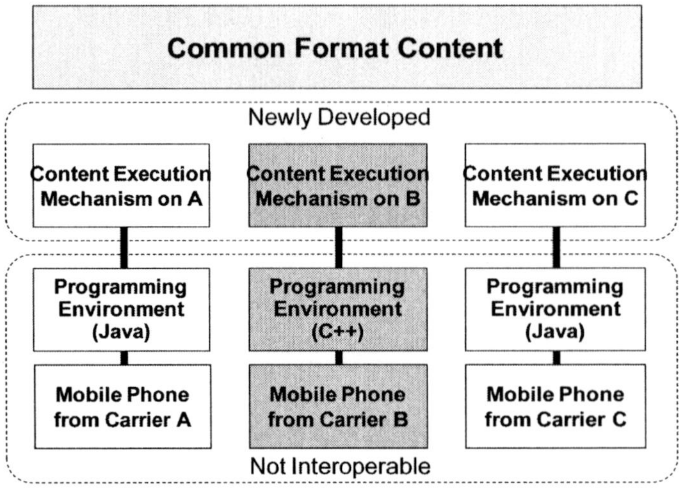

SYSTEM IMPLEMENTATION

A mobile learning environment was developed based on the design policies described in the previous section. The system's configuration is shown in Fig. 3.

The SCORM 2004 LMS was developed based on the open source SCORM 2004 engine (Nakabayashi et al., 2006). Learners log on to the system using personal computers or mobile phones through the protocol transformation server, and the server checks what type the client device is. If the device is a personal computer, it will directly communicate with the LMS during the following session. Manifest files and learning results are common to both mobile phones and personal computers, but learning resources (SCOs or Assets) to be sent to the client need to be prepared for either mobile phones or personal computers. The details on each system's components are described in the following sections.

Extension of Manifest Files

The SCORM 2004 manifest-file format was extended for mobile learning. One of the extensions

Figure 3. System configuration

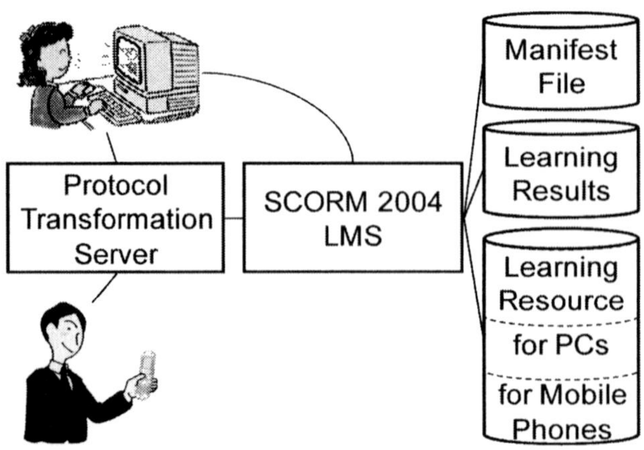

Figure 4. Identification of learning terminal and downloading of content in manifest file

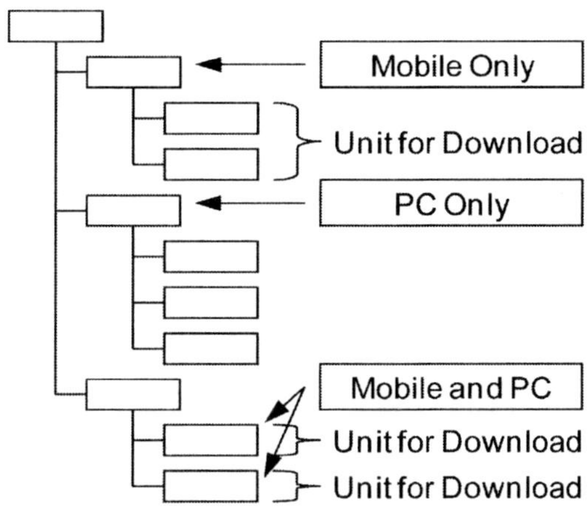

was a specification for suitable terminal devices and units for download. Another extension was a specification for multiple resources for one activity.

The first extension is summarized in Fig. 4. This extension of manifest files makes it possible to specify any given hierarchical activity as "mobile only", "PC only", or "mobile and PC". This allows the content designer to declare which part of the content is to be executed on which terminal device. For example, the designer can declare that a pre-test can only be accessed from a mobile phone or that a lecture session should only be accessed from a personal computer. Content accessible from a mobile phone will also become a unit of a download. If a leaf activity is specified as "mobile only" or "mobile and PC", then the learning resource associated with the activity becomes a unit of the download. If a parent activity is specified, then the unit of the download will consist of the part of the manifest file that corresponds to the specified parent activity and its whole descendant sub-tree as well as all the learning resources associated with the leaf activities in the sub-tree.

The second extension is to enable multiple learning resources to be associated with one activity. This is in contrast to the original SCORM 2004 specification, which only allows one learning resource to be associated with one activity. A type of terminal device and/or mobile phone carrier is specified for each learning resource that is associated with the activity. When the activity is accessed, the system can select a suitable resource depending on the type or the client device's carrier.

Content-Execution Mechanism and Content for Mobile Phones

As described in the previous section, content downloaded to mobile phones consists of learning resources suitable to be rendered on them and the sub-tree of a SCORM 2004 manifest file. The following sections discuss content specifications and the execution mechanisms for these components. The mechanism to input external information using mobile-phone devices is also described.

Learning-Resource Specification and Browser for Mobile Phones

The content browser is designed to render downloaded learning resources. The learning resource format is shown in Figure 5. A simplified version of HTML is used. A learning resource consists of multiple HTML pages and associated bitmap images.

Figure 5. Example of learning-resource format

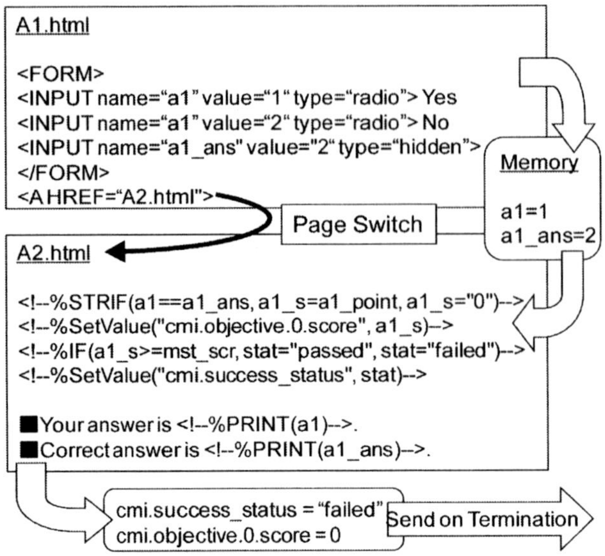

The content browser is capable of executing simple script embedded in HTML pages to provide score, feedback, and communication functionalities. With this script, it is possible to describe behaviors such as addition, comparison, retrieval, and the setting of values for the RTE data-model elements. This script is evaluated when the HTML page is switched to another page. In the example in Figure 5, a script in A2.html will be evaluated when the page is switched from A1.html. The learner's input value to the <INPUT> tag will be held in memory. By referencing this value, the content browser scores, generates feedback, and communicates with the SCORM 2004 RTE data model.

SCORM 2004 Sequencing Engine for Mobile Phones and Synchronization of Tracking Information

To execute the sub-tree of a manifest file consisting of a SCORM 2004 sequencing rule, a SCORM 2004 sequencing engine with full functions has been implemented on mobile phones. This enables a meaningful chunk of learning activities to be executed such as adaptive learning material providing a pre-test followed by remedial learning and a post test. It is possible for recent mobile phones with a few mega-bytes of memory space for application programs to implement the SCORM 2004 sequencing engine.

To implement learner-adaptive behavior according to the sequencing rules defined in the sub-tree of the manifest file, it is necessary to synchronize tracking information between the LMS and mobile phones. As seen in Figure 6, the tracking information associated with each activity and learning objective is downloaded together with the content before offline learning on the mobile phone. When the offline learning is finished, the updated tracking information on the mobile phone is uploaded to the LMS (Figure 6 (b)), and a rollup process is executed to update the tracking information of the whole activity tree to reflect the offline learning results.

Input of External Information Using Mobile Phone Devices

A mechanism to read QR code using mobile-phone cameras is implemented as an external-information input function. When the URI as-

Figure 6. Synchronization of tracking information

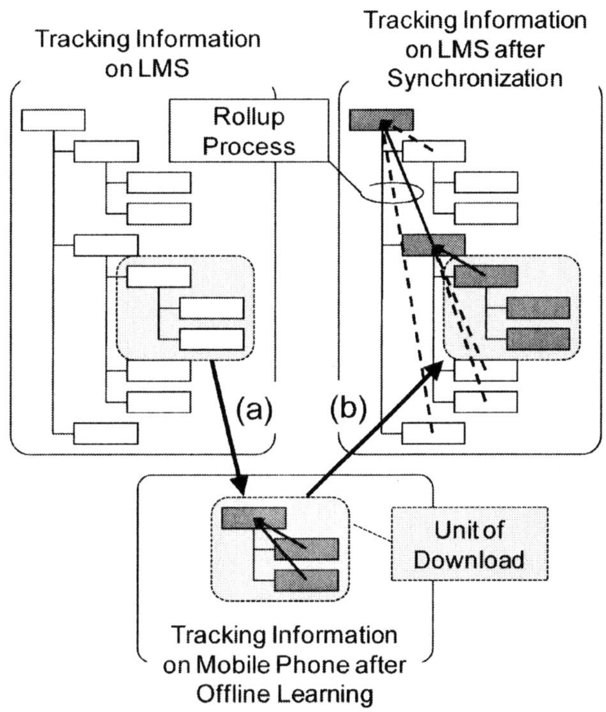

sociated with the QR code is read by the camera, the system automatically issues a SCORM 2004 choice command to jump to the learning activity specified by the URL; then, the predefined RTE parameters associated with the activity are set. Since the values of the RTE parameters affect the sequencing behavior, it is possible to make the system aware of external information input from mobile-phone devices.

TRIAL EXPERIMENTS

Two different trial experiments were conducted with the system we developed. The first involved adult users checking basic system functionality according to defined learning scenarios such as pre-testing, post-testing, and remedial situations. The second was intended to use the system we developed in much more "authentic" situations. The experiment particularly involved small children and their parents to enhance their interest and understanding of "foods" by providing them with an environment where they could touch "real" foods and interact with the system exploiting mobile-phone camera input.

First Experiment

The first experiment was conducted to check system functionality. Two types of content were prepared. The first consisted of four sections with pages from a lecture and a final test. The second consisted of a "mobile only" pre-test followed by seven sections of "PC only" lecture pages selectively presented based on the results of the pre-test. Four learners participated in the experiment. They checked the behavior of the system with these two types of content using personal computers and mobile phones from three different Japanese mobile-phone carriers. A Java version of the sequencing engine and content browser were implemented for mobile phones from two of the carriers, and a C++ version for the remaining carrier.

Seven different learning scenarios were prepared to check the system. For example, one of the learning scenarios checked whether the pre-test results from mobile phones were reflected in the subsequent learning behavior from the personal computers so that the mastered content could be skipped. Another utilized QR code input to check whether the sequencing behavior was altered by the input code.

Although we demonstrated that the system worked correctly according to our design intentions, the experimental learners encountered several usability issues. One was the lack of tables of contents. Most learners wanted this capability. Another frequent comment was that the images were occasionally too small. A function should be equipped that will enable images to be freely enlarged.

Second Experiment

The second experiment allowed learners to use the system rather freely in an informal but "authentic" situation. The topic to be learned was "foods", which was intended to enable small children and their parents to enhance their interests and understanding of foods. To achieve this goal, the experiment was conducted in a large shopping center where the learners had the daily "authentic" experience of looking, selecting, and purchasing various food items. In the experiment, the learners had a chance of looking at actual vegetables while perusing learning content with their mobile phones. The content on their mobile phones contained quizzes on vegetables, such as the various kinds, their nutritional value, and ways of cooking them. To answer the questions, the learners could handle the vegetables, examine their characteristics closely, and input answers to the system with their mobile phones by reading the QR code tags attached to the vegetables.

Figure 7 illustrates an example of the content provided on the mobile phones. Learners were given instructions on the "Preparation" screen about how to distinguish a particular kind of vegetable from other similar looking vegetables. Learners were asked to find a particular kind of vegetable on the "Question" screen, spinach in this case, from the vegetables displayed at the shopping center and to input the QR code tags attached to it. They were given feedback on their answers on the "Feedback and Lecture" screen and some explanation and background knowledge concerning the question. The learning results acquired at the shopping center were stored in the LMS so that learners could later access the learning content through their personal computers at home. This encouraged communication between parents and children to improve their interest and understanding of foods.

The experiment involved pairs of small children around five or six years old and their parents. The families who visited the shopping center were randomly chosen. A total of 59 pairs participated four times in the experiment on two days. The system was stable during the whole experiment and we found that even small children could easily operate it. Figure 8 has photographs of the experiment with vegetables on the left and a subject manipulating her mobile phone on the right.

The parents answered a questionnaire after the experiment. The results are listed in Tables 1 and 2. Forty-nine out of 59 parents responded to the first question in Table 1 that they thought that mobile-phone learning about vegetables was a worthwhile experience for their children. Responding to the second question in Table 2, 48 parents indicated their intent to again participate in an experiment if they had the chance. Thus, most participants gained positive impressions of the system, the content, and the entire experiment.

Parents provided free responses to the questionnaire one week after the experiment had been conducted. These included positive comments that met our intent in the experiment to enhance

Figure 7. Content for mobile phones

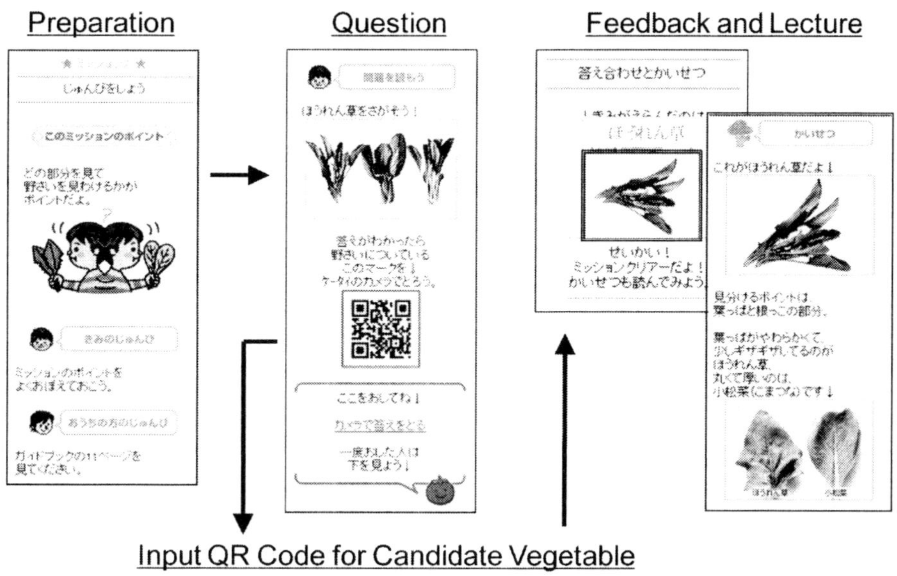

Figure 8. Photographs of vegetables (left) and subject manipulating her mobile phone (right) in experiment

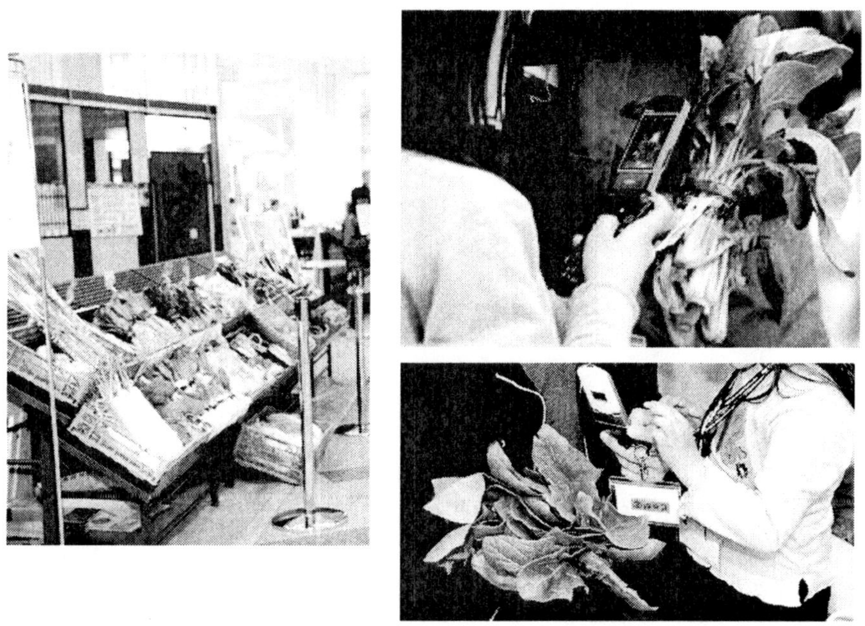

the learners' interest and knowledge of vegetables such as:

- My child now prefers vegetables more than he/she did before.
- In the supermarket, my child demonstrated an interest in the shapes and colors of vegetables.
- There has been a great deal of conversation in our family about nutrition in foods.

Table 1. Answers to question: "Was it a worthwhile experience for your children to participate in the experiment?"

Answer	Strongly Agree	Agree	Neutral	Disagree	Strongly Disagree	No Answer
# of people	36	13	2	0	0	8

Table 2. Answers to question: "Would you like to participate in a similar experiment with different content?"

Answer	Strongly Agree	Agree	Neutral	Disagree	Strongly Disagree	No Answer
# of people	38	10	3	0	0	8

However, there were comments concerning the usability of the system and the learning scenarios such as:

- The only feedback was in the form of character texts, which was difficult for children to understand.
- It was difficult to answer the questions.
- It would have been preferable to have more time to discuss the learning content in terms of the differences between various vegetables.

FUTURE TRENDS

Several issues that will need to be resolved were identified during the system's design, development, and trial experiments.

- The usability of the content browser must be improved. It should provide tables of contents and be able to enlarge images as discussed in the previous section.
- It should be possible to automatically generate client-side content. It is very time consuming to separately prepare the same learning resources for personal computers and mobile phones. A mechanism that automatically generates learning resources for both types of devices from one original set of content would save a great deal of time.
- It is necessary to design learning scenarios that will utilize input devices much more. Recent mobile phones have various input devices, i.e., not only cameras but contactless smart-card readers and global positioning systems (GPSs). Such devices could be incorporated to enhance adaptive functionalities to support "authenticity" and "situatedness" in learning so that the system could take into account learners' current learning contexts as well as their knowledge or level of understanding.
- Investigations into learning scenarios other than self-learning are also an important issue. Interaction between teachers and students or groups of students is essential to enhance learning activities. It would be worth investigating the design of a learning platform based on mobile devices and personal computers that support such group-learning in addition to self-learning activities.
- We need to establish a unified learning platform that is independent of devices where not only content but also application programs work without being dependent on these. To establish such interoperability, it is necessary to investigate new

software-platform technologies supporting Web2.0 and SaaS as well as new mobile-phone technologies such as Android from Google, Symbian, or i-Phone OS. A recent proposal has been for new learner-adaptive specifications (Nakabayashi et al., 2008) dealing with object-oriented architectures for learner adaptive environments in which various objects with certain educational functionalities are integrated to work together. The objects in these environments are implemented not only as program modules in the LMS but also as parts of Web services and widgets on client terminals. This may be a possible framework for a unified learning platform that is independent of devices with learner-adaptation capabilities.

CONCLUSION

A learner-adaptive self-learning environment in which both mobile phones and personal computers can be used as client terminals has been developed. The SCORM 2004 specifications were used to implement a learner-adaptive function with an extension to offline learning using mobile phones. The specifications were extended to enable offline learning using mobile phones. To enable content-platform interoperability despite variations in application-programming environments between mobile-phone carriers, a common content format was specified and content-execution mechanisms were developed for the environments of three carriers. This mechanism synchronized the latest learning results achieved using mobile phones with the latest ones on the server-side sequencing engine so that the learner-adaptive function was also available from personal computers. The system could provide adaptive courses such that the results of a pre-test taken on mobile phones were reflected in the lecture content on personal computers, fitting them to each learner's level of knowledge and understanding. The functionality and usability of the system was evaluated through two trial experiments, which indicated that the intended functionality was successfully implemented and a certain level of usability was achieved. Future work includes introducing the system we developed into actual learning settings. It is also necessary to incorporate new mobile-phone technologies in terms of both hardware facilities and software-application environments, as well as to investigate unified mobile-learning platforms that support group-learning and self-learning activities.

ACKNOWLEDGMENT

This work was supported by the Japanese Ministry of Internal Affairs and Communication's 2005-2006 program of "Development and Experiments with Ubiquitous Learning Infrastructures".

REFERENCES

Advanced Distributed Learning Initiative. (2006). *SCORM® Shareable Content Object Reference Model SCORM 2004 3rd Edition.*

Brown, J. S., & Burton, R. R. (1978). Diagnostic models for procedural bugs in basic mathematical skills. *Cognitive Science, 2*, 155–191.

Brown, J. S., Collins, A., & Duguid, P. (1989). Situated cognition and culture of learning. *Educational Researcher, 18*(1), 32–42.

Brusilovsky, B. (2003). Developing adaptive educational hypermedia system. In T. Murray, S. Blessing, & S. Ainsworth (Eds.), *Authoring Tools for Advanced Technology Learning Environments* (pp. 337–409), Dordrecht, Kluwer Academic Publishers.

Carr, B., & Goldstein, I. P. (1977). Overlays: a theory of modeling for computer-aided instruction. *AI Lab Memo 406 (Logo Memo 40)*. Massachusetts Institute of Technology, Cambridge, MA.

Chen, Y. S., Kao, T. C., Sheu, J. P., & Chiang, C. A. (2002). A mobile scaffolding-aid-based bird-watching learning system. In *Proc. of the IEEE International Workshop on Wireless and Mobile Technologies in Education 2002* (pp. 15–22), IEEE Computer Society.

Cortez, C., Nussbaum, M., Santelices, R., Rodrigues, P., & Zurita, G. (2004). Teaching Science with Mobile Computer Supported Collaborative Learning (MCSCL). In *Proc. of the IEEE International Workshop on Wireless and Mobile Technologies in Education 2004* (pp. 67–74), IEEE Computer Society.

De Bra, P., & Ruiter, J.-P. (2001). AHA! Adaptive hypermedia for all. In *Proc. of the World Conference of the WWW and Internet 2001* (pp. 262-268), AACE.

Fallon, C., & Brown, S. (2003). *E-Learning Standards*. Boca Raton, St. Lucie Press.

Fletcher, J. D. (1975). Modeling of learner in computer-based instruction. *Journal of Computer-based Instruction, 1*, 118–126.

Hsi, S., & Fait, H. (2005). RFID enhances visitors' museum experience at the Exploratorium . *Communications of the ACM, 48*(9), 60–65. doi:10.1145/1081992.1082021

Kazi, S. (2004). A conceptual framework for Web-based intelligent learning environments using SCORM 2004. In *Proc. of the IEEE International Conference on Advanced Learning Technologies 2004* (pp. 12–15), IEEE Computer Society.

Lave, J., & Wenger, E. (1991). *Situated learning legitimate peripheral participation.* Cambridge, MA: Cambridge University Press.

Lemlouma, T., & Layaida, N. (2004). Context-aware adaptation for mobile devices. In *Proc. of the IEEE International Conference on Mobile Data Management* (pp. 106–111), IEEE Computer Society.

Leuhn, C., & Chan, Y. (2003). Mobile learning: a new paradigm in electronic learning. In *Proc. of the IEEE International Conference on Advanced Learning Technologies 2003* (pp. 76–80), IEEE Computer Society.

Mukherjee, D., Delfosse, E., Kim, J.-G., & Wang, Y. (2005). Optimal adaptation decision-taking for terminal and network quality-of-service. *IEEE Transactions on Multimedia, 7*(3), 454–462. doi:10.1109/TMM.2005.846798

Murray, T., Blessing, S., & Ainsworth, S. (Eds.). (2003). *Authoring tools for advanced technology learning environments.* Dordrecht, Kluwer Academic Publishers.

Nakabayashi, K. (2004). E-Learning technology standardization – make it converge!! In *Proc. of the International Conference on Computers in Education 2004* (pp. 33–39), Asia-Pacific Society for Computers in Education.

Nakabayashi, K., Hoshide, T., Hosokawa, M., Kawakami, T., & Sato, K. (2007a). Design and implementation of a mobile learning environment as an extension of SCORM 2004 specifications. In *Proc. of the IEEE International Conference on Advanced Learning Technologies 2007* (pp. 369–373), IEEE Computer Society.

Nakabayashi, K., Hosokawa, M., Kawakami, T., & Sato, K. (2007b). Mobile learning environment with content download capability as an extension of SCORM 2004 specification, In *Suppl. Proc. of the 15th Intentional Conference on Computers in Education, Vol.1.* (pp. 45–47), Asia-Pacific Society for Computers in Education.

Nakabayashi, K., Koike, Y., Maruyama, M., Touhei, H., Ishiuchi, S., & Fukuhara, Y. (1995). A distributed intelligent-CAI system on the World-Wide Web. In *Proc. of the International Conference on Computers in Education 1995* (pp. 214–221), Asia-Pacific Society for Computers in Education.

Nakabayashi, K., Moroimoto, Y., & Hada, Y. (2008). Investigation into object oriented architecture for extensible learner-adaptive environment. In *Suppl. Proc. of the 16th Intentional Conference on Computers in Education,* (pp. 161–165), Asia-Pacific Society for Computers in Education.

Nakabayashi, K., Nakamura, A., Kosaka, Y., & Nagaoka, K. (2006). Design and implementation of SCORM 2004 execution engine and its performance evaluation. In *Proc. of the 2006 International Conference on SCORM 2004* (pp. 31–35).

Ogata, H., & Yano, Y. (2004). Context-aware support for computer-supported ubiquitous learning. In *Proc. of the IEEE International Workshop on Wireless and Mobile Technologies in Education 2004* (pp. 27–34), IEEE Computer Society.

Schilit, B. N., Adams, N. I., & Want, R. (1994). Context-aware computing applications, In *Proc. of the IEEE Workshop on Mobile Computing Systems and Applications* (pp. 85–90), IEEE Computer Society.

Shih, T. K., Lin, N. H., Chang, W., Wang, T., Wen, H., & Yang, J. (2005). The hard SCORM: reading SCORM courseware on hardcopy textbooks. In *Proc. of the IEEE International Conference on Advanced Learning Technologies 2005* (pp. 812–816), IEEE Computer Society.

Sosnovsky, S., Dolog, P., Henze, N., Brusilovsky, P., & Nejdl, W. (2007). Translation of overlay models of student knowledge for relative domains based on domain ontology mapping. In *Proc. of the 13th Int. Conf. Artificial Intelligence in Education* (pp. 289–296), IOS Press.

Stone, A. (2004). Mobile scaffolding: an eiment in using SMS text messaging to support first year university students. In *Proc. of the IEEE International Conference on Advanced Learning Technologies 2004* (pp. 405–409), IEEE Computer Society.

Thornton, P., & Houser, C. (2004). Using mobile phones in education. In *Proc. of the International Workshop on Wireless and Mobile Technologies in Education 2004* (pp. 3–9), IEEE Computer Society.

Vetro, A., & Timmerer, C. (2005). Digital item adaptation: overview of standardization and research activities. *IEEE Transactions on Multimedia, 7*(3), 418–426. doi:10.1109/TMM.2005.846795

Wang, C., Liu, B., Chang, K., Horng, J., & Chen, G. (2003). Using mobile techniques in improving awareness to promote learning performance. In *Proc. of the IEEE International Conference on Advanced Learning Technologies 2003*, IEEE Computer Society.

Wenger, E. (1987). *Artificial intelligence and tutoring systems: computational and cognitive approaches to the communication of knowledge.* San Fransisco, CA: Morgan Kaufmann.

Xie, X., Liu, H., Ma, W.-Y., & Zhang, H.-J. (2006). Browsing large pictures under limited display sizes. *IEEE Transactions on Multimedia, 8*(4), 707–715. doi:10.1109/TMM.2006.876294

Yang, J., Chiu, C., Tsai, C., & Wu, T. (2004). Visualized online simple sequencing authoring tool for SCORM-compliant content package. In *Proc. of the IEEE International Conference on Advanced Learning Technologies 2004* (pp. 609–613), IEEE Computer Society.

Yang, S. J. H. (2006). Context aware ubiquitous learning environments for peer-to-peer collaborative learning. Journal of*Educational Technology and Society, 9*(1), 188–201.

Chapter 11

Towards Mobile Learning Applications Integration with Learning Management Systems

Marc Alier Forment
Universitat Politècnica de Catalunya, Spain

María José Casany Guerrero
Universitat Politècnica de Catalunya, Spain

Jordi Piguillem Poch
Universitat Politècnica de Catalunya, Spain

ABSTRACT

ICT in education innovators are creating new kinds of learning applications using all sorts of new technologies available: Web 2.0, Mobile, Gaming platforms and even Virtual Worlds. Mobile learning applications (m-learning) take advantage of the ubiquitousness of the mobile devices to explore new kinds of ways of learning. Learning Management Systems (LMS) are a consolidated kind of Web based learning software that over the last 15 years have evolved to meet the needs of the learning institution to basic, common online educational platforms. The LMS creates a Web based space for every course (Virtual classroom) that can be used to complement the presence learning activities (Blended Learning) or to fully deliver the course contents (Online Learning). Nowadays most learning organizations have integrated a LMS with their information systems (back-office, academic management, etc.) to a point where all learning activities (virtual and non virtual) have a counterpart (syllabus, assessments, scheduling, etc.) in the LMS virtual classrooms. M-learning is not destined to replace the current Web based learning applications, but to extend it, that is why Mobile Applications will need to be able to integrate with the LMS. It also makes sense to be able to access some of the services of the LMS Virtual Classroom from the mobile device. But, to accomplish this goal might not be a simple task. This chapter analyzes the complexities involved to achieve that goal, and describes some standard interoperability architectures and related research and development projects that will allow this kind of interaction between the LMS and the m-learning applications.

DOI: 10.4018/978-1-60566-703-4.ch011

THE NEW GENERATION LEARNING APPLICATIONS

The present time is characterized by the unstoppable technological change. New technologies such as Web 2.0 or affordable and connected mobile devices have enabled the re-conceptualization of learning spaces. This technological explosion has many implications. Today the classroom is not the only space where learners can learn, mobile devices enable the possibility of learning anywhere and anytime.

For example, game-based learning has a huge potential in the learning process of children, adolescents and even grownups (for example Big Brain Academy), and has been an important field of research since late 1970s (de Aguilera & Mendiz, 2003). More recent studies (Prensky, 2001; Prensky, 2008) explore the potential of using game consoles and other portable devices such as Nintendo DS or PlayStation Portable for education purposes. Such technologies with which children spend so much time. Game players can learn to do things such as driving a car, but deeper inside, they learn thinks such as take information from many sources and make decisions quickly, deduce the games rules rather than being told, create strategies, overcome obstacles or learn to collaborate with others though the Network.

Other learning applications use portable technology such as digital cameras, mobile phones, MP4 players, or GPS devices to enhance the learning process. These applications are often called mobile learning (m-learning) applications. Although m-learning is in its infancy, there are many experiences using mobile technology (Brown-Martin, 2008).

Blogging, wikis, podcast, screen-cast, contents from youtube, Google Maps, pictures in Flickr, and social interaction in Facebook or Twitter, are common sources of information used by students while they learn or work in their assignments.

The consumers of these applications are the 'digital natives' (children who have lived all their lives with technology). Studies have tried to define the preferred learning approaches of this generation (Bradley, Haynes, & Boyle, 2005; Kennedy, Krause, Judd, Churchward, & Gray, 2006). Digital natives learning style can be characterized by: preference for receiving information quickly and the ability to process it quickly, a bias towards multitasking and non linear access to information, a heavy reliance on ICTS for information access and communication active involvement (Cao, Tin, McGreal, Ally, & Coffey, 2006).

THE NEED FOR INTEGRATION OF LEARNING APPLICATIONS WITH LMS

Current Web based Learning Management Systems are focused on meeting the needs of the institution in providing a basic, common educational platform. Most of universities worldwide have successfully integrated the use of a LMS where all the academic information services, online contents and learning application are centralized and managed. LMS are a consolidated online learning environment already adopted by learners, teachers and institutions.

Right now we can find lots of learning applications, like the ones described in the previous section, living outside the LMS ecosystems (Mobile applications in particular). Teachers willing to innovate are using applications and technologies not supported by their institution LMS, and by doing so they are taking their students outside the virtual campus. Thus the students need to go to several different sites (using different usernames and passwords) in a scrambled learning environment. This may cause confusion and frustration to students.

We need to allow the use of these new kinds of learning technologies inside the LMS. To keep a coherent learning environment for the learners without limiting the kind of applications to use. In addition to this, the good practices of the innovators could be relayed to the rest of the docents. We need to find a way to easily get the upcoming new generations of learning applications inside the LMS.

That is why the LMS need to incorporate the ability to integrate new applications from all these variety of new technologies, and the developers of Mobile learning applications need to know how to connect their systems with the LMS easily. To achieve this goal the interconnection middleware has to be defined, it needs to be based on sound standards and we need reference implementations with Open Source.

In the next sections we will discuss some promising standards that are being defined, some implementations in real world applications and the Moodbile application, which takes advantage of them.

LEARNING INTEROPERABILITY STANDARDS

In the early days of audio recording, you had to buy music content and the device to play it from the same manufacturer. As the industry evolved, standards created a new market based on the fact that you could buy music in a standard format that could play on devices from more than one manufacturer.

The same problem has been addressed successfully in different ways regarding the delivery of educational content through the Web, and its integration in Web based Learning Management Systems. There was a clear interest for the industry to port all the contents being created for CD-ROM to the online world. Standards such as Advanced Distributed Learning (ADL http://www.adlnet.gov/) SCORM (http://www.adlnet.gov/scorm/) have been widely implemented and adopted. So there are standards implemented to create, share and use educational contents.

But education is not only about content, as the last trends in online pedagogy models make explicit connectivism (Arina, 2008), social constructionism (Alier, 2007). Thus the goals of the interoperability we seek are not bound only to content interoperability but to a wider scope of features and services that learning applications can offer. Integration is defined by the Oxford English dictionary as *"the act of making two systems work together to achieve a functional goal, regardless of how difficult or expensive the task might be"*. Interoperability is defined by IEEE as *"the ability of two or more systems, or components to exchange information and to use the information that has been exchanged"*. The IEEE definition for interoperability is 16 years old, and nowadays software systems can do more things together than just exchange information, for example share functionality. So Merriman (2008) from the Open Knowledge Initiative (OKI) offers a new definition for interoperability: *"the measure of ease of integration between two systems or software components to achieve a functional goal. A highly interoperable integration is one that can easily achieved by the individual who requires the result"*. According to this definition, interoperability is about making the integration **as simple and cost effective as technologically possible**.

Source Code is not Enough

Free Libre Open Source Software (FLOSS http://en.wikipedia.org/wiki/FLOSS, from now on Open Source) projects often assume the ease of integration with the software can be achieved simply through providing the source code to a community, regardless of architecture or design considerations. In this scenario interoperability can be achieved assuming that the organization requiring integration will have access to the necessary expertise to achieve it.

Let us assume a hypothetical organization called the ACME organization that wants to integrate the Open Source software XXX with its legacy systems YYY. ACME needs to build up a team (or hire it) with expertise about the software platforms of XXX and YYY, with deep knowledge about the way XXX and YYY are designed and a good idea of the technologies to bind platforms such as RPC (Remote Procedure Calls), CORBA, SOA or - as seems to be the common tendency last years: - Webservices. Thus it is technically possible to achieve. Let us suppose that ACME spends the kind of money and time necessary to make this integration happen. With most proprietary systems this could not be done at all, and that is often the killer reason to adopt a Open Source system instead of a proprietary solution (sometimes at a greater cost in customization effort), even abandoning a yet implanted proprietary solution. But soon enough the XXX Open Source community will come up with a new revision or upgrade of the XXX software. The integration performed at ACME will likely not work with the new version of XXX without a thorough revision and testing. So, with each new version of XXX, ACME needs to decide if upgrading to the new version of XXX is worth revising all the integration code all over again.

This is not a hypothetical case, is happening in lots of learning organizations that need to integrate their enterprise applications with LMS.

Integration Technologies

To integrate two systems without access to the source code of both systems can be very tricky. Several years ago this could only be accomplished by hacking the data files, sometimes by reverse engineering the file format (just like OpenOffice opens the Microsoft Office's files), or accessing directly the database engine tables. Fortunately during the last 15 years the software industry has developed, and widely adopted, several technologies to ease the exchange of information between applications: such as Microsoft OLE (Object Linking and Embedding, well known for being the technology that allowed originally the *Cut & Paste* between Windows Applications), DCOM (Distributed Compound Object Model and its successor ActiveX), CORBA (Compound Object Request and Broker Architecture). In the last years the Webservices empowered by XML related technologies (WSDL, XSLT, XML-RPC, SOAP, REST) are widespread and promise the arrival of real interoperability frameworks.

But interoperability is not only about having a language and a channel to exchange information and services (like XML and Webservices), but something that need to be considered in the very design of the software. For software developers and project managers, building highly interoperable software is often harder than not doing so, and usually more expensive. To build a interoperable system involves several tasks such as understanding a standard, engaging with a community or refining the standard. The immediate value to a project of easing future integration is usually not highly regarded.

Merriman (2008) states that interoperability *"is something that need to be addressed thinking in terms of system's architecture, at the very beginning of the design of the system"*. But as he also admits it is more difficult to follow a highly interoperable approach, following standard specification or best practices is more complex and does not have immediate benefit. And we need solutions that not only consist in good practices to build the systems of the future, but that provide a strategy to adapt existing systems to be interoperable.

The **Service Oriented Architecture** (SOA) is a software engineering approach that provides a separation between the interface of a service, and its underlying implementation. Such that consumers (applications) can interoperate across the widest set of service providers (implementations), and providers can easily be swapped on-the-fly without modification to application code. SOA preserves the investment in software development as underlying technologies and mechanisms evolve and allow enterprises to incorporate externally developed application software without the cost of a porting effort to achieve interoperability with an existing computing infrastructure.

In the service oriented approach data synchronization between systems and data exchange are not problems because information is kept in one place. On the other hand, the SOA approach requires redesigning software.

The Open Knowledge Initiative OSIDs

The *Open Knowledge Initiative* (http://okiproject.org) was born in 2003 with the purpose of creating a standard architecture of common services that learning software systems need to share, such as Authentication, Authorization, Logging etc. The OKI project has developed and published a suite of interfaces know as Open Service Interface Definitions (OSIDs) whose design has been informed by a broad architectural view. The OSIDs specifications provide interoperability among applications across a varied base of underlying and changing technologies. The OSIDs define important components of a SOA as they provide general software contracts between service consumers and service providers. The OSIDs enable choice of end-user tools by providing plug-in interoperability. OSIDs are software contracts only and therefore are compatible with most other technologies and specifications, such a SOAP, WSDL. They can be used with existing technology, open source or vended solutions.

Each OSID describes a logical service. They separate program logic from underlying technology using software interfaces. These interfaces represent a contract between a software consumer and a software provider. The separation between the software consumer and provider is done at the application level to separate consumers from specific protocols. This enables applications to be constructed independently from any particular service environment, and eases integration (see Figure 1).

For example, services such as authentication are common functions required by many systems. Usually each application has built this specific function. As a result the authentication function is implemented in many ways and this results in information being maintained in different places and being unable to easily reuse. OKI would separate the authentication function from the rest of the systems and provide an central authentication service for all the applications.

Figure 1. An OSID is a description of a logical service between a server provider and a consumer, with independence from the communication framework or data definition language

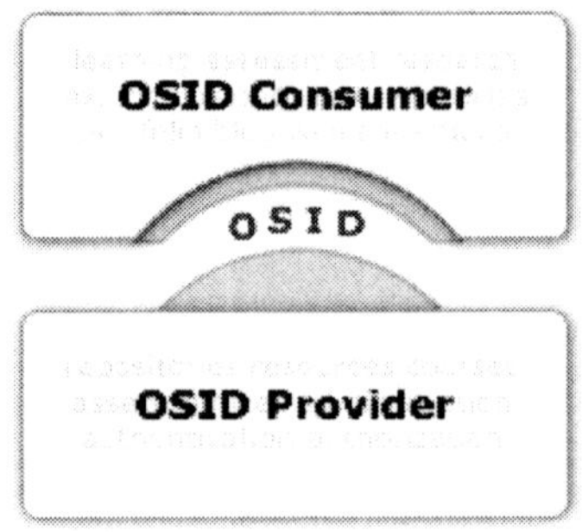

OKI describes with OSIDs the basic services already available in e-learning platforms. Among others, these basic services used by many e-learning platforms are described in the following OKI OSIDs:

- The **authentication** OSID is used to register a new user or to know if the user is connected to the system. This is a basic service in any software system.
- The **authorization** OSID is used to know if a user has rights to access a service or function. This service is necessary in any system using roles.
- The **logging** OSID is used to capture usage information. It is useful to know how the system is working for system diagnostics and performance.
- The **internationalization** OSID is used to change the language of the application or add new languages.
- The **configuration** OSID is used to change configuration parameters.

Thus using the OKI OSIDs has the following advantages:

- Ease to develop software. The organization only has to concentrate in the part of the problem where they can add value. There is no need to redo common functions among most of the systems.
- Common service factoring. OKI provides a general service factory so that services can be reused.
- Reduce integration cost. The current cost of integration is so high that prevents new solutions from being easily adopted. OSIDs are a neutral open interface that provides well understood integration points. This way there is no need to build a dependency on a particular vendor.
- Software usable across a wider range of environments, because OKI is a SOA architecture.

But OKI still has a long way to go before becomes a de facto standard of interoperability. Nowadays up to 75 projects have implemented the OSIDs and given feedback to the OKI community process.

The IMS Global Learning Consortium Initiatives for Interoperability in Learning Systems

The IMS Global Learning Consortium is also working since 2005 in standards towards interoperability and integration of learning services and systems.

The IMS Abstract Framework is set of (abstract) specifications to build a generic e-learning framework, which might be able to interoperate with other systems following the IMS AF specifications. IMS AF describes a e-learning system as the set of services that need to be offered (IMS, 2003). IMS AF is a standard that can be complemented by the OKI OSIDs because OKI provides more specific information about the semantics of the services, how to use them and in what kind of situations they could be used.

IMS also defines the IMS Learning Technologies for Interoperability. While IMS AF and OKI work on the exchange of information and services, IMS LTI developed under supervision of Dr. Charles Severance, focuses on the process on how a remote service is installed on a Web based learning system (IMS, 2006).

The OSIDs tells us how to exchange information between the LMS and an external learning application, but how will the teacher and the student reach the application form the LMS? These kinds of proxy bindings are described by the IMS LTI 1.0 and 2.0 standards.

IMS Learning Technologies for Interoperability

The basic idea of IMS LTI is that the LMS has a proxy tool that provides an endpoint for an externally hosted tool and makes it appear if the externally hosted tool is running within the LMS. In a sense this is kind of like a smart tool that can host lots of different content.

The proxy tool provides the externally hosted with information about the individual, course, tool placement, and role within the course. In a sense the Proxy Tool allows a single-sign-on behind the scenes using Web services and allows an externally hosted tool to support many different LMS's with a single instance of the tool.

The IMS LTI 2.0 architecture focuses on the launch phase of the LMS-to-tool interaction. The launch accomplishes several things in a single Web service call:

- Establish the identity of the user (effectively like a single sign-on).
- Provide directory information (First Name, Last Name, and E-Mail address) for the user.
- Indicate the role of the current user whether

Figure 2. With a very little customization the server side of a mobile learning application can be fully integrated with an LMS

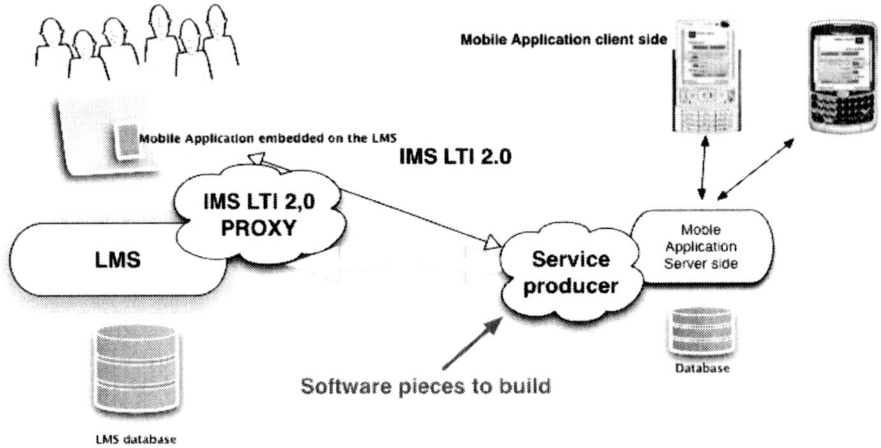

the user is an Administrator, Instructor, or Student.

- Provide information about the current course that the Proxy tool is being executed from such as Course ID and Course Title.
- Provide a unique key for the particular placement of the Proxy Tool.
- Securely provide proof of the shared secret.
- Hints as to display size.
- An optional URL of a resource, which is stored in the LMS – which is being provided to the external tool as part of a launch.

Mobile Learning and Interoperability Standards

Most learning mobile applications consist of a mobile client software complemented with a server side software, often Web based that is usually in charge of managing the users, the contents to be delivered to the mobile client, the pacing of learning activities, assessment and activity logging. Using standards like OKI, IMS AF and IMS LTI the server side software of mobile learning applications can be integrated as a native activity of the LMS, using the users, roles and contexts (courses) defined in the LMS and bound to the academic syllabus. See Figure 2.

Since September of 2008 there are two reference implementations for the IMS LTI 2.0 proxy for Moodle and Sakai, the two major Open Source LMS in the market. Proprietary LMS vendors (like Blackboard and Microsoft) are participating in the IMS community in the definition of the IMS LTI 2.0 standard and have announced the compliance of their LMS with this standard.

A CASE OF STUDY: THE CAMPUS PROJECT

The Campus Project (http://www.campusproject.org), promoted by the Government of Catalonia's Secretaría de Telecomunicacions i Societat de la Informació (STSI), was the initiative of several Catalan universities (including UOC and UPC), which came together to create a virtual open source campus infrastructure. The Campus project had to bind in the same Open Source Enterprise Virtual Learning Environment up to 23 different educational existing applications developed by the project partners. So the responsible of the Campus project decided to use the OSIDs as framework to integrate all these applications in two well known

Open Source LMS: Moodle and Sakai.

Moodle and Sakai implement most of the services described in the OSIDs but they do not comply with the semantics and the interfaces defined by the OSIDs. For this reason the Campus project focused in developing a plug-in for Moodle and Sakai that offers the interface of the OSIDs based on the implementation of the host system. This layer was named the Campus Gateway.

The Campus Gateway implements several OSIDs in their native languages: PHP for Moodle and J2EE for Sakai. Then these OSIDs APIs where transformed in a SOAP Webservices Layer (OKI-BUS) that offers to the Learning Applications developers a framework in their native languages (PHP, J2EE, .NET, etc) to integrate with a Campus Gateway implementation. Thus the integration of a Learning Application can be done with an estimate effort of two weeks of development time (see Figure 3).

The OSIDs framework allows learning systems to share information and services. For example if we have an online collaborative paint application we can use the OSIDs to authenticate the users from a LMS, check if they are authorized agents to act according to a concrete role, to log their activity and even store the drawings in a repository.

If we want this cool collaborative paint application to be really integrated in a Moodle or Sakai course as a native-like module, this would go beyond integration and become what Severance, Hardin & Whyte (2008) call functionality mashup.

To enable the mashup of the Campus applications inside Moodle or Sakai, the Campus team decided to adapt the IMS Learning Tools for Interoperability 1.0 specification combined with the OSIDs. IMS LTI 1.0 covers the needs for installation and launch of remote mashups,

Figure 3. The campus project architecture combining OKI OSID's and IMS LTI has been implemented under Moodle, Sakai and the UOC (http://www.uoc.edu) proprietary LMS, to make available to these LMS up to 24 learning applications (Web and mobile)

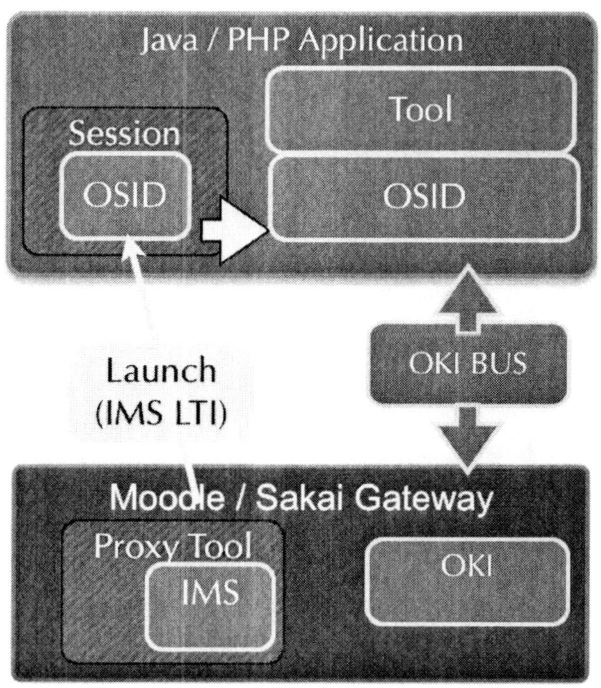

however it does not support runtime operations, which are covered by the OSIDs.

After three years of work the Campus project has succeeded in developing the OKI Gateway for Moodle and Sakai, the OKI Bus and the installation proxy applications (based in IMS LTI 1.0). Up to 23 educational applications have been adapted to be OSIDs consumers and are able to be installed trough the IMS LTI 1.0 proxies in both Moodle and Sakai. All this work is available and licensed under General Public License (GPL) 2.5.

So the first lesson learned is that using sound architectural principles the integration of learning applications is possible. But the Campus Gateways are developed *ad hoc* for concreted versions of Moodle and Sakai. Both LMS have evolved and we should need to modify a lot of code (and test a bit more) in order to have the Gateways working on the last releases.

We are back in the problem of ACME with the project XXX that we discussed before. In order to be effective the developments in interoperability need to remain stable in future versions of the LMS.

DESIGNING A WEBSERVICES ARCHITECTURE FOR MOODLE: MOODLE-DFWSS

Given the experience of the Campus Project, the Moodle lead developer and founder Martin Dougiamas, assigned in early 2008 to the team in UPC (http://www.dfwikilabs.org) the task of developing a new API to access the services of the Moodle core system, with independence of its implementation (in kind of parameters, environment, session etc.) that may remain stable in the following versions of Moodle. This task is described in the Moodle tracker (http://tracker.moodle.org/browse/MDL-12886) and in Moodle Docs (http://docs.moodle.org/en/Development:Web_services). It consists on a set of PHP functions that encapsulate most of the services that an external (and even internal) application shall need from a Moodle server. In October 2008 this Webservices layer has been integrated in the Moodle standard distribution for Moodle 1.9.3 and is going to be the standard interoperability subsystem for the future versions of Moodle.

This layer is intended to be useful for all developers who want to build applications for Moodle, because this development can lead to a documented and stable API to hack into Moodle that should overcome new versions of Moodle.

This API is the base to develop a set of Webservices served by Moodle: Moodle-DFWSs.

Moodle needs to be accessible using any transport protocol present or future. So it cannot depend on a concrete Webservices protocol, name it XML-RCP, SOAP, REST etc. Moodle-DFWSs be implemented in the present version of Moodle (Moodle 1.9) and in the future versions as well (appearing as a core feature in Moodle 2.0 expected early 2009).

Moodle-DFWSs architecture need to be extendable, so each Moodle Module can be a service provider. The proposed architecture consists in 3 layers described in Figure 4:

- **Connect Layer:** Contains the connectors that implement services to local or remote applications. Each component manifests the services implemented using the "*info*" function, and implement the *Connect* and *InOut* components.
- **Integration layer:** This layer consists on THE API (being implemented) that provides a one point access to the Moodle plus *contrib* functionalities.
- **Services Layer:** Is where real things happen. THE API knows how to deal with the Moodle core, and in future posts we will deal on how the activity modules, course formats and plugins can offer their services to the clients.

The Connect layer can implement connectors adjusted, without hacking inside Moodle and

Figure 4. Moodle DFWSs architecture

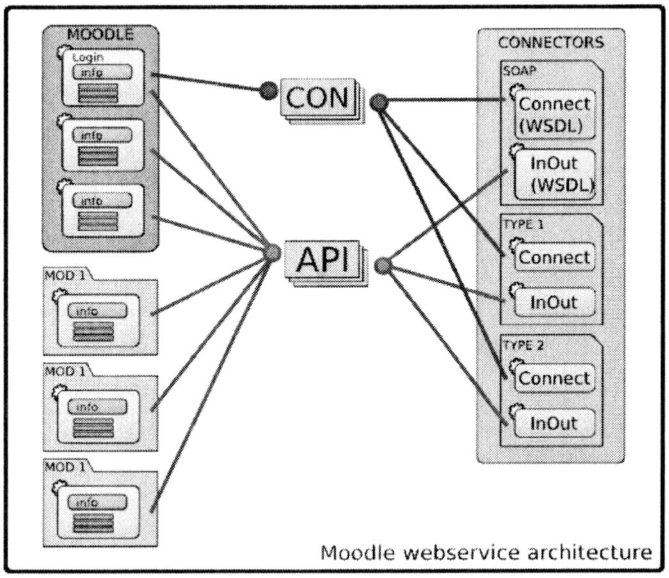

creating code that will survive the new releases of Moodle for some years, to behave according to different standards. One of these connectors will be a Campus Gateway clone, so Moodle will implement the OKI OSIDs v3 on 2009.

MOODBILE: THE MOODLE MOBILE CLIENT

The customization of LMS to be accessible from Web browser to mobile devices, has been an interesting problem, which researchers and Web developers have come up with possible solutions with different degrees of success. But what happens if a student wants to read the latest forum posts while she/he is traveling on the underground train without wireless access? Does the student need to pay to the wireless operator every time she/he wants to access to contents from the virtual campus?. And what if she/he prefers to use her wifi access in the cell phone to get all the data for free while she has free connection and review these data while she/he is on the go, afterwards?.

The point is that the students might want to access the data from the LMS when they are offline and synchronize whenever they want. And this is not possible in a Web-based scenario, which requires being online to work. One possible way to overcome this problem would be the use of Web caching tools or *RSS feed readers*. But the data in the LMS is password protected and many issues can appear, even if we do not consider the security problems. In the best case only data would we available, no services.

For the previous reasons, the UPC team started a research project which leads to the development of a specific mobile client to access the Free Open Source LMS Moodle, which constitutes an extension of this LMS to the mobile scenario: the Moodbile project (http://potato.lsi.upc.edu/projects/moodbile).

On the client side a rich mobile client application with persistent storage capabilities is needed, because we want to provide the user with offline functions when either no network coverage is available or the user does not want to pay for it.

Moodbile is the test drive application developed by the authors and their team in the UPC that implements this kind of mobile client application for Moodle 1.9 (Alier & Casany 2008; Alier, Casany & Casado 2007).

The general system architecture consists of the following parts:

- **The *Moodle* LMS** that runs on the server: Moodle (but other systems such as Sakai could be adapted).
- **The Moodle interoperability extension (Moodle-DFWS)** described in the previous section. This part runs on the same server as Moodle does. Using Webservices as transport implementing both the XML-RPC and SOAP standards. However the mobile client will use only the XML-RPC protocol because -theoretically - will be more efficient in this kind of scenario. The analysis of this issue is material for another eventual research.
- **The *Moodbile* Client**. Through the interoperability layer the *Moodbile* client syncs the data with the *Moodle* server. The mobile user can work offline using the mobile device with the same data he can get through the Web interface. The user can even contribute to the *Moodle* course while offline. All the modifications will be stored in the local database and sent to the server in the next sync.

Moodbile can work online as well as offline. When working online the mobile client application uses the interoperability extension to access the new information originated in the *Moodle* server. This new information is sent to the mobile client and stored persistently for further or offline access. When working offline the mobile user will be able to access the information stored on the mobile device in the last synchronization. The mobile user will also be able to do some update information from the mobile device. When the user updates an activity the changes will be stored locally on the mobile device database and sent to the Moodle server database when the user decides to synchronize.

Considering that the user will use the mobile device to access very specific information about recent events in short connections or extend the learning process on the move, we selected the following activities to be access from the user's mobile device: 1) forums, 2) wiki contents, 3) glossary entries, 4) internal mail messages and 5) calendar from the virtual classroom. The selected activities are the Moodle core activities. In these activities the following tasks were selected to be extended to the mobile client:

- Read posts in a forum and add a specific post to a forum
- Read the wiki contents and add a comment to a wiki page
- Read and answer internal mail messages
- Read the glossary entries
- Synchronize the mobile calendar with the user's calendar on the Moodle server
- Access to the grades and qualification

The previous tasks mainly involve reading information. Updates are very limited especially because mobile phones are not adapted to enter large amounts of data.

For the development of Moodbile the following technologies were selected: J2ME, Android and iPhone. Java 2 MicroEdition (J2ME) technology was the first platform to implement the Moodbile client because it is an open development platform and because it is independent from the hardware and the operating system. Because J2ME lacks persistent storage capabilities specific software for this purpose had to be developed (Casany, Alier & Casado 2007).

Android is a software stack for mobile devices that Google Inc. released on fall 2007. Android includes an operating system (linux likw), middleware and key applications. The Android SDK provides the tools and APIs necessary to begin developing applications on the Android platform using the Java programming language. Android looks like the answer that the mobile developers have been waiting for because it provides solutions to the main drawbacks of J2ME.

In summary, we did not want to develop a full Moodle client able to perform all the tasks performed from the Web interface, because we considered that mobile devices are only suitable for specific tasks of the learning process. Instead, we considered that the mobile device could be useful to do short connections to the Moodle system to access specific information and to do limited updates that do not require large amounts of data entries from the mobile device

CONCLUSION

Moodbile is a foothold of the Moodle LMS inside the m-learning world. It also sets the necessary technological grounds so other m-learning applications or Web 2.0 applications can be integrated with Moodle or other LMS: SOA, OKI OSIDs, IMS LTI 2.0 and related technologies.

Let us imagine a teacher with a group of students that is going to visit the city of Barcelona in Spain. The students carry a GPS and digital camera enabled mobile connected device. When they are close to, let us say, *La Sagrada Familia* temple they have to take a picture and send it to a flickr group tracked by their Moodle course. They receive also the thumbnails of the pictures of other students as they arrive to the server, and they can track the location where the best pictures have been taken. When the teacher decides that there are enough pictures of one kind, he can text all the students with a new assignment. Later they can engage in a discussion about the visit and the history related. They can blog about it and they can go to Google Maps and add their pictures, share the visit to Barcelona with their friends in Facebook. And all these activities can relate to the Moodle course as learning activities because all these systems can interoperate together.

The standards and projects described in this chapter, provide the necessary technologies and development best practices so the mobile application developers can create this kind of mobile systems and get them working inside the current IT infrastructures of the learning organizations, not as something outside of the system.

REFERENCES

Alier, M. (2007). A social constructionist approach to learning communities: Moodle. In M. D. Lytras (Ed.), *Open Source for Knowledge Management: Strategies beyond tools* (pp. 369-381). Hersley: Idea Group Publishing.

Alier, M., & Casany, M. (2008). Moodbile: Extending Moodle to the Mobile on/offline Scenario. In I. Arnedillo & P. Isaías (Ed.), *IADIS International Conference Mobile Learning* (pp.11-18). IADIS PRESS.

Alier, M., Casany, M., & Casado, P. (2007). A Mobile extension to a Web based Moodle virtual classroom. In P. Cunningham & M. Cunningham (Ed.), *Expanding the Knowledge Economy: Issues, Applications, Case Studies, 4*, 1169-1176. IOS Press.

Arina, T. (2008). Learning Zeitgeist: The Future of Education is Just-in-Time, Multidisciplinary, Experimental. In *Masternewmedia*. Retreived July, 20, 2008 from http://www.masternewmedia.org/news/2008/02/13/learning_zeitgeist_the_future_of.htm

Bradley, C., Haynes, R., & Boyle, T. (2005). Adult multimedia learning with PDAs - the user experience. *Proceedings of the 4th World Conference on mobile learning (mLearn)*. Retrieved June 3, 2008, from http://www.mlearn.org.za/CD/papers/bradley-1.pdf

Brown-Martin, G. (2008). It's the learning, stupid! In I. Arnedillo & P. Isaias (Ed.), *IADIS International Conference Mobile Learning* (pp. XVII-XIX). IADIS PRESS.

Cao, Y., Tin, T., McGreal, R., Ally, M., & Coffey, S. (2006). The Athabasca University Mobile Library Project: Increasing the Boundaries of Anytime and Anywhere Learning for Students. *International conference on Wireless Communications and Mobile Computing* (*IWCMC '06)* (pp. 1289-1294).

Casany, M., Alier, M., & Casado, P. (2007). JDMEMicroDB: A new Open Source Lightweight Database Engine for J2ME Mobile Devices. In S. Kim et al. (Ed.), *Multimedia and Ubiquitous Engineering* (pp. 247-252). IEEE Computer Society.

de Aguilera, M., & Mendiz, A. (2003). Video games and education: (education in the face of a "parallel school"). [CIE]. *Computers in Entertainment, 1*, 1–10. doi:10.1145/950566.950583

IMS Global Learning Consortium. IMS Abstract Framework: White Paper. (2003). Retrieved June 10, 2008 from http://www.imsglobal.org/af/index.html

IMS Global Learning Consortium. IMS Tools Interoperability Guidelines. (2006). Retrieved June 10, 2008 from http://www.imsglobal.org/ti/index.html

Kennedy, G., Krause, K., Judd, T., Churchward, A., & Gray, K. (2006). *First Year Students Experiences with Technology: Are they really Digital Natives*? Retrieved May 5, 2008, from http://www.bmu.unimelb.edu.au/research/munatives/natives_report2006.pdf

Merriman, J. (2008). Redefining interoprability. *The Open Knowledge Initiative (OKI*). Retrieved May 5, 2008, from http://www.okiproject.org/view/html/node/2916

Prensky, M. (2001). *Digital Game-Based Learning*. New York, NY: McGraw-Hill.

Prensky, M. (2008). *Mark Prensky writing*. Retrieved Juny 3, 2008 from http://www.marcprensky.com/writing/default.asp

Severance, C., Hardin, J., & Whyte, A. (2008). The coming functionality mashup in personal learning environments. *Interactive Learning Environments (Journal)*, *16*(1), 1–15. doi:10.1080/10494820701772645

Section 3
Innovative Tools

Chapter 12
Using Mobile and Pervasive Technologies to Engage Formal and Informal Learners in Scientific Debate

Dawn Woodgate
University of Bath, UK

Danaë Stanton Fraser
University of Bath, UK

Amanda Gower
BT Innovate, UK

Maxine Glancy
BBC Research & Innovation, UK

Andrew Gower
BT Innovate, UK

Alan Chamberlain
University of Nottingham, UK

Teresa Dillon
Polar Produce, UK

David Crellin
Abington Partners, UK

ABSTRACT

In a climate of concern in the United Kingdom about a perceived loss of interest in science among schoolchildren and the general public, we consider the relationships that exist between science education and public engagement in science, and "formal" and "informal" learning contexts. The authors

DOI: 10.4018/978-1-60566-703-4.ch012

move on to describe four case studies drawn from our research, where mobile technologies have been used in ubiquitous ICT-based science-related learning activities. Three of these studies were of school based activities which took place in timetabled science lesson time. The fourth was set in Kew Gardens in London, during a holiday period, and involved leisure-time visitors of all ages. Finally, they describe a planned integrated trial, which will draw together "formal" and "informal" learners in environmental and scientific debate, scaffolding previous mobile learning experiences towards a genuinely multiplatform e-learning system.

INTRODUCTION

Maintaining school pupils" enthusiasm for STEM subjects (Science, Technology, Engineering and Mathematics) can be problematic. Too often, these subjects are perceived to be more difficult than many of the others on offer, and science in particular often tends to be seen as remote from young people's everyday lives and experiences. There is evidence too, that this ambivalence about science is of a wider nature, extending beyond the classroom to the adult community. This has led to concerns in the UK about levels of what has been termed "scientific literacy" (Bybee 1997; Murphy et al., 2001), and prompted a number of initiatives intended to "engage" people (both schoolchildren and the general public) with science. Promoting a wide-scale interest in science is seen as essential, not only because of the economic need for a workforce equipped with sufficient scientific and technical skills to secure the nation"s competitiveness in the global marketplace, but also because science is an important part of our culture (Osborne & Hennessy, 2003). People who lack a measure of basic scientific knowledge run the risk of being excluded from taking a full part in debates on the social, economic, legal and ethical implications of new scientific and technical developments that affect all of us.

The reasons for this seemingly widespread lack of interest in science amongst the general public are likely to be complex and multidimensional, but one unintentional contributory factor may be the science education system itself. During the early years of primary education in the UK, most young children are enthusiastic about their science lessons. There is at this stage an emphasis on constructivist, "learning by doing" methods, where they are engaged in practical investigative activities. However, by the later primary years and the transition to secondary schooling, there is a move away from constructivist principles towards more factual and theoretical forms of learning, in response to the perceived demands of the National Curriculum and the system of formal assessment linked with it (Hacker & Rowe, 1997; Murphy, 2003; Wadsworth, 2000). This switch of emphasis has been implicated in pupils" disengagement, and changes are currently being implemented in the curriculum to introduce a greater number of practical investigations for older children, and foster in them more of an understanding of how "real" science works.

One way in which curricular changes of this type could be supported is through the use of new technologies. In particular, the potential of emerging mobile technologies has excited a great deal of interest, because of their portability and relatively low cost. These small devices can be used in any classroom, which contrasts with the traditional scenario of expensive desktop computers sited in school IT suites, where access is necessarily limited, due to timetabling demand. Furthermore, mobile technologies can be taken outside for fieldwork, accompany pupils on school trips to museums, or even be taken home to help with homework, thus blurring the boundaries between what have been termed "formal" and "informal" learning contexts.

Our aims in this chapter are firstly to consider

the relationship between "formal" and "informal" learning settings. We will argue that this distinction is not clear cut, and predict that the adoption of emerging mobile technologies for learning will render it still more ambiguous. We will describe four case studies drawn from our research, where mobile technologies have been used in ubiquitous ICT-Based Educational activities. Three of these studies took place in what could broadly be termed "formal" educational settings, in that they were school-based activities which took place in timetabled science lesson time, though in the interests of accuracy, it should be stated that pupils, teachers and technologies moved in and out of the confines of the physical classroom as appropriate to the activities concerned. The fourth was set in an unequivocally "informal" learning context; that of Kew Gardens in London, during a holiday period, where visitors of all ages took part in a series of activities where information normally available in the Gardens was augmented by additional content provided by means of specially configured mobile phones. We will conclude by describing a planned integrated trial, which will draw together "formal" and "informal" learners in environmental and scientific debate, scaffolding previous mobile learning experiences towards a genuinely multiplatform e-learning system. This trial is scheduled to take place towards the end of 2008.

BACKGROUND

As others have suggested (Scanlon et al 2005; Sharples et al., 2005; Traxler, 2005), there is a need to focus less attention upon the mobility of the *technologies* concerned, and more upon that of the *learners*. This is because their mobility has important implications for the organisation of learning. In the traditional model, "formal" learning takes place in specific places and at set times, with teacher and pupils usually co-present. Mobile learning on the contrary, occurs (or can occur) at any time, and takes place across, as well as within specific contexts (Roschelle et al 2005; Sharples 2006). It can also occur remotely. Hartnell-Young (2007) suggested that, even in these relatively early stages of research and implementation, there is a need to consider the effects of the changes in the nature of time and space brought about by mobile learning. In respect of the primary age children to whom Hartnell-Young referred, this is expressed mainly in terms of the relationship between home (parents) and school. With older students, these changes are potentially much broader, to encompass offline friendship groups outside of the family, and contacts made through online social networking, as well as family relationships. This raises the possibility at least, of building learning communities that extend far beyond the confines of the traditional classroom, and challenges the legitimacy of conventional distinctions between "formal" and "informal" learning.

The difficulty in respect of defining what is meant by "formal" and "informal" learning is well known and well documented. For example, does a school trip to a museum count as "formal" or "informal" learning, and is it significantly different from a trip to the same museum organised by parents or a youth group, particularly where the trip is instigated by the interest of a child who has previously visited the facility with her school? Sefton-Green (2004) suggested that the settings in which learning takes place should be thought about in terms of a continuum, from formal *settings*, such as schools and universities, to *social structures* such as friendship groups, and it does indeed seem useful to move away from thinking about this distinction in terms of a dichotomy. In any case, as Scanlon et al (2005) suggest, insufficient work has so far been carried out on the intersection between informal (and "formal" learning for that matter), mobile learning and science for a strict separation to be meaningful. This approach is useful in respect of our own work, which attempts, among other things, to cut across the boundaries between science education, science practice and public engagement in science (Woodgate & Stanton Fraser, 2005, p.48).

Reporting on ubiquitous learning with hand-held computers in schools, Ng & Nicholas (2007) pointed out that learning with mobile devices is in reality "blended" learning. This is because mobile devices tend to have limitations of functionality and computing power. Typically therefore, a range of mobile and other learning materials and technological tools are used together. In the examples we describe below, mobile devices such as phones, GPS, cameras and sensors are used alongside PCs, videoconferencing technologies and the internet, within and across formal and informal learning situations. Our research in schools (some of which is described in the first three case studies below), builds upon a body of work including that of Roy Pea and his colleagues (eg, Edelson et al 1995; Gordin et al., 1994; Gordin et al., 1995; Gordin & Pea, 1995; Pea, 2002). Pea"s team used the technologies available during the early 1990s to show the potential of adapted versions of the types of data visualization tools used by professional scientists, along with communication technologies, to engage and enthuse schoolchildren. This was achieved by facilitating collaboration over dynamically rendered scientific data within individual science classrooms, across schools, and with professional scientists. We have added a personalised and mobile dimension, where children can collect their own scientific data locally, using tailored sensors, sometimes alongside other devices such as mobile phones and cameras. In some instances, the data collection devices have been co-designed with the young users. These mobile technologies are juxtaposed with visualization and collaboration tools to provide a realistic eScience – like experience for school students from the age of around 10 years (Woodgate, & Stanton Fraser, 2005), to help facilitate a hands-on approach to learning science, to aid their understanding, and to motivate and enthuse them.

Our fourth exemplar shows how the wider public too, outside of the classroom situation, can become involved in this type of experience, with a view to promoting learning, discussion and sharing of experiences on science-related topics, in this instance, botany and horticulture. These four studies trace what we believe to be a coherent progression in our thinking on the topic. All involve participants in a range of technology-augmented activities based upon scientific or environmental themes, such as monitoring the local environment using specialized sensors and digital cameras, carrying out (and digitally documenting) environmental improvement projects such as clearing rivers and ponds, or merely recording or commenting upon artefacts in the environment. All of this activity results in user generated content (UGC) of various types; data sets, written comment, audio files, films, still images and posters, which are uploaded to a digital repository so that others can view and comment upon the items. Also, there is often a call to action, encouraging others to contribute their own material to produce a picture of the wider situation. In each case study, we have employed different combinations of tools for data collection, content creation, collaboration and visualization. In the following sections, we briefly describe 4 research projects: The Sense Project, Mobile Phones, and The Schools Trials and Stories@Kew trials which formed part of the Participate project. We will conclude by outlining the integrated study which is planned to bring the Participate project to conclusion, where participation in a range of environmentally themed activities will be possible across mobile phone, internet and digital TV platforms.

CASE STUDY 1. THE SENSE PROJECT: INTRODUCING ESCIENCE TO THE CLASSROOM

SENSE was a collaboration between researchers at the Universities of Nottingham and Sussex, and began to explore the potential of sensor technologies and within- and across -school collaboration on science activities and scientific data, to support a hands-on approach to school science education (Stanton Fraser et al., 2005). A particular emphasis was placed upon promoting understanding of the scientific process, and the use of video to aid children"s understanding of self-collected scientific data in context. The project aimed to initiate and support collaborative activity within individual schools, between different schools and between schools and professional scientists. The focus of inquiry was carbon monoxide (CO) pollution from road traffic, and a series of activities based around this was carried out with pupils at two schools. Firstly, the pupils were encouraged to hypothesize about where this pollution might occur in the areas surrounding their schools, by creating maps and counting traffic from webcam recordings. Some low-tech prototyping was then carried out, where pupils used cardboard and Vaseline to make their own low tech "sensors". These were placed in locations where they had previously hypothesised there would be particularly high or low pollution levels, and after a period of time, the results were examined. Finally, the children helped to design and trial high-tech pollution sensors within their local environment.

The technology consisted of a PDA and pollution sensor. Each school"s sensor was slightly different, reflecting their own design ideas. In the case illustrated in Figure 1, the sensor was coloured differently on each side so that the direction in which the sensor was facing would be evident when the children later inspected the video data of their sensor in use. Groups of pupils captured their own sensor data using these devices, and at the same time videoed the data collection process. Visualization software displayed the data as graphs which ran in time sequence with the video footage, to help them analyse and understand their data. The interface is shown in Figure 2.

They then shared and compared their data across the two schools, using an identical interface. They also discussed their data with a pollution expert remotely. Results of video analysis of the sessions and interviews with the teachers suggest that this context-inclusive approach is significant for three key reasons. Firstly, it allows individuals to reflect upon scientific method as part of the data collection process. Secondly it provides an aide-memoir to groups who have collected data together, in interpreting their results. Thirdly, it allows new participants who have engaged in similar processes elsewhere (or on other occasions) to understand new perspectives on their own and others" data.

This early exploration of the potential of eScience tools and methodologies to engage children in science learning prompted us to take stock of the extent of current and past educational eScience activities in the UK and beyond, to see what we could learn from them. To this end, we carried out a review exercise. At this stage, not only did we find that relatively few examples of hands-on collaborative eScience activities for schools existed, but it was necessary first of all to scope and define exactly what we understood by educational eScience. We have defined eScience in the context of education as: "The use of ICT in education, to enable local and remote communication and collaboration on scientific topics and with scientific data" (Woodgate & Stanton Fraser, 2005). Although most of the projects that featured in our review were schools-based, others, such as the BBC's Springwatch, whose topic was seasonal change, were not specifically confined to schools, but aimed at any interested members of the general public. Again, this suggests a blurring of boundaries between formal science education and informal learning and engagement in scientific topics.

CASE STUDY 2. "MOBILE PHONES IN SCHOOLS"

Returning to the classroom, the "Mobile Phones in Schools" (Towards a National Scale eScience and Education) project took place during late 2005

Figure 1. A group of children and a teacher collecting sensor data. One pupil (second from the left in the group) is capturing video footage of the data collection process.

and early 2006, and was focused around a school-based Participatory Design (PD) exercise aimed at raising awareness of local environmental issues among the young participants, and designing in collaboration with them, an environmental sensor that could be used with a mobile phone. One of our aims was to lay the foundation for a larger project which would explore how a combination of eScience methodologies, mobile and personal technologies could lead to exciting new kinds of educational projects that could involve many schools across the UK. We worked with a class of approximately 30 Year 9 students (aged 13-14 years) and their science teacher at a secondary school in the South West of England, during six of their timetabled science lessons. To set the activity in context, we began by carrying out a series of exercises to familiarise the children with the issue of environmental pollution. Using paper maps of the local area, we brainstormed questions such as: What types of pollution are likely to occur in the area round the school? Where and when would the pollution occur? What might cause it? How would we know it was there? Although a number of potential pollutants were identified, there was particular interest in noise and light pollution, probably because these issues had recently received media coverage.

The second session consisted of a demonstration of datalogging and sensors in the classroom using off the shelf equipment manufactured by a local company called Science Scope. This was followed by a simple hands-on activity where the pupils used the equipment to measure light levels in various parts of the school grounds, and a demonstration of ways in which sensor data can be displayed. During the third session, we carried out a “Bluetooth challenge” (an exercise in using Bluetooth connectivity with mobile phones), and carried out some low tech prototyping activities using craft materials. To introduce this, we went back to the ideas generated during session 1, and asked groups of children to draw on these in designing sensors that could be used with mobile phones. The groups then presented their ideas to the rest of the class.

Figure 2. The SENSE data analysis tool interface with annotated CO graph, space for notes and video of data collection context

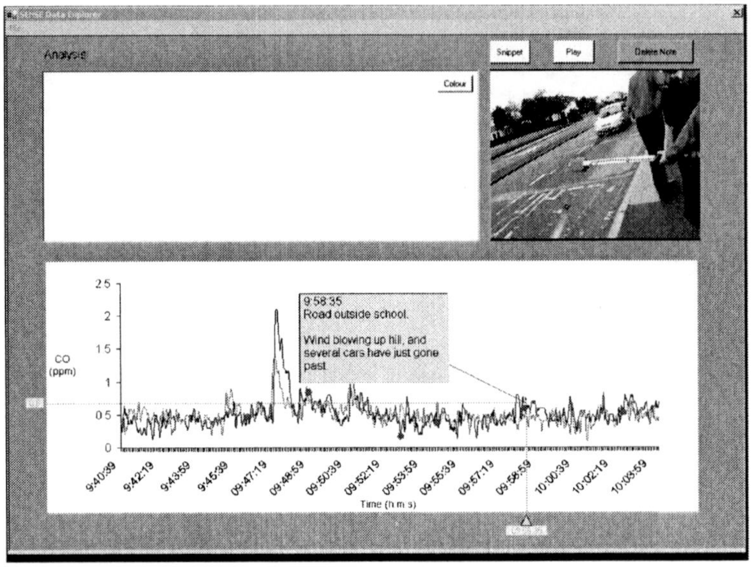

The fourth session took place after an interval of around four weeks, to allow time for the development of a functioning prototype. We started by giving feedback on some of the children"s design ideas. In some cases, to the students" surprise, similar technologies were already in commercial production, though not necessarily available in the UK. We then introduced our prototype. This first iteration comprised software that enabled a Nokia 66 Series mobile phone to connect with a Science Scope Logbook datalogger via Bluetooth, which enabled the phone to be used to collect a range of sensor data within Bluetooth range. The data could then be downloaded to a PC for visualization and analysis. The children tried out temperature, light and velocity sensors using this device. Although they enjoyed trying out the equipment around the school buildings and grounds, they were not particularly impressed by the idea of attaching extra sensors to the phone. Many expressed the view that, although the system might be good for providing fixed sensors in the environment, they questioned its suitability for mobile work. They didn"t like the idea of carrying this quantity of equipment around with them; A typical comment was: *"What's the point of having the phone when you still need all the other stuff?"* We next carried out an interface design session, using paper templates of the mobile phone screen. Pupils were asked to sketch out the screens they would like to see at various stages of the process of using a mobile phone sensor. We then facilitated a class discussion on what had been achieved so far, and ideas for further development.

Figure 3. Students working with phones in the classroom

Figure 4. An example of low tech prototyping output, showing design ideas for mobile sound sensors

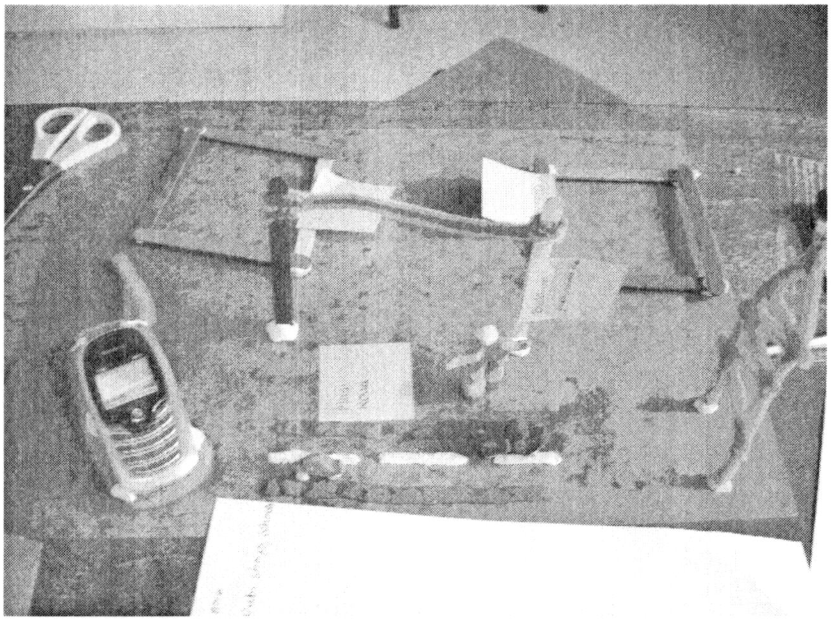

Our final session again took place after some weeks, to allow time to develop a stand-alone sound sensor to work on mobile phone only, using the phone's microphone. Technical information on the design of both of these prototypes is available in Kanjo et al (2007). We demonstrated this second prototype, and tried it out by asking the students to hypothesize whereabouts in the school and grounds it would be more (or less) noisy. Groups were then sent to different parts of the school campus to collect sound data on the phones. Back in class, each group presented their data, displayed as Excel graphs, and told their classmates about the locations and circumstances in which they had been collected. All sessions were videotaped, and all physical artifacts (such as notes, designs and models) were collected to aid our analysis.

The sessions were "quick and dirty" in that only the 1 hour lesson period was available for each. As a result, not all activities were completed. More time would have been extremely useful, but as we were working within the constraints of a real-life school context, we were fortunate to have as much time as we did. We focused initial analysis on the Participatory Design (PD) approach, reflecting upon how this work, carried out in school with a whole class of around 30 students of mixed ability and motivation, relates to much previous PD work with children which has tended to focus upon small numbers of carefully chosen children in a much more controlled, laboratory situation. We concluded that "quick and dirty" studies such as this, carried out "in the wild" in everyday classrooms, are potentially useful as a design technique. Despite problems such as the limited time available and large numbers of students, such studies have value both in terms of generating a lot of ideas quickly, and for the rigorous testing of prototypes of educational technologies in the situation in which their use is intended, particularly if used alongside other methods such as ethnographic studies or more controlled laboratory design sessions.

CASE STUDY 3. THE PARTICIPATE PROJECT SCHOOLS" TRIALS

Following on directly from *Mobile Phones,* and this time involving work with groups both inside and outside of formal education, *Participate* is a large scale collaborative project which aims to use pervasive technologies to inform environmental debate, among groups such as school pupils, computer gamers and community groups. Project participants are encouraged to actively generate their own media (user generated content, or UGC), in the form of scientific data, text, images and video, as opposed to being passive consumers of professionally produced material. Project partners are the Universities of Bath and Nottingham, the BBC, British Telecom, Microsoft Research and Science Scope. The project is still in progress at the time of writing. Initially, Schools, Gaming and Community trials were carried out independently, though there was inevitably some cross-over, with some "schools" studies being carried out with young people outside of the classroom in informal learning contexts. For example, a small trial was carried out at the World Scout Jamboree, which was held in the UK in 2007. Initially, most activities were based around the collection, analysis and visualization of environmental data. More recently, a series of curriculum relevant "missions" for schools has been developed by project team members, or in some cases, contributed by participating teachers. Some of the "missions" continue with the theme of self-collected sensor data, while others are less dependent upon specific technologies, opening up participation in activities based around topics such as energy use, recycling and environmental conservation, to younger age groups (i.e. in primary schools which may not have sensor technologies available), a wide range of abilities, and extracurricular groups.

An early trial was centred around the idea of journeys; the daily journeys that children make between home and school. Classes of 13-15 year old pupils in two schools were loaned a laptop PC with Google Earth™, and Science Scope's Datadisk graphing software installed, and five sets of data collection equipment. These comprised a Science Scope Logbook datalogger with a selection of sensors from which the pupils could choose, and a Nokia 66 series mobile phone with sound sensor software which was a further iteration of that developed under the Mobile Phones project described above. The phone connected via Bluetooth to a GPS unit, the idea being that all the Latitude, Longitude and sound data would be saved in the phone's memory to a time-stamped KML file, which could be displayed as trails on high resolution 3D maps in Google Earth™. Sensor data from the Logbooks were to be displayed separately as conventional line graphs. Disposable cameras and notebooks were also provided. Once pupils had collected and downloaded their data, they then had one or more teacher-led sessions to work with the data.

Pupils took turns to take a set of data collection equipment on their journeys, collecting data as they went, on parameters such as carbon monoxide (CO), sound and temperature. The idea was to produce a snapshot of the conditions that they experienced on a daily basis, to promote discussion about how their personal journeys, whether by car, bus, bike or on foot, impacted on the environment and quality of life locally, and how the environmental conditions that they encountered on their journeys may in turn affect them. The pupils were briefed that the trial would include new technologies that had not previously been tested in schools, and that consequently, they might experience technical problems. The only notable problem, however, was an intermittent loss of connectivity between the phones and the GPS, due to a software issue. Despite this, pupils succeeded in collecting short sequences of simultaneous sound and GPS data with the phones. These sequences were then manipulated by the project team to visualize them as data trails in Google Earth™, showing the sound levels along the routes taken on a 3D map. Data from the Logbooks were downloaded to Science Scope's graphing software (Datadisc Pt), displaying as coloured date and time stamped line graphs. The pupils collected a range of materials during the two week trialling period: a large number of printouts of line graphs showing levels of parameters such as CO, temperature, and light levels; a few sequences of sound data visualized in Google Earth™ ; some photographs and some handwritten notes taken during the data collection.

The pupils were very engaged by the Google Earth™ visualizations. The data trails provoked considerable discussion about the routes taken, and possible causes of the data peaks. They also raised other interesting issues, such as how this type of technology could potentially be used for surveillance purposes, and even possible implications for personal safety if technology of this nature were used inappropriately. An example of a data trail produced in this trial is shown in Figure 5. Perhaps more surprisingly, an almost equally high level of engagement was elicited by the other materials that the pupils had collected, even though these seemed quite bland in comparison to the Google Earth™ visualizations. Despite this, pupils were nevertheless motivated to spend considerable time examining them, attempting to make sense of their results, we suggest because the material was personal to them, and enabled them to reflect upon their own activities (Woodgate et al 2008). Pupils at one school decided to make posters with the "low tech" materials that they had gathered, to record what they had done and display their results.

As a further aid to reflection, BBC colleagues concluded this trial by running a one day "60 second scientist" film-making workshop at each school. Groups of pupils were helped to make 60 second short films centred around the trial activities and findings. Each group was given a topic or question upon which to base their ideas, and shown how to storyboard, shoot and edit their own short film. The day finished with a general viewing of all the films. This activity was intensive and engaging, prompting pupils to reflect upon both their own activities, and environmental issues more generally. They spent a

Figure 5. A data trail in Google Earth™ from the pilot trial

lot of time discussing their experiences, and looking for additional information on their topic. We have included an adapted version of both the poster task and "60 second scientist" in later trials.

Currently, around 15 schools, at varying levels of engagement, are involved in a further, ongoing trial. As before, dataloggers, sensors and GPS are used, though some changes have been made to the technology based upon revised research requirements, feedback from participants, and the need to render the activities more appropriate for the involvement of multiple schools. We have retained the compelling Google Earth™ visualizations, but now Google Maps™ can also be used if preferred. This time, data from the loggers and GPS are downloaded to software called JData3D, produced by members of the project team. This program automatically displays the time and location stamped data as trails in Google Earth™. Additionally, pupils' digital photographs can be incorporated with the data, and opened by clicking on placefinders along the data trails. Examples of both types of visualizations can be seen in Figures 6 and 7.

To support storage and sharing of data, a secure website has been developed within the Participate project (www.participateschools.co.uk). Teachers can control the setting up of pupil groups, access to different areas of the site, and the upload of both data trails, and class work in the form of digital posters and short "films" which are easily created as Microsoft Photostory presentations. Instructions for creating these materials are available on a resources area on the website. Some carefully moderated items are available to view on the site's public page, but most are password protected. The site thus enables the controlled sharing of data and other materials between participating schools, while still maintaining the security and privacy of children's personal data. Observations indicate that the combination of data visualizations and pupil generated material is compelling as tool for learning and sharing, engaging pupils and provoking lively discussion.

Our final example moves beyond the classroom, to engage visitors of all ages with a popular tourist and heritage site, which also has a commitment to education and research. This is the type of facility visited by individuals, interest groups

Figure 6. A Google Maps™ trail to show conductivity along part of the course of a river

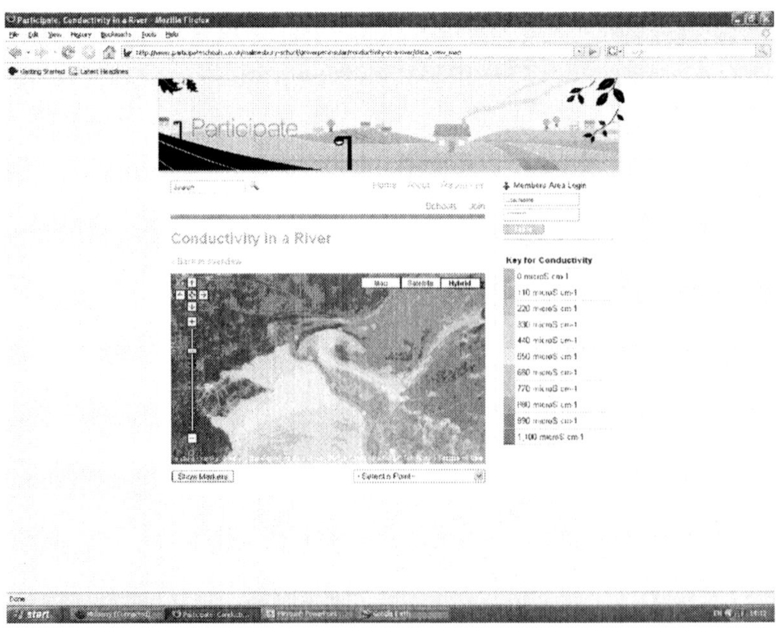

Figure 7. Google Earth™ visualization showing carbon monoxide (CO) levels along a city street, with associated photograph

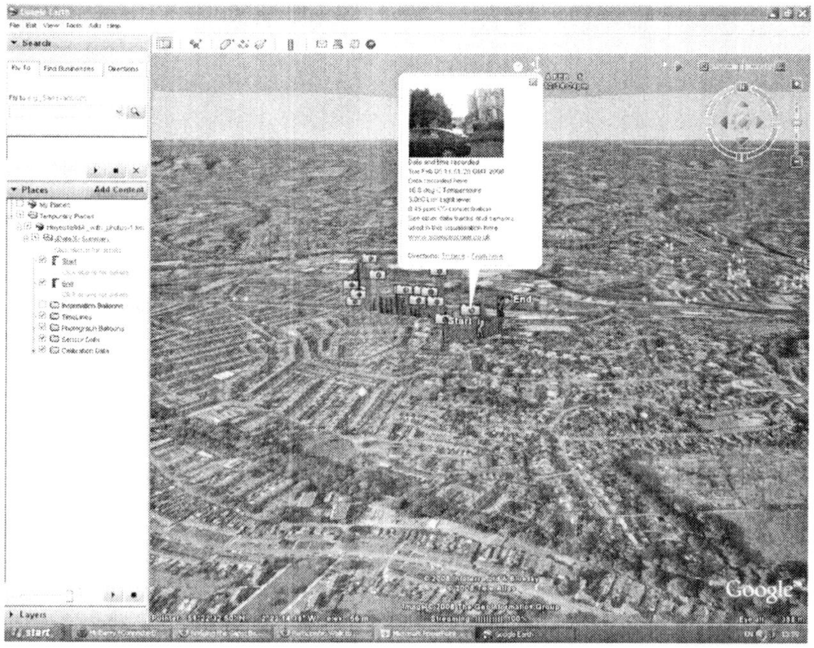

and families, as well as organised parties from schools, colleges and universities. It begins to explore how techniques similar in some respects to those described above, which were trialled in "formal" educational contexts, have the potential also to engage learners at the "informal" end of the spectrum.

STORIES@KEW

Also undertaken within the Participate project, Stories@Kew was a location-based mobile experience which took place over a five day period during the Easter break from the 5-9 April 2007. Stories@Kew enabled the discovery and creation of located content by visitors to the Royal Botanic Gardens, Kew, London, and was led by researchers from the BBC and BT. Within the experience at Kew, specific locations (Points of Interest or POIs) were augmented with "hidden" information (media bundles), as a catalyst to stimulate participants of all ages to record their own contextual stories in video format, which could then be viewed and added to as more people took part. The trial aimed to develop both tools and applications for participatory campaigns or events, and new models for user participation.

To direct participants to locations, two types of location based mobile devices were used during the study, representing both ends of the technology spectrum. The two options provided the opportunity to explore different methods to structure way-finding. At the low-tech end of the spectrum, a physical paper map and signage placed at relevant locations alerted the user to a POI. Users would then key in a number displayed on the signage into a Nokia 6630 mobile phone to unlock the relevant media bundle. At the high-tech end of the spectrum, the second mobile device was a location aware system using GPS tracking for outside locations, and Bluetooth enabled positioning for inside locations. The device concerned was a Nokia N73 paired with a TomTom wireless

GPS MKII dongle. This used an on-screen map, and alerted the user, via a ring tone, vibrate and visual indicator on the map, when a POI was nearby. The media bundle could then be viewed using the menu options. This device also created data logs of when and where users participated, which could be viewed using Google Earth™. All the devices were loaned; for various reasons it was not possible during this trial for users to use their own devices.

To run the trial, a Stories@Kew base was set up within the gardens at a high footfall location called The Orangery. Participants were recruited both on the day and in advance. Some were members of the general public who happened to be visiting Kew, and others came specifically for the experience. They were invited to explore Kew Gardens with one of the mobile device options and discover media bundles virtually located at 34 POIs in the 300 acre environment. The bundles contained a variety of "prompt content" and a specific question which would provoke people to record a response at each point. At 13 of the points. the bundles contained an archive editorial video clip, a text file, occasionally an audio file, a selection of user generated videos and a prompt question. The "editorial" video clips consisted of material from the BBC archives, including clips from the popular "Year at Kew" TV programme, and news items, which provided factual information about features of Kew. These points were accessible using both devices. At an additional 21 locations, bundles contained a recently recorded interview video clip, a selection of user generated videos and a prompt question. The "interview" footage included locally produced video interviews with staff and volunteers at Kew, telling stories about the place, the plants and their memories of Kew especially for the experience. These more widely dispersed points were accessible only by the GPS enabled device. This combination of material served to augment key features of the gardens with contextual information that otherwise would be difficult to obtain.

Once a POI was discovered, participants could view the information provided, and contribute their own content to the location. The prompt question would ask for opinions, thoughts, ideas and stories relevant to the location. The resulting video based user generated content (UGC), once moderated, was placed into the system for other participants to see in context, and also made available on a public display at The Orangery and online at the Stories@Kew website. All content (moderated and unmoderated) was made available within a secure area of the website, which the authors could ac-

Figure 8. BBC's (low tech) interface and map

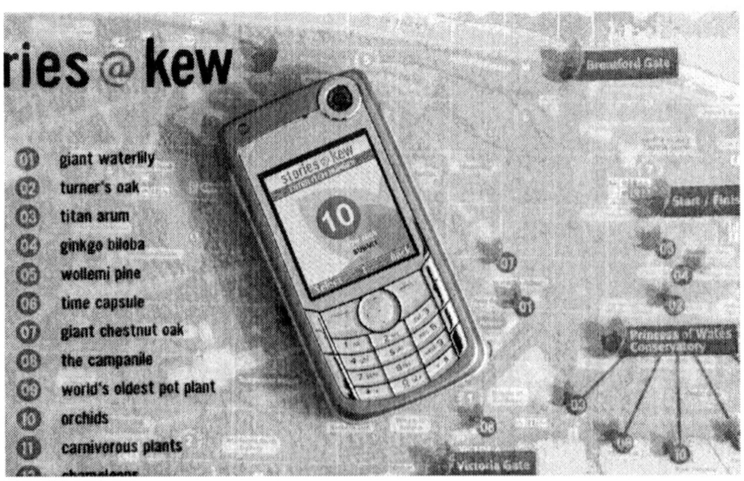

Figure 9. BT's GPS (high tech) interface

cess using a unique personal identification code. The experience was designed to be playful and engaging, to appeal to different age groups, and to be flexible for groups or individuals to take part. The prototype applications developed have the potential to be installed and used in any number of locations and communities.

Participants were of all ages, from as young as 6 years old and the oldest in their seventies, and could take part as individuals, in pairs or in family or friendship groups. Most of them took part in the Stories@Kew experience for periods of 2 hours or more, and on average accessed 8 media bundles. Data from pre- and post test questionnaires and user logs indicated that users found the experience very engaging. Some content items proved more engaging than others. The most popular had one or more of the following features: it was interesting and distinctive, it taught the user something, it had a personal or familiar aspect, it was reflective or touching, unexpected, had a "cliff-hanger", or was funny. Although the experience was not specifically intended as a learning or educational activity, some users spontaneously recalled information they had picked up during their visit, such as facts relating to plant species on display, gardening tips, descriptions of places they had not been aware of previously, and historical anecdotes. Many users wished to extend their experience and their engagement further, requesting that more media bundles be made available around the gardens, and some (particularly Kew members) wanted more in-depth information, while others requested "specialist" information grouped by themes such as botanical, historical, architecture, separate adults" and children's material or alternative language options. It was noted by some parents that their teenage children had later viewed and recalled factual content that otherwise may not have interested them. The playful aspect of discovering POI's and creating video responses was highly motivating for this age-group within the experience. There was a clear tendency for participants to record their videos in the location they were prompted, maintaining a strong contextual link to the prompt content. Post event it is estimated that seventy percent of participants visited the website to view and download their own videos. The desire to extend the experience beyond the day of participation, to share videos with family and friends, and take time to see videos created by other participants was very strong.

THE PARTICIPATE PROJECT INTEGRATED TRIAL: MOVING TOWARDS A MULTI-PLATFORM SYSTEM

We have begun to explore how the use of "blended" mobile and internet technologies can provide an eScience-like learning experience for schoolchildren, increasing motivation and interest in science lessons, and promoting understanding of the scientific process, by giving them an authentic experience of scientific inquiry. Feedback from teachers indicates that streamlined versions of these methods would fit well with the UK's revised National Curriculum. There have also been unforeseen benefits: in one school, it has been reported that the activities have played a key part in the initial training of student teachers on placement, and have also impacted on the Continuing Professional Development (CPD) of newly qualified teachers. Alongside this, using the example of Stories@Kew, we have also shown how related activities can engage people of all ages in "informal" learning contexts.

Material produced by school trials participants, and insights gained by the Stories@Kew study, will contribute to an integrated trial currently in preparation, which will mark the culmination of the *Participate* project. This will take the form of a campaign which will use the construct of a dysfunctional family residing at a fictional address known as "Bicker Manor" as a means to deliver playful and thought provoking "Missions" to participants. Elements from the schools, gaming and community strands of the project will be combined in a multi-platform experience for all ages, based upon the theme of the environment. Participants will be able take part in their own homes, in public venues and on the move, alone or with friends and family, consuming and contributing content as appropriate via the internet, their own mobile phones, and IPTV. Users will sign up to receive "Missions" that cover topics such as energy use, transport and recycling, delivered via their choice of platform. There will also be a pool of additional missions available on a website, from which users can choose if the designated missions do not appeal, or if they want to do additional activities. Some missions will be simple and quick to complete, such as providing the answer to a multiple choice question. Feedback will be provided to respondents in return for their contributions, for example in the form of a tailored response, or a summary of all the contributions so far received. Other missions will require more effort from users, and will vary in the amount and type of input required. Rather than simply rating something or answering a multiple choice question, these missions will typically involve a number of stages, and may require participants to carry out a task or set of related tasks and record the results, creating content in the form of uploaded text, audio, still images or video, which after moderation, will be available for viewing by other participants. The changing dynamics of the Bicker family is a wrapper to this purpose, and will provide closure to the end of the trial. Each member of the family has their own point of view and motivation regarding environmental issues, which is echoed in the different types of missions they provide to participants. However, these apparently divergent missions ultimately "work together" to provide a "big picture" at the end of the campaign. At the end of the trial a personal reflection will be provided to participants, which summarises what they have done in the trial, and provides a global view of the total data collected. These may be presented as graphs, a piece of text commentary or an image as appropriate. This integrated trial will provide an insight into how a multi-platform system might function, as well as how people participate, to inform future design, and provide detailed information on how this type of system could be leveraged for formal and informal learning purposes.

CONCLUSION

In conclusion, we have reviewed some of our past, current and future work with blended technologies across the continuum of formal and informal learning situations, and from quite rigorous curriculum-relevant science learning activities, to popular engagement with environmental themes. In doing so, we have brought to light what we feel are some important insights for research, teaching and learning with these types of technologies. The SENSE project highlighted the importance of providing schoolchildren with information on the context of scientific data collection, to facilitate their understanding of the data's significance. It also demonstrated the potential for eScience methodologies, currently more familiar in "big" science contexts such as physics and genomics than in education, to engage pupils in science learning by providing authentic hands-on activities and adding value by allowing children to collaborate on those activities across schools and with professional scientists, as well as within individual classrooms. The Mobile Phones in Schools work used adapted Participatory Design (PD) methods in an ordinary classroom, to encourage children to reflect upon issues within their local environment, and engage them with science and technology problems. In doing so, we have contributed to debates on Participatory Design (eg, see Druin, 1999; Guha et al., 2005; Scaife et al 1997), and produced early working prototypes of sensor devices based upon mobile phones.

In the Participate Schools Trials we advanced our understanding still further of the importance of contextual information to facilitate children's grasp of the significance of scientific data. Rather than the video footage running in time sequence with graphed data that we used in SENSE, context information in this instance, has ranged from low tech analogue photographs and printouts of graphs, to high-end data trails in Google Earth™ or Google Maps™, which show dynamically the routes taken, and the levels of the parameters measured along the paths followed. Still more contextual information can be provided by means of linked digital photographs, and data trails can now be animated if required. All of this has raised a number of interesting questions about the best type and quantity of contextual information to provide for optimum learning. This will vary according to circumstances such as the age and ability of the students, and the learning topic. Apart from the issue of context, indications are that personalization of the data, and providing interesting activities to help pupils to reflect upon what they have learned, are also significant. Pupils are keen to take ownership of their data, and this appears almost equally true of bland data forms such as line graphs, as of richer material such as high-end computer visualizations. When pupils collect their own data, they are motivated to make a much greater effort to grasp its meaning than they would in the case, for example, of similar material shown in a textbook. Finally, the importance of reflection in learning is well known, and is a key factor in professional training in various disciplines (Schon, 1983; Schon, 1987). Our observations indicate that opportunities for reflection can be provided by various means, such as discussion, within small groups, a whole class, or cross schools, working with and interpreting self-collected data, and creating and sharing user-generated material such as posters and films based upon the activities.

We do not claim that this type of research will directly and immediately improve science teaching and learning, though we do hope that some of the enthusiasm that we have encountered along the way, even in children whom teachers reported as prone to exhibiting disaffected behaviour during science lessons, will have made a small contribution to their ongoing interest in the topics covered. If we genuinely wish to engage and motivate children in science education, whether our intention in doing so is to produce the next generation of scientists, or more prosaically, to ensure that they will be equipped to participate in informed debate on scientific topics, we do feel that work of this kind has the potential to do so. Moving beyond the trialling situation in which we currently find ourselves, any national implementation of such opportunities would require a large scale rethinking of how science classes and ICT facilities in schools are organised. However, it is fair to say that, prompted in some respects by Government initiatives, some progress is already being made on addressing issues of access to technology and its integration into subject teaching.

Other activities carried out within the *Participate* project such as Stories@Kew, broaden our thinking about learning in science to encompass ways in which technology can engage people at the informal end of the learning spectrum. The rigorous demands of the curriculum do not feature here, but the problems of engagement are not dissimilar. Although the focus in this instance is more on using technology to facilitate fun activities and collaboration on popular interests, we believe that these can work usefully alongside scientifically valid classroom- based study to raise awareness and debate on some of the big issues for science and society, and to begin to break down some of the barriers that exist between science practice, science education and public engagement in science.

ACKNOWLEDGMENT

We would like to thank our supporters, the Engineering and Physical Sciences Research Council, the Technology Strategy Board and the Joint Information Systems Committee (JISC) of HEFCE. We acknowledge the contributions of all partners in the Participate project; the Universities of Bath and Nottingham, The BBC, BT, Microsoft Research and ScienceScope. Also those of the researchers at the Universities of Nottingham and Sussex who were involved in the SENSE project. Finally and importantly, we extend our thanks to all the schools, teachers, pupils and members of the public who took part in the studies.

REFERENCES

Bybee, R. W. (1997). *Achieving Scientific Literacy: From Purposes to Practices*. New York: Heinemann.

Druin, A. (1999). Cooperative Inquiry: Developing New Technologies for Children with Children. *Proc. CHI '99* (pp. 592-599). Pittsburgh, PA, USA: ACM

Edelson, D. C., Pea, R. D., & Gomez, L. (1995). Constructivism in the Collaboratory. In B.G. Wilson (Ed.), *Constructivist learning environments: Case studies in instructional design.* Englewood Cliffs, NJ: Educational Technology Publications.

Gordin, D., Polman, J., & Pea, R. D. (1994). The Climate Visualizer: Sense-making through scientific visualization. *Journal of Science Education and Technology, 3*, 203–226. doi:10.1007/BF01575897

Gordin, D. N., Edelson, D., & Pea, R. D. (1995). The Greenhouse Effect Visualizer: A tool for the science classroom. *Proceedings of the Fourth American Meteorological Society Education Symposium.*

Gordin, D. N., & Pea, R. D. (1995). Prospects for scientific visualization as an educational technology. *Journal of the Learning Sciences, 4*(3), 249–279. doi:10.1207/s15327809jls0403_1

Guha, M. L., Druin, A., Chipman, G., Fails, J. A., Simms, S., & Farber, A. (2005, January). Working with Young Children as Technology Design Partners. *Communications of the ACM, 48*(1), 39–42. doi:10.1145/1039539.1039567

Hacker, R. J., & Rowe, M. J. (1997). The impact of National Curriculum development on teaching and learning behaviours. *International Journal of Science Education, 19*(9), 997–1004. doi:10.1080/0950069970190901

Hartnell-Young, E. (2007). Making the Connections: Theory and Practice of Mobile Learning in Schools. In A. Norman & J.Pearce (Eds.), *Proceedings of MLearn Melbourne 2007: Making the Connections* (pp. 87-95). Melbourne: University of Melbourne Press.

Kanjo, E., Benford, S., Paxton, M., Chamberlain, A., Woodgate, D., & Stanton Fraser, D. (2007). MobGeoSen: Facilitating Personal GeoSensor Data Collection and Visualization using Mobile Phones. *Personal and Ubiquitous Computing Journal*. Springer ISSN 1617-4909 (Print) 1617-4917.

Murphy, C., Beggs, J., Hickey, I., O'Meara, J., & Sweeney, J. (2001). National Curriculum: compulsory school science – is it improving scientific literacy? *Educational Research, 43*(2), 189–199. doi:10.1080/00131880010021294

Murphy (2003). *Report 7: Literature Review in Primary Science and ICT*. Bristol.

Ng, W., & Nicholas, H. (2007). Ubiquitous Learning with Handheld Computers in Schools. In A. Norman & J. Pearce (Eds.), *Proceedings of MLearn Melbourne 2007: Making the Connections* (pp.186-193). Melbourne: University of Melbourne Press.

Osborne, J., & Hennessy, S. (2003). *Report 6: Literature Review in Science Education and the Role of ICT: Promise, Problems and Future Directions*. Bristol: Nesta Futurelabs. www.nestafuturelab.org

Pea, R. D. (2002). Learning Science through Collaborative Visualization over the Internet. Nobel Symposium (NS 120), 'Virtual Museums and Public Understanding of Science and Culture. May 26-29 2002, Stockholm, Sweden. http://nobelprize.org/nobel/nobel-foundation/symposia/interdisciplinary/ns120/lectures/pea.pdf

Roschelle, J., Sharples, M., & Chan, T. W. (2005). Introduction to the Special Issue on Wireless and Mobile Technologies in Education. [Blackwell Publishing.]. *Journal of Computer Assisted Learning, 21*, 159–161. doi:10.1111/j.1365-2729.2005.00123.x

Scaife, M., Rogers, Y., Aldrich, F., & Davies, M. (1997, March 22-27). Designing For or Designing With? Informant Design for Interactive Learning Environments. *Proc. CHI '97* (pp. 343-350), Atlanta, GA, USA: ACM.

Scanlon, E., Jones, A., & Waycott, J. (2005). Mobile Technologies: prospects for their use in learning in informal science settings. In A.Jones, A. Kukulska-Hulme, & D. Mwanza (Eds.), *Journal of Interactive Media in Education, special issue on Portable Learning: Experiences with Mobile Devices*. [jime.open.ac.uk/2005/25]

Schon, D. (1983). *Reflective Practitioner*. Basic Books.

Schon, D. (1987). *Educating the Reflective Practitioner: Towards a New Design for Teaching and Learning in the Professions*. Jossey-Bass Inc.

Sefton-Green, J. (2004). *Report 7: Literature Review in Informal Learning with Technology Outside School*. Bristol: Nesta Futurelabs. www.nestafuturelab.org

Sharples, M. (2006). *Big Issues in Mobile Learning: Report of a Workshop by the Kaleidoscope Network of Excellence Mobile Learning Initiative*. University of Nottingham 2006.

Sharples, M., Taylor, J., & Vavoula, G. (2005). Towards a Theory of Mobile Learning. In *Proceedings of mLearn 2005 Conference, 4th World conference on mLearning: Mobile technology: The future of learning in your hands,* Capetown.

Stanton Fraser, D., Smith, H., Tallyn, E., Kirk, D., & Benford, S. Rowland, D., Paxton, M., Price, S., & Fitzpatrick, G. (2005). *The Sense Project: a context-inclusive approach to studying environmental science within and across schools*. www.equator.ac.uk/index.php?module=upload&func-download&fileID-227

Traxler, J. (2005). Mobile Learning: it's here, but what is it? *Interactions, 9*(1). University of Warwick.

Wadsworth, P. (2000). What's happening to science? *Primary Science Review, 61*, 31–32.

Woodgate, D., & Stanton Fraser, D. (2005) *eScience and Education: A Review*. (Review report commissioned by JISC). http://jisc.ac.uk/uploaded_documents/ACF2B4.pdf

Woodgate, D., Stanton Fraser, D., Paxton, M., Crellin, D., Woolard, A., & Dillon, T. (2008, March 23-26). Bringing school science to life: personalization, contextualization and reflection of self collected data. *Proceedings of the Fifth IEEE International Conference on Wireless, Mobile and Ubiquitous Technologies in Education, WMUTE 2008*, Beijing, China. IEEE Computer Society.

Chapter 13
Tools for Students Doing Mobile Fieldwork

Mattias Rost
Göteborg University, Sweden

Lars Erik Holmquist
Swedish Institute of Computer Science, Sweden

ABSTRACT

Students are not always sitting at their desk when learning new things – they are also out in the world. The authors present a set of tools they developed to support groups of students who are doing field studies. Initially, the authors gave the students a Wiki for gathering field notes and their group work material. Based on observations on how they used it and collaborated, they developed additional tools to run along with the Wiki. These include a mobile application for capturing data (photo, video, audio, and text) and automatically uploading to the Wiki, and a set of Web tools which run on top of the Wiki for increasing the awareness between students, and for browsing the captured data. They describe the implementation of these tools and report on the experience from having students using them on their own equipment during the course.

INTRODUCTION

Students often work at a desk, either reading a book or listening to a lecture. But there are also many forms of activities where students are actually out in the real world. When being mobile, it is not always suitable to bring a laptop computer even if they need the capabilities that these devices offer. Instead they inhibit their freedom of movement, and can also serve as an obstacle when interacting with other people at the same time. However, it might be that students are actually out gathering observations and experience about a phenomenon or practice, and therefore need to take notes or capture data which they have to bring back to their desktop for reflection and discussion. This poses various problems.

We report from a course teaching ethnography and design at the IT University of Göteborg, where students work in groups studying a workplace of their choice. They start by getting access to the workplace, and then spend two weeks out in the

DOI: 10.4018/978-1-60566-703-4.ch013

field. During this time they have to take notes and collect data - taking photos, recording videos or audio. The students themselves have reported that just deciding what kind of notebooks to bring into the field is hard (as it may affect how they are treated by the people they observe) (Brown, Lundin, Rost, Lymer, & Holmquist, 2007), suggesting that using a laptop computer is out of the question. At the end of the day however, they need to get what they have found into their computers to be able to share with their friends in the group to analyze the data. We were therefore interested in building tools to support the students in this endeavor.

In previous years we have experimented with having a *Wiki* – easily editable web pages - to support the students. They used the Wiki to type in their field notes, put up their work plans, and upload other material gathered in the field, such as photos and drawings. The Wiki thus served as a group repository, allowing the individuals to collect their own material as well as get access to their group's material (for more details, see (Lymer, Lundin, Brown, Rost, & Holmquist, 2007)). Having your material in one place was highly beneficial compared to having it spread out on the group members' personal computers. Users could access the material anywhere as long as they had access to a web browser, and they could link the material directly to individual Wiki pages and discuss it. Even if the Wiki supported the collaborative aspect of their work it did not support them when they were actually mobile. They still had to type in their notes when they got home, and upload any photos or videos after getting that data of their cameras. We therefore decided to build a mobile tool to easily take photos, record video and audio, and write short notes, and automatically get these into the Wiki.

When we studied the usage of the Wiki, it became apparent that the students found it very beneficial to look at each other's texts, and that they would benefit from an increased knowledge about the others' work - what in the field of computer-supported cooperative work is known as *awareness* (Dourish & Bellotti, 1992). We therefore decided to provide an extension to the Wiki to provide this awareness, that would tell students at a glance what others had been doing, without forcing them to install any special new software.

In this paper we present three addons for Wikis; an awareness extension, a mobile application for capturing data in the field (photo, video, audio, text) and uploading the data to the Wiki, and an extension to the Wiki which let you browse through captured material on the Wiki. When capturing the data we also store where the data is gathered, using *cell IDs*. The cell ID is the ID of the current GSM base station that a mobile phone is communicating with. Thus taking for instance two photos at the same location would result in them both carrying the same cell ID and can therefore later be found together if organized by location.

RELATED WORK

ZoneTag (Ahern, Davis, Eckles, King, Naaman, Nair, et al., 2006) is an application for mobile phones that automatically uploads photos to the photosharing site Flickr (www.flickr.com). It uses cell IDs to tag the photos with location, and to suggest tags that the user might want to use, based on current location and previously used tags. If the location is not known, the user can specify the location on the ZoneTag web site. The location specified will propagate through the network of ZoneTag users so that other photos from the same location (identified by cell ID) will be named. Unlike ZoneTag, our intended use of cell IDs is not to simplify tagging, but rather to simplify the organization of material.

Meneses and Moreira investigated how cell IDs can be used to find a phone's location (Meneses & Moriera, 2006). Instead of just the current cell ID, their algorithm uses a set of last seen cell IDs and their time stamp. In this way they are able to get a more precise location than by just assigning a cell ID with an area, since cells in the network topology will usually overlap. Furthermore, they use this to determine when a phone is stationary and to find familiar locations. We use a similar scheme to determine whether two pieces of data (e.g. photos) were created at the same location or not.

Several researchers have explored how people organize and identify photos. Rodden and Wood (Rodden & Wood, 2003) showed that people find it beneficial to have their collections of digital photos in chronological order, as it is easier to remember when an event occurred relative to other events, rather than to remember its absolute occurrence. Rodden investigated how visual similarities between images can be used to browse through photos (Rodden, 2001). Cooper et al. used time as a way of clustering the images so that the clusters formed events (Cooper, Foote, Girgensohn, & Wilcox, 2005).

A number of systems have been presented that visualize awareness in distributed workgroups. For instance, *AwarenessMaps* (Gross, Wirsam, & Graether, 2003) supports awareness by visualizing activities in a web-based shared workspace system. The system consists of two parts. The first is *PeopleMap,* which showed the activities of users. The second is *DocumentMap,* which shows the current status of the content of the workspace. AwarenessMaps only shows activities within the last twenty-four hours; when a document is changed its representation is changed for twenty-four hours and is then changed back. Thus it does not give any sense of the history of the document. Another example is *YeTi* (Yamada, Shingu, Churchill, Nelson, Helfman, & Murphy, 2004), an information sharing system for informal digital sharing over distances, which includes a history view for showing when and how information has been accessed by people at different places. The history view is a timeline showing the time and place where a piece of information has been accessed.

Awareness is also an important issue in software development. This practice usually has a high degree of cooperation and the need to know the work of others is especially important. Storey, Čubranić, and German, (2005) presented a framework for how to evaluate visualization tools that aim to support awareness in software development. One notable system is *Jazz* (Hupfer, Cheng, Ross, & Patterson, 2004) (not to be confused with the zooming graphics toolkit of the same name (Bederson, Meyer, & Lance, 2000)), a software development environment where an existing system, Eclipse (www.eclipse.org) was extended with functions for *contextual collaboration* (Fontana, 2003). The idea was to add functions and tools to the existing environment that the programmers were already using, in order to support collaboration unobtrusively. The added functions included both support for awareness and active communication channels such as chat.

An example of visualizing Wiki activity is *history flow visualizations,* which were used to analyze the evolution of pages in WikiPedia (Viégas, Wattenberg & Dave, 2004). History flow visualizations produce a visual map that shows how a page has been edited and by whom at what time. It was used to analyze the collaboration within WikiPedia and to understand what makes it successful. While history flow visualizations does give a good indication to what has happened to a page historically, it does not convey any information to what is going on in the Wiki as a whole or support awareness of what other contributors are doing.

SYSTEMS

As a starting point we created a Wiki, for which we used the popular Wiki engine *TikiWiki* (tikiWiki.org). TikiWiki supports numerous features in addition to the basic Wiki functionality, including file and image galleries, blogs, discussion boards, etc. Our configuration had the Wiki and the galleries enabled, and allowed comments on Wiki pages. Access was restricted so that users had to log in with a username and password in order to read any text, and the only ones given access were the students and the teachers of the course.

One of the strengths with a Wiki is the accessibility of it. As long as you have a web browser you can access whatever is in it from anywhere. We wanted to incorporate this strength as far as we could, and so we wanted to implement the Wiki-extensions as simple Rich Internet Applications, running inside the web page using standard APIs (with Ajax techniques (Garett, 2005)).

We will now talk about the tools we built.

Awareness Tool

In order to support awareness for groups working in our Wiki, we wanted to design a visual representation of the activity, which clearly showed what had been added or changed, and by whom at what time. Whenever a person creates or edits a page, or uploads an image or a file, it should be visible to anyone else without being intrusive, when they visit the Wiki. In this way users would be able to keep track of each other's work, and follow the progress of the Wiki content.

Design

The result is an interactive zooming graphical timeline at the bottom of each Wiki page, as shown in Figure 1. The timeline is split horizontally in two parts, providing both overview and detail. The bottom part shows the total number of *events* each day for the last thirty days represented as a histogram. An event here is an action within the Wiki, such as creating or editing a page, or uploading an image or a file. The user can zoom in on a time interval by dragging two sliders to choose specific dates. When dragging the sliders to zoom in or out, the visualization is animated in real-time, creating a smooth animation as objects in the upper view gets more spread out or more compact, much like other zoomable interfaces, e.g those created with the zoomable UI toolkit Jazz (Bederson, Meyer, & Lance, 2000) (not to be confused with the software development system of the same name (Hupfer, Cheng, Ross, & Patterson, 2004)).

The upper part of the timeline shows the detail view, i.e. the events in the chosen time interval. The events are represented by short text strings stating which object (Wiki page, image, etc.) is concerned and who caused it. If the same user causes many events on the same object in a short amount of time, they are grouped together. The events are spread out vertically to give more space for events occurring close in time. If the user lets the mouse pointer hover above the text string, a box (similar to a tool-tip) will appear to describe the event in more detail, giving information about exact date, what type of object it is, etc. To go to the corresponding page in the Wiki for the object, the user simply clicks on the text string.

Usage Scenario

To illustrate how the users interact and experience this we present the following scenario. When a user first visits the Wiki he or she is presented with the start page. This is intended to be the starting point of all pages and there should be no orphan pages (pages not linked to). The integrated awareness view then shows all activities within the last thirty days (Figure 2, top). The user can then zoom in to see what happened on a specific day to get more details and a less cluttered view (Figure 2, middle). By clicking on one of the events, for instance the one called 'mitra_artikel' the browser is redirected to the page for the Wiki page named 'mitra_artikel', and the web browser loads the page. The events shown in the awareness module will now only include pages that are accessible from this page (Figure 2, bottom). The events for pages that are outside the scope of 'mitra_artikel' will then disappear. Thus when reading someone's field notes for instance, only

Figure 1. A collaborative Wiki page with our awareness extension visible at the bottom

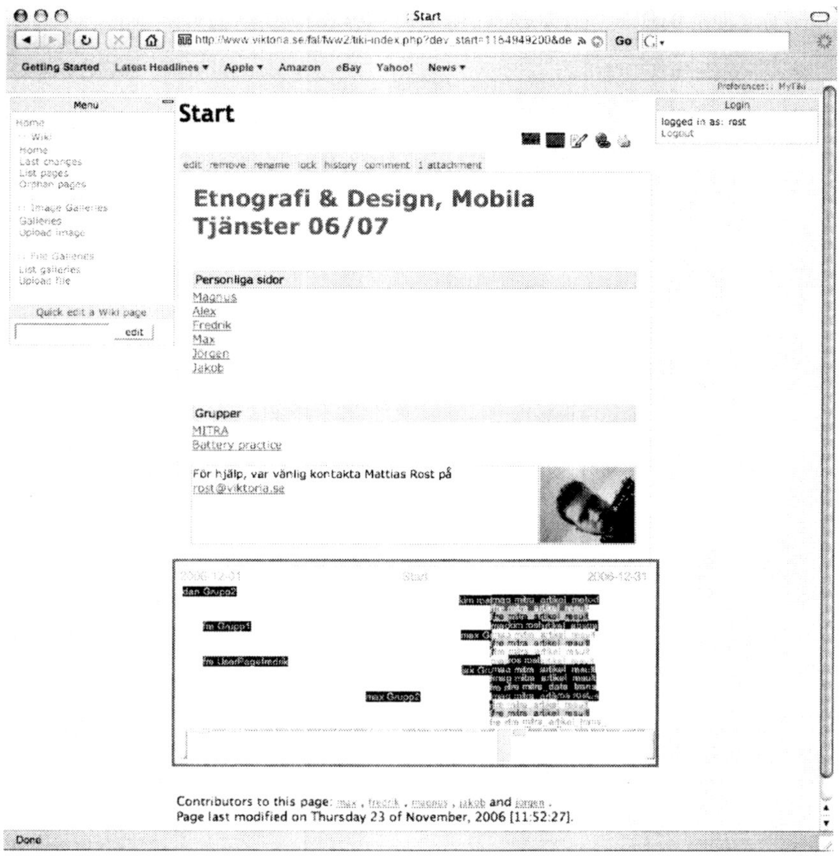

changes done to the field notes will be seen as events in the awareness module.

Implementation

The interactive timeline was implemented using *Ajax* techniques (Garett, 2005). This means that javascript on the client side is used to fetch data asynchronously from the server without having to reload the web page. The resulting application runs in the web browser without the need for any special applications or extra plug-ins, such as Java or Flash. This gives a significant advantage for material that is accessed on-line from several different computers and sites, as the only requirement is a web browser.

In order to render the timeline, the client needs data about the events. The data is fetched with an HTTP GET request. The response is XML-formatted data, which is easy to parse on the client. There are two types of data: *histogram data,* and *event data.* The histogram data gives the number of events for the last thirty days. This is typically only fetched once, when the page is being loaded. The event data is a list of all information about the events within a time interval. This data is fetched when the page is loaded, and whenever the user changes the time window. The server part is implemented in PHP to fetch the data.

Page Structure

All pages on a Wiki are typically on the same level and thus the structure is flat. They are connected through links between the pages. Some Wikis offer namespaces which allow you to structure the Wiki content hierarchically by creating pages within

Figure 2. Example views of the awareness extension. (top) View from the start page. A lot of edits have been made around the 21st of December. (middle) Zoomed in view around the dates shown at the top, and in the timeline. (bottom) The view from the 'mitra_artikel' page, showing only those event that belong in that hierarchy

namespaces and namespaces in other namespaces to create a tree structure of pages. This however forces the content creators to manually organize and manage this structure and can be a burden. The Wiki engine we used, TikiWiki, does not support namespaces and hence the structure of our Wiki was flat. However, in order to only display events for the currently viewed Wiki page and subsequent pages, we had to artificially impose such a structure on the set of pages.

In the course, the students were divided into groups, and each group created their own group page. The group pages would contain links to other pages with more information such as each group member's own field notes, project plans, etc. Thus the way students entered information into the Wiki and linked pages generated an implicit structure. We therefore wanted to find this hierarchical structure to find out which events to show.

The algorithm for finding the hierarchy works by following links in a breadth-first search order starting at the first page of the Wiki. The algorithm works by collecting all links from the first page and puts these in a link queue. It then creates a root node for our tree structure representing the start page, and creates a child node for all links on this page. It then continues with the next link in the queue. By ignoring any subsequent links to pages already visited, no cycles will be formed, and we will get a nice tree structure. This does not necessarily yield the most optimal (or logical) structure, but from experience of how pages are linked together it does give a good enough result.

We used this hierarchy to filter the events shown in the timeline, so that when a user looks at a certain group's page, the only events shown are those that have taken place on this page or pages linked from it and so forth. The result is that users will see only the specific group's activities. But it is also more general such that when looking at a specific report page for instance, a change to any page linked from it (not linked from a page further up the tree) will be shown, recursively.

Mobile Capturing Tool

We wanted to give the students a simple tool for capturing photos, videos, audio, and text when in the field, and also to be able to easily upload that material directly to the Wiki. Since many mobile phones today have all of those capabilities – they have a camera, microphone, a keypad for text input, and also have the treat of being a communication device rendering them able to send data to a server – we decided to write an application for mobile phones as our tool. This would also make the tool useful to students as all students carry mobile phones. During the deployment of the application we therefore made sure the students would be able to use the phones as their primary phone.

Design

The application allows a user to collect photos, videos, texts, and audio recordings (collectively referred to as *data* or data *objects*). The application makes sure that the material is automatically uploaded to the Wiki. Apart from uploading the material automatically we also wanted to save information about the location of where the data was collected. The application therefore stores a set of the last seen cell IDs seen prior to taking for instance a photo. As phones today (having GPRS, UMTS, and WLAN network capabilities) are in practice always capable of uploading data over an internet connection, we wanted to give some control to the user how much data it actually sent over the network as there might be charges for data traffic. We limit this by only automatically uploading data objects below a certain size. We set this size so that photos would always be uploaded instantly, whereas video or audio recordings would only be uploaded if the length is short enough. If the size of the object is too big it will instead be put in a separate queue for large files. In order to have the large files uploaded the user has to manually start the process and select which data connection on the phone to use. The idea here is that the user can upload big files when there is access to WLAN to avoid the cost.

The application has four views, one for each function. The user navigates between the views using left/right on the joystick. The information shown is only the most basic information required, such as how long a video recording is and how much longer it is possible to record before the memory runs out. It is also possible to set the application to be started instead of the regular camera application when pressing the camera button, which enables the user to quickly start the application.

Implementation

The mobile application is built for smart phones running Symbian 3rd Edition operating system. The application was implemented in C++ using Carbide.C++. The targeted phone model was the Nokia E70 which has a 2 megapixel camera, and a folded keyboard which makes it suitable for writing texts (see Figure 3).

Most current mobile phones already have applications for recording video and audio, taking photos, and writing text. Initially, we intended to use the standard built-in applications, and extend them by for instance monitoring the directories where the applications store data, and when discovered upload the file together with the cell IDs.

Figure 3. Mobile application. Writing text (left), and taking a photo (right)

This would make the development more simple and rapid, and would allow the users to use the programs they already knew. However, it proved to be inefficient to start and switch between existing applications and users were often required to wait a considerable amount of time before a needed function was ready for use. Instead, we built an application from scratch containing the four functions of collecting each type of media which significantly optimizes the startup time for each function.

The data is uploaded over a HTTP connection using POST. The information sent, besides the actual object data, is cell ID information, timestamp, and IMEI number (unique identification number for all mobile phones) of the originating phone. In order to improve the results of using the cell ID as a metric of location, a list of cell IDs seen in the last couple of minutes is sent. In order to do this, the application has to track as the cell ID changes over time before the user captures any data. Thus the application actually consists of two programs; one background process which monitors cell movement, i.e. keeps track of when the cell ID of the base station changes, and records this in a database; and one GUI application which exposes the core functionality to the user, capturing the data. The background application also handles the scheduling and uploading of the data objects, which results in that the user does not have to wait until an object is uploaded before another can be captured. Also the uploading can be done in the background even if the GUI application is closed.

A Symbian GUI application is recommended to follow certain architectures, which are supported with different base classes for application logic and GUI components. Our application uses a *view architecture* in which each view or function of an application are separated and easily invoked when needed. This fit well with our intended application, as we wanted four distinct functions.

The background application automatically starts when the phone is powered on, and is always running. It registers a callback to be acknowledged whenever the cell ID changes and stores this. This process also serves as an upload server that attempts to upload anything that is put on queue. The GUI application then issues commands through a custom API to get status information about uploads, and to put data in the queue for uploading.

The two applications communicate using a client/server model where the background application implements the server part and the GUI application issues commands as a client. This is the recommended style for Symbian programming.

Tool for Browsing Captured Data

Although the photos and videos etc. are uploaded automatically to the Wiki, you still need to manage and organize them. The standard way is to have a file gallery where all data is put, and you have to refer to the object in the gallery in some way. The organization within such a gallery is usually very linear and simplistic showing the filenames and the time of upload, and it can be hard to find what you want, especially if there is a lot of data. We therefore wanted to create a different mean of managing the data in the Wiki, by building a browser which shows how the data objects are related to each other. Again we wanted to implement it as a web application to use the strength of being able to use it from any computer.

The web application lets the user browse objects arranged after either *time* or *location*. The two views differ in principle and are therefore explained separately below. The application is built using Ajax techniques in the same way as the Awareness addon and is run on top of the Wiki itself. A button is added to each Wiki page which invokes the Browser. The Browser is then brought up on top of the Wiki page, occluding the Wiki, and can be hidden by hitting a close button to bring back the Wiki.

View by Time

The web application that shows objects on a timeline is shown in Figure 4. The bottom of the view shows a number of bars. There are thirty bars representing the last thirty days, where the height indicates the number of objects collected on that day. This part also contains a narrow horizontal bar indicating the time interval for which the objects are shown above. The objects are arranged horizontally according to time, and spread out vertically to increase the visibility. On the very top of the application is a toolbar of options.

The browser is highly interactive and the aim was to make it easy to navigate through objects

Figure 4. The browser application showing pictures ordered by time

and quickly get an overview of what the collection contains to find objects of interest. This is done by letting the user zoom in and out by simply dragging and clicking with the mouse. In order to change the view, the user can either zoom out (enlarge the time interval) by pressing the right mouse button or a left click while holding the ctrl-button down. In order to zoom in, the user can simply drag the mouse horizontally while holding down the left mouse button and select the time interval to show. Alternatively, by double clicking on one of the bars at the bottom, the view will zoom in on that particular day. The zooming is animated to create a smooth user experience, but the animation can be disabled if desired.

Even though it is very easy to zoom in, it is equally easy to zoom in by mistake or over an incorrect time interval. To mitigate this, the zoom levels are put on a stack when zooming in. When zooming out, the application will first check if the stack is empty, and if not it will zoom out to the zoom level on top of the stack. Thus, to recover from a zoom mistake the user can easily zoom out to the last zoom level. The result is that when browsing large amounts of objects, the user can still find what he/she wants due to the simplicity of zooming in and out over different time intervals.

Object View

By holding the mouse pointer over an object, the name of the object and when it was created is shown. By clicking the object representation, the object view is shown (see Figure 6). In the object view, the object is shown to the right and information about it is shown on the left. For video and audio the object is loaded in a QuickTime plug-in (requires QuickTime to be installed on the computer). This allows the object to be previewed and examined easily, however if lacking QuickTime support this feature will not be accessible. As a way integrate the browser with the Wiki, we added support for adding the viewed object to the currently active Wiki page. Depending on the nature of the object, the object is added in different ways. For an image, video clip, or sound clip, a thumbnail is added with a link to the object in the file gallery. For a text note, the text is simply added to the page. This allows the students to write stories based on and around the material gathered in the field, by simply adding it from the browser, and then write additional text around it.

View by Location

As the collected data objects are tagged with a set of recent cell IDs, they can be related to each other in terms of location. In order to find objects from the same location as another object the view can be changed to a location view (shown in Figure 5). In the location view, thumbnails representing the objects are put in chronological order on the left, scrollable by dragging. By choosing an object (clicking the thumbnail), all objects considered to be captured "close" to it are loaded in a view to the right of it.

There are two ways to go to the location view. The user can switch to the view by pressing a link in the top menu bar. Or, in the object view, the user can click a link in the menu going to the location view and automatically highlight the object in the chronological list, and see all objects from the same location below. In this way it is possible to simply locate a photo, and then find all other photos taken at this location, but at the same time find photos related in time (maybe taken by other team members) as well. (see Figure 6)

Using Cell IDs to Decide Location

For the purpose of the browser, all we need is to be able to decide whether two objects relate to the same location or not. The actual geographical location is of secondary interest and not resolved in our system.

To decide if two objects are associated with the location using only cell IDs is not trivial and the reason for this is twofold. First, cell IDs only give

Figure 5. Location view

coarse-grained details about the location, as cells in the network topology can be rather big. Secondly, different algorithms for jumping between cells are used by network operators and a phone is seldom connected to one base station for too long, even when the phone is stationary. This means that if the phone is near three base stations when at a particular location, the current base station when capturing one object may differ from when capturing another object at a later time but at the same location, and hence also the cell ID. Care must therefore be taken when dealing with cell IDs as location.

We use a similar algorithm as (Meneses & Moriera, 2006). By, instead of comparing just the most recent cell ID, comparing a sequence of the last seen cell IDs for two objects using a similarity measure, we get a value for how similar the locations are. Defining a threshold value for how similar two such sequences must be, allows us to make a decision on whether they are the same location or not. When looking for objects collected at the same place as another object, the similarity is simply calculated for all objects and the ones above the threshold are presented as the nearby objects.

Figure 6. Object view

USER EXPERIENCE

The awareness visualization was designed based on our studies of the students' work practices during the first year the Wiki was used. The second time the course was run (with a new group of students) we installed the awareness visualization and the browser extensions, and gave them each a mobile phone (to be used as their ordinary phone). Two student groups used the Wiki and mobile phones to upload ethnographic field notes and other material.

As for the awareness visualization we found that the students mainly used it at the moment when they logged in, from the front page, to quickly get aware of what had happened since they last logged in. However, they rarely used when they had navigated deeper into the Wiki. This is probably because the number of pages they created (and also the number of students) was significantly lower, and their history could easily be grasped in the timeline on the front page of the Wiki. Thus after looking at the awareness module from the front page they already had a good sense of what had happened since last time in the Wiki, and where the most action had taken place. This might not have been the case had there been more students, and more pages created (and hence more activity in the Wiki).

They mainly used the phone application for taking photos and some videos, and the browser was used to review, find, and discuss photos and videos that they needed for their analysis. They reported that they found the mobile phone and application a very simple but powerful and helpful tool and appreciated the way material automatically got to the Wiki without having to manually upload it. They also liked how it was visualized in the browser in the Wiki, however they did not use the location view at all. The reason for this is most probably because of the way they collected the data, such that the timeline served as an implicit location divider, as they knew where they had been at different times. One important note here is that they always went out in groups, and never individually, meaning that all group members' data were gathered over the same time interval, and at the same location.

FUTURE TRENDS

Since our deployment of the tools presented we have started to work toward even more accessibility of the tools. A Java application similar to the mobile tool has been written which includes most functionality, however due to restrictions in the Java APIs it cannot track the cell ID. The Wiki has been made more accessible from mobile devices as well by making a version especially adapted for mobile screens, with the ability to upload phone data directly from the web browser in the phone.

The tools we have built and used are to a high degree run on the students own equipment, equipment that the students bring with them anyway. Laptop computers are as common for the students as are their mobile phones, and they bring the mobile phone where ever they go. Therefore we believe one possible direction for research in mobile learning, is to see how the students own technology can be used in education, as a contrast to the common agenda of bringing technology to the classroom.

DISCUSSION AND CONCLUSION

We have presented an awareness tool, a mobile capturing tool, and a data browser, to aid students who are doing mobile fieldwork. The tools were deployed to students in a course at a local university. A major feature of this work is that the stationary applications actually run inside the web browser using Ajax techniques, and that the capturing tool runs on their mobile phones. It thus does not require any extra effort from the students besides learning to use the tools, rather than having to install and manage special software on their computers, or carrying around extra equipment when in the field – something that can be seen as intrusive by the people they observe.

We believe that there will be a widespread use of mobile devices to create and share collections of digital media. This project represents one approach to organizing such collections, and turned out to be useful in an educational setting, when students created and analyzed field notes and data captured in the field. At the same time as the prices for data charges are dropping and flat rate data plans are becoming more common, we can see an increasing bandwidth capacity, with transfer speeds higher than most people had in their homes just a couple of years ago. In the future, we hope to generalize our application to other application areas, to support emerging practices such as mobile photo-blogging and other user-created content.

ACKNOWLEDGMENT

Thanks to our colleagues Johan Lundin, Barry Brown and Gustav Lymer. We also want to thank Kim Grahn Dahlberg for assisting in development. This work was part of the iDeas project, a collaboration with Stanford University supported by the Wallenberg Global Learning Network.

REFERENCES

Ahern, S., Davis, M., Eckles, D., King, S., Naaman, M., Nair, R., et al. (2006, September). ZoneTag: Designing context-aware mobile media Capture to increase participation. Paper presented at *Workshop on Pervasive Image Capture and Sharing at the Eight International Conference on Ubiquitous Computing*, Orange County, CA.

Bederson, B., Meyer, J., & Lance, G. (2000) Jazz: An extensible zoomable user interface graphics toolkit in Java. In *Proceedings of the 13th Annual ACM Symposium on User Interface Software and Technology* (pp. 171-180). ACM press.

Brown, B., Lundin, J., Rost, M., Lymer, G., & Holmquist, L. E. (2007) Seeing ethnographically: Teaching ethnography as part of CSCW. In L. J. Bannon, I. Wagner, C. Gutwin, R. Harper, & K. Schmidt (Ed.), *Proceedings of the 10th European Conference on Computer-Supported Cooperative Work* (pp. 411-430). Springer.

Cooper, M., Foote, J., Girgensohn, A., & Wilcox, L. (2005) Temporal event clustering for digital photo collections. In *ACM Transactions on Multimedia Computing, Communications, and Applications* (pp. 269-288). ACM press.

Dourish, P., & Bellotti, V. (1992) Awareness and Coordination in Shared Workspaces. In *Proceedings of the 1992 ACM conference on Computer-supported cooperative work* (pp. 107-114). ACM press.

Fontana, J. (2003). *Collaborative software ages slowly*. Retrieved July 30, 2008, from http://www.networkworld.com

Garett, J. J. (2005). *Ajax: A new approach to Web applications*. Retrieved on July 30, 2008, from http://www.adaptivepath.com/ideas/essays/archives/000385.php

Gross, T., Wirsam, W., & Graether, W. (2003). AwarenessMaps: Visualizing awareness in shared workspaces. In *CHI '03 Extended abstracts on Human factors in computing systems* (pp. 784-785). ACM press.

Hupfer, S., Cheng, L., Ross, S., & Patterson, S. (2004) Introducing collaboration into an application development environment. In *Proceedings of the 2004 ACM conference on Computer supported cooperative work* (pp. 21-24). ACM press.

Lymer, G., Lundin, J., Brown, B., Rost, M., & Holmquist, L. E. (2007). Web based platforms in co-located practice: The use of a Wiki as support for learning and instruction. In *Proceedings of Computer Supported Collaborative Learning 2007*. Lawrence Erlbaum Associates, p.

Meneses, F., & Moriera, A. (2006) Using GSM CellID positioning for place discovering. Presented at *First Workshop on Location Based Services for Health Care in the International Conference on Pervasive Computing Technologies for Healthcare, (Locare '06)* Innsbruck, Austria.

Rodden, K. (2001). *Evaluating similarity-based visualizations as interfaces for image browsing.* Unpublished doctorial dissertation, University of Cambridge Computer Laboratory.

Rodden, K., & Wood, K. (2003). How do people manage their digital photographs? In *Proceedings of the SIGCHI conference on Human factors in computing systems* (pp. 409-416). ACM press.

Storey, M.-A., Čubranić, D., & German, D. (2005). On the use of visualization to support awareness of human activities in software development: A survey and a framework. In *Proceedings of the 2005 ACM symposium on Software visualization* (pp. 193-202). ACM press.

Viégas, F., Wattenberg, M., & Dave, K. (2004). Studying cooperation and conflict between author with history flow visualizations. In *Proceedings of the SIGCHI conference on Human factors in computing systems* (pp. 575-582). ACM press.

Yamada, T., Shingu, J., Churchill, E., Nelson, L., Helfman, J., & Murphy, P. (2004) Who Cares? Reflecting who is reading what on distributed community bulletin boards. In *Proceedings of the 17th annual ACM symposium on User interface software and technology* (pp. 109-118). ACM press.

Chapter 14
SMART
Stop-Motion Animation and Reviewing Tool

Peter Byrne
Trinity College, Ireland

Brendan Tangney
Trinity College, Ireland

ABSTRACT

Animation shares many of the educational advantages of digital video production. However, both activities can be time consuming, are non-trivial to implement as whole class activities and there are aspects of the process that are not well scaffolded by currently available software tools. The design, implementation, and evaluation of a mobile learning application called the Stop-Motion Animation and Reviewing Tool (SMART) are described. The application enables users to create animations on a mobile phone and is part of a larger generic suite of open-system software we are developing to facilitate the development of cross platform applications in the area of digital narrative production.

INTRODUCTION

Digital video production can provide many opportunities for learning (Buckingham, 2003; Buckingham, Harvey, & Sefton-Green, 1999; Hofer & Swan, 2005; Kearney & Schuck, 2005; Posner, Baecker, & Homer, 1997; Reid, Burn, & Parker, 2002). Animation is a related, yet simpler, activity that shares many of the educational advantages of digital video production (Madden, Chung, & Dawson, 2008). However, both activities can be time consuming, involve using a diversity of devices and are non-trivial to implement as whole class activities. This chapter advocates developing a dedicated application for mobile phones that uses the cameras, communications facilities, and ready-at-hand nature of mobile phones to help overcome these problems.

The specific focus of this chapter is the design, implementation, and evaluation of a mobile learning application called the Stop-Motion Animation and Reviewing Tool (SMART). The application enables mobile phone users to create animations using the stop-motion animation technique. SMART adheres to the constructionist, collaborative, contextualized and constructivist approach to developing learning

DOI: 10.4018/978-1-60566-703-4.ch014

applications for mobile devices argued for by (Patten, Arnedillo Sánchez, & Tangney, 2006). SMART allows users to capture images, sequence the images using a filmstrip paradigm, insert title cards, and view the completed movie, all on the mobile phone. From a technical perspective, an XML document represents the animations, which can be transferred to a PC with Bluetooth for further editing by third-party applications if required. SMART is part of a larger generic suite of open-system software, called Mobile Unified Storytelling Environment (MUSE) (P. Byrne, Arnedillo Sánchez, & Tangney, 2008), which we are developing to facilitate the development of cross platform applications in the area of digital narrative production. MUSE includes a middleware that implements a service-oriented architecture, which provides a reliable platform to support collaborative applications on mobile phones, PCs and the internet. MUSE includes several services to support digital narrative production, including services to generate video files from still images and sound. The Digital Narrative Tool (DNT) (Arnedillo-Sánchez, 2008) is a tool, built on MUSE, to support users creating digital narratives. The DNT includes shared workspaces on both the PC and the mobile phones comprising collaborative concept-mapping tools to scaffold the digital narrative process, and a collaborative timeline to edit the digital narrative.

SMART is evaluated according to the framework described by (Sharples, Lonsdale, Meek, Rudman, & Vavoula, 2007), and further expanded on in (Vavoula, 2007; Vavoula & Sharples, 2008), which advocates evaluating mobile learning projects according to three levels of granularity, the micro level (usability), the meso level (educational), and the macro level (organisational). This evaluation will focus on the micro and meso levels from this framework, with the macro level being outside the scope of the research. At the micro level, the question asked is it possible to design an application to allow users to create animations on mobile phones? Further questions examine the usability and utility of the application. At the meso level the question asked is does the application enable constructionist, collaborative, contextualized and constructivist approaches to learning?

The current trend in mobile and software development is towards generic open-systems that use the service-oriented architecture paradigm. This chapter concludes by acknowledging this trend, and considers the advantages of integrating SMART to use MUSE and interoperate with the DNT.

BACKGROUND

There is a growing body of evidence in the literature that digital video production can facilitate powerful learning experiences. Digital video projects support collaborative learning, problem solving, critical thinking, and creativity, while encouraging development of media literacy, communication, and presentations skills e.g. (Buckingham, 2003; Buckingham et al., 1999; Hofer et al., 2005; Kearney et al., 2005; Posner et al., 1997; Reid et al., 2002; Swain, Sharpe, & Dawson, 2003). Furthermore, digital video presents opportunities for student-centered, inquiry–based projects (Hofer et al., 2005). Digital video production and animation, and more generally moving image media, are familiar even to preschool children (Marsh & Thompson, 2001) and "learning activities which incorporate them may help to connect school life with the wider world" (Madden et al., 2008) (p. 901).

Animation is an analogous process to digital video production that shares many of the potential educational advantages while being a simpler activity. Collin et al. consider animation a subset of video, of which they recognise three such divisions: live action; animation; and talking heads, e.g. face-to-face video conferencing (Collins, Neville, & Bielaczyc, 2000). The important distinction between live action and animation is that live action records real life events as they occur whereas animation creates the illusion of life and motion from static frames. This allows users to depict scenes not possible with live action, e.g. making objects appear to move by themselves or illustrate dynamic or scientific processes like blood moving around the body (Collins et al., 2000).

Earlier studies of animation for learning largely focused on studying how viewing animations of dynamic systems helped users to understand complex systems. For example, one study of the use of animation on student understanding of computer algorithms (M. D. Byrne, Catrambone, & Stasko, 1999) notes that their "experiments show a trend towards a benefit of animations and predictions on students' ability to solve procedural problems about algorithms". On the other hand, recent research has supported interactive and constructionist approaches to using animation. (Tatar, Roschelle, Vahey, & R.Penuel, 2003) describe a project to animate scientific processes dynamically using Sketchy, an animation and drawing tool for PDAs. An advantage of this system is that users engage in a constructivist process, for instance exchanging drawings and animations to uncover misunderstandings of scientific phenomena (Tatar et al., 2003). In a different subject domain (Zagal, Piper, & Brukman, 2004) use animation to support storytelling by children aged 11 – 12 using software called Alice. Alice is a 3D programming environment and teaching tool for introductory computing that enables users to tell stories using animation. To control the animation, and on-screen characters, the users drag-and-drop graphical objects, which represent statements in a programming language. While (Zagal et al., 2004) describe this animation activity as a success they note that a supportive social context is important for children to become authors of multimedia in an educational context, e.g. collaborative skills are necessary as is providing structure to scaffold the animation activity.

There are several different animation techniques, for instance, cell animation or computer animation; however this study uses the stop-motion animation technique because it is "concrete and easy to approach for the beginner" while supporting development of additional skills including hand-eye coordination (Hämäläinen, Lindholm, Nykänen, & Höysniemi, 2004). Traditional stop motion animation involves shooting a movie one frame at a time, changing drawings of characters slightly between each, thereby creating the impression of movement. Stop-motion animation is not limited to drawings, with variations of the technique using clay models, Lego® bricks, everyday household objects, and people. Animaatiokone (Hämäläinen et al., 2004) is a system for creating clay animation and learning about stop-motion animation. The Animaatiokone installation consists of a desk to stage the animations, a mounted camera to capture the images, and a mounted screen to view the animations and timeline. In addition, Animaatiokone supports collaboration by allowing users to share clay models and extend previous users' animations. One limitation is that the Animaatiokone installation is large and fixed to one location therefore animations can only include objects and drawings that can fit into the machine.

Digital video production and animation are not without their problems. Although the cost of digital cameras continues to decrease they are not yet ubiquitous devices and more importantly animation and digital video production are time consuming activities (Burden & Kuechel, 2004). Using a digital camera means that images have to be transferred to a desktop machine and loaded into another application to create the final animation. For a whole class activity access to the desktop may prove to be an issue, images are processed in a different physical location to which they were captured and the learner is required to master two pieces of technology. Carrying out the image capture and animation editing on the single mobile phone device means that the ready-at-hand nature of the technology is being exploited, only a single application needs to be mastered and where a large number of learners are involved it is much easier for different groups to work in parallel. Furthermore, the in-built communication facilities of a phone mean that the learners can easily exchange the images, and completed animations.

Mobile learning or mobile computer supported collaborative learning (MCSCL) is an emerging area and while there is "no single overarching theory of mobile learning" (Naismith, Lonsdale, Vavoula, & Sharples, 2004) several taxonomies have been put forward for classifying mobile learning applications, e.g. (Roschelle, 2003) and (Naismith et al., 2004). The functional-pedagogical framework for mobile learning proposed by Patten et al. suggests the best examples of mobile technology for learning are informed by constructionist, collaborative, contextualised and constructivist learning theories (Patten et al., 2006).

SMART is evaluated according to the framework described by (Sharples et al., 2007; Vavoula, 2007; Vavoula et al., 2008) which advocates evaluating mobile learning projects according to three levels of granularity, the micro level, the meso level, and the macro level. The micro level examines activities of the users and assesses usability and utility of the technology used. The meso level examines the learning experience of the activity and technologies used. The macro level relates to the longer-term impact of the technology on educational and learning practice.

This framework was developed in the context of the MyArtSpace (Sharples et al., 2007; Vavoula, Meek, Sharples, Lonsdale, & Rudman, 2006) project, which is an attempt to support structured inquiry learning with mobile technology to connect learning in the classroom with learning in museums. MyArtSpace enables users to take pictures, record sound, and write comments using the supplied multimedia mobile phones in order to reflect upon and share their experiences upon returning to the classroom. In addition, this framework was used by (Spikol, 2007) for evaluating their mobile game "Skattjakt", a collaborative treasure hunt game using mobile phones with GPS to navigate through the game. They found that this framework helped to identify problems, understand the learning processes, and identify further requirements. This evaluation will focus on the micro and meso levels from this framework, with the macro level being outside the scope of the research.

Synthesising what has just been said about the potential advantages of digital video and animation production as vehicles for learning and the approach to developing mobile learning applications advocated by (Patten et al., 2006), this chapter describes the design, implementation and initial evaluation of The Stop-Motion Animation and Reviewing Tool (SMART), an application for creating animations on mobile phones.

STOP-MOTION ANIMATION AND REVIEWING TOOL

Animation supported by mobile devices is an activity that lends itself to the argument of (Patten et al., 2006) that an MCSCL tool should encourage elements of a constructionist, collaborative, contextualized and constructivist approach to learning. Animation is an inherently constructionist activity with the learner required to create scenes and characters at the physical level and the actual animation itself at a higher level of abstraction. The activity lends itself to collaboration since there are easily separated tasks, which different participants can undertake. Requiring the learners to display their finished animation to their peers and to reflect on the product, and the process used to create it, promotes a constructivist approach to learning. A level of contextualization is achieved through the choice of topic the animation addresses and the mobility facilitated by the phones, which enables the users to incorporate elements from their surroundings in their animations.

Mobile phones are ready-at-hand devices (Soloway, Norris, Blumenfeld, Fishman, Krajcik, & Max, 2001) which users typically have with them at all times, the users are familiar and feel comfortable using them (Geser, 2004) and mobile

Figure 1. SMART graphical user interface

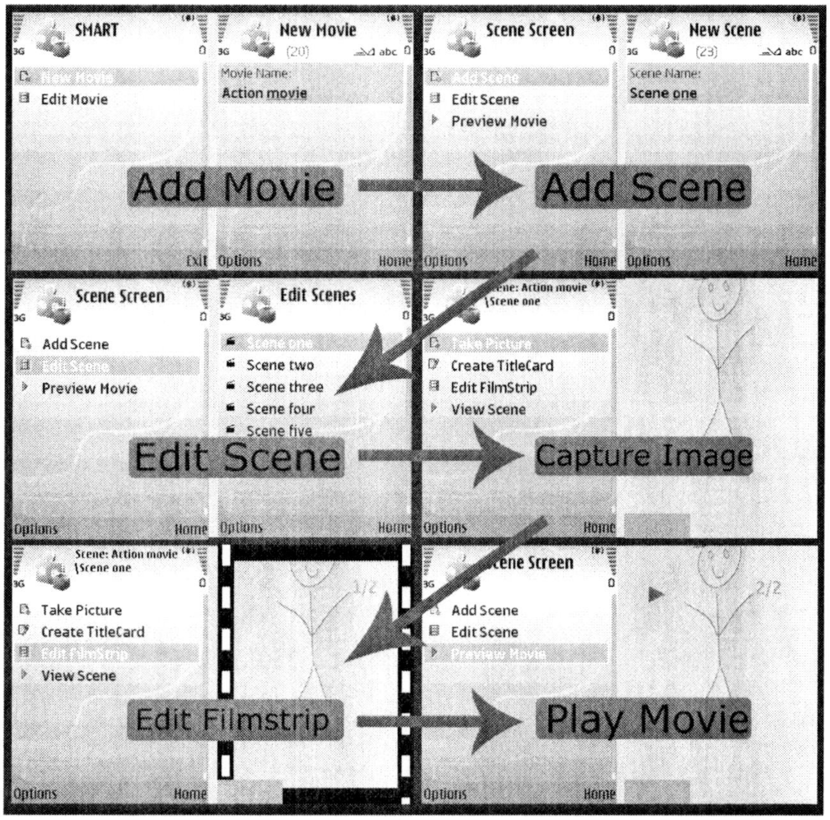

phones generally include support for J2ME. By using J2ME, it is possible to create sophisticated applications that use the multi-media, image capture and communications features of the devices. In addition, the communications facilities enable users to collaborate and exchange animations, thereby supporting a collaborative, and constructivist approach to animation.

The Application

SMART, supports the shooting of small animated-movies using the stop-motion animation technique. A user can capture frames, containing images from their surroundings, drawings or clay models (created by themselves or others) and additional real world objects. Several frames form a scene, and an animated movie can contain numerous scenes. When users select a scene to edit, the application displays a filmstrip. Users can use the filmstrip screen to add, reorder, or delete frames. Similarly, users can add, reorder, or delete scenes. They can then review their work by playing the full movie or previewing individual scenes. SMART also includes a facility to add text frames to the filmstrip, which is a concept borrowed from silent movies and helps the users tell the story.

Described here is a typical interaction and usage of the system from starting a project to creating the final movie. Initially the users in a group create a simple storyboard, typically containing three to six story elements, to represent the story for the animation. Then they create and gather the artefacts to use in the animation, for example, drawing images, or sculpting clay models. The users then switch to SMART to cre-

Figure 2. Process for using SMART

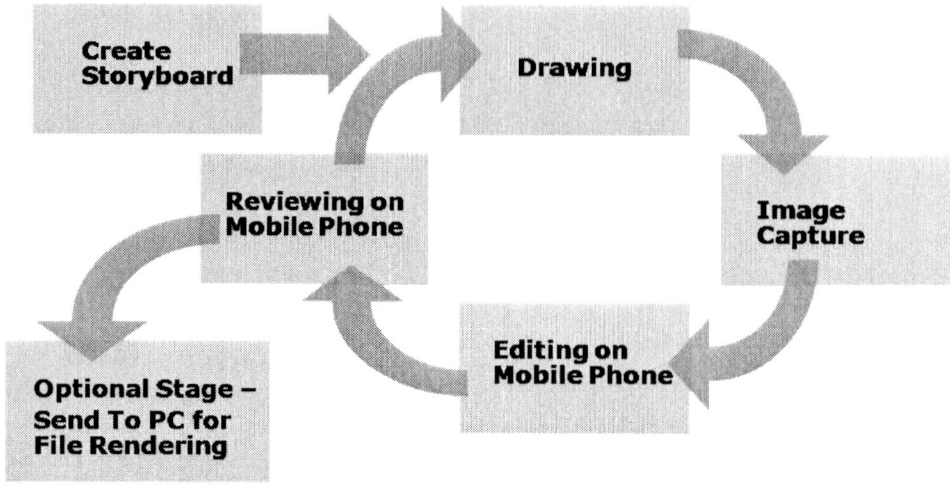

ate the animation. First, the users create a new movie, the top-level unit in SMART. The users then create a number of scenes that will comprise the completed animation. Once a scene is selected a series of frames for the scene can be captured using the integrated camera. When the users are happy that they have captured the appropriate frames, they can edit the scene, reordering or removing frames as desired. The users can then preview the scene or view the complete movie on the mobile device. When the movie is complete, users have the option of sending the movie to a server where it can be rendered into a format that allows it to be distributed to non-SMART users. In addition, on the server the users can add sound to the animation if desired.

The application was developed on Sony-Ericsson™ K750s, Nokia™ N73s, and Nokia™ N95s mobile phones, all of which have integrated cameras and support Bluetooth, MMS, and J2ME. SMART can be ported to any other similar mobile phones. SMART is deployed as a J2ME MIDlet using the Connected Limited Device Configuration 1.1 (CLDC) and the Media Information Device Profile 2.0 (MIDP). In addition, SMART requires access to the Java APIs for Bluetooth, the PDA optional packages and the Wireless Messaging API 2.0. SMART uses XML to represent the animations. XML is used as it is an open format and third-party developers can easily parse the XML to create tools to manipulate the animations, for example, adding a soundtrack.

However, the J2ME platform has several technical constraints, which include a lack of memory and processing power for multimedia manipulation, the restrictions of the J2ME security model, and difficulties ensuring that applications are portable on multiple mobile phones.

Currently the processing power and memory limitations on small mobile devices are a constraint on performing advanced image and video editing tasks. Additionally, the lack of support for the Java Media Framework APIs is an obstacle to converting a series of JPEG images into a portable movie file format. There is also a deficit of third party tools supporting file conversion operations on mobile devices. Therefore, if the user wishes to convert the completed animation into a portable movie file format, for example 3GP, they must send it to a server for further processing. 3GP is a multimedia container format defined by the Third Generation Partnership Project (3GGPP) for use on 3G mobile phones. As this is a standard file format, its use supports portability of the application and allows users who do not have J2ME, or SMART, installed on their mobile phone to view the animations. In addition, it was impossible to implement an onion-skinning function on the mobile phones because they lacked the processing power, memory, and advanced media manipulation facilities. Onion skinning is an animation technique that displays the current frame overlaid on the previous frames to help the animator to align the next frame.

Java MIDlets use a sandbox model similar to Java Applets and therefore require that the application be signed with a valid digital certificate before permission to use many for the functions of the phone can be granted, e.g. local file system access, camera functions, and MMS facilities. This is a practical barrier to deployment because each phone model and each mobile network operator can require a different set of valid certificates. Each certificate costs between €200 and €400 per year, so it can be expensive to buy the correct certificates for wide deployment on users' mobile phones.

The graphical user interface (GUI) for the application (see Figure 1) is built predominantly using the high-level API of the MIDP. This has the advantage that as the GUI components are defined in a device independent manner it simplifies porting the application to other mobile devices. A supplementary advantage is that menus and buttons in the application will operate using the same interface paradigm as the native interface on the mobile device, thereby simplifying the task of learning to use the application. However, some of the unique requirements of the application necessitated creating specific interfaces using the low-level Java API of the MIDP, for instance, the filmstrip screen. These APIs control drawing directly to the display on the phone. The user interface design and development must consider this when dealing with the varying display sizes on different mobile phone models

Evaluation Setting

A social outreach programme run within our university provides the setting for the evaluation of SMART described here. The programme provides students (typically 16 years of age) from designated disadvantaged second-level schools with an innovative technology-mediated learning experience. Volunteer third-level students provide the mentoring for the programme.

The configuration, furnishings, and decoration of the learning space, promote a creative and collaborative learning environment. The space contains two sections; the first half contains a display space and open-plan layout with movable desks and chairs to support group work, brainstorming, discussion, and presentation of project work. The second half of the space contains six *"pods"*, which are bays surrounding a computer and desk-space to support small groups working around a PC.

The animation activity that is the focus of this evaluation contains four stages. In the first stage, the mentors introduce the activity and software. Then the participants break-up into groups of two or three people. In the second stage, each group creates a storyboard on paper, outlining the story elements that will form the basis of their animation. They create the drawings, backgrounds and other artefacts that they will need to tell their story and create the animations using SMART. During the third stage, the participants upload their animations to a server in order to present their animations to all the participants. The fourth and final stage is a discussion involving the participants, allowing them the chance to reflect on the activity.

This evaluation workshop described here lasted for three hours at the end of which participants were asked to complete a short questionnaire. The questions fell into three broad categories that examined the participants' technology and multimedia background, their reflection on the activity, and solicited their feedback on the application. All participants completed the survey.

The group in question consisted of fourteen participants between the ages of thirteen and sixteen, 4 of whom were female and 10 of whom were male. The participants worked in four self selected groups containing three participants and one group containing two participants. Each group was supplied with a Nokia N73 mobile phone with SMART installed on it.

The participants were familiar with technology, spending on average over eleven hours per week using computers. Typical activities they engaged in included YouTube™ (thirteen participants); playing games (eleven participants); using social networking websites (nine participants); editing photographs (four participants) and editing movies (two participants). From a mobile technology perspective, all participants had their own mobile phone with built-in camera, while twelve participants had a phone potentially able to run SMART.

To ascertain if they had engaged in similar activities before the participants were asked if they had created animations or digital videos previously, thirteen of the participants had recorded digital videos while seven participants had created animations on their PC. The most popular activity with digital videos was to share the video with friends (six participants) or share them online (two participants uploaded them to YouTube).

Results

As previously mentioned the evaluation follows the framework for evaluation described by (Sharples et al., 2007; Vavoula, 2007; Vavoula et al., 2008) and focuses on the micro and meso levels.

Taking the micro level first, this section discusses the usability of SMART. Questions at this level include, were the participants able to create animations, what did they like and dislike about the application under investigation and were there any problems with the application.

Firstly, the application did prove usable. All five groups were able to produce a short animation ranging from 10-15 seconds long during the short workshop session. When asked *"what did you like about the animation activity today?"* seven of the participants liked the hands-on artistic aspect to the activity, for example making the characters for the movies with modelling clay. Five of the participants mentioned that it was fun, and one participant liked the novelty of the activity. Two of the participants liked the fact it was simple while two others liked the social aspect of working in teams of the activity. Finally, one participant enjoyed the storytelling aspect.

In related responses to the question "how *would you improve the software?"*, three participants suggested making the application faster with two being so specific as to suggest that more memory should be added to the phones to improve the software. A common complaint from the participants was that the final animations were moving too quickly. SMART fixes the default frame-rate at four frames a second. This proved too fast in practice and resulted in jumpy animations as the participants took too few images for this frame rate. While four frames a second can produce smooth animations, it requires many images to advance the story. Therefore, to improve SMART the frame rate should be configurable from the SMART user-interface, allowing the users to experiment with different frame rates to suit their animation.

At the meso level, the focus was upon the degree to which constructionism, collaboration, contextualisation, and constructivism were encouraged by the activity. The users were observed engaging in a constructionist process by creating the physical objects for the animation and creating the actual animations themselves. All of the groups managed to produce an animated movie during the three hours of the activity. The users were observed collaborating to create the objects, the storyboard, and the movie. In addition, while creating the animation, the users often employed a division-of-labour strategy, with one user holding the phone to capture the images, while the other users moved the characters in the movie. Again when one of the mentors asked, *"What are you working on?"* the participant replied, *"He is doing the characters and I'm doing the storyboard".*

The setting of this evaluation provided opportunities for users to contextualise their work by using tools and artefacts available in their surroundings to improve their animations. There were computers and printers available in the room and three of the groups took advantage of these facilities to draw characters and backgrounds that they could print out and use in their animations.

In order to gauge the participants' reflections on their movies and the animation activity two further questions were asked, *"If you had more time, how would you improve your movie?"* and *"if you were starting again with a new movie, how would you make it better?"* Eleven responses advocate making better drawings and characters. Six responses suggest either more planning from the start or better storyboards would improve the animations. Both sets of responses show that the participants came to realize if they gave more consideration to the story to be told, before any artefacts or animations were created, it would have resulted in an improved animated story. Several of the movies that the participants created were not very smooth and contained jumpy animation. Three of the responses indicate cognisance of that fact, as they recommend adding more frames. Similarly, one response suggests reshooting the movie. As the movies were typically 10-15 seconds in duration, four responses favour creating longer movies if they had another chance at this activity.

This workshop, and others we have carried out, indicates that the use of SMART does lead to successful learning experiences and offers a practical way in which mobile technology can be used to support animation as a whole class activity.

Future Trends

Generic systems have the potential to overcome the limited reuse potential of specific tailored systems and the general trend in computer supported collaborative systems (CSCL) has been towards generic systems (Lonchamp, 2006). Dimitracopoulou, while describing current trends for the design of collaborative learning systems, mentions that it is "important to provide flexible architectures and customisable tools" (Dimitracopoulou, 2005). In addition, current trends challenge software developers "to provide a uniform and integrated user experience across the desktop, web, and mobile platforms" (p. 51) (Bosch, Friedrichs, Jung, Helbig, & Scherdin, 2007). The architectural goal is to minimize the platform-specific parts of the software while maximizing the commonalities across the platforms, with both parts separated by simple interfaces. (Bosch et al., 2007) recommend a service-oriented architecture (SOA) as a technical solution to achieving this architectural goal. A SOA consists of loosely coupled reusable components and provides a flexible, reconfigurable platform on which to develop applications. Service-oriented architectures provide a logical way to design software systems that provide services to end-user applications or services distributed over a network (Papazoglou, Traverso, Dustdar, & Leymann, 2007).

There have been some previous attempts to bridge the gap between users on PCs and mobile devices in mobile learning. For instance, MoULe (Arrigo, Di Giuseppe, Fulantelli, Gentile, Novara, Seta et al., 2007) is a Mobile and Ubiquitous Learning system, which enables users edit and share location based documents, concept-maps and wiki pages using PCs and mobile phones with GPS and built-in cameras. In addition, the system includes a learning management system so that users can access these documents online using moodle software. Another relevant study (Hwang, Hsu, & Huang, 2007) describes an application called "StudentPartner", an integrated multimedia forum, which allows users to capture media files with a mobile device and upload them to a shared database that users access through PCs and mobile devices in order to create a shared discussion forum.

SMART is a standalone mobile application, which the users can use to create and view animations all on the mobile device, however, communication between the devices and the server is achieved manually using Bluetooth. The principle drawback of such a manual process it that it is not transparent to the users. With the result that it is difficult for the users to generate the finished 3GP files, or take advantage of the facilities for adding sound to the movies without assistance or technical support.

To overcome this drawback and to improve the reliability and functionality of the application SMART needs access to better networking functionality and to integrate better with third party tools. Furthermore, given the trend in software development towards open-systems, we have developed a reliable platform called the Mobile Unified Storytelling Environment (MUSE) (P. Byrne et al., 2008) that uses the SOA paradigm to provide a simple API on which to build mobile computer supported collaborative applications in the area of digital narrative production. It is designed to provide support to applications operating in a heterogeneous technical environment containing mobile phones, PCs, and differing networking technologies. MUSE uses the Java platform and XML is the main data-representation format. For communication over different networks, it contains a transparent network layer, abstracting from the client applications the differences between TCP/IP, HTTP, Bluetooth, and MMS. The MUSE middleware includes several services to support collaboration and digital narrative production in general, including a service to managing data-structures representing, concept-maps, timelines, a service to enable text-chat, a service to generate movie files from still images and sound, and a service to manage user groups.

The Digital Narrative Tool (Arnedillo-Sánchez, 2008) (DNT) is the first application to make use of the services exported by MUSE. The tool provides integrated facilities that include a collaborative story script, shared storyboard, timeline-editor, and communication tools to help users maintain a shared understanding during the creation of a digital narrative. The performance and functionality of the tool changes as users move from high performance PCs to mobile devices, replacing the rich graphical and communication support on PCs with alternative communications tools and simpler graphical user interfaces on mobile devices.

It is relevant to ask if integrating future versions of SMART to use MUSE as a middleware platform and to use the DNT to edit the animations would improve the software and further support learning. According to the original design considerations for SMART, the software should support a constructionist, collaborative, contextualized and constructivist approach to learning. This section will take each design consideration in turn and discuss the benefits of adopting the MUSE open architecture to provide common support across the desktop, web, and mobile platforms and using the DNT shared workspaces to edit the animations.

If SMART used MUSE to communicate with the DNT application then the DNT GUI would provide an interface for editing animations on the PC, in addition to the current mobile only interface. This configuration would support and simplify further 'post-production' of the animations, which is not technically feasible on mobile devices now. The first activity that would benefit from this approach is the ability to simplify adding and editing a soundtrack to the animation. This additional facility would enhance the finished animation while continuing to support a constructionist approach to animation.

Increasing opportunities for collaboration is an advantage of using an open system like MUSE. While a central benefit of SMART is that the application is self-contained on the mobile device, supporting stop-motion animation from start to finish all on the device, additional collaboration involving MUSE and DNT application on the PC would provide new opportunities for collaboration. Firstly, the users could split up with some users creating characters for the animation and capturing images, while other users would use the DNT to edit the animation. A second option is to increase collaboration by enabling users on multiple PCs to edit the animation. This is possible as the DNT contains shared-workspaces that allow users on separate PCs to edit the filmstrip simultaneously.

Integration with MUSE and the DNT would present further options for the users to contextualise their work, as there would be a more diverse range of platforms from which the users could interact with the animation. Besides the DNT editor would enable the users to include not only elements from their surrounding in their animations but previous animations and other media.

MUSE can send any images captured by SMART automatically to the DNT editor if required. This enables faster feedback between users creating the animations and users editing the animation on the PC. This feedback together with the other communications facilities, including text chatting tools, available within MUSE enables the users to engage a constructivist process including the negotiation of meaning between the collaborators, even if they are not co-located.

As mentioned earlier, SMART operates as a standalone application with images and files manually transferred to a PC to generate 3GP movie files if required. However due to the limitation of this process SMART is currently being integrated to work with the DNT via the MUSE platform. In this case, the DNT GUI will provide an interface for editing animations on the PC, in addition to the current mobile only interface. This configuration supports further 'post-production' of the animations, which is not technically feasible on mobile devices now. In particular, this gives the ability to add and edit a soundtrack for the animation. The DNT's shared workspaces support additional levels of collaboration. Therefore, integrating SMART with MUSE and the DNT could benefit and enhance each of the pedagogical underpinnings of SMART. However, SMART should also retain the ability to operate as a stand-a-lone application on a mobile phone to satisfy an original design goal to support animation without the need for access to computers.

CONCLUSION

This chapter argues that stop motion animation can provide the potential educational advantages of digital video production while also being a simpler and less expensive activity in which to engage. However, both digital video production and animation are time-consuming activity with problems regarding access to expensive equipment. This chapter recommended using mobile technology to overcome these issues. Therefore, this chapter described SMART, an application designed to support a collaborative, contextual, constructionist and constructivist approach to making animations on mobile phones. A noted advantage of which is that as the application is implemented on a mobile phone it has, among others, the benefit of being relatively inexpensive, can exploit the ready-at-hand nature of the device, and it is a familiar technology.

This chapter presented the results from a small user study, with some recommendations for improving the software. The participants in the study were familiar with technology and most of them had mobile phones capable of running SMART. The evaluation used the framework for evaluation described by (Sharples et al., 2007), and further expanded on in (Vavoula, 2007; Vavoula et al., 2008), which advocates evaluating mobile learning projects according to three levels of granularity, the micro level, the meso level, and the macro level. At the micro level (usability level of the framework), the question asked is it possible to design an application to allow users to create animations on mobile phones? Further questions examine the usability and utility of the application. The application proved usable, as all the participants were able to produce animations during the three-hour animation activity. However, the study revealed one usability problem with SMART because the participants complained that the final animations were moving too quickly. The fixed frame rate of SMART was too fast resulting in jumpy animations as the participants took too few images for this frame rate. Therefore, to improve SMART, the frame rate should be configurable from the SMART user-interface, allowing the users to experiment with different frame rates to suit their animation. At the meso level (the educational level of the framework), this study observed the participants engaged in constructionist, collaborative, contextualized and constructivist approaches to learning. The macro level is concerned with institutional and organisational implications of mobile technology and is outside the scope of this study.

In addition, this chapter noted the trend in software development towards open-systems and introduced a open-system called MUSE (P. Byrne et al., 2008) that we are developing to facilitate the development of cross platform applications in the area of digital narrative production. The chapter noted some current limitations with SMART and described how integrating SMART with MUSE and the DNT would provide additional opportunities to support constructionist, collaborative, contextual, and constructivist approaches to animation. Furthermore, the participants liked to share multimedia with their friends, witness their usage of social networking websites and YouTube. In addition, when they previously created digital videos or animations, the most popular activity was to share it with friends. Therefore, the extra collaboration support provided by MUSE would support users who wish share the finished animations in future.

REFERENCES

Arnedillo-Sánchez, I. (2008). The mobile digital narrative tool. In [Lisbon, Portugal.]. *Proceedings of IADIS Mobile Learning, 2008*, 77–83.

Arrigo, M., Di Giuseppe, O., Fulantelli, G., Gentile, M., Novara, G., Seta, L., et al. (2007, October 16-19). A Collaborative mlearning environment. In *Proceedings of mLearn* (pp. 13-21). Melbourne, Australia.

Bosch, J., Friedrichs, S., Jung, S., Helbig, J., & Scherdin, A. (2007). Service orientation in the enterprise. *Computer*, *40*(11), 51–56. doi:10.1109/MC.2007.399

Buckingham, D. (2003). *Media education: Literacy, learning and contemporary culture.* Cambridge, UK: Polity Press.

Buckingham, D., Harvey, I., & Sefton-Green, J. (1999). The difference is digital: Digital technology and student media production. *Convergence: The Journal of Research into New Media Technologies*, *5*(4), 10–21. doi:10.1177/135485659900500402

Burden, K., & Kuechel, T. (2004). *Evaluation report of the Teaching and Learning with Digital Video Assets Pilot 2003-2004: Becta.* Retrieved from http://www.becta.org.uk/page_documents/research/evaluation_dv_assets03.pdf.

Byrne, M. D., Catrambone, R., & Stasko, J. T. (1999). Evaluating animations as student aids in learning computer algorithms. *Computers & Education*, *33*(4), 253–278. doi:10.1016/S0360-1315(99)00023-8

Byrne, P., Arnedillo Sánchez, I., & Tangney, B. (2008, October). A mobile computer supported collaborative learning tool for digital narrative pProduction. In *Proceedings of the MLearn2008 Conference: The Bridge From Text To Context* (pp. 66-73). UK: University of Wolverhampton.

Collins, A., Neville, P., & Bielaczyc, K. (2000). The role of different mMedia in designing learning environments. *International Journal of Artificial Intelligence in Education, 11*, 144–162.

Dimitracopoulou, A. (2005). Designing collaborative learning systems: Current trends & future research agenda. In *Proceedings of Computer Support for Collaborative Learning: Learning 2005: The Next 10 Years*! (pp. 115-124). Taipei, Taiwan.

Geser, H. (2004). *Towards a sociological theory of the mobile phone.* (University of Zurich Release 3.0).

Hämäläinen, P., Lindholm, M., Nykänen, A., & Höysniemi, J. (2004, April). Animaatiokone: An installation for creating clay animation. In *Proceedings of CHI* (pp. 24-29). Vienna, Austria.

Hofer, M., & Swan, K. O. (2005). Digital moviemaking—The harmonization of technology, pedagogy and content. *International Journal of Technology in Teaching and Learning & Leading with Technology, 1*(2), 102–110.

Hwang, W.-Y., Hsu, J.-L., & Huang, H.-J. (2007, October 16-19). A study on ubiquitous computer supported collaborative learning with hybrid mobile discussion forum. In *Proceedings of mLearn* (pp. 101-109), Melbourne, Australia.

Kearney, M., & Schuck, S. (2005). Students in the director's seat: Teaching and learning with student-generated video. In *Proceedings of World Conference on Educational Multimedia, Hypermedia and Telecommunications* (pp. 2864-2871), Montreal, Canada.

Lonchamp, J. (2006). Supporting synchronous collaborative learning: A generic, multi-dimensional model. *International Journal of Computer-Supported Collaborative Learning, 1*(2), 247–276. doi:10.1007/s11412-006-8996-7

Madden, M., Chung, P. W. H., & Dawson, C. W. (2008). The effect of a computer-based cartooning tool on children's cartoons and written stories. *Computers & Education, 51*(2), 900–925. doi:10.1016/j.compedu.2007.09.008

Marsh, J., & Thompson, P. (2001). Parental involvement in literacy development: using media texts. *Journal of Research in Reading, 24*(3), 266–278. doi:10.1111/1467-9817.00148

Naismith, L., Lonsdale, P., Vavoula, G., & Sharples, M. (2004). R*eport 11: Literature review in mobile technologies and learning, (No. 11): Nesta Futurelab*. Retrieved from http://www.nestafuturelab.org/download/pdfs/research/lit_reviews/futurelab_review_11.pdf.

Papazoglou, M. P., Traverso, P., Dustdar, S., & Leymann, F. (2007). Service-oriented computing: State of the art and research challenges. *Computer, 40*(11), 38–45. doi:10.1109/MC.2007.400

Patten, B., Arnedillo Sánchez, I., & Tangney, B. (2006). Designing collaborative, constructionist and contextual applications for handheld devices. *Computers & Education, 46*(3), 294–308. doi:10.1016/j.compedu.2005.11.011

Posner, I., Baecker, R., & Homer, B. (1997). Children learning filmmaking using multimedia tools. In *Proceedings of ED-MEDIA/ED-TELECOM*, Calgary, Canada.

Reid, M., Burn, A., & Parker, D. (2002). *Evaluation report of the Becta Digital Video Pilot Project: Becta.* Retrieved, from http://www.becta.org.uk/research/reports/digitalvideo/

Roschelle, J. (2003). Keynote paper: Unlocking the learning value of wireless mobile devices. *Journal of Computer Assisted Learning, 19*(3), 260–272. doi:10.1046/j.0266-4909.2003.00028.x

Sharples, M., Lonsdale, P., Meek, J., Rudman, P., & Vavoula, G. (2007, October 16-19). An evaluation of MyArtSpace: A mobile learning service for school museum trips.In *Proceedings of mLearn* (pp. 238-244), Melbourne, Australia.

Soloway, E., Norris, C., Blumenfeld, P., Fishman, B., Krajcik, J., & Max, R. (2001). Log on education: Handheld devices are ready-at-hand. *Communications of the ACM, 44*(6), 15–20. doi:10.1145/376134.376140

Spikol, D. (2007, December). Designing mobile games that explore novel learning practices with co-design. In *Proceedings of Research Methods in Informal and Mobile Learning (MIL-RM)* (pp. 41-47), Institute of Education, University of London, UK.

Swain, C., Sharpe, R., & Dawson, K. (2003). Using digital video to study history. *Social Education, 67*(3), 154–158.

Tatar, D., Roschelle, J., Vahey, P., & Penuel, R. W. (2003). Handhelds go to school:Lessons learned. *SRI International* (pp. 30-37).

Vavoula, G. (2007). *Evaluating mobile learning*.

Vavoula, G., Meek, J., Sharples, M., Lonsdale, P., & Rudman, P. (2006). *A lifecycle approach to evaluating MyArtSpace*.

Vavoula, G., & Sharples, M. (2008, October 7-10). Challenges in evaluating mobile learning. In *Proceedings of the MLearn2008 Conference: The Bridge From Text To Context* (pp. 296-303), University of Wolverhampton, UK.

Zagal, J. P., Piper, A. M., & Brukman, A. S. (2004). *Kids telling fables through 3D animation: Georgia Institute of Technology.* Retrieved, from http://hdl.handle.net/1853/3732

Section 4
Innovative Cases

Chapter 15
A Multiplatform E-learning System for Collaborative Learning
The Potential of Interactions for Learning Fraction Equivalence

Siu Cheung Kong
The Hong Kong Institute of Education, Hong Kong

ABSTRACT

A multiplatform e-learning system called the "Graphical Partitioning Model (GPM)", with the separate versions for desktop computers and mobile devices, was developed for learning knowledge of fraction equivalence. This chapter presents a case study on the use of the mobile version GPM for the learning of the targeted topic in a mobile technology supported environment. The interactions between a dyad of Primary 5 students and the GPM were analyzed in order to understand the feasibility of the design of the mobile version e-learning system. The results show that the interactions between the students and the GPM have the potential to enhance the learning effectiveness of the targeted topic. The mobile version GPM demonstrated a possibility to integrate with collaborative learning strategies such as reciprocal tutoring and peer discussion. The case study also reveals that there is a potential for the flexible use of the dual-version GPM to foster deep learning.

INTRODUCTION

Knowledge of fraction equivalence is a fundamental element in the learning of the mathematics topic "mathematical fraction". This knowledge, which comprises the concept of fraction equivalence and knowledge of the computation of equivalent fractions, both of which are of equal importance, is a prerequisite for the further conceptual development of the targeted topic such as the procedural knowledge about the operation of mathematical fractions (Kong & Kwok, 2005). Researchers suggest that computer-supported learning environments that provide graphical supports facilitate the knowledge generation about fraction equivalence (Ohlsson, 1991; Steffe & Olive, 1996). In this regard, a desktop version of a web-based e-learning system for learning the knowledge of fraction equivalence was

DOI: 10.4018/978-1-60566-703-4.ch015

designed (Kong, 2008a; Kong & Kwok, 2003).

Previous evaluation study shows that the interactions among learners, for example in the form of reciprocal tutoring, in a collaborative learning environment have the potential to increase learning effectiveness in this domain (Kong, 2008a, 2008b; Kong & Kwok, 2005). As the portable and versatile nature of mobile devices offers the opportunity to promote reciprocal tutoring in a mobile technology supported environment, the desktop version e-learning system for comprehending knowledge of fraction equivalence was adapted to create a mobile version for collaborative learning (Kong, 2008b). This chapter presents a real case on analyzing the interactions of learners who used the mobile version e-learning system in a mobile technology supported environment for the collaborative learning of fraction equivalence.

THE MULTIPLATFORM E-LEARNING SYSTEM

Researchers have suggested that visualization plays an important role in learning mathematics. Visualization is a "cognitive technology" using visual means, such as visual representations in terms of diagrams or graphs, to "see" abstract concepts and ideas (Arcavi, 2003; Borba & Villarreal, 2005). Visualization encompasses four elements, namely mental images, external representations, visualization processes and visualization abilities. In mathematics, visualization is a process requires the ability to interpret and understand figural information and the ability to conceptualize and translate abstract relationships and nonfigural information into visual terms.

Visualization is considered as a helpful tool for mathematical comprehension because many concepts and processes in school mathematics can be tied to visual representations. By virtue of the concreteness of visual representations, visualization becomes an essential factor for learners to create a sense of self-evidence and immediacy (Arcavi, 2003; Borba & Villarreal, 2005). The integration of visualization with e-learning, which refers to the use of computer technology to access digital resources on the Internet for learning purposes (Holmes & Gardner, 2006), plays a relevant role in this educational context because computer is a rich source of visual and computational images that makes the exploration of mathematical concepts possible. It is suggested that e-learning systems in the nature of computer-based graphical tools are able to support the dialectic reasoning of learners in the mathematics classroom by providing opportunities for exploring hypothetical queries and making mental manipulation of concepts easier (Sedig & Liang, 2006).

Researchers suggest that computer-based graphical tools are particularly suitable for the topics that emphasize visualization, such as mathematical fractions. Early studies on computer-supported learning environments for learning fractions used graphical presentations or representations, or operators in a micro-world to help learners to develop the conceptual understanding and procedural knowledge of fraction equivalence (Ohlsson, 1991; Steffe & Olive, 1996). The purpose of the graphics was to link fraction symbols with pictorial presentations or representations in order to coordinate the internal mental models of learners with the external visual representations, thereby increasing understanding of the concepts (Sedig & Liang, 2006). With the aim of providing graphical supports for the knowledge generation about fraction equivalence, a desktop version of a web-based e-learning system called the "Graphical Partitioning Model (GPM)" was developed for individual learning of the targeted topic (Kong, 2008a; Kong & Kwok, 2002, 2003, 2005).

The GPM is a graphical model of a rectangular bar for representing fractions, of which each fraction being represented by displaying shaded fractional parts of an equally partitioned rectangular bar according to the value of the fraction. This e-learning system was designed as a model of affordances to support the learning of mathematical fractions. Gibson (1979) introduces the notion of affordances and suggests that the perceptual task of human beings is to detect environmental aids that could be used in their attempts to interact with the environment to meet their needs. In human-computer interaction, affordances refer to the interface cues that help learners know the inherent functions and operation methods of specific features of the computer technologies (Sedig & Sumner, 2006). Based on the rationale that the means of instruction should not be predetermined because each learner constructs knowledge in a unique way, Akhras and Self (2000) suggest that the pedagogical role of e-learning systems should provide profitable spaces for interaction to the learner based on some model of the affordances of potential situations. The function of such model of affordances is to make available profitable spaces, or provide the necessary scaffolding (Clark, 1997). It is found that the provision of appropriate models of affordances has a constructive effect on the learning outcomes of learners (Wijekumar, Meyer, Wagoner & Ferguson, 2006).

According to the findings of a qualitative pilot evaluation study of the use of the GPM (Kong & Kwok, 2002), learners with varying mathematical abilities exhibit diverse needs for the functionalities of this e-learning system in the learning process. To enhance the effectiveness of the desktop version e-learning system to support the learning of mathematical fractions, the desktop version GPM was therefore designed as a model of affordances with three spaces to give learners the means to interact in a way that meets their needs (Kong, 2008a; Kong & Kwok, 2003, 2005). The first space was the feature for partitioning (see Figure 1). This space allows a choice to be made between an intentional slowed down animation that shows the partitioning or regrouping process and an instantaneous change that shows the results of the partitioning or regrouping process. The simulation of the partitioning strategy by the slowed down animation addresses the lack of intention of representing fractions to compare their equivalence and the failure to recognize the inverse relationship between number of parts and the size of a part of a unit. This space enables learners to interact with the GPM according to their needs, with capable learners being able to generalize the knowledge by rapidly calling up the results of

Figure 1. The interface of the desktop version GPM for comparing equivalence state of two fractions

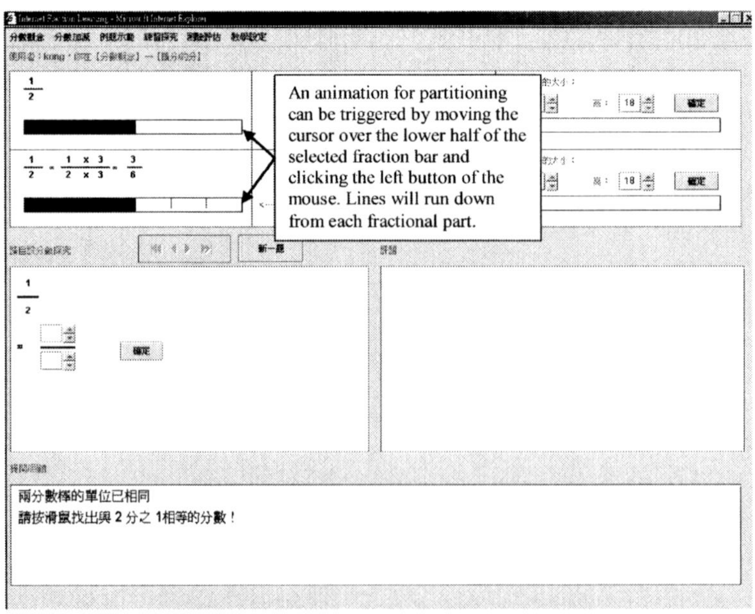

partitioning and less capable learners being able to pick up the idea by activating the slowed down animation of the process of partitioning.

The second space was the feature for the comparison of the equivalence of fractions in response to the difficulties that learners who have no intention of representing fractions with the same unit to compare their equivalence encounter (see Figure 2). An animation that shows the direct comparison of the equivalence of two fraction bars, which is triggered by dragging a fraction bar and dropping it onto another bar, is designed to allow an extra comparison of fraction equivalence in addition to the visual inspection of two separate fraction bars. This space gives learners multiple opportunities to compare fraction equivalence in an interactive way.

The third space was the feature that consists of a hypothesis-testing interface $a/b = a \times c/b \times d$ to address the problem of learners who lack the ability to find equivalent fractions systematically (see Figure 2). The hypothesis-testing interface asks learners to test possible fraction equivalent states by adjusting parameters c and d, and allows them to compare fraction equivalence using the aforementioned comparison animation by dragging the fraction bar of the hypothesized fraction and dropping it onto the fraction bar of the original fraction.

The desktop version GPM was evaluated by a quasi-experimental evaluation study with a pre-test/post-test control group design (Kong & Kwok, 2005). The results of this evaluation study indicate that the model of affordances allows learners of varying learning abilities to develop a concept of fraction equivalence. It is also found that the interactions among learners, for example in the form of reciprocal tutoring, in a collaborative learning environment have the potential to increase learning effectiveness in the targeted topic.

Collaborative learning, a process that encourages learners to participate in coordinated and synchronous learning activities with a number

Figure 2. The interface of the desktop version GPM for computing equivalent fractions

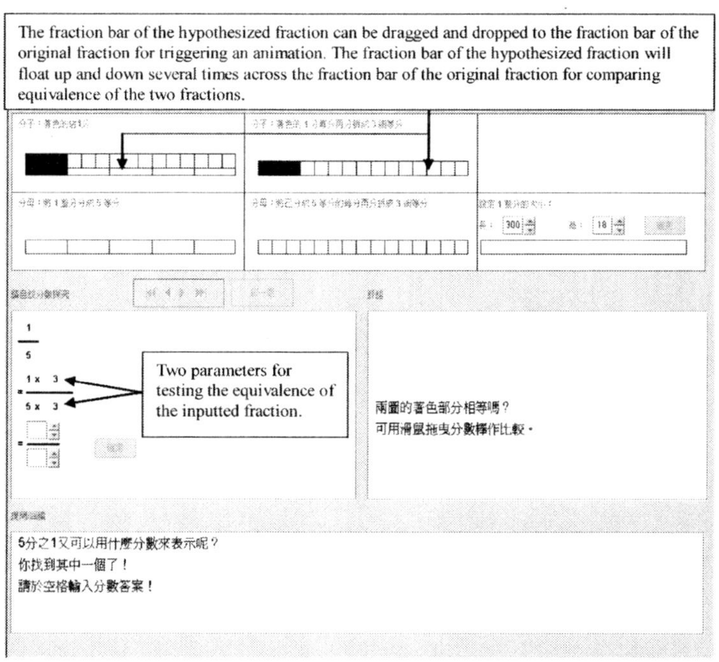

of other learners, has been found to be effective for learners at all learning achievement levels (Roschelle & Teasley, 1995; Slavin, 1996). Reciprocal tutoring is a good strategy for engaging learners in classroom interactions for learning purposes, in which learners take turns to tutor each other in a collaborative learning environment (Chan & Chou, 1997; Wong, Chan, Chou, Heh & Tung, 2003). Learners are placed in some form of grouping within classrooms throughout their schooling. The majority of their learning time in school is spent on classroom interactions, such as explorative and elaborated talks, in the presence of their peers. The strategy of reciprocal tutoring enables learners to learn from one another through the verbal elaboration of the new knowledge in a group learning context, thus allows learners who have gained insight into the new concept to reinforce their knowledge by providing explanations to others who need more opportunity to comprehend the knowledge.

Previous studies find that the integration of the use of computer technology into collaborative learning not only increases learning enjoyment of learners, but also deepens their cognitive understanding of various basic subjects including mathematics (Kutnick, Ota & Berdondini, 2008; Scott, Mandryk & Inkpen, 2003). Mobile learning, which refers to the use of mobile technology for learning and teaching, is an emergent learning approach that has the potential to facilitate interactions and foster deep learning in a collaborative learning environment. Portability and visualization capability make mobile devices a powerful medium for learning (Roschelle & Pea, 2002; Sharples, Taylor & Vavoula, 2005). The former attribute of existing mobile technologies offers learners a sense of ownership of individual mobile devices, and thus helps to provide incentives to learners to actively participate in collaborative learning activities. The latter attribute of currently available mobile devices enables learners to visualize and manipulate abstract concepts through visual representations, and therefore helps to promote the sharing and communication among learners on their ideas and knowledge with visual support in collaborative learning activities. As the portable and versatile nature of mobile devices offers the opportunity to promote reciprocal tutoring in a mobile technology supported environment, the desktop version GPM for comprehending knowledge of fraction equivalence was adapted to create a mobile version for collaborative learning of the targeted topic (Kong, 2008a).

A mobile platform is established for the mobile version GPM for interaction between learners. The learners interact in pairs through a server. The server acts as a communication coordinator of synchronous interactions between paired learners. Adaptations of display layout and cognitive artifacts are made from the three spaces of the model of affordance of the desktop version GPM to enhance the capability of the mobile version GPM to address diverse learning needs of individual students. First, the space for partitioning is modified to a learner-controlled animation of partitioning (see Figure 3). This feature allows learners to partition fractions by clicking the graphical representation of the fractions for an iterative display of the initial blank fraction bar, the calibrated fraction bar and the original fraction bar with shaded fractional part. This cognitive artifact is incorporated in all of the fraction bars displayed on the GPM. It aims to return control of the learning process to the learners under two approaches: when learners work with only one fraction bar, the stepwise design helps them to develop the part-whole concept; and when they work with both fraction bars, the stepwise design helps them to understand the inverse relationship between the number of parts in a unit and the size of a part.

Second, the space for comparison of the equivalence of fractions undergoes a twofold modification. On the one hand, the positioning of the two fraction bars for comparison, in that the fraction bars in the desktop version are arranged in a row, are arranged in a column because of the

Figure 3. The interface of the mobile version GPM for comparing equivalence state of two fractions

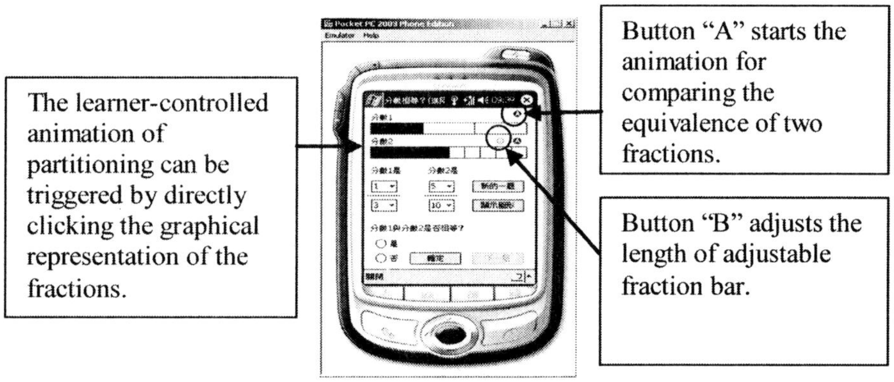

Figure 4. The interface of the mobile version GPM for computing equivalent fractions

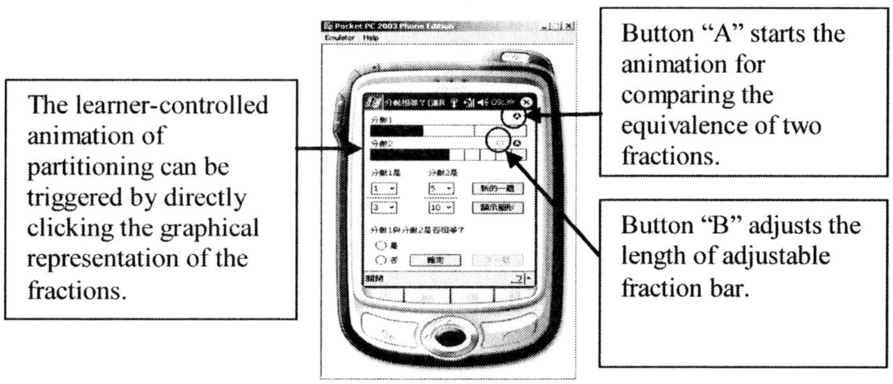

relatively narrower screen of the mobile device. A button "A" is designed for learners to start a comparison animation of which the fraction bar of the selected fraction rolls over the other fraction bar when learners press this button to obtain a graphical representation of a fraction (see Figure 3). On the other hand, an additional feature of an adjustable fraction bar is included to act as a random alert to strengthen the concept of representing fractions using common units to compare their equivalence. An adjustable bar that is 50% to 70% of the length of the bar of the other fraction is displayed occasionally. A button "B" is designed for adjusting the length of the bar in between its elongated and original length (see Figure 3).

Third, the space for hypothesis-testing is modified to a time-keeping hypothesis-testing competition (see Figure 4). There is a similar hypothesis-testing interface $a/b = a \times c/b \times d$ in the desktop version GPM, for adjusting parameters c and d to test a possible fraction equivalent state, but the interface in the mobile version is improved by the use of competition between learners to stimulate their learning interest.

THE CASE STUDY

This chapter reports a case study on the use of the mobile version GPM in a mobile technology supported learning environment. The learning activities take place in a wireless-networked environment. The learners are provided with a personal digital assistant (PDA) which is pre-installed the GPM for learning fraction equivalence. The learners interact in pairs through a server that is connected to an SQL database.

There are two learning activities designed for the abovementioned mobile technology supported environment. The first one is a true-false learning activity for developing concept of fraction equivalence. The learners are arranged in pairs, in which one learner is the question-setter and the other the respondent. The learners are placed in a situation in which they have to decide the equivalence of two fractions (see Figure 3). The learners alternate between the two roles.

The learning activity involves three steps. Step 1 is the process of question-setting, in which the learners who are playing the role of question-setter set and send out questions about the equivalence of two fraction expressions. The question-setters have to state whether the two fraction expressions that they have chosen are equivalent by graphically representing the two fraction expressions at the top of the interface. Once the learners are satisfied with the question that they have set, they can click the "Confirm" button to send the question to their partners through the server. Step 2 is the response process. In this step, the learners who are playing the role of respondent receive the question in the form of two fraction expressions from their partner, and then have to decide whether the two expressions are equivalent with the help of graphical representation. After indicating their decision, learners click the "Confirm" button to send their answer to the server. Step 3 is the process of judgment. In this step, the computer system assesses the correctness of the questions that are set by the question-setters and the answers that are provided by the respondents. The computer system then sends the messages in the form of the words "Correct" and "Incorrect" for right and wrong questions or answers, respectively.

The second learning activity is a time-keeping learning activity for developing knowledge of the computation of equivalent fractions. Learners are grouped into pairs to engage in a competition to find an equivalent fraction of a fraction that is assigned by the computer system (see Figure 4). This learning activity comprises two steps. Step 1 is the process of finding the equivalent fraction. In this step, learners require to find an equivalent fraction with the use of a hypothesis-testing interface $a/b = a \times c/b \times d$. Learners are asked to adjust parameters c and d to test a possible fraction equivalent state. Graphical representations are generated to help learners to compare the equivalence of the two fractions by the instant change that takes place following the adjustment of parameters c and d. Once learners have decided on their answer, then they can click the "Confirm" button to send the answer to the server. Step 2 is the process of judgment. In this step, the computer system measures the response time and judges the correctness of the answers that are provided by learners. For quick responses that are correct the words "Correct" and "Yeah!" are displayed on the screen; for slow responses that are correct, the words "Correct" and "Cheer up!" are displayed; and for incorrect answers the words "Incorrect" and "Cheer up!" are displayed regardless of the response time.

A study was conducted to investigate the remedial effect of the designed mobile technology supported environment on supporting at-risk students to learn the targeted topic. A dyad of Primary 5 students, one is male (hereafter "S1") and one is female (hereafter "S2"), was invited to participate in this study. These two students completed the learning of the targeted topic in the grade of Primary 4 and had relatively low mathematical ability. In this study, the students were asked to perform the abovementioned learning activities along with the completion of a set of guiding worksheets under an iterative process of replying questions based on prior knowledge, exploring questions with the use of the GPM, and answering questions according to computer-generated results.

In view of the importance of interactions in the learning of the targeted topic, this study focused on the human-computer interactions and student-student interactions in the learning process. The learning scenarios were video-recorded. Analysis on the transcribed dialogues among the participants of this study was conducted to study the characteristics and effects of the interactions between the students and the GPM during the learning process in the designed mobile technology supported environment.

RESULTS AND DISCUSSIONS

The results of this case study show that the interactions between the students and the GPM could induce constructive changes in the learning of the targeted topic. Two pieces of findings are drawn from the case study. The first one relates to the gain in subject knowledge of learners from interacting with the graphical support of the GPM in the mobile learning environment. The second one is related to the facilitation of reciprocal tutoring between learners by interacting with the graphical support of the GPM in the mobile learning environment. This section depicts the scenarios that illustrate how the mobile version GPM exhibited constructive effects on the learning of the targeted topic via the human-computer interactions and student-student interactions in the learning process.

Gain in Subject Knowledge of Learners from Interacting with the GPM

The interactions between the students and the GPM facilitated the individual learning of the targeted topic via self-exploration. One of the evidences comes from the interaction process of S1 in the learning task on completing a guiding worksheet "Concept of Fraction Equivalence". The question involved was about the comparison of equivalence state of two designated fractions 1/3 and 3/9.

Before working with the mobile version GPM, S1 applied his prior knowledge about fraction equivalence to answer the question. This student could recall the correct procedural knowledge about comparing the equivalence state of the two designated fractions by describing the algorithm in computing equivalent fraction. However, his incorrect drawing of graphical representation for the designated fraction 3/9 indicates that this procedural knowledge was not in line with the meaning of the graphical representations (see Table 1). S1 exhibited one of the four typical inadequacies in the learning of mathematical fractions – the failure to understand the part-whole concept, that is, parts of the whole are equal (Kong & Kwok, 2002). The work of S1 reflects the separation of procedural knowledge from its underlying meaning of his knowledge of fraction equivalence.

After working with the mobile version GPM, the student consolidated and furthered his acquired knowledge of fraction equivalence. He

Table 1. A scenario that shows a learner compared the equivalence state of the fractions 1/3 and 3/9 before working with the mobile version GPM

Dialogue translated from Cantonese into English (R stands for researcher and S1 stands for Student 1)		
R	:	You think they (i.e. the fractions 1/3 and 3/9) are equivalent. Can you explain your answer?
S1	:	Since 3 times 3 equals to 9 (pointing to the graphical representation for the denominator of the fraction 1/3) and 1 times 3 equals to 3 (pointing to the graphical representation for the numerator of the fraction 1/3), they are equivalent.
R	:	Equivalent? How about the graphs that you have drawn? Can these graphs indicate the equivalence state of these fractions?
S1	:	This is (pointing to the shaded part of the graphical representation for the fraction 1/3, seen in Figure 5) a portion of the three sub-parts, and this is (pointing to the shaded part of the graphical representation for the fraction 3/9) a portion of the nine sub-parts.

Figure 5. Remark: This is an incorrect graphical representation for the fraction 3/9, in which the length of the shaded part was not the same as the one for the fraction 1/3, was drawn by S1.

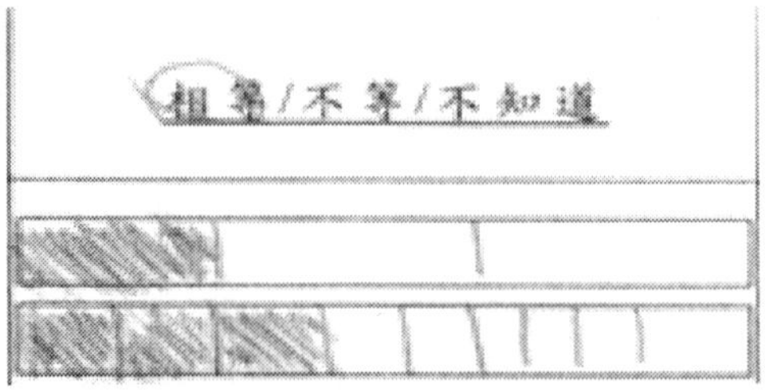

Table 2. A scenario that shows a learner compared the equivalence state of the fractions 1/3 and 3/9 after working with the mobile version GPM

Dialogue translated from Cantonese into English (R stands for researcher and S1 stands for Student 1)		
R	:	This time the answer is "Yes" again. Can you explain again?
S1	:	For the fraction 1/3, divide it into three sub-parts and select one of them. For the fraction 3/9, divide it into nine sub-parts and select three of them. The selected parts of these two fractions are the same. (see Figure 6)

confirmed the correctness of his answer with the aid of the graphical support. The student described the procedure of comparing the "length" of the colored portions of the fraction bars displayed on the GPM for finding solutions. The student also gave a proper graphical representation for the designated fractions after working with the mobile version GPM (see Table 2). Although it is uncertain whether the student grasped the relevant concepts or simply a memorization of the graph from the display on the GPM, the explanation given by S1 in Table 2 supports the former guess.

The above scenarios show that the graphical support of the GPM not only made the student more conscious of the proper representation of fractions graphically, but also enabled him to give a more clear and justified explanation on the equivalence state of two fractions. The interactions between the student and the GPM are considered meaningful for the student on learning about the meaning of fractions.

Facilitation of Reciprocal Tutoring between Learners by Interacting with the GPM

The interactions between the students and the GPM facilitated the collaborative learning of the targeted topic via reciprocal tutoring. There is evidence in the interaction process of S1 and S2 in the learning task on completing a guiding worksheet "Computation of Equivalent Fractions". The question involved was about the computation of equivalent forms for the designated fraction 1/4.

Before working with the mobile version GPM, both students exhibited misunderstanding in finding equivalent fractions of the designated fraction. For S2, there was a misunderstanding that the inverse of the given fraction was its equivalent form. This student recalled two types of procedures in completing this task. The first one was the algorithmic method. This is the correct procedural knowledge for finding equivalent fractions, which led S2 to give a correct set of answer and graphical representation. The second one was the procedure for finding the inverse of

Figure 6. Remark: This is a correct graphical representation for the fraction 3/9, in which the length of the shaded part was nearly the same as the one for the fraction 1/3, was drawn by S1

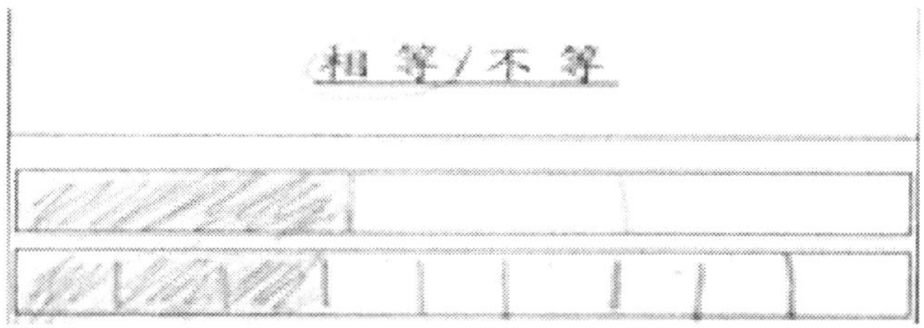

Table 3. A scenario that shows a learner found an equivalent fraction of 1/4 before working with the mobile version GPM

Dialogue translated from Cantonese into English (R stands for researcher and S1 stands for Student 1)		
R	:	I want to know why 4/1 and 4 are equivalent to 1/4.
S1	:	4 is equal to 1/4.
R	:	You write 4/1 equals to 1/4. What does 4/1 represent?
S1	:	4/1 equals to 4 (pointing to the upper graphical representation in his work). 4 equals to 1/4 (pointing to the lower graphical representation in his work).
R	:	To represent 4/1 graphically, you think it should be this graph (the upper graphical representation in his work). That graph (the lower graphical representation in his work) represents 4, doesn't it?
S1	:	(It) also represents 1/4.
R	:	This space is for you drawing the graphical representation for the fraction 4/1 after you write down the fraction. I would like to know whether you think this graph (the upper graphical representation in his work) represents 4/1. (See Figure 7a)
S1	:	Yes.
R	:	Does this graph (the lower graphical representation in his work) represent 4? (See Figure 7b)
S1	:	(It) also represents 1/4.
R	:	In this situation, you think 4 and 1/4 are equivalent?
S1	:	Yes.

a fraction. This is a wrong procedural knowledge for computing equivalent fractions, which led S2 to give another unacceptable set of answer and graphical representation (see the work of S2 before working with the GPM as shown in Table 3).

The student S1, who seemed to know the procedural knowledge for comparing the equivalence state of two given fractions in the above task, showed misunderstanding again in this task of finding an equivalent fraction of 1/4. He gave the inverse in two forms (see the work of S1 before working with the GPM as shown in Table 3) as the equivalent fractions. He demonstrated difficulty in applying relevant procedural knowledge for finding equivalent fractions (see Table 3).

After working with the mobile version GPM, the students rectified their misunderstandings in finding equivalent fractions of the designated fraction. For S2, a clear understanding of the concept was established through her observation of the relationship between the changes in the value of multiplier for the denominator and numerator in the hypothesis-testing bed and the length of the colored portion of the fraction bars displayed on the GPM. This student confirmed the correctness of her answer guess from the graphical support in

Figure 7. a) Remark: This is the work of S1 before working with the GPM; b) Remark: This is the work of S2 before working with the GPM

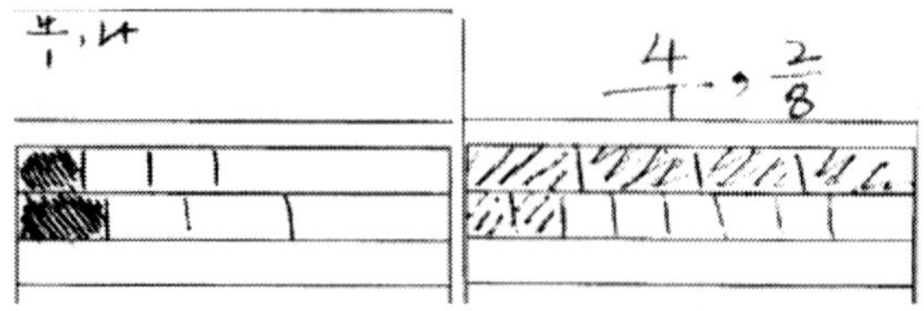

(a) (b)

the GPM and gave the correct graphical representation on the guiding worksheet (see the work of S2 after working with the GPM as shown in Table 4). With the encouragement of the researcher, the student could confidently point out the mistakes made by her partner and engage in reciprocal tutoring to clearly explain the procedures for giving her correct answer to S1 with the support of the GPM (see Table 4). S2 gained an experience of explanation to her partner.

Table 4. A scenario that shows learners found an equivalent fraction of 1/4 after working with the mobile version GPM

Dialogue translated from Cantonese into English (R stands for researcher; S1 stands for Student 1; and S2 stands for Student 2)		
R	:	... How many sub-parts are there now?
S2	:	Eight.
R	:	How many sub-parts are selected?
S2	:	Two.
R	:	... Can you two discuss the results?
S2	:	He (i.e. S1) produces three sub-parts more.
R	:	How come three sub-parts more result non-equivalence? Can you explain? Does he make a mistake?
S2	:	Yes, he does (make a mistake).
R	:	What is the mistake?
S2	:	It is a mistake that 4 times 1 will produce four sub-parts (pointing to the denominator of the hypothesis-testing bed on the PDA of S1), and 1 times 4 means four sub-parts are colored (pointing to the numerator of the hypothesis-testing bed on the PDA of S1). In this way, the fraction is equal to one. [Remark: S1 input 1 as the denominator and 4 as the numerator in finding an equivalent fraction of 1/4. Please refer to the interface in Figure 4 for these inputs.]
R	:	That is, the fraction equals to 1?
S2	:	Yes.
R	:	So what is the correct one? Give me an answer.
S1	:	(It) could be 2/8.
R	:	Why should it be 2/8?
S2	:	It is because 2/8 means two sub-parts are selected from eight sub-parts (S2 compared each dyadic sub-part of the graphical representation of the fraction 2/8 on her PDA to each single sub-part of the graphical representation for the fraction 1/4). Here two sub-parts, two sub-parts, two sub-parts, two sub-parts, fits perfectly (with each sub-part of 1/4). Denominator should be the same as the above. [Remark: The meaning of S2 was the input of the parameter for the denominator should be the same as the input to the numerator on the top.] (see Figure 8)
R	:	Denominator should be the same as the above (fraction)?
S2	:	Actually it is my guess. Since here two sub-parts, two sub-parts, two sub-parts, two sub-parts. This is then the same.

Figure 8. a) Remark: This is the work of S1 after working with the GPM; b) Remark: This is the work of S2 after working with the GPM

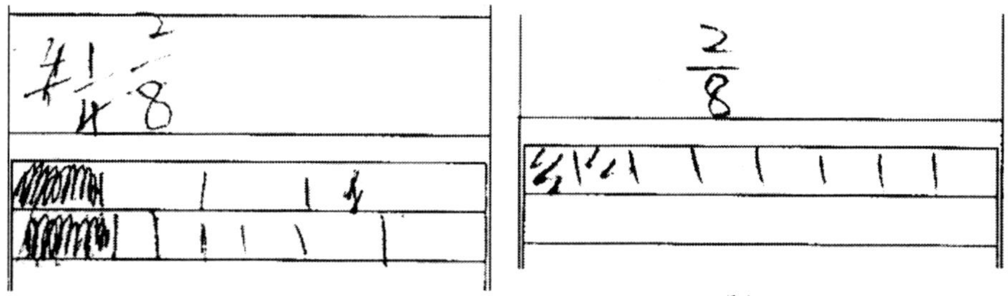

(a) (b)

For S1, a realization of his wrong answer guess and improper graphical representation was made after working with the GPM and listening to the explanation from S2. The student gave the correct answer (see Table 4) and graphical representation (see the work of S1 after working with the GPM as shown in Table 4) after listening to the explanations from his partner who referred to the graphical support of the GPM in the explanation. S1 gained another experience in relating the procedure for finding equivalent fraction with its graphical meaning.

The above scenarios show that the graphical support of the GPM not only assisted students to deepen their conceptual understanding of the targeted knowledge via the interactions with their own GPM, but also facilitated students to discuss their procedural skills in the targeted topic via the interactions with other's GPM. The interactions among the students with the use of the GPM were thus constructive to students in the collaborative learning of the targeted topic.

In summary, the mobile version GPM was productive for the learning of the targeted topic. The GPM provided visualization support to foster the development of conceptual knowledge and acted as a learning authority to facilitate the practice of explanation skills and the learning of targeted knowledge in the reciprocal tutoring process. The mobility of the GPM also enhanced the involvement of the students in learning the subject knowledge by offering them a sense of ownership of the mobile device.

IMPLICATIONS

Two implications in connection with the use of the mobile version GPM are drawn based on the results of this case study. The first concerns the pedagogical value of the mobile version GPM. The results of this case study show that the use of the mobile version GPM could greatly realize the potential of visualization in learning fraction equivalence. The scenarios reported in the previous section reflect that after interacting with the mobile version GPM, the two at-risk students could make use of visual artifacts to rectify common misunderstandings, grasp necessary knowledge and explain algorithmic procedures in the learning of the targeted topic. It is also found that the mobile version GPM was conducive to reciprocal tutoring because students were allowed to interactively take the graphical displays on their GPM as evidences to explain their ideas about the targeted topic. In this regard, the mobile version GPM will be further refined to enhance its capability to facilitate student-student interactions in the collaborative learning in a mobile technology supported environment.

The second implication relates to the pedagogical use of the mobile version GPM. The results of this case study reveal that teacher mediation was required to facilitate student-student interactions with the use of the mobile version GPM. From the scenarios depicted in the previous section, the dyad of senior primary students was not spontaneous to discuss the targeted topic. They engaged in a small degree of interaction upon the provision of external mediation from the researcher. In this respect, teacher mediation for the trigger to student-student interactions should be provided for students of junior grades in the collaborative learning activities involving the use of mobile version GPM. Three approaches are worthy of consideration in future implementation of the GPM in the mobile technology supported classroom. First, teachers may act as the mediator during the collaborative learning activities to encourage students to make use of the graphical displays on the GPM for group discussions about the targeted topic. Second, teachers may invite a number of "Student Teachers", who are capable of completing learning tasks with the use of the GPM, to act as the facilitator to lead group discussions based on the graphical displays on the GPM. Third, teachers may embed collaborative learning elements, such as the instructions for students to compare answers with their partners by referring to the graphical displays on their GPM, in the design of learning activities and guiding worksheets.

FUTURE WORK

The encouraging findings of this case study shed light on the scaffolding of the existing mobile learning experience towards multiplatform e-learning systems. The e-learning system GPM, with the desktop version and mobile version, is designed to support the learning of mathematical fractions with the use of various access devices. Previous evaluation studies have already attested to the effectiveness of the desktop version GPM on the individual learning of the targeted topic (Kong & Kwok, 2002, 2005; Kong, 2008a). The study reported in this chapter reveals the potential of the mobile version GPM for the collaborative learning of the targeted topic. With the latest development of relatively low-cost mobile computing devices, such as the subnotebooks Eee PCs, that combines the functionalities of traditional desktop computers and the portability of traditional mobile devices, the mixed use of the dual-version GPM in a computing device in a mobile technology supported classroom is worthy of consideration.

Owing to its web-enabled capability and portable-size design, it is possible for students to use the Eee PCs to access the dual-version GPM in a mobile technology supported classroom for an alternative engagement in individual learning and collaborative learning of the targeted topic without seat constraints. Investigations into appropriate pedagogical designs for the use of the dual-version GPM under these two learning modes are worth consideration in future research studies. Regarding the individual learning, it is anticipated that each of the students can use their own Eee PC to access the desktop version GPM, of which the computer system acts as the learning authority. The major type of interactions involved will be the student-GPM interactions. The students will follow the instructions on the guiding worksheets and interact with the computer system to formulate knowledge of the targeted topic. It would thus be worthwhile to study whether teachers should take the role of observer and assessor to offer students ample opportunities to self-explore the knowledge of the targeted topic with the use of the GPM in this case.

Concerning the collaborative learning, it is anticipated that each of the students can use their own Eee PC to access the mobile version GPM, of which the computer system acts as the learning authority. The interactions in the corresponding learning activities will include both student-GPM interactions and student-student interactions. On the one hand, the students, again, follow the instructions on the guiding worksheets and interact with the computer system to explore knowledge of the targeted topic. On the other hand, the students refer to their answers on the guiding worksheets and interact with their partners to discuss knowledge about the targeted topic. It would then be worthwhile to investigate whether teachers should take the role of facilitator to provide students with necessary encouragement and mediations in this situation.

CONCLUSION

This chapter reports a case study on the use of a mobile version e-learning system called the "Graphical Partitioning Model (GPM)" for the learning of fraction equivalence in a mobile technology supported environment. The findings of this case study indicate that the mobile version GPM had constructive effect on the remedial learning of the participating students who were less able in mathematics learning. Through their interactions with the mobile version GPM and their partners, the students were not only stimulated to deepen their knowledge about fraction equivalence, but also fostered to engage in reciprocal tutoring about the targeted topic. With the appropriate external mediation, there is a potential to integrate the use of the mobile version GPM with collaborative learning strategies such as reciprocal tutoring and peer discussion to facilitate deep learning of the targeted topic.

The findings of this case study reported in this chapter reveal the value of the scaffolding of the existing mobile learning experience towards multiplatform e-learning systems. With the availability of the desktop version GPM and the low-cost mobile computing devices, one of the future research foci will target at the mixed use of the desktop version GPM for individual learning and the mobile version GPM for collaborative learning. To maximize the integrated use of the dual-version GPM for fostering deep learning in the classroom context, field studies will be conducted to look into implementation characteristics of these two platforms and decide on integration approach for the mixed use of these two platforms to facilitate human-computer interactions and student-student interactions in the collaboration-based mobile learning of mathematical fractions.

REFERENCES

Akhras, F. N., & Self, J. A. (2000). Modeling the process, not the product, of learning. In S. P. Lajoie (Ed.), *Computers as cognitive tools, volume two: No more walls* (pp. 3-28). Mahwah, NJ: Lawrence Erlbaum Associates.

Arcavi, A. (2003). The role of visual representations in the learning of mathematics. *Educational Studies in Mathematics*, *52*(3), 215–241. doi:10.1023/A:1024312321077

Borba, M. C., & Villarreal, M. E. (2005). *Humans-with-media and the reorganization of mathematical thinking: Information and communication technologies, modeling, visualization and experimentation*. New York: Springer.

Chan, T. W., & Chou, C. Y. (1997). Exploring the design of computer supports for reciprocal tutoring. *International Journal of Artificial Intelligence in Education*, *8*, 1–29.

Clark, A. (1997). *Being there: Putting brain and world together again*. Cambridge, MA: MIT press.

Gibson, J. J. (1979). *The ecological approach to visual perception*. Boston: Houghton Mifflin.

Holmes, B., & Gardner, J. (2006). *E-learning: Concepts and practice*. London: Sage.

Kong, S. C. (2008a). The development of a cognitive tool for teaching and learning fractions in the mathematics classroom: a design-based study. *Computers & Education*, *51*(2), 886–899. doi:10.1016/j.compedu.2007.09.007

Kong, S. C. (2008b). Collaborative learning in a mobile technology supported classroom. In J. Lumsden (Ed.), *Handbook of research on user interface design and evaluation for mobile technology* (pp. 269-284). Hershey, PA: Information Science Reference.

Kong, S. C., & Kwok, L. F. (2002). Modeling a cognitive tool for teaching the addition/subtraction of common fractions. *International Journal of Cognitive Technology*, *1*(2), 327–352. doi:10.1075/ijct.1.2.07kon

Kong, S. C., & Kwok, L. F. (2003). A graphical partitioning model for learning common fractions: Designing affordances on a web-supported learning environment. *Computers & Education*, *40*(2), 137–155. doi:10.1016/S0360-1315(02)00118-5

Kong, S.C., & Kwok, L. F. (2005). A cognitive tool for teaching the addition/subtraction of common fractions: A model of affordances. *Computers & Education*, *45*(2), 245–265. doi:10.1016/j.compedu.2004.12.002

Kutnick, P., Ota, C., & Berdondini, L. (2008). Improving the effects of group working in classrooms with young school-aged children: Facilitating attainment, interaction and classroom activity. *Learning and Instruction*, *18*(1), 83–95. doi:10.1016/j.learninstruc.2006.12.002

Ohlsson, S. (1991). Knowledge requirements for teaching: The case of fractions. In P. Goodyear (Ed.), *Teaching knowledge and intelligent tutoring* (pp. 25-59). Norwood, NJ: Ablex Publishing Corporation.

Roschelle, J., & Pea, R. (2002). A walk on the WILD side: How wireless handhelds may change computer-supported collaborative learning. *International Journal of Cognitive Technology*, *1*(1), 145–168. doi:10.1075/ijct.1.1.09ros

Roschelle, J., & Teasley, S. D. (1995). The construction of shared knowledge in collaborative problem solving. In C. O'Malley (Ed.), *Computer-supported collaborative learning* (pp. 145-168). Berlin: Springer-Verlag.

Scott, S. D., Mandryk, R. L., & Inkpen, K. M. (2003). Understanding children's collaborative interactions in shared environments. *Journal of Computer Assisted Learning*, *19*(2), 220–228. doi:10.1046/j.0266-4909.2003.00022.x

Sedig, K., & Liang, H. N. (2006). Interactivity of visual mathematical representations: factors affecting learning and cognitive processes. *Journal of Interactive Learning Research*, *17*(2), 179–212.

Sedig, K., & Sumner, M. (2006). Characterizing interaction with visual mathematical representations. *International Journal of Computers for Mathematical Learning*, *11*(1), 1–55. doi:10.1007/s10758-006-0001-z

Sharples, M., Taylor, J., & Vavoula, G. (2005). Towards a theory of mobile learning. *Proceedings of the mLearn 2005 Conference* (9 pp.). Cape Town: mLearn 2005.

Slavin, R. E. (1996). Research on cooperative learning and achievement: What we know, what we need to know. *Contemporary Educational Psychology*, *21*(1), 43–69. doi:10.1006/ceps.1996.0004

Steffe, L. P., & Olive, J. (1996). Symbolizing as a constructive activity in a computer micro-world. *Journal of Educational Computing Research*, *14*(2), 113–138.

Wijekumar, K. J., Meyer, B. J. F., Wagoner, D., & Ferguson, L. (2006). Technology affordances: the 'real story' in research with K-12 and undergraduate learners. *British Journal of Educational Technology*, *37*(2), 191–209. doi:10.1111/j.1467-8535.2005.00528.x

Wong, W. K., Chan, T. W., Chou, C. Y., Heh, J. S., & Tung, S. H. (2003). Reciprocal tutoring using cognitive tools. *Journal of Computer Assisted Learning*, *19*, 416–428. doi:10.1046/j.0266-4909.2003.00046.x

Chapter 16
Mobile Interactive Learning in Large Classes:
Towards an Integrated Instructor-Centric and Peer-to-Peer Approach

Kin-Choong Yow
Nanyang Technological University, Singapore

Boon-Chong Seet
Auckland University of Technology, New Zealand

ABSTRACT

This chapter aims at describing a new platform for mobile and interactive learning targeted as an effective communication medium between the professor and students during lectures. In this system, students and professors will be equipped with a Multimedia Messaging Service (MMS) capable device (which may be PDAs, Laptops, or Tablet PCs) that is connected on the campus-wide Wireless LAN. During lectures, students can ask questions, response to questions or give immediate feedback on the lecture simply by composing a MMS message and sending it to the professor. The main advantage of this learning system is that MMS messaging is easily extensible to the mobile GSM networks, so students are not restricted to use it only on campus. In addition to enabling better interaction between students and instructor, an approach to facilitate student-to-student interaction during a lecture for peer-to-peer learning is proposed, which can be easily integrated into our existing system.

INTRODUCTION

This is an era of global mobile communication, in which instant communication and information transfer are the major driving forces of the society. In keeping up with the advancement of technology, the learning process has also undergone through rapid changes. The introduction of the Internet and e-education (including e-learning and e-teaching) have changed the way knowledge and education are being transferred to students all over the world. In this chapter, a new learning platform developed in Nanyang Technological University (NTU), which utilizes the latest technologies to bring a mobile interactive learning environment into the class-

DOI: 10.4018/978-1-60566-703-1.ch016

rooms in a Computer Engineering course will be described. This learning platform is targeted as an effective communication medium between the professor and the students, in an attempt to enhance the quality of the learning process.

BACKGROUND

The Computer Engineering Course at NTU

The Computer Engineering course at Nanyang Technological University (NTU) is a 4-year direct honours degree program. Students will have to read subjects such as Electronics, Engineering Mathematics, Programming, Algorithms, etc., in their lower years and achieve specialization in their final year in such areas as Embedded Systems, Computer and Communication Networks, Computer Vision and Graphics, Intelligent Systems, Software Engineering, Information Management, etc. Each subject consists of 3 one-hour lecture and 1 one-hour tutorial session per week, and certain subjects have a laboratory component, which demands a two-hour session on alternate weeks. Typical class sizes are 450 for lectures, 35 for tutorials, and 35 for laboratory sessions.

Problems in Large Class Learning

There are two key problems identified with the current large-class learning system in NTU. They are: (1) the lack of interaction; and (2) the need to be physically present in the classroom.

- Lack of interaction
  - The education community has long discussed the challenges of facilitating student-instructor interaction in large classes (Geske, 1992; Gleason, 1986). Several primary factors that inhibit student initiated interaction in large classes are feedback lag, student apprehension, and single-speaker paradigm (Anderson et al., 2003). The current learning systems are mostly one-way communication, in which the professor is giving lectures to hundreds of students in a class (Barajas et al., 1998). Even in a tutorial or laboratory session, there is seldom feedback, comments or questions arise during classes. As a consequence to the one-to-many relationship between the professor and the students, there is lack of interaction between the students and the professor (Miner, 1992; Mortera-Gutiérrez, 2002)
- The need to be physically present
  - Distance learning is the main model where the need to be physically present can be eliminated (Brown, 2001; Bullen, 1998). In this model, there is the opportunity for student-to-student interactions and student-to-instructor interaction and the faculty do not change their role significantly from the traditional classroom, although presentations will have to adapt to the technology used (Andronico et al., 2004). However, in the ordinary classroom learning model, a student may be, say, 10 minutes late for a class, and that might cause the student to have difficulty following the rest of the lecture. So it will be desirable to apply the distance learning model to this scenario so that the student will still be able to follow the lessons and participate in student-to-instructor interaction.

In the vision of the 21st century classroom, students are equipped with portable wireless devices connected to an infrastructure, which enables polling, question queue, slide synchronization, and remote access (Dominick, 2002; Kramer &

Strohlein, 2006). However, such interactivity restricts the students to be connected to the same infrastructure. Students connected to other infrastructures may still be able to access information from the server but will lose the opportunity to participate interactively in the class.

Several wireless classroom projects have been implemented (Singh & Baker, 2007; Choi et. al, 2007; Petropoulakis & Flood, 2008) and tested with encouraging results. However, the implementation is restricted to a Local Area Network (LAN) coverage, making it impossible to carry out such interactive learning when a student is outside of the LAN.

NTU'S MOBILE AND INTERACTIVE LEARNING SYSTEM

Objectives

To overcome the above mentioned problems, i.e. the lack of interaction and the need to be physically present, a wireless mobile interactive learning solution is proposed with the following objectives in mind:

1. To provide interactive learning capabilities in classroom environment by allowing instant communication between the professor and the student through messaging services.
2. To extend the mobile learning opportunities in NTU to include the Wide Area Network (WAN) coverage provided by telephone companies.
3. To make wireless mobile interactive learning available to all students both on and off campus without having to incur costly hardware.

Proposed Approach

The new solution is based on two new wireless technologies, namely (1) Multimedia Messaging Service (MMS), and (2) Internet Naming Service (iName). Multimedia Messaging Service (MMS) is a new way of mobile communication and is believed to be one of the key driving forces of mobile data service business for 2.5G and 3G. Major advances in technology of instant messaging and the rapid evolution of the capabilities of mobile devices has made it possible to provide multimedia rich messaging application to mobile users. The project described in this chapter made use of the MMS technology and wireless LAN infrastructure, to build a mobile learning platform.

The key idea is to make use of MMS service as the interactive communication medium during a class. Professors can check the understanding of students by posting a question during a class and the students will reply by MMS over the University's WLAN network. An application on the professor's console will then receive and be able to parse these responses and summarize them for the professor. When satisfied with the response, the professor can then proceed to the next part of his/her lecture.

Students can also post feedback messages to the professor during the class (e.g. to inform the professor that he/she is too fast and needs to slow down on his explanation). These messages will appear on the professor's console and the professor can decide whether or not to respond to them. One of the advantages of using MMS is that this interaction can easily be extended to students who are not physically present in the classroom. The same data format can be used over WLAN or GPRS.

Live audio streaming of the lectures can be performed which allows the student to participate in a class even though he/she is not physically present in the classroom. The student can be on-campus and connected via WLAN, or be off-campus and connected via GSM/GPRS. The wireless devices used can even be a MMS-enabled mobile phone, thus reducing the need to depend on expensive laptop or tablet PCs.

The iName technology serves to provide address abstraction for the whole service. With iName

Figure 1. The overall architecture of the wireless mobile interactive learning framework

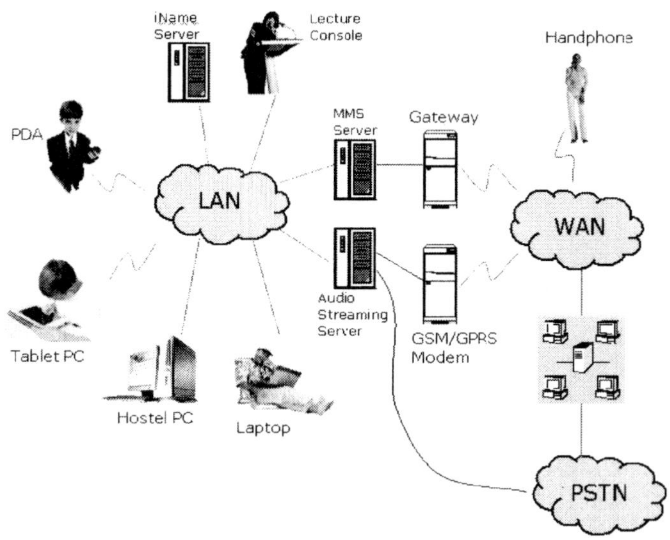

service, the student only need to enter an intuitive text-based name such as "CE101Lecture" and the iName server will map it to a valid IP address. Similarly, the professor can instantly communicate with all his students no matter whether they are physically present in the classroom. The iName server will host the current dynamic IP address for connected laptops or PDAs, or phone numbers for mobile phone connections so that all students can be reached via the MMS messages.

Overall Architecture

The overall architecture of the NTU Wireless Mobile Interactive Learning framework is shown in Figure 1. Client devices such as PDAs, Tablet PCs, will be connected to the NTU LAN via wireless access points. The MMS sent by the students (e.g. lecture feedback) during a lecture will go through the iName server to determine the forwarding address of the professor. An application residing on the lecture console can then pick up the MMS and display it to the professor.

The professor may send out a pop quiz to its students via MMS. The MMS sent by the Lecture console will go through the iName server to determine the forwarding address of the students' client devices for all the students registered for his/her class. If the clients are on the WLAN, the iName server will return the client's dynamic IP address. If the clients are outside the WLAN, it will return a handphone number and the MMS server will send it out to the student's handphone on the GSM/GPRS telephone network (WAN).

The professor's vocal presentation is recorded live on an Audio Streaming server and will be streamed out to a conference call centre hosted by the telephone company. Students who are not in the campus WLAN may dial in to the conference call centre via a PSTN line or even through their GSM handphones. In this way, students who are outside campus can listen and interact with the class during a lecture.

Key Components

Multimedia Messaging Service

MMS (Ericsson, 2005), as its name suggests, is the ability to send and receive messages comprising of a combination of text, sound images and

Figure 2. Transaction flow diagram of the system

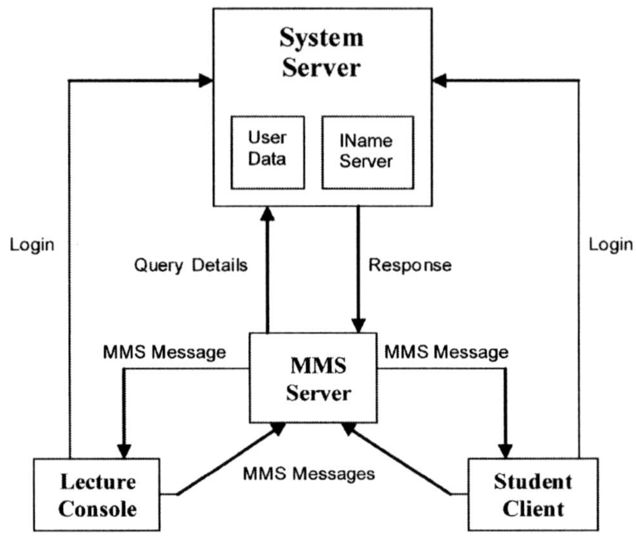

video to MMS capable handsets. The WAP-forum (Open Mobile Alliance, 2006) and the 3GPP (3GPP, 2003) are the groups responsible for standardizing MMS. MMS employs the Wireless Application Protocol (WAP) and therefore it is bearer independent – supporting either Circuit Switched Data or General Packet Radio Service (GPRS). MMS should also eventually support bearers such as Enhanced Data rates for GSM Evolution (EDGE) and 3G too.

MMS presentations are different from email presentations. MMS provides advanced layout and timing for multimedia contents in the message, which is not provided by email message. Furthermore, neither IMAP3 not POP3 provide a standard technology to notify a non-connected client of an incoming mail.

The MMS-enabled device (e.g. PDAs, laptops, tablet PCs) must be able to compose, send, receive and play an MMS message over the 802.11b WLAN. The MMS server sitting in the infrastructure network will serve two purposes: (1) MMS Centre (MMSC) that stores and forward the MMS messages, and (2) MMS Proxy/Relay that will send/receive MMS messages to the Wide Area network through the gateway.

Internet Naming Service

iName is an internet naming service which translates a name query into network addresses such as IP address or mobile phone numbers. iName is designed to be a physically independent way of referencing any user in the system. All the components in the project address a particular user in the system with their respective iName. This frees the system from the need of knowing where and how the user is logged on to the system. To consolidate information regarding connectivity, each student's connection is passed through an iName server to identify current network connection type. The iName server then serves as a host to store the current dynamic IP address for connected laptops or PDAs, or phone numbers to handphone connections. Hence, if the established connection is through a handphone, the e-slides/video or audio transmission will be modified to MMS format and transmitted through the telecommunication network; while for LAN devices, the relevant information would be transmitted through the net.

The iName server consists of several protocol layers to communicate with the infrastructure

Figure 3. Screenshot of the integrated lecture console

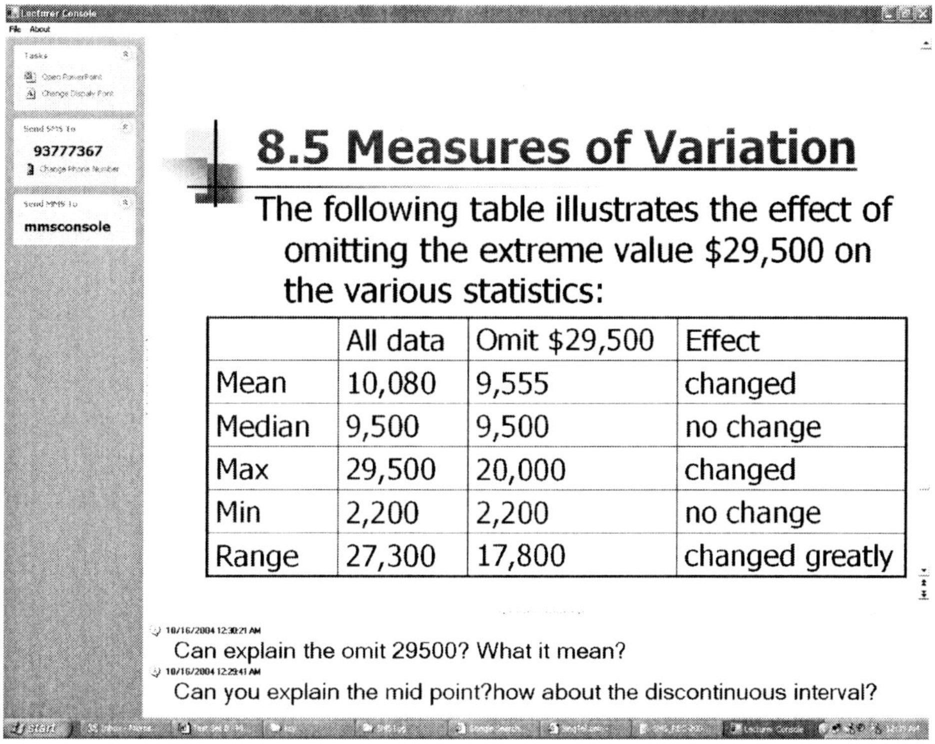

network. It accesses an SQL database to store as well as to retrieve the current dynamic IP addresses of all its registered clients.

Figure 2 shows the message transaction flow schema of the overall system. The communication between the professor and the student will be done through MMS-formatted messages. Both the Lecture Console and the Student Client are the client-type device in the system. The MMS Server is sitting in between the clients and serves to store and route messages among clients. The iName Service serves to provide address mapping to all entities in the system.

System Evaluation

The first version of the system has been rolled out and tested on students in real classes. The test groups consist of two groups, with size of 150 students, comprising of first year Computer Engineering students, in the subject of Engineering Mathematics. The system ran during the one-hour lectures in Engineering Mathematics, 3 times a week, for a duration of 2 months. All received messages were logged and displayed anonymously on the lecture console. Figure 3 shows a screenshot of the integrated lecture console.

On average, in a typical one-hour lecture, 57 messages were received. These messages can be categorized into the different group headings with a mean distribution as shown in Table 1:

From the messages received, a lot can be understood about the students' understanding and their needs. Because the system is anonymous, the students were not intimidated to speak freely. Also, the students lend support to one another to reinforce a request that many of them want but is too shy to ask. This is something that could not have being possible without this interactive system.

Table 1. Messages received in a typical one-hour lecture

Test Messages	**6**
Questions about subject	**10**
Requests for additional information / actions	**10**
Supportive messages	**13**
Feedback on Lectures	**11**
Error messages	**2**
Irrelevant Messages	**5**
Message Received	**57**

Here are the descriptions of the different message categories and what can be understood about the students:

- **Test messages**: These are messages sent by the students to test that their messages are received properly by the system. Examples include "*Test*", "*testing...*" and "*Hi*", Some students send these test messages before sending their real messages for the first time, and these messages are sent even when they can see other students messaging successfully in the system. This shows the level of belief (or rather disbelief) they have that the system actually works.
- **Questions about subject**: These are messages that directly pose a query on the content of the subject. Examples include "*Sir, what do i and j in the formulas stand for?*" and "*why n-1 in this eg.Pls sir explain again*". The number of such messages shows directly the level of understanding that the students have in the subject.
- **Requests for additional information / actions**: These are messages that students send to request for information / action that is not directly related to the content. Examples include "*Sir wats not examinable*", and "*Tell a story first*" and "*any in between breaks?*". These messages are the most important ones that portray the needs of the students. Many messages are related to what topics will be examined – an indication that the students are more interested in the exam than in the learning.
- **Supportive messages**: These are messages that students send to support a request made earlier by another student. When one student messages "*no 1 can understand anything in the prog lect*", several other students messages in unison with "*agree:)*" and "*Agree =*".
- **Feedback on lectures**: These are messages that give feedback on the lectures. Examples include "*Please lower down your volume!*" and "*Havent finish looking..*"
- **Irrelevant messages**: These are messages that are totally irrelevant to the lecture or the course. This is sent because the student may be bored, or just wants to attract attention. Examples include "*Upper half, middle column, 4th row, girl in white jacket... ^ look cute... Asl please ?*" (internet chat lingo for age, sex and location) and "*There's a lizard on the ceiling.* " These messages are actually very disruptive to the class and it affects both the professor as well as the students. However, there are many other students who do their part to respond "*Stop abusing the system....*". and " *IF U ARE EDUCATED ENOUGH, PLS POST QNS CONCERNING THE LECTURE. THANK U*". Upon seeing these messages, the culprit does stop. (see Figure 4)

In summary, some of the direct advantages observable from the test results are:

- Students are more active in giving feedbacks. From the test conducted, the increment in the level of participation of the students in class can be easily seen. Students who are normally too shy to interact are also able to participate.
- The professor can get a feel of students'

Figure 4. Examples of messages received in a 1 hour lecture

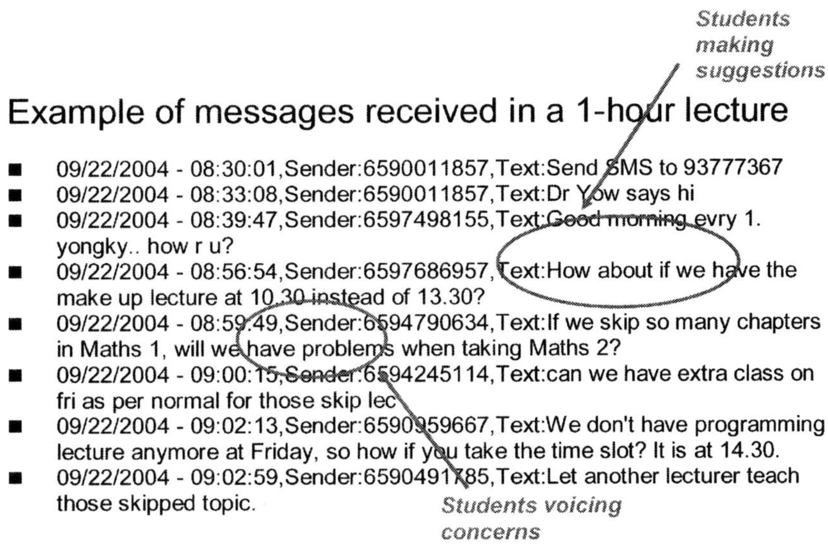

needs and the their understanding level. The comments posted indirectly reflect the understanding level of the student, and therefore based on the comments, the professor can adjust the pace of learning.

- Whenever the professor receives multiple and different responses, he/she can query the class to get a consensus.

However, there are also some disadvantages noticeable from testing the system:

- An overwhelming number of replies, which makes it difficult for the professor to address all of the student's needs.
- The professor will have to be able to split his concentration between the subject and the responses coming from students.
- Disruptive remarks.

The overall responses shown favorable results. Messages from the student includes "*Nice system, well done !*" and "*Agree..! Asik2.. This program is cool..*". Students in general finds this beneficial and are eager to make use of the system. The main advantage of this learning system is that MMS messaging is easily extensible to the mobile GSM networks, so students are not restricted to use it only on campus. Sending MMS messages on the Wireless LAN will be free to students on campus, and only at a low cost outside. This system is particularly beneficial in engineering education, since engineering students are generally tech-savvy, and therefore can easily adapt to this medium. This learning system will encourage students to be more participative in the learning process, since usually students tend to be shy to speak up in large groups.

Integrating Peer-to-Peer Learning

While the original intent of the developed system is mainly to facilitate the interaction between the instructor and students during a lecture in a large class setting, in this section, we propose to extend the use of the system beyond this traditional instructor-centric approach of learning to encompass some form of concurrent peer-to-peer based learning in the background, by using the same system to facilitate the communication of knowledge between the students during a lecture. Research into peer-to-peer learning (e.g. Soller et. al, 1999; Wagner & Gansemer-Topf, 2005; Resta & Laferrière, 2007) has shown that more effective learning can be achieved when learners interact, communicate their thoughts, and share their knowledge with each other than when they work in an individualistic and competitive manner. Such practice of students giving and receiving advice and assistance from one another will foster a greater sense of belonging, communal learning, and group support among the learners.

The proposed peer-to-peer learning takes the form of providing an avenue for students to offer an answer or share their thoughts about some of the questions sent to the professor by other students through MMS during the lecture. This brings about two advantages, one of which addresses an issue discovered with the system mentioned earlier:

- It relieves the professor from having to respond to all questions from students sent during the lecture by himself/herself
- It harnesses the intelligence and knowledge of stronger students to aid the understanding of the weaker students on the subject, and in the process reinforces the understanding of the stronger students themselves.

Proposed Integration

The peer-to-peer learning feature can be incorporated into the present system with no changes required in the hardware setup and only a software update to the application residing in the lecture console is expected. In the current system, the professor has the option of either making the MMS messages received by the console visible to the students through the screen projector, or allowing them to be seen only by the professor on the console display. Here, we assume the latter, which is the choice of most professors who used the system, as it avoids the risk of the lecture being disrupted unnecessarily by irrelevant messages.

We also assume there is a team of students who register themselves with the system as 'perceptors' for each specific course subject. The term 'perceptors' refer to students who are concurrently enrolled in a course in which they serve as 'in-class' teaching assistants. This is a term first coined by researchers at the University of Arizona (Larson et. al, 2001) in their proposed 'just-in-time' model for student-assisted teaching, which contrasted with the conventional use of senior undergraduates or graduate students as teaching assistants (TAs) for 'outside-of-lecture' assistance (e.g. in labs and tutorials). Due to the fact that perceptors are co-enrolled peer assistants in the course, their ability to be content experts may be limited to their 'just-in-time' command of the course material. However, their status as peer learners often makes them more approachable, and their in-class presence means greater opportunities to mentor and offer immediate advice to their fellow classmates than are possible with TAs. Thus, they represent an essential in-class resource that should be harnessed for more effective classroom teaching and learning.

Figure 5 illustrates the modified transaction flow diagram of the system in which a new perceptor console group has been introduced for peer-to-peer learning. The following explains the sequence of transactions (or events) that will occur when a student sends a MMS message to the professor. The sequence follows an alphabetical order as shown in the figure.

a. Prior to the start of the lecture, both perceptors and students will login to the system to register their presence (or attendance) for the lecture.
b. During the lecture, Student A sends a MMS message to the professor.

Figure 5. Modified transaction flow diagram of the system

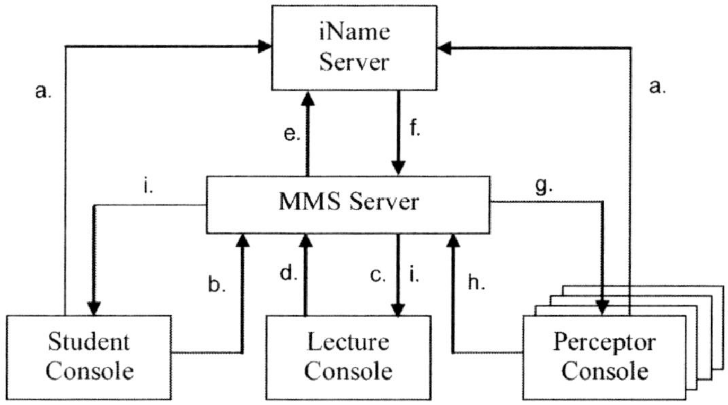

c. The message arrives at the Lecture Console through the MMS Server.
d. The professor reviews the message, determines that it is about the topic of the lecture which may be answered by other students, and decides to delegate the question to the perceptor group.
e. The message reaches the MMS Server, which in turn queries the iName Server for the IP addresses of perceptor members who are present.
f. The iName Server responds with the required addresses.
g. The MMS server forwards the message of Student A to the perceptor group. The message should be forwarded without revealing the identity of the student in order to avoid potential bias or discrimination, and also to better encourage participation from students who may feel shy or embarassed to ask questions.
h. Each perceptor who receives the message is then requested to respond within a certain time limit. Likewise, only the reply but not the identity of the perceptor should be sent to the enquirying student. When a perceptor responds, the reply message is first received by the MMS Server.
i. The MMS Server in turn forwards the reply message to the student as well as sending a copy to the professor.

It is essential to inform the professor of any responses received from the perceptors, so that in the event that no response is received within the timeout of the message, the professor can still be notified by the Lecture Console, and he/she may respond directly to the question during the lecture. The professor can also evaluate the quality of the perceptor responses by having each response (not the perceptors themselves) rated by the student immediately after receiving, and sent to the Lecture Console. The responses may be rated in a way similar to that of current Internet Q&A portals such as (Answers, 2005, Answers, 2005Yahoo! Answers, 2005!) and (Amazon Askville, 2006).

Discussions

As with the original intent of our learning system, the proposed peer-to-peer learning sub-system is not restricted to be used by only 'in-class' students who are physically present in the lecture. Instead, any 'in-lecture' student, including students who are accessing the lecture remotely via MMS-enabled handphones, can participate and benefit from the proposed system, as its underlying communication architecture is already able to support interaction between any participants 'in' the lecture.

There is an open issue on how professors can get students to participate as perceptors. In a way, the 'service' rendered by the perceptors to their peers is similar to that by existing student teaching assistants. Hence, the forms of incentives to encourage participation such as the award of bonus credits or monetary units upon satisfactory completion of their service as perceptors, can still be effective.

While the proposed system opens a new avenue for students to seek peer advice in their learning, there is also a potential for abuse by some students who have become over-reliant on their peers for quick answers to questions that they might have answered for themselves. To discourage such behavior, a possible strategy is to enable the perceptors to similarly rate the questions they received from others, and those with a tendency of making 'low-rated' queries may have their access to the system restricted in future.

FUTURE TRENDS

In the future, technology-aided learning will bring about changes in the way students learn. The following lists the key situations where students will benefit from such technology-enhanced learning systems. These are situations where one or both identified problems of learning in large classes, i.e. the lack of interaction; and the need to be physically present in the lecture, can be addressed by the proposed system.

Embarrassment-Free Instant Lecture Feedback

During the course of the lecture, students can send feedback in the form of MMS messages to the professor to indicate that they do not understand certain part of the lecture or to request the professor to slow down or speed up. This could be initiated by the student or as a response to the professor's query. The professor can see all these messages on the console and he/she can determine whether or not to respond to them. Students with questions that need the help of a diagram can pull up a figure from the lecture slides, annotate on it and send it to the professor.

Catch a Lecture in a Bus When Late for Class

Suppose a student is late for class and he/she is still on the bus journey when a particular lecture starts. Instead of missing the first part of the lecture, the student can use his/her handphone and dial in to the lecture. The student only needs to type in the name of the lecture he/she wants to listen to and the iName server will map it to the appropriate conference call number. While listening to the lecture over the handphone, the student may view the lecture slides from a hardcopy printout that is printed earlier.

Avoid Wastage of Waiting Time at Official Functions

Students (especially student leaders) sometimes have to miss lectures due to official functions in the University (e.g. visits by International Accreditation Panel, etc.). These students usually have to report to the meeting venue well ahead of time and thus lose valuable time waiting for their guests. Since these students are still on campus they can participate in the lecture over their WLAN-enabled devices. They can listen to the professor's voice, view lecture slides and even send their queries over WLAN.

Lectures Will Not be Missed Due to Illness

Students who are ill and had to stay at home will not need to miss their lectures. From their home, they can dial in to the lecture from their PSTN phone and listen to the lecture. They can still interact with the class through their MMS-enabled handphones.

Learning Support can Come from Beyond Your Social Network

With the proposed peer-to-peer learning, students can benefit from the knowledge of anonymous peers, not just from their own study mates and close friends. Since the query and response are anonymous (but known to the system), one no longer has to feel awkward to seek support in their learning from unknown peers.

CONCLUSION

In this chapter, a new mobile interactive learning system developed at Nanyang Technological University is presented. A system that allows instant feedback on teaching has been developed using MMS on the campus-wide Wireless LAN. The system made use of two new wireless technologies, namely MMS and iName. The proposed system enables instant lecture feedback to be delivered and interactive quizzes to be carried out during the lecture. This learning system is intended to encourage students to be more participative in their learning process. Sending MMS is free on campus, and its use can be easily extensible to GSM networks. The current learning experience through the system can be further enriched by a proposed integrated support for peer-to-peer learning to facilitate knowledge sharing among students during the lecture. A design for the support of such peer-to-peer learning within our current system architecture is presented, and we reserve the implementation and investigation on its impact on the students' learning for future work.

ACKNOWLEDGMENT

The work in this chapter is supported by the Hewlett-Packard Philanthropy Grant for applying mobile technologies in a learning environment.

REFERENCES

Amazon Askville. (2006). Retrieved 24 July, 2008, from http://askville.amazon.com

Anderson, R. J., Anderson, R., Vandergrift, T., Wolfman, S., & Yasuhara, K. (2003). *Promoting Interaction in Large Classes with Computer Mediated Feedback*. Paper presented at 5th International Conference on Computer Support for Collaborative Learning (CSCL), Bergen, Norway.

Andronico, A., Carbonaro, A., Colazzo, L., & Molinari, A. (2004). Personalisation services for learning management systems in mobile settings. *International Journal of Continuing Engineering Education and Lifelong Learning, 14*(4/5), 353–369. doi:10.1504/IJCEELL.2004.005726

Barajas, M., Chrysos, A., Bosco, A., Fonollosa, M., Alvarez, I., & Sancho, J. M. (1998). *Virtual Classrooms in Traditional Universities: Changing Teaching Cultures Through Telematics*. Paper presented at the World Conference on Educational Multimedia and Hypermedia, Freiburg, Germany.

Brown, R. E. (2001). The process of community-building in distance learning classes. *Journal of Asynchronous Learning Networks, 5*(2), 18–35.

Bullen, M. (1998). Participation and critical thinking in online university distance education. *Journal of Distance Education, 13*(2), 1–32.

Choi, Y. B., Abbott, T. A., Arthur, M. A., & Hill, D. N. (2007). Toward a future wireless classroom paradigm. *International Journal of Innovation and Learning, 4*(1), 14–25. doi:10.1504/IJIL.2007.011472

Dominick, J. (2002). Ready or Not – PDAs in the Classroom, *Campus Technology Magazine*, Retrieved 24 July, 2008 from http://www.campustechnology.com/article.asp?id=6705.

Ericsson, (2005). MMS Services Moving Forward. *Ericsson News*, Retrieved 24 July, 2008 from http://www.ericsson.com/solutions/news/2005/q3/20050725_mms.shtml

Geske, J. (1992). Overcoming the drawbacks of the large lecture class. *College Teaching, 40*(4), 151–154.

Gleason, M. (1986). Better communication in large courses. *College Teaching, 34*(1), 20–24.

3GPP. (2003). 3GPP TS 23.140: MMS Functional Description. Retrieved 24 July, 2008, from http://www.3gpp.org

Kramer, B. J., & Strohlein, G. (2006). *Exploring the Use of Cellular Phones for Pervasive eLearning*. Paper present at 4th IEEE International Conference on Pervasive Computing and Communications Workshops, Pisa, Italy.

Larson, H. P., Menke, R., Tollefson, S. J., Harrison, E., & Berman, E. (2001). The University of Arizona Teaching Teams Program: A "Just-in-Time" Model for Peer Assistance. In J. E. Miller, J. E. Grocia, & M. S. Miller (Eds.), *Student-Assisted Teaching: A Guide to Faculty-Student Teamwork* (pp. 27-33). Bolton, MA: Anker Publishing.

Miner, R. (1992). Reflections on teaching a large class. *Journal of Management Education, 16*(3), 290–302. doi:10.1177/105256299201600302

Mortera-Gutiérrez, F. (2002). Instructor interactions in distance education environments. *Journal of Interactive Learning Research, 13*(3), 191–209.

Open Mobile Alliance. (2006). *Wireless Application Protocol*. Retrieved 24 July, 2008, from http://www.openmobilealliance.org/Technical/wapindex.aspx

Petropoulakis, L., & Flood, F. (2008). Interactive student engagement using wireless handheld devices. *International Journal of Engineering Education, 24*(1), 46–55.

Resta, P., & Laferrière, T. (2007). Technology in support of collaborative learning. *Educational Psychology Review, 19*(1), 65–83. doi:10.1007/s10648-007-9042-7

Singh, D., & Baker, Z. A. (2007). Wireless implementation of a mobile learning environment based on students' expectations. *International Journal of Mobile Learning and Organisation, 1*(2), 198–215. doi:10.1504/IJMLO.2007.012678

Soller, A., Lesgold, A., Linton, F., & Goodman, B. (1999). *What Makes Peer Interaction Effective? Modeling Effective Communication in an Intelligent CSCL*. Paper presented at the AAAI Fall Symposium: Psychological Models of Communication in Collaborative Systems, North Falmouth, MA, USA.

Wagner, M., & Gansemer-Topf, A. (2005). Learning by teaching others: A qualitative study exploring the benefits of peer teaching . *Landscape Journal, 24*(2), 198–208. doi:10.3368/lj.24.2.198

Yahoo! Answers (2005). Retrieved 24 July, 2008, from http://answers.yahoo.com

Chapter 17
The "Trigger" Experience
Text Messaging as an Aide Memoire to Alert Students in Mobile Usage of Teaching and Learning Resources

Joan Richardson
RMIT University, Australia

John Lenarcic
RMIT University, Australia

ABSTRACT

This case study chapter will outline the results of a 2006 pilot test into the use of Short Message Service (SMS) to augment the provision of student administrative services currently available through a university website. The pilot conducted utilised an SMS Prototype Tool Trigger that enabled dynamic information transfer between staff and students. Trigger facilitated live update reminders that assisted students to schedule their time and better organise themselves. Specifically, SMS technology was used to deliver physical class locations, availability and web addresses of iPod resources, important events, alerts for multimedia, examination schedules, and, assessment feedback by 'pushing' information to students. Trigger also provided students with pull access to study schedules and requirements. The aim of the test was to evaluate student response to the use of Trigger to improve the learning environment. The case study will identify student responses to the 2006 pilot and describe a current project that has extended the number of students participating in the study.

INTRODUCTION

The product of contemporary civilization's symbiotic relationship with technology is the prospect of being immersed in data smog (Shenk, 1997), an experience that can lead to a state of information anxiety (Wurman, 1989) in some individuals.

DOI: 10.4018/978-1-60566-703-4.ch017

External aids to memory have served humanity as mnemonic devices since the first person tied a piece of string around their finger to remember something important. In this era we have digital watches, PDAs and mobile phones to act as memory prompts and prosthetic aids to time management. In the case of mobile phones, these devices virtually eliminate the need to recall phone numbers with their electronic storage capacity capable of holding hundreds of personal contact details (this being a conservative estimate). An emerging adjunct to the mnemonic facility of the mobile phone is the use of text messaging to emit and receive alerts. SMS has rapidly become a commonly employed communications protocol amongst mobile phone owners (Faulkner & Fintan, 2005). Mobile phones permit users to almost communicate in parallel modes, these being voice and text. An itinerant technology that has spawned a cult dialect, text messaging is the austere equivalent of conventional e-mail: Economical in practice and terse in delivery with standard messages capped at 160 characters. It is the fragmented lingo of students the world over and as such an excellent candidate for a personal aide memoire. The *Trigger* system outlined in this chapter describes one approach to using an existing means of communication (namely SMS) to function as a mnemonic construct to enable better personal organisation of a student's scholastic activities.

BACKGROUND: MOBILE COMMUNICATION

Mobile phone penetration of the Australian population is high and expected to remain that way. In 2006, 950 million mobile phones were expected to be sold, a figure that far exceeded the 234 million PC's. (Arvind & Hicks, 2006). The number of mobile phones owned has increased from 8.1 million in 1999-2000 to 19.8 million in 2008 (The Age, 6/04/2008). This recent explosion in SMS use for the purpose of communication is global in nature, although not in all countries. However, the 20.5 billion messages sent in the UK in 2003, (Faulkner & Fintan, 2005) substantiate SMS as the 'killer' application of mobile phones. Australian Mobile Market statistics indicate that Australians sent over eight billion SMS messages in the 2005/6 financial year, an average of at least 300 messages for each subscriber. Figure 1 shows the exponential increase in the volume of text messages that occurred in Australia from 2005/2006.

The growth in general usage has exceeded expectations primarily due to the low cost of messages and acceptance in the youth culture as a means of scheduling social events. Reminders and the sense of control provided by the asynchronous nature of the communication mode enable users to reflect before sending a reply at their leisure. SMS enable generation Y to receive and send private messages (Markett & Sanchez & Weber & Tangney, 2006). Steve Boom, Yahoo's senior vice president for broadband and mobile, views the mobile internet today as entering an era where the PC-based Internet was in '96 or 97 (The New York Times, 9/01/2007). This strong sales market for technology has made mobile phone ownership ubiquitous amongst university students. A survey conducted by the University of Technology Sydney (UTS) showed that 95% of students owned mobile phones, 73% owned MP3 players or iPods, 23% had their own games console and 15% had a PDA (Bachfischer & Lawrence & Litchfield &Dyson & Raban, 2008).

The New Media Consortium 2007 Horizon Report stated that the higher education sector was facing a growing expectation to deliver services, content and media to mobile and personal devices (The New Media Consortium & EDUCASE Learning Initiative, 2006). Mobile technology provides the ability to deliver content, but also to efficiently provide important alerts, warnings and instructions over a greater range of times and locations than any other device, a demand of the general marketplace. For applications where information is concise and timeliness and ease of access are important, smaller capacity mobile devices will have a major role to play in education, as they do in other parts the community. Connectivity is increased to almost 100% of the time for those who carry a mobile phone. Whilst the potential of mobile technology has been demonstrated, ways to effectively deploy it more broadly in the education sector have yet to be established.

As a result, Pearson Education Australia and RMIT University designed and developed a prototype of a Short Messaging Service (SMS) system

Figure 1. Text messages, Australia mobile market statistics (2008)

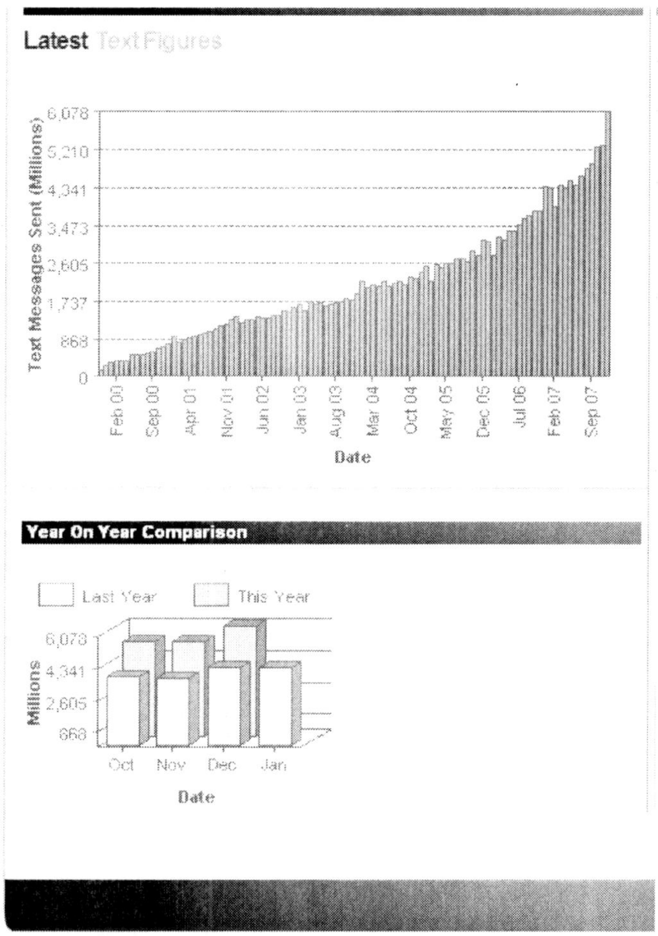

to improve student uptake of learning resources outside traditional classroom boundaries; further develop the *Trigger* prototype to demonstrate fast one-to-many communication and immediate access to facilitator instructions, in real-time. (Platts, 2004) Teachers were able to engage students outside the classroom through the use of targeted SMS reminders for work to be undertaken and alerts for events that provide opportunistic augmentation of learning. Specifically, SMS technology was used to facilitate the delivery of, physical class locations, availability and web addresses of iPod resources, important events, alerts for multimedia, examination schedules, and, assessment feedback by '*pushing*' information to students. The *Trigger* application also provided students with *pull* access to study schedules and requirements that enabled them to manage their time.

Learning management systems (LMS) are pervasive. Subject content and information can be read as printed or electronic texts or on the web, students can listen to lecture material on their iPod, and timetabling, enrolment, assessment schedules and grades are all available online. All of this University information is readily available though the Internet. As such, students now use a number of technologies to access a range of services. Access to educational Information Systems (IS) generally involves the ICT *pull technique*, which typically employs a desktop computer with

Internet access. This may involve communicating with multiple IS that require disparate log on processes. A frequent request by students (Platts, 2004) is to avoid this replication. Students' value proactive information that includes: important announcements and significant events; their preference is for Email or SMS.

A 2007 survey of 442 students at the University of Technology Sydney provided the following valuable insights, which supported the findings of the RMIT 2006 and 2008 pilot studies. Some students could see SMS as useful for administration tasks, such as room or timetable changes, but also believed that email backup was essential. They I expressed interest in receiving SMS notifications, but as an addition to e-mail rather than a replacement. This seems to be the most preferred option, with many other students requesting that SMS be used to alert them to important information by providing them with basic information, advising them to check their e-mail to obtain an announcement, or more detailed information. SMS was favoured for alerts and reminders; however, it is not seen as being particularly suitable for mobile learning tasks. (Lawrence & Bachfischer & Dyson & Litchfield 2008)

Issues, Controversies, Problems

There are challenges in effective integration of SMS technology with existing educational sector ICT systems. It is not just a matter of ensuring that the technology application is robust. Trials and subsequent review is needed to identify the most appropriate use of SMS in the educational environment. Reminders for assessment due dates and alerts for additional available learning activities have been included in the prototype's functional design. The next problem experienced during the prototype pilot was the use of language elements to convey the messages to the student participants.

The current project is examining the quality and appropriateness of the information made available in this manner. How language can most effectively be used within technological and socially constructed constraints is important to the uptake of digital curriculum products by the student population. SMS can facilitate reminders and alerts where only the information requested is delivered which has the potential to reduce the impression of information overload and scheduling confusion. Acquisition and exchange of data and information was designed to be as simple and efficient as possible. Scheduling information and in particular assessments due dates were made available to students whilst reminders for assignments and feedbacks were sent. This ensured that only information deemed necessary or relevant to the user was exchanged (Huang & Pulli & Rudolph 2005).

Universities, publishers and authors have invested heavily in the development of suites of resources that use a range of new technologies to augment learning in traditional educational environments, homes, workplaces and more recently in transit among these diverse settings. This variety of delivery modes for learning materials provides choice for academics and students' but also causes fragmentation of the learning narrative, information overload, confusion about activity scheduling and difficulty in the provision of alerts or warnings to all students in a timely fashion. A profound gap has emerged between the expected improvements to learning and teaching expected when technology is used and the real impact (Head, 2008). How to install and use the technology is not difficult as institutions have technology services staff and students are adept at Googling to enable cutting and pasting, defining terms using Wikipedia, entertaining themselves with content from YouTube, iTunes and MySpace and scheduling their life using SMS before they enter university (Head, 2008). How to remind students using technology to increase the uptake of curriculum resources outside the traditional classroom is at the centre of the problem for academics.

The complexity of the environment generates

a potential to reduce rather than improve learning, despite the time, funds and expertise directed to creating and delivering readings, videos, exercises, case studies, assessment tools, lectures and simulations. A simple and direct method for advising students of services available, suitable delivery modes and a sensible order for undertaking learning activities is a necessary but complex task due to technologies' removal of boundaries around learning environments (Stone, 2004).

Generally textbooks currently contain theoretical knowledge, self-assessment exercises, business scenarios that highlight ethical issues, problem solving and group based activities that require discussions and reflective practice questions to help students illuminate key issues. CDs contain case-study problems and step-by-step guides to the acquisition of skills that ensure attainment of 'work ready' capabilities. Facilitator's guides are disseminated using web-based LMS's and CDs and provide suggestions for appropriate uses of subject resource components to support classroom activities and assessment tasks. Web based Learning Management Systems also generally provide student choices in the completion of self assessment tasks, immediate feedback in the form of staged solutions, schedules for learning modules and the capacity to work collaboratively. SMS reminders to complete teacher directed activities utilising existing resources associated with texts outside traditional classrooms may improve uptake of available resources. (Richardson & Lenarcic, 2007a). Currently a major problem for resource construction and delivery, using technology is the re-integration of the learning narrative despite technologically supported deliveries in an environment where the boundaries of the traditional classroom have been removed. Re-building the storey is vital to produce students that are technically proficient within their discipline and able to effectively share and integrate discipline-specific knowledge (Corsini & Crittenden & Keeley & Trompter & Viechnicki, 2000; Wheeler, 1998).

The pilot of the SMS trigger prototype enables text message responses specifically generated (triggers) words for each subject without the addition of technical adjuncts to the students' phone which enables the application to function as if there was an invisible client on the mobile phone that provides *pull* access to information without direct access to teachers. Widespread ownership of mobiles (The Age, 6/04/2008) creates the potential of SMS as a simple source of important one-to-many *blast* messages from the disparate university and school systems to specific student cohorts, via Email accounts, SMS or web interfaces. *The ability of students' to obtain reminders changes administrative information systems (IS) from primarily information dispersal TO students, to an information acquisition FROM the institution using an informal communication style commonly used by students and accessible to parents.*

SMS Alerts and Reminders

The 2006 *Trigger* fulfilled the reminder and alert provision potential of SMS technology through the creation of a prototype that enabled '*push*' and '*pull*' instructions for students to use existing resources to complete homework, that:

- Enable reflection and review of important concepts;
- Provide recommended schedules for learning activities and assessments;
- Headline Web addresses for downloadable television, radio, newspaper articles journals papers, magazines and events and
- Encourage the completion of preparatory exercises.

SMS has the potential; to be used as an utilitarian tool, that provides access to information in real-time, as there is often no impact during times of heavy traffic or adverse conditions that can overwhelm other wireless networks, which also makes the tool effective as a link of last resort for crisis communications (McAdams, 2006). SMS systems utilise the ability of text messages to reach their destination even when a cell phone signal may be too weak to sustain a spoken conversation (Boyd, 2008). The fact that a text message will queue if the destination is unavailable rather than requiring a redial like a traditional voice phone call enables asynchronous communication.

The type of communication for the digital native generation would appear to favour short and simple messages: – urgent SMS alerts, as mentioned earlier, seem very suitable. The April 2007 case (Hauser & O'Connor, 2007) of a student shooting staff and other students at Virginia Technical University provided a graphic example of how university management could have warned students that a gunman was on campus had such an alert system been in place. *Trigger's* functionality reminds and alerts students about work requirements to be completed outside class time and also utilises the ability of mobile services to free people from committing to physical presence and commitment to a pre-determined schedule in order to be accessible to another (McClatchey, 2006).

SMS to Remove Geographical Boundaries

McClatchey (2006) described how innovative utilisation of available technologies such as the Internet, e-mail and iPods has enabled Universities to respond to generation Y, fee-paying student expectations driven by the marketplace. *Trigger* utilises SMS to respond to students, guardians and staff expectations for immediate responses to questions asked especially when reminders for schedules, alerts for important events or warnings for disasters are requested. The shift towards mobile technologies highlights the importance of allowing individuals to access private information whilst in a public space (Faulkner & Fintan, 2005). Customers expect messages to be transmitted and responded to without the necessity for a face-to-face interaction or both parties being at either end of a telephone simultaneously. In the past students' excuses for late submission of assessment tasks could include statements like "I tried to call but you did not answer" or "I left a message on your answering service you but you did not call" or "I was out when you tried to call" or "Sorry I forgot to leave the call back number" (Reisman, 2006). None of these reasons not communicating are effective anymore due to advances in mobile telephones, like SMS.

A study conducted at Kingston University in the UK, identified SMS as a technology that students already used (Stone, 2004). This influenced the choice to utilise it, to *blast* first year undergraduate students with reminders for assignment due dates and available learning resources (Stone, 2004). Information made available to students using *Trigger* at RMIT provided time-sensitive information that was also disseminated using Blackboard, the Internet, e-mail and hardcopy. SMS technology reflects attributes of Email in that it is asynchronous and enables automatic reply without having to recall an address or phone number (Faulkner & Fintan, 2005).

SMS Provides Information Access at a Minimal Cost to Students

Cost factors are another major impediment to universities embracing m-learning. The main cost barriers appear to be usage charges billed by telecommunications providers and the price of mobile hardware. If anything, universities are probably more afraid of usage charges than hardware costs. The contained cost of hardware versus the unknown, and potentially escalating cost of usage, poses challenges to university budgeting and existing practices (Richardson & Lenarcic, 2007a).

The cost of mobile phone technologies is minimal compared with desktop computing facilities, which increases accessibility to information for students and guardians, without adding to social equity issues in the education sector. Purchase of hardware, such as kiosks and networks that enable information access in corridors and cafes is also more expensive to institutions than uploading information using existing Internet infrastructure that then becomes available using SMS technology. Institutional choice between kiosks and mobile technologies represents a choice between ownership (of the Internet access device) and access on demand. (Slack & Rowley, 2002) The relative price of each of these options for different levels and types of use might be a significant factor in determining the respective roles of each channel.

Trigger: Solutions and Recommendations

Work currently underway extends the RMIT University and Pearson Education Australia 2006 pilot and research undertaken at Kingston University in the UK, the University of Technology Sydney (UTS) and the MIT and Nokia Research Centre (Huang & Pulli & Rudolph & 2005). The current system is ubiquitous in that once a student is registered the system recognises the individual and provides information tailored to them. This innovative application of mobile technology enables access to information, at a minimal cost, irrespective of geographical location, using a limited vocabulary of requests.

As the students' provided an overwhelmingly positive response to the 'usefulness' of the technology in the 2006 trial (Richardson & Lenarcic, 2007a) it was extended to evaluate technology effectiveness with a larger and more diverse group of students in a 2008 pilot. Work undertaken at UTS also illustrated that educational institutions across the sectors now exist in a cultural environment that favours SMS (Bachfischer & Lawrence & Litchfield & Dyson & Raban, 2008) usage for reminders and organising time. The provision of real-time access to logistical information by means of SMS, removes barriers to student/guardian and staff communications imposed by geography and memory. All responses to student triggers are sourced from a temporary database built by specialist staff, via password protected *Trigger* web interfaces embedded in existing organisational systems. The emphasis is on enabling one-stop specialist staff data input.

Participants of the 2006 pilot were drawn from staff and students at RMIT University. The aim was to inform a broader rollout of SMS usage and its integration with existing infrastructure. In order to create an application that supports 'best-practice' the pilots included a semiotic analysis of existing dispersal of administrative information intended to assist interested parties in logistical scheduling and completion of learning activities offsite. The frequency of business service requests for information suited to SMS dissemination will inform the creation of Triggers and associated accessible data. Analysis of current student service communication was also conducted to establish what needed to be integrated into the extended *Trigger* system. The principle aim of the project was to make innovative use of SMS mobile technology, as a vehicle to augment uptake of digitised learning resources for use outside traditional classrooms. The system was designed to effectively and efficiently disseminate information between educational institutions and students to improve usage of educational resources. The value of a range and combination of mobile technologies to learning through the provision of 'on-demand' student access to quality information (such as their study schedules, assessment performance, and institution's provision of information to students) was investigated.

Extensive review of pilots was conducted to provide guidance for effective educational deployment of SMS to support staff–to-students, students-to-staff, guardians-to-staff and staff-to-guardians communication. The different requirements of students undertaking higher education subjects were factored into the design of the prototype and pilots. The use of SMS to access data services reflects an identified trend away from mobile voice usage, particularly amongst University students.

McClatchey, (2006) stated that 3G service availability would reinforce these current trends and increase the variety of hedonistic functional uses awaiting the new technology. Social networking and group formation will be facilitated as youths' download music and games and utilise Web blogs, chat and FaceBook facilities.

Case Study Pilot

The project used a 5 phase mixed methodology approach which included:

- Phase 1 – Scoping;
- Phase 2 – SMS Development and Prototyping;
- Phase 3 – Trial Design and Conduct;
- Phase 4 – Trial Evaluation Stakeholder Survey; and
- Phase 5 – Trial Evaluation Focus Groups.

The scoping, prototype construction and review were underpinned by concepts drawn from the Technology Acceptance Model (TAM) (Davis, 1989; Davis, 1993) and the Theory of Planned Behaviour (TPB) (Ajzen, 1991). The two-way communication process enabled by the SMS technology application is illustrated in Figure 2: *Trigger Enabled Multidirectional Flow of SMS Text Messages*.

In the initial pilot study the information made available to students was restricted to that input to the user interface illustrated in Figure 3. An academic course coordinator using the prototype, could provide a 'blast' service and send assessment feedback and reminders to the entire student cohort when required. Consider instructor-initiated e-mails to an entire class.

It is argued here that text messaging is analogous in this respect and consequently ideal for impulsive consumers seeking immediate informa-

Figure 2. Trigger enabled multidirectional flow of SMS text messages

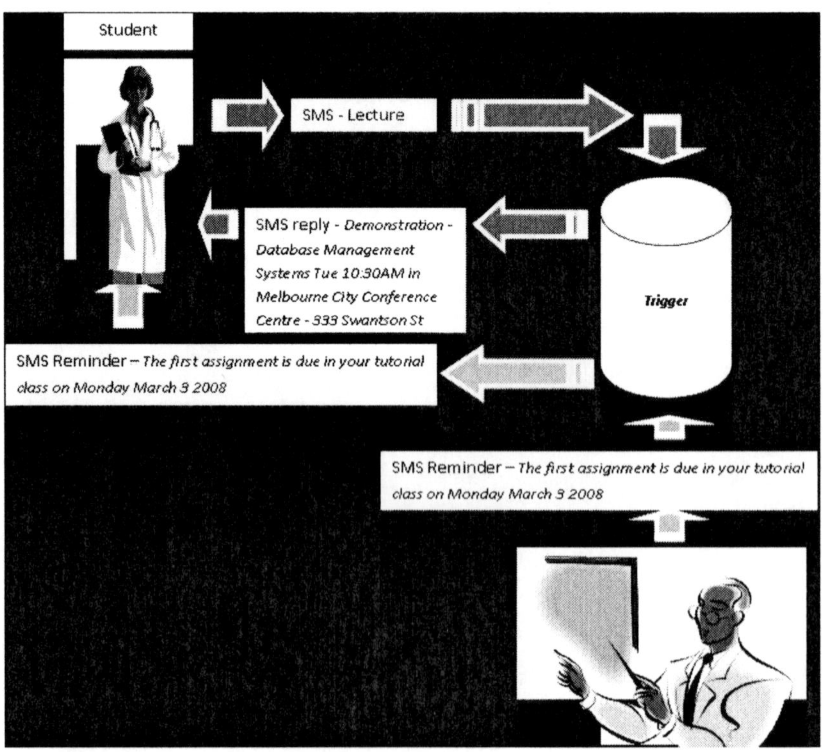

tion interchange where brevity is the aim. Figure 4 displays the Web based prototype feature that enables academics to send messages to all registered students in their subject.

Students were given the opportunity to trial the SMS prototype to receive reminders and alerts and access scheduling and assessment information. To register they provided their student number, password, Email, subject, lecture times, tutorial class and most importantly their mobile phone number. For one semester registered students were pushed reminders and results given access to schedules and assessment requirements. At the end of the specified semester the transient database containing class and student information was removed.

A combination of qualitative and quantitative data was gathered to evaluate the potential for adoption and improved uptake of available learning resources outside the traditional learning environment. Both the 'usefulness' and the student's perception of the impact of SMS reminders, alerts and factoids on their use of available resources was evaluated to build on the dataset generated by the RMIT University/Pearson Education Australia 2006 pilot (Richardson & Lenarcic & McKay & Craig, 2008). A project-based methodology was utilised to scope, develop, prototype, test, and review. The methodology supported development of the technology application in conjunction with the identification of innovative uses of mobile technologies and a holistic and rigorous evaluation.

Project Scoping

Trigger applied an existing technology in an innovative manner within the education sector to enable 'on-demand' information exchange between teachers and students. The SMS communication trial was limited in breadth to enable an in-depth investigation of the impact of technology usage on students' use of available resources. Students were able to interact with the educational institution remotely using mobile technology. Generation Y are typically more technologically sophisticated than any other generation. They are educated and highly employable. They have dispensed with five-to-five and they are in control of their schedules. (Fragiacomo, 2005)

Figure 3. The academic information input screen

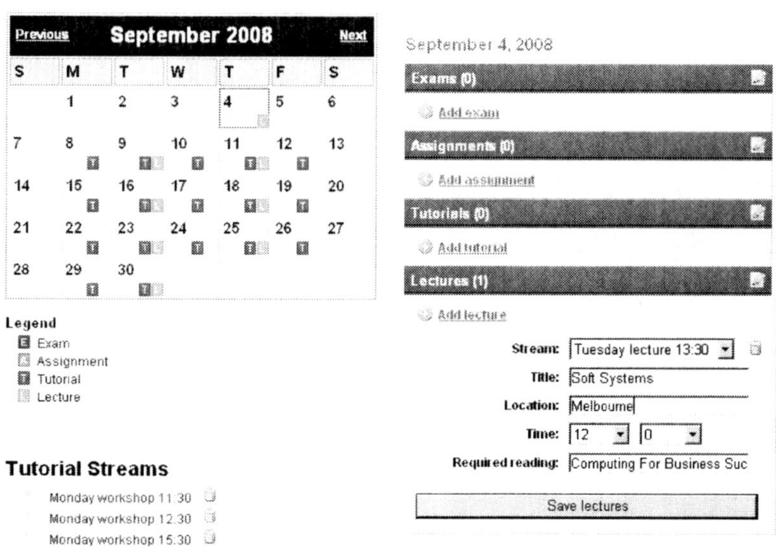

Figure 4. Academics 'push' information to the student cohort

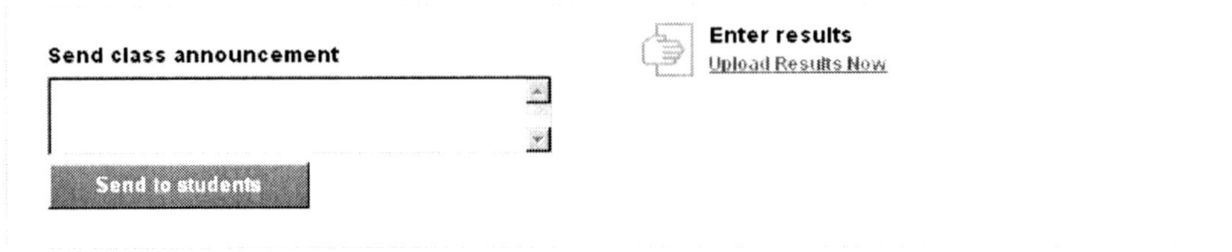

SMS Application Development and Prototyping

The choice of Triggers and accessible information was underpinned by an analysis of current business service requests and web-based dissemination. Evaluations of information quality, timeliness and language use was undertaken with respect to existing communications. Appropriate information for SMS access and delivery of weekly prioritised homework reminders, multimedia event alerts and assessment feedback was identified. The existing *Trigger* prototype was extended to manage 6 subjects across three Faculties in the 2008 pilot. Partners were involved throughout the project to determine requisite additions and deletions to the application that represented improvements to the quality of the information available.

Trial Design and Conduct

The information disseminated using SMS included:

- Reminders for deadlines for assessment;
- Time and location information about lectures and workshops;
- Time and location information about examinations and assessment tasks; and
- Assignment and exam marks.
- Guidelines for augmentations of staff/student/guardian communications utilising SMS technology

The trial provided the means to evaluate the students' response to the technology would be consistent for pilot studies conducted across the institution.

Trial Evaluation

The TAM is used to explain and predict how users come to accept and use novel forms of technology. The model suggests that when users are presented with a technological system, the perceived usefulness (PU) and the perceived ease-of-use (PEOU) influence their decision about how and when they will use it. Hypotheses were generated to test PU and PEOU. The study then used a survey tool comprised of questions based on the TAM to identify whether the purpose and operational features of the technology led to acceptance of the SMS system.

All stakeholders were surveyed to establish the 'usefulness' of the technology combinations for removing location barriers to uptake of digitised resources. This required examination of '*push-pull*' information exchange between students and teachers. '*Push*' refers to initiating information exchange by the institution and academics (sending information like reminders for work to be completed prior to class). '*Pull*' refers to students requesting information (receiving). The project will contained two distinct activities subject to review:

- *The pushing phase: All students' received a welcome message, prioritised homework reminders and assessment;*

- *The pulling phase: Each message sent by a student was recorded and tallied against the associated trigger-word.*

As the volume of SMS in a social domain has increased the communication channels between educational institutions and students have altered. Cultural behaviour impacts on educational stakeholder acceptance of the SMS service technology within work and study contexts as well as the social. The TAM model can be used to predict SMS adoption. The model recognises that there are circumstances in which technology usage behaviour might be expected to result in positive consequences (or net benefits), yet not be undertaken due to a perceived lack of ability to control the execution of the behaviour (Fragiacomo, 2005).

Outcomes

There are challenges involved in the provision of a safe, sustainable, positive culture in which students and staff can interact comfortably. Innovations in communications technologies enable information dispersal to assist access to information required to learn. The aim of this project was to develop the *Trigger* SMS tool to provide students with pragmatic assistance to make each class and exam easy to attend. The project aimed to capture requisite information services, from each stakeholder's perspective, and then use this knowledge of existing systems to build SMS capacity.

Outcomes of the projects will guide the application of SMS mobile technology use in the education sector and enable systems integration driven by end-user work practice and usage. These projects are required to further refine the technology application and ensure establishment of effective SMS communications that increase resource uptake, in the higher education sector. This removes barriers to innovation adoption at an institutional level where the typical approach of experimentally deploying new technologies on campuses does not include processes to quickly scale them up to broad usage when they work. The lack of implementation planning that incorporates scalability often creates its own obstacles to full deployment. (The New Media Consortium & EDUCASE Learning Initiative, 2007).

Results 2006 Pilot

Data collected informed an understanding of the impact of the mobile technology application's functionality on the students' transition into the higher education social network, overall impact on the learning experience and individual time-management capacity. The initial pilot used a large undergraduate subject delivered to the entire business cohort of students to trial the application of SMS technology. The survey response dataset used to evaluate the application of the technology was relatively small. Student responses were collected on a four point scale. Missing responses were also recorded and a quantitative analysis was conducted.

The overwhelming positive response to the technology application provided a basis for additional testing and evaluation to build on the dataset. Normal methods of information provision to the students were maintained. Web based access to class timetabling, location and scheduling information, as well as assessment requirements was provided in addition to the SMS. Web based access to class timetabling, location and scheduling information, as well as assessment requirements were provided in addition to the SMS.

Student Uptake of the SMS Application

During the pilot students were informed about the application during lectures, by email, via Blackboard and provided with Trigger control cards created by Pearson Education Australia. These cards enabled them to have ready access to the online registration address and possible text triggers at any time. At the end of the trial

180 students (which represented over half of the students) had voluntarily registered with Trigger. Interestingly, 45 students or 25% registered in the last month of the pilot that could indicate the students' positive response to fast access to assessment results as a key driver for registration to the system. Spikes in registration could be observed prior to release dates of assessment marks. This indicated that access to fast assessment results was a key driver for registration on the system. Word of mouth around the successful delivery of results for Assignment 1 in October and reminders to students that were registered that they would be receiving their results by means of SMS also explains the late spike in registrations numbers prior to the November release of the final two assessment results. The relatively slow initial uptake of the system and poor survey response rate was mirrored in (Stone, 2004) study findings at Kingston University.

Table 1. *Triggers Used in the Pilot* illustrates all of the triggers available to the student body, the content of database responses and the number of requests for information during the pilot. As the application was only used for one subject rather than all of the subjects in which a student was enrolled it is postulated that some of the figures are lower than expected.

An online student survey was conducted at the end of the pilot. Although the response rate for the initial trial was disappointing (13%) the results affirmed findings identified in the literature and at other universities. At Orange University 62% of students surveyed were in favour of getting their grades sent directly to their mobile phones. 96% of student respondents in the trial agreed or strongly agreed when asked "are you satisfied with assessment results being sent to you via an individual message?'.

Students were satisfied with the ease of use of the application as 92% of survey respondents agreed or strongly agreed that the application was easy to use. The popularity of SMS in the marketplace as a means of communication amongst Generation Y' *ers* evidences the ease with which students' input *Triggers* and perceive the relevance and usefulness of the output information quality.

Student Perceptions of the Effectiveness and Usefulness of the SMS Application

On the basis of the study, embedding the use of SMS will improve the student experience by increasing the effectiveness of 'student to student', 'student to staff', 'staff to student' and 'university to student' communication. The impact of mobile technology adoption on students' first year experience, will underpin University system

Table 1. Triggers used in the 2006 pilot

Trigger word	Trigger sample response	Number of Trigger requests
Lectures	Demonstration - Database Management Systems Tue 10:30AM in Melbourne City Conference Centre - 333 Swantson St	106
Tutorials	Database - Designing for data Wed 01:30PM in 108.09.003	8
Latest Results	98%	37
My Progress	11.25% (with 75% of assessments still to be released)	22
Next Assignment	Presentation and Spreadsheet Assignment due Mon 12:00PM worth 25	10
Next Exam	End of Semester Exam conducted during the University Official Examination period Mon 12:00AM in TBA	3
Due this week	Assignment 1- ISYS2056	5

and process recommendations for use, changes to practice and uptake.

The Trigger application demonstrates an innovative use of the technology in enhancing students' experience of administrative service. It enables organisations to support the students' capacity to manage their learning environment in an individual manner. Students were able to interact with the University remotely using mobile technology to obtain scheduling and assessment details on-demand. Questions based on the TAM model's evaluation of effectiveness as an indicator of students' adoption of the technology provided an overwhelmingly positive response as displayed in Table 2. *Students' Rating of the Effectiveness of SMS.*

Although the response rate for the initial trial was disappointing the results affirmed findings identified in the literature and at other universities. The student cohort found the use of the technology effective and a convenient source of information.

Table 3. *Expected Frequency and Relevance of Trigger Use* indicates the percentage of students that stated they would use a particular trigger frequently or very frequently and found it relevant or very relevant. This indicates the students' perception of the usefulness of particular *Triggers* that enables evaluation of PU.

There was an appreciable difference between the expected use of trigger words to elicit information relating to exams at the end of semester and the next assignment due. However, this study did not enable causal linkages to be made suggested explanations include timing or the appropriateness of the trigger language or the students' need for immediate access and delivery of information upon request.

The student perception of the usefulness or relevance of the lecture and tutorial trigger responses was marginally higher than their evaluation of the usefulness of the actual triggers. In relation to the responses to questions about assessment the students' felt more strongly than the triggers used to elicit information. The timing and feedback in relation to assessment is critical to the students' experience and they recorded higher rates of response to *'Very Effective'* as compared to *'Very Frequently'*. Students overwhelmingly supported the relevance of the information delivered using SMS to their needs. An alignment between the trigger words and the standard designed responses was obtained.

The students' response to the survey also demonstrated a positive correlation between 'on-demand' access to lecture and tutorial location and topic information as 88% found SMS to be effective or extremely effective to obtain this information. A model to estimate costs to the student cohort and the institution was developed during the pilot. Estimates generated on the basis of student use of *Trigger* indicated that the cost was minimal. Cost and transaction time minimization alongside geographic convenience have

Table 2. Students' rating of the effectiveness of SMS

Survey Question	Extremely ineffective	Ineffective	Effective	Extremely Effective	N/A
How effective did you find the SMS provision of assessment results?	4%	8%	44%	44%	0%
How effective did you find the SMS service for the provision of assessment results?	4%	12%	36%	44%	0%
How covenient was the ability to access assessment information anytime, anyplace?	0%	4%	48%	48%	0%
Did the system improve your ability to schedule assessment task work?	4%	24%	44%	28%	0%

Table 3. Expected frequency and relevance of trigger use

Trigger Words	Frequently	Very Frequently	Relevant	Very Relevant
Lectures	52%	12%	68%	12%
Tutorials	40%	16%	60%	24%
Latest Results	44%	24%	44%	32%
My Progress	44%	20%	20%	40%
Next Assignment	52%	16%	36%	32%
Next Exam	40%	16%	60%	20%
Due this week	44%	16%	36%	40%
Due next week	40%	20%	44%	32%

Table 4. Students perceptions of the information quality of the SMS application

Survey Question	Very Inaccurate	Inaccurate	Accurate	Extremley Accurate	N/A
How accurate were the SMS responses to your SMS requests from *Trigger?*	0%	5%	90%	5%	0%

the potential to positively impact on perceived usefulness of the application.

Student Perceptions of the Information Quality of the SMS Application

Students overwhelmingly supported the privacy and information quality characteristics of the application used to provide class and work scheduling details and assessment results. The quality of information was determined by questions that ascertained the students' perceptions of the information accuracy, ease of understanding, timeliness and availability of the SMS service. Student responses are displayed in Table 4: *Students Perceptions of the Information Quality of the SMS Application*.

The data collected from the 25 survey respondents provided support that the students were positive or very positive (P=54.1%, VP=29.8%) about the key aspects of this SMS mobile technology. Moreover, those replying to the survey gave the Pearson's *Trigger* tool a 100% rating (positive and very positive) for the system registration, availability, information, quality and response clarity. High ratings (>90% V and VP), were also received for system security, usability and convenience, and information accuracy.

FUTURE APPLICATIONS FOR TRIGGER IN EDUCATION

In extending student use of popular mobile ICTs for social involvement into the learning environment the pilot sought to improve student engagement with their learning outside the traditional classroom boundaries and thus to provide a more realistic context. Student pilot participants suggested useful additions to the SMS prototype. Qualitative comments received supported the use of SMS technology to provide examination location and individual seating details and to provide fast instructions to the student body in the case of emergencies. This use of SMS technology alongside the current reminder and alert system implements the University strategic priority of creating "*Work ready graduates*" (Academic Plan 2006-2010, 2006: p6) as the connection between the internal and external environment of the University becomes seamless. It also contributes to the ability of graduates to adapt to, and indeed potentially lead, the process of change in the workplace. In summary, the intersection between the cultural expectations of the workplace, the internal University demands and the students' behaviour is re-enforced.

Planned additions to Trigger to include multiple access modes for students are in accord with the University's academic plan of developing infrastructure that provides a single point of access to student support services, learning content, interactivity with peers, subject and program/course information, and student administration (Academic Plan 2006-2010, 2006) in the form of an electronic portal. The student electronic portal will provide students with a choice between communicating with the university by means of Email or SMS where the information required suits the brevity and anytime anywhere nature of text messaging.

Emergencies

Fast communication and a high hit-rate in terms of message and instruction receipt, irrespective of physical location are imperatives to ensure staff and student safety during disaster situations. Strengths of SMS technology are its dependability when other communication networks like voice calls fail and its ability to relay logistical information. High ownership of mobile phones amongst University staff and students enhance the technology's potential as an early warning and disaster recovery device that will be trialled in higher education institutions during this project. The type of communication chosen for mobile devices in learning needs to be very carefully considered, and, at the moment, would appear to favour short and simple messages: - urgent SMS alerts seem very suitable (Lawrence & Bachfischer & Dyson & Litchfield, 2008).

Cascio (2008) describes the importance of fast efficient communication to effective disaster recovery and the usefulness of SMS technology due to the widespread ownership of mobiles and the need to hit only a percentage of mobiles to spread the word. McAdams (2006) described the role of SMS as a link of last resort for crisis communications. As SMS messages traverse an isolated channel there is often no impact during times of heavy traffic or adverse conditions that can overwhelm other wireless networks. During the aftermath of hurricane Katrina SMS was used by members of the coast guard to inform loved–ones of individual safety, to direct life-saving helicopter rescues and to relay messages to hospitals. SMS systems utilise the ability of text messages to reach their destination even when a cell phone signal may be too weak to sustain a spoken conversation (Boyd, 2008). The fact that a text message will queue if the destination is unavailable rather than requiring a redial like a traditional voice phone call enables asynchronous communication, another useful feature during a crisis.

An emergency warning password protected interface will be embedded in the *Trigger* application to enable restriction of generation of warning messages to authorised staff and to enable emergency logistical organisation and dissemination of information to be the province of trained staff. One-to-one SMS messages to instruct and inform emergency and public relations staff will be facilitated. The application will also enable information to be directed to Email accounts and web interfaces. Anecdotal evidence suggests that the technology will be useful in crisis situations. This project intends to mimic a crisis situation and rigorously evaluate the effectiveness of SMS to provide a quick way to get information dispersed to a large number of recipients and to assist in the logistical communications required to effect recovery. McAdams (2006) described the unplanned use of SMS during a crisis at Tulane University where staff on-site messaged personnel offsite to report damage assessments, medical updates, continuity-of-operations plans and communication with hospitals.

CONCLUSION

The application of mobile technology outlined in this paper enables students to access information, at a minimal cost, irrespective of geographical location, using a limited request lexicon. Text messaging represents an opportunistic agent for the mutual exchange of information between the administrative sector of a university and its first year undergraduate clientele. The primary intention of this project was to alter the students' behaviour in relation to access to timetabling information. The emphasis on text to interface via Trigger was an example of fostering system adoption through an existing persuasive channel, thus allowing technology itself to shape the behaviour of a user.

The value of a SMS mobile technology to student academic learning through the provision of 'on-demand' student access to quality academic information (such as their study schedules, assessment performance, and institution's provision of information to students) was validated.

The SMS application reduced the need for students to access university or home computer systems to find subject timetables and locations, assessment schedules and feedback or marks. In addition students can use SMS to check what reading, events or tasks are scheduled off-campus. If this facility increases the uptake of digitised learning resources learning outcomes will improve. The SMS application changes the nature of the administrative system from primarily being seen as a dispersal apparatus that is automatic in function to that of a dynamic acquisition process that is purposely governed by students. SMS use also removes barriers for students in a new environment where accessing staff to ask questions, and information provided on the Web, can be difficult.

The Trigger system piloted at RMIT University was ubiquitous in that once a student was registered the system recognised the individual and provided information tailored to be relevant to them, This system extended the use of the current web-based infrastructure and usage by enabling receipt of Internet based information on mobile phone devices on-demand.

Survey data collected in the initial pilot combined with a literature survey indicated that the use of the technology in the sector would potentially improve student administrative service. Engagement of students can be assisted by the technology as it is a fast way to get important information about the students' university life and workload without booting computers or logging into a Content Management System. The system enables dynamic information transfer with live updates and potentially allows students to better schedule and organise themselves. The cost to students is minimal as all that they require is a mobile phone.

REFERENCES

Ajzen, I. (1991). The theory of planned behaviour . *Organizational Behavior and Human Decision Processes*, *50*, 179–211. doi:10.1016/0749-5978(91)90020-T

Arvind, A., & Hicks, J. (2006). A Mobile phone Ecosystem: MIT and Nokia's Joint Research Venture . *IEEE Intelligent Systems*, *21*(5), 78–79. doi:10.1109/MIS.2006.84

Atun, R., Greene, S., Morris, S., Franklin, V., & Samyshkin, Y. (2006, March). *Use of mobile technologies to enhance control of type 1 diabetes in young people: economic evaluation*, (Paper No. 4), The Vodafone Policy Paper Series.

Atun, R., & Sittampalam, S. (2006, March). *The upward trend in healthcare spend, The role of mobile phones in increasing accessibility and efficiency in healthcare* (Paper No. 4), The Vodafone Policy Paper Series.

Bachfischer, A., Lawrence, E., Litchfield, A., Dyson, L., & Raban, R. (2008). Student Perspectives about Using Mobile Devices in Their Studies, *IADIS International Conference Mobile Learning* (pp. 8-18). Algarve, Portugal.

Boyd, C. (2008). *Text Messages and Disaster Recovery, World a BBC Service and WGBH-Boston co-production.* Retrieved 23 March, 2008, from http://newsbbc.co.uk/2/hi/technology/4149977.htm.

Budde, P. (2008). *Australian Mobile Market Statistics,* [REMOVED HYPERLINK FIELD] *Mobile Data Association*. Retrieved 12 Oct 2008 from http://www.themda.org/Page._

Cascio, J. (2008). *World Changing.* Retrieved 23 March, 2008, from, http://www.worldchanging.com/archives/00128.htm.

Chauhan, D., Pritchard, C., & Sussex, J. (2008, March). *The upward trend in healthcare spend, the role of mobile phones in increasing accessibility and efficiency in healthcare*, (Paper No. 4), The Vodafone Policy Paper Series.

Compeau, D., Higgins, C., & Huff, S. (1999). Social Cognitive Theory and Individual Reactions to Computing Technology: A Longitudinal Study. *MIS Quarterly*, *23*(2), 146–158. doi:10.2307/249749

Corsini, L. S., Crittenden, V. L., Keeley, R. C., Trompter, G. M., & Viechnicki, B. (2000). Integrating cross-functional undergraduate teaching and learning: A work in progress. *Marketing Education Review*, *10*(3), 1–17.

Danesi, M. (2007). *The quest for meaning: A guide to semiotic theory and practice*. Toronto Studies in Semiotics and Communication, University of Toronto Press, Toronto,Canada.

Davis, F. D. (1989). Perceived usefulness, perceived ease of use, and user acceptance of information technology. *MIS Quarterly*, *13*(3), 319–340. doi:10.2307/249008

Davis, F. D. (1993). User acceptance of information technology: system characteristics, user perceptions and behavioural impacts. *International Journal of Man-Machine Studies*, *38*, 475–487. doi:10.1006/imms.1993.1022

Desouza, K. C., & Hensgen, T. (2003). 9 -11 information failures: A semiotic approach. *IT Professional*, *5*(2), 61–63. doi:10.1109/MITP.2003.1191793

Faulkner, X., & Fintan, C. (2005). When Fingers do the talking: a study of text messaging. *Interacting with Computers*, *17*, 167–185. doi:10.1016/j.intcom.2004.11.002

Fogg, B. (1999). Persuasive technologies: Now is your chance to decide what they will persuade us to do- and how they'll do it. *Communications of the ACM*, *42*(5), 26–29. doi:10.1145/301353.301396

Fragiacomo, L. (2005). Capturing the Y, MIS Managing Information Strategies (www.misweb.com), *Communications of the ACM, 42(*5), 54-59 Australia, Summer 2004/2005.

Gough, N., & McCulloch, B. (2006). *The role of mobile phones in increasing accessibility and efficiency in healthcare, Moving the debate forward* (Paper No. 4), The Vodafone Policy Paper Series, March 2006.

Hartenian, S., Schelienger, M., & Frederickson, P. (2001). Creation and Assessment of an Integrated Business Course: One College's Experience. *Journal of Education for Business*, *76*(3), 149–159.

Hauser, C., & O'Connor, A. (2007). Virginia Tech Shooting Leaves 33 Dead. *The New York Times Online*. Retrieved 20 Feb 2008 from http://www.nytimes.com/2007/04/ 16/us/16cnd-shooting.html.

Head, B. (2008, September 1-3). Who's Teaching who? How the gap between students – the digital natives – and teachers – the digital immigrants – can be closed. *Education Review Technology Guide*, (www.campusreview.com.au).

Huang, A., Pulli, K., & Rudolph, L. (2005). *Kimono: Kiosk-Mobile Phone Knowledge Sharing System*, Massachusetts Institute of Technology, Nokia Research Centre, HIT Lab NZ, University of Canterbury, Christchurch, New Zealand.

Leong, K. C., Chen, W. S., Leong, K. W., Mastura, I., Mim, O., & Sheikh, M. A. (2006). The use of text messaging to improve attendance in primary care: a randomized controlled trial. *Family Practice*, *23*(6), 699–705. doi:10.1093/fampra/cml044

Markett, C., Sanchez, A., Weber, S., & Tangney, B. (2006). Using Short Message Service to Encourage Interactivity in the Classroom. *Computers & Education*, *46*, 280–293. doi:10.1016/j.compedu.2005.11.014

McAdams, U. (2006). *SMS does SOS.* Retrieved 23 March 2008. from http://www.fcw.com/12_11/news/92790- 1.htm, FLW staff.

McClatchey, S. (2006). The Consumption of Mobile Services by Australian University Students . *International Journal of Mobile Marketing*, *1*, 35–42.

New York Times. (2007). Yahoo Introduces Mobile Service Software, late Edition New York Times, New York, N. Y. (9/01/2007)

Platts, B. (2004). *ATN Online Learning Surveys Project 2004*. Online Learning Student Survey Report, Combined ATN Universities: Institute for Interactive Media and Learning, University of Technology Sydney. Lawrence, E., Bachfischer, A., Dyson, L. E., & Litchfield, A. (2008). Mobile Learning and Student Perspectives: An mReality Check! *7th International Conference on Mobile Business*, (pp. 287-295). Barcelona, Spain.

Reisman, S. (2006, January/February). The Ivory Tower – Failing to Communicate in an Educational Environment. New York: IEEE Computer Society. *IT Professional*, *8*(1), 62–66.doi:doi:10.1109/MITP.2006.53

Richardson, J., & Lenarcic, J. (2007a, December 5-7). Trigger iz gr8 4 Gen Y: mobile Student Administration via a Text Messaging System. *Australasian Conference on Information Systems (ACIS)*, (pp. 991-998), Toowomba, Australia Richardson, J, & Lenarcic, J. (2007, June 4-6). E-inclusion through text messaging: The emergence of an admisnistrative ecology within a university student population via the use of a mobile academic information delivery system. *20th Bled eConference*, Bled, Slovenia.

Richardson, J., Lenarcic, J., McKay, E., & Craig, C. (2008, March 17-19). Pull with Pearson's Trigger Word Tool to Enable Simplexity: Student Empowerment via a Text Messaging System for Mobile Administration. *IASTED International Conference on Web-Based Education*, Innsbruck, Austria

Sebeok, T. A. (2001). Signs an introduction to semiotics, *Toronto studies in semiotics and communication*. University of Toronto Press, Toronto, Canada.

Shenk, D. (1997). *Data smog: surviving the information glut*. San Francisco, CA: Harper Edge.

Slack, F., & Rowley, J. (2002). Online kiosks: the alternative to mobile technologies for mobile users . *Internet Research: Electronic Networking Applications and Policy*, *12*(3), 248–257. doi:10.1108/10662240210430928

Stone, A. (2004). Mobile Scaffolding: An Experiment in Using SMS Text Messaging to Support First Year University Students. *Fourth IEEE International Conference on Advanced Learning Technologies* (pp. 405-409) (ICALT'04), Joonsuu, Finaland.

The Age (2008). *Life & Times*. 6/04/2008.

The New Media Consortium & EDUCAUSE Learning Initiative, The Horizon Project 2006.

The New Media Consortium & EDUCAUSE Learning Initiative, The Horizon Project 2007.

Wheeler, B. C. (1998). The state of business education: Preparation of the past? *Selections*, *14*(2), 19–21.

Wurman, R. S. (1989). *Information Anxiety*. New York: Doubleday.

Chapter 18
Use of Mobile Technology at Montclair State University

Patricia Kahn
Montclair State University, USA

Edward Chapel
Montclair State University, USA

ABSTRACT

Educators strive to develop innovative teaching strategies to meet the expectations of digital natives that are accustomed to social networking environments. The Campus Connect project at Montclair State University, which began in the fall 2005 semester, provided an innovative mobile technology service, in order to meet these expectations. The program, which included a custom designed, high speed, rich media and GPS (location based services) capable cellular network as well as a rich array of cell phone based applications enabled students to customize their mobile phone for 24/7 access to the University's teaching and learning, information, and administrative resources. This chapter will describe the growth and evolution of the Campus Connect program and the applications that were frequented by the student population on mobile technology through this innovative program. In addition, a description of how these applications enhanced the learning environment will be provided as well as the changes the program underwent in order to best suit the demands of the changing population of students. Quantitative and qualitative survey results are offered to describe the student's reaction to using mobile technology in a learning environment as well as identify those applications that students utilized most often. Based on these results, recommendations for future iterations of the Campus Connect program will be provided, which can be used as a guide for administrators who may be contemplating comparable mobile technology programs at their institutions.

INTRODUCTION

When reflecting upon creative uses of technology for the University campus an inspiring story told by Renee Dubos in his classic work, A God Within (Dubos, 1972), which explores humanity's stewardship of the planet comes to mind. He tells of the Inuit people of the far North who are known for carving very intricate and beautiful figurines from the ivory extracted from whales. When Dubos

DOI: 10.4018/978-1-60566-703-4.ch018

inquired of these artists how they decide what to carve from the pieces of ivory they worked with, he was surprised to learn that they do not make a decision about what to carve at all. Rather, they see the figurine that will result from their carvings as latent in the raw media and their artistry is to remove the excess to release the "god within." These artists have a unique insight about their media. In some way they comprehend all the facets and nuances of the undifferentiated material in hand and with precision and care they are able to fully realize it's potential. What a fitting metaphor for our efforts to introduce new technologies into the fabric of the living and learning environment that is a University campus. We would do well to remember that the campus is a delicate ecosystem. We should take care to consider the multitude of subtle forces that define a campus community and treat technology as the carving knife, so to speak, that refines the campus environment in a way that all members of the community are able to fully realize their potential.

It is this campus community that has subtly changed over the years, which reaps the benefits of new technologies or the integration of existing technologies into the learning environment in new ways. In today's higher education environment, campuses face multiple challenges and opportunities—from keeping students safe, to building stronger campus communities, and to designing technology enriched courses that improve learning. Universities therefore need to remain sensitive to the various ways in which people can learn.

The benefits of these new technologies have also accelerated the pace of change at colleges and Universities throughout the nation, specifically in the area of online education. The percentage of distance learners as a proportion of all those pursuing higher education has increased significantly over the past decade and a half according to the United States Department of Education (2003). This rise has encouraged educators to rethink many facets of their teaching strategies. Distance-based delivery systems significantly alter the dynamic relationship among the instructor, student and content; these alterations must be anticipated in the design of instruction. Foremost among the factors that must be considered is the need to accommodate the various learning styles that are characteristic of the 21st century learner. Today's students who are broadening their knowledge are coming from diverse backgrounds demanding curriculum changes in higher education. Demographic changes such as differences in family status and age of population are requiring educators to "adjust their curriculums to match the diversity of their students" (Tunstall, 1995, p. 2) and establish learning environments that foster individualization and accommodate diversity.

In response to this diverse student culture, postsecondary educators respond to these varied backgrounds by supporting individual strengths and weaknesses. Dunn and Griggs (2000) describe how instructors "need to become aware of the variety of ways in which individual learners process new and difficult information" (p. 134). The research performed by Dunn and Griggs indicates how there exists significant differences in learning style preferences among a variety of cultural groups such as Native Americans, Hispanic Americans, African Americans, Asian Americans, and European Americans. Therefore, instructors are discovering that a single instructional method is not effective when educating the 21st Century learner whose differences in age, culture, experience, and language skills have become so apparent in the classroom.

The typical college student has grown up with computers, and the Internet has become an essential ingredient in their everyday lives (Jones, 2002). The student of the past, who broadened their knowledge by attending traditional lectures and researched material in the library, has been replaced by the computer savvy learner relying on an infrastructure that promotes the use of technology meeting the demands of this new generation. Newly designed courses should promote student learning through active engagement, resulting in knowledge produced from experience, as opposed to the passive receiving of knowledge (McGriff, 2001). Teachers face the problem of finding instructional technologies that support the digital learner but do not require a high level of technical proficiency. Research has shown that technology can be especially useful when introducing complex subject matter and support collaborative and interactive learning activities consistent with constructivist learning principles (McLaughlan & Kirkpatrick, 2004; Valdez, McNabb, Foertsch, Anderson, Hawkes, & Raack, 2004). Instructors seek a low threshold technology that promotes a constructivist approach to teaching and learning that customizes the curriculum to the learner.

At Montclair State University, in order to address the challenges associated with the learning style of the 21st century learner as well as take advantage of advancements in technology that could better serve these learning styles, a solution was sought in cell phone technology. Montclair State University, instituted a cell phone program called Campus Connect, in order to enhance the student's learning experience within a social learning context. These 21st century learners, referred to by many as "digital natives," adopt technologies at a furious pace and use these technologies in different ways. As described by Prensky (2004), students are not just using technologies but have adapted every day practices to these activities. An example of this can be seen in the proliferation and adoption of cell phones, where today's college student has selected them as their main communication gateway and a necessity to their existence. Universities face the unique challenge of not only using this device as a method of communicating with students, but also as a teaching and learning device that supports the method of learning that complements the learning style of the 21st century learner who are programmed to and have the desire to keep up with technology.

This chapter will describe the Campus Connect cell phone program that enabled students and faculty to step well beyond traditional cell phone services by offering a variety of applications that could be used to enhance the learning environment. Survey results will provide the student's reaction to the learning activities that were phased into the learning environment over a two year period as well as detail those applications that students used most frequently. Pedagogical and administrative challenges will also be addressed, which will determine future implementation and instructional design strategies for Montclair State University's mobile phone initiatives. Finally, solutions and recommendations will be provided for those administrators and decision makers who are considering integrating mobile technology into their University's future goals and objectives.

BACKGROUND

Montclair State University (MSU), New Jersey's second largest and fastest growing University is committed to maintaining a campus community that reflects the diversity of New Jersey. The campus is comprised of more than 17,000 students, of which 75 percent are commuters. The University's strategic plan outlines technology-related goals that include providing technical and design support for faculty who incorporate technology into courses and providing incentives for faculty to embrace new pedagogies made possible by technology (Montclair State University Board of Trustees, 2002). Montclair State supports the partnerships with faculty as described by Albright and Nworie (2008), where commitment and services provided by information technology units support these challenges associated with curriculum redesign and encourages the integration of technology into teaching and learning.

The mission of Montclair State University as a multipurpose public institution is to develop educated persons of inquiring, creative, and disciplined intelligence to be competent in careers that are fulfilling and to be socially responsible contributors to society. The University strives to graduate people on the bachelor's and master's level who have had sound education in the arts and sciences and relevant, specialized, training built upon that base (Montclair State University Mission Statement, 2002).

In order to effectively support its mission, Montclair State University is committed to using applications and technologies that foster and encourage a strong campus culture. Since many students have lives that focus off-campus, in order to keep them engaged, the University strives to provide this segment of the population with technology enabled solutions that create a "virtual" campus. This environment encourages student participation in campus activities, creates a safe and secure learning environment for all students, and provides easy access to faculty and educational tools regardless of the student's location.

In support of this "virtual campus," Montclair State University is the nation's first institution to implement a revolutionary new mobile phone program that enables students and faculty to step well beyond traditional cell phone services such as voice, IM and text messaging, affording them the ability to integrate interactive activities into the teaching and learning environment and foster a social and dynamic learning community regardless of the student's location. The award winning program, referred to as "Campus Connect" (Campus Technology, 2007) resulted from a collaboration among the University, a tier one cellular service provider, and a mobile services software development company. It was our hope that the "Campus Connect" program if used effectively would foster a learning environment outside of the classroom walls by facilitating innovative use of academic applications and timely access to available resources. The objectives of this innovative approach to student communications are increased academic participation, improved student retention rates, and stronger student participation in a more well-defined campus culture.

CAMPUS CONNECT PROGRAM

In 2005, the Montclair State University administration began looking at new ways it could harness today's ubiquitous cell phones to communicate with students. This action was precipitated by the reality that landline usage on campus has become obsolete. The University was failing in its efforts to communicate effectively with our student body. These challenges, if left unchecked, threatened to fracture the sense of community across campus. Montclair's implementation of mobile applications shows that the University recognizes the need to stay one step ahead of student technology adoption. Students were opting to use their personal cell phones as their primary mode of communication, without even sharing their mobile phone numbers with the University; therefore, it became increasingly difficult to reach students for basic messaging (i.e. class cancellations, grade postings, or other academic and community information.) In order to address these challenges the Campus Connect program was developed.

Campus Connect enables students to step beyond traditional cell phone services such as voice, IM and text messaging and enables students to participate in a rich array of cell phone based applications by customizing their mobile phone for access to the University's teaching and learning and administrative resources. Partnering with Rave Wireless, a company that offers a complete suite of academic, safety, community, and social applications specially designed for students' mobile phones (Rave Wireless, 2008) and Sprint/Nextel, a tier one cellular carrier company which offered specific technologies that aligned with the features and functionality envisioned for our virtual campus solution, the Campus Connect program was launched.

The first iteration of the Campus Connect program included access to core phone-based mobile applications for communication, collaboration, safety and academics. These included broadcast text alerts for campus-wide communications, including emergency security situations, severe weather, and class cancellations. In addition, the application supported a social networking peer location and communication tool called "entourage" with one to many messaging and location capability to locate and communicate with selected groups. This group feature fostered group collaboration through its opt-in feature using formats similar to Facebook.

Although the "entourage" capabilities were useful, the real backbone of the application resided in the ability for students to use the cell phone to view course announcements and grades that were posted in the course management system (Blackboard). This allowed students to receive alerts related to class cancellations, homework updates, and other relevant coursework without having to access a computer. Students also received real-time alerts, directories, event information, shuttle bus information, and dining menus, as well as accessing mobile guardian, where students traveling around campus can be monitored by the campus police.

Providing this application suite was an excellent first start, but the University's primary objective for introducing cell phones into the fabric of the campus was the intent that cell phone technology could be used to enhance and enrich teaching and learning. This objective was integral to subsequent iterations of the Campus Connect program, which was implemented in two phases. During Phase I, the various features and functions described earlier remained in place; however, these were augmented with an enhancement that enabled students to view video podcasts of recorded lectures from their mobile phone. The outcomes from Phase I paved the way for further curriculum redesign, which went into effect during Phase II of the program. These curriculum redesign efforts afforded students with interactive and social learning activities such as blogging, polling, and field work assignments. The following describes in detail a description of these activities that were implemented during both phases as well as the student's reaction to these activities as detailed in the post survey.

Phase I

With the start of the fall 2006 semester, the Campus Connect program subscribed all sophomore students in residence as well as all entering undergraduate students for academic year 2006-2007 representing nearly 3,500 subscribed campus connect cell phones. This number provided the critical mass of students required for the Office of Information Technology to introduce academic uses of this technology. In order to accomplish this, a pilot was ran during the fall 2006 semester, which allowed for 60 students from a Contemporary Business and College Writing classes to access recorded course lectures from their mobile phone. These classes were chosen as they represent subject matter with differing pedagogical flavor in order to determine whether cell phone technology was suitable for all courses, or if subject matter should be considered before integrating mobile technology across all disciplines.

Content in the business course included practical aspects of how a business is organized, which included various functional areas of business such as finance, management, production, and marketing. As part of this course, students learned material not only through traditional lecture and textbook readings, but also participated in an online simulation program, which provided for role based scenarios. On the opposite side of the spectrum was the college writing class where the material was more abstract in nature where students gained a better understanding of the writing process by establishing proficiency in a range of writing assignments that included essays and journal entries. Students learned material through lecture and in class writing exercises. The method used for demonstrating knowledge of content also differed between these two courses, where the business course's primary evaluation tool consisted of traditional assessments (multiple choice and true/false exams) as opposed to the writing class where the students were primarily evaluated on their essay submissions and journal entries.

The goal of the stage I pilot was to determine whether learning activities that used mobile technology enhanced the learning environment as well as provided easy access to relevant course information. Primarily, this would be accomplished by providing course content as podcasts that students could review from the provided cell phone. In addition, students would be able to access course announcements and gradebook information that was posted in the course management system. Although these postings were accessible by logging onto the learning management system, it was our hope that students would have a preference to accessing this information from their mobile phone, since it was more in tune to their style of learning as well as affording greater flexibility. This assumption was based on past research, which defined the digital learner as one who emphasizes a sense of immediacy and faster access to content because they are no longer tied to a physical space or hardware (Oblinger & Hawkins, 2005). In addition to this sense of immediacy, today's learner also has an exploratory style of learning and "prefer to learn by doing rather by being told what to do" (p. 2.6).

In keeping this in mind, a meeting with faculty was held to discuss the instructional design and pedagogy behind the cell phone pilot, where recommendations were made on how to integrate mobile technology into their pedagogy. Firstly, besides posting announcements that provided assignment due date information, test reminders or class cancellations, it was suggested that announcements also be used to ask thought provoking questions to promote critical thinking based on content that was covered in the course lecture. Secondly, podcasts would be used to record "snippets" of information as a supplement to the material covered in class. These "snippets" would provide content that was a supplement to the course lecture, promote critical thinking, as well as provide a mechanism that reinforced the "announcement" question. Instructors would use recording/narrating software that worked in conjunction with PowerPoint presentations to capture these "snippets," which was accessible from the student's cell phones as a video podcast.

Despite these recommendations, instructors in both groups succumbed to traditional methods of instruction when integrating mobile technology into their pedagogy. For example, instead of recording podcasts that supplemented the course lecture as recommended in the instructional design, the instructor would record the entire lecture when the course met during face-to-face meetings. In addition, course announcements posted in the learning management system were informative in nature and did not initiate any critical thinking by offering thought provoking questions. Figure one displays a typical announcement generated from a course within Blackboard as it appeared on students' mobile phones, reminding them when their final exam will be occurring. A similar screen would also show a grade that was posted on Blackboard.

The goal of Phase I was to evaluate whether accessing course lecture "snippets" via cell phone technology as well as using the cell phone features as an integral component of the University's learning management system satisfied today's student and enhanced learning. Out of the 60 students from the business and college writing courses who participated in the pilot, 44 students responded to our questionnaire probing their experience with these cell phone innovations as previously described in the instructional design methodology. Each class submitted 22 responses (see Figure 1).

The quantitative data from Phase I showed disparate results for the two groups tested. In response to the question whether students liked using cell phones in class, survey results indicated that 59% of the business students responded favorably, as compared to their counterparts in the English curriculum where only 23% responded favorably. The same disparity in findings persisted

Figure 1. Instructor's announcement posted on Blackboard

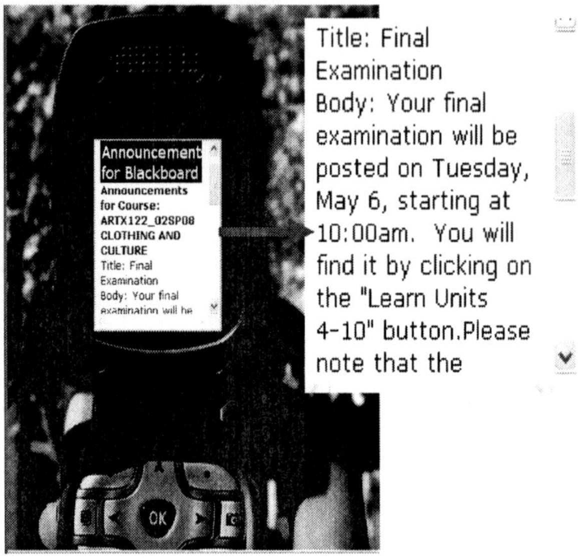

with regard to student's assessment of whether cell phones helped them learn better in class. Forty six percent (46%) of the business students responded favorably; as compared to only 19% favorable responses from the English students.

It was apparent from these results that subject matter was a motivating factor as to whether the use of cell phones assisted in the learning process. This was evident in the greater percentage of learners in the business class responding favorably as to whether they liked using cell phones in class as compared to the English counterpart. However, in response to the question whether cell phones helped students learn better, despite the greater number of business students responding more favorably as compared to the English counterparts, more than half (54%) of the learners did not feel that mobile technology helped in the learning process. This could be attributed to the fact that the instructors did not adhere to the instructional design recommendations and created podcast recordings that repeated the course lecture. In addition, the announcement postings were administrative in nature and did not bridge the podcast with critical thinking processes by invoking questions that motivated the learner to review the podcast as recommended in the instructional design.

One additional observation can be made in both classes, which corroborate the research conducted by Oblinger and Hawkins (2005), are that students prefer participatory activities in the learning environment. In both groups, learners participated in either role based scenarios or writing assignments, which seemed to be the preferred learning activity as opposed to the passive use of mobile technology. Paradoxically, despite the seemingly different reviews the cell phone technology received between these two classes, there was general consensus among all of the students that mobile phones had the potential to help them achieve their educational objectives. Seventy three percent (73%) of the business students as well as 68% of the English students responded favorably to the help in education dimension of our inquiry (see Figure 2).

The results displayed in Figure 2, suggest agreement about the potential value of the cell phones to support education but clear differences in the perceived value of the devices for

Figure 2. Student quantitative survey results for Phase I

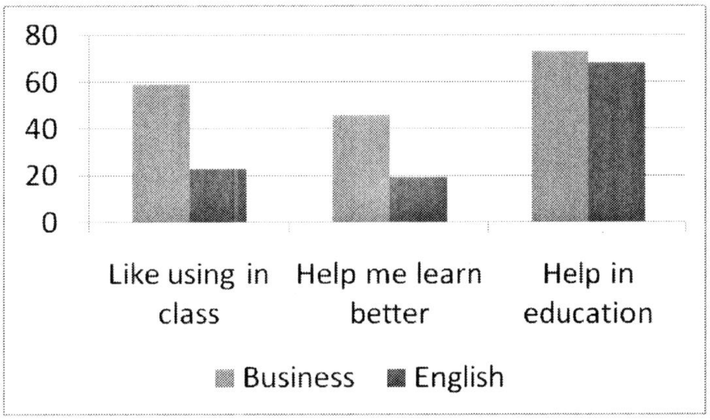

classroom use and to reinforce learning. The pattern discerned in Phase I of our cell phone experimentation presented the possibility that the differential perceptions of students were more so attributable to the nature of the subject matter of the courses involved as well as the method in which the activities were presented. This was a key factor in whether students were motivated to use cell phone technology with the business students displaying a marked preference for the cell phones as classroom and learning tools than their English class counterparts.

Further analysis of qualitative data collected in conjunction with our research supported this speculation. Anecdotal comments obtained from student focus groups as well as written comments provided from the post survey were evaluated. These results indicated that students were not motivated to view their instructor's recorded lectures from their cell phones for several reasons. Firstly, the same material was covered in class; therefore, if the student was a good listener in class or took excellent notes, there was no reason to revisit the course lecture on the phone. Secondly, because the instructor was recording the entire lecture, the podcast was lengthy. This made it difficult for the student to fast forward to the part of the podcast that he or she wished to review. In addition, the cell phone screen size posed a challenge for viewing the podcasts. This was particularly expressed by the business students who had difficulty viewing income statements from the cell phone screen. In addition, the subject matter did not warrant student's use of the cell phone to view the content. As noted previously, the English students participated in writing assignments and in class exercises that did not warrant cell phone usage. Finally, the preference for direct activities was apparent, especially for the Business student's who expressed how the real world experience that they were exposed to in the online simulation activity provided a deeper learning experience. Unfortunately, the suggested use of mobile technology to bridge the course content with these activities did not occur. Table 1 provides a sample of the comments provided by the English and business students on the post survey and during the focus sessions.

In addition to the pedagogical challenges the students described in Phase I, there were various administrative policies enforced that students were not in favor of, resulting in negative feedback about the program. For example, as part of the Campus Connect program, students were required to purchase a Montclair provided phone and service agreement. The charges associated with these phones and plans were included with their telecommunication fee. Although the phone and service plan was cost effective (in many in-

Table 1. Phase I Focus Group Student Comments

Favorable	Unfavorable
Able to obtain reminders from the instructor without having to logon to Blackboard Enjoyed listening to the lectures while on the train or driving to and from school Rewrote lecture notes while reviewing the lesson for a test	Same material covered in class; therefore, no motivation to view lectures on the phone Required assignments were in-class written exercises Material for assignments was obtained from in-class lectures

stances, better than the student's personal plan), the students were resistant and did not want to give up their personal phone. In fact, many students carried two phones – one that was used for the University activities and applications and the second phone, which was used for personal communication.

An additional administrative challenge that resulted in less than favorable reactions to the program was related to the support for the mobile phones. Although the students received an orientation for the phone set up and use, the instructions provided during these orientations were not thorough enough for the students to begin using the phone for class assignments. Supplementary in-class instruction was provided by the University Information Technology staff, which took away from class time requiring the instructor to modify the course's lesson plan. Needless to say, these interruptions resulted in negative feedback from all parties involved.

Despite these mixed reviews, as indicated in both the quantitative and qualitative data, students in both groups did indicate that the learning process could be enhanced with mobile technology as long as they felt motivated to use the tool. It became evident from these results that the primary factor for a positive and enriching learning experience resided in the types of activities that were being offered as well as the particular curriculum where these activities were introduced. These two factors would strongly influence whether students felt compelled to use their mobile device in order to enhance their learning experience. These conclusions informed the research protocol for Phase II of the Campus Connect program.

Phase II

Based on the results from Phase I, several modifications were included in the Campus Connect program, specifically towards the design of academic activities that were facilitated through mobile technology. These changes, implemented in Phase II, were an attempt to address student concerns around cell phone activities not serving a clear purpose in a particular course; thereby causing students to have no incentive to use the technology. In addition, based on past research, which described the 21st century learner as Internet savvy and preferred learning through active engagement, (Jones, 2002; McGriff, 2001; Oblinger & Hawkins, 2005) changes were made to the activities facilitated by the mobile phone in order to support these findings.

Firstly, the appropriate courses were chosen in order to guarantee that the learning activities facilitated by the cell phone were applicable to the objectives of the course. These courses included the New Student Experience course, which involved in class interactive exercises as part of the curriculum. In addition, a linguistics, language, and history course were selected, since the learning activities as described in the course syllabi could be easily modified to accommodate cell phone use, promote a social and collaborative learning environment, and facilitate "learn by doing" activities.

As part of the redesign process that occurred in Phase II, an instructional designer met with the faculty member and reviewed the course syllabus and content in order to design activities that were suitable for the cell phone. Again, suggested activities were designed as a classroom supplement, which required students to use the cell phone outside of class for fieldwork exercises in order to foster the social learning and collaborative environment the program was designed to achieve. The cell phones were also used as an assessment tool, in order to evaluate student comprehension of material. Finally, when the cell phone was used during class time, it was used as a teaching tool to stimulate classroom discussion.

The cell phone applications, which allowed students to view Blackboard announcements, grades, and University email, continued to be part of the student's academic agenda. However, during Phase II, based on the student responses from Phase I, a variety of activities that reinforced course content by direct student involvement designed around mobile technology were also added. These learning activities reinforced not only social and constructivist learning methodologies, but also included assessment tools, which provided the students and the instructor immediate feedback regarding the learner's understanding of the course material. Examples of mobile phone activities included students answering multiple choice questions from the mobile phone, texting and capturing images from their mobile phone to a blog for peer and instructor review, scavenger hunts, polling, and field trips. In addition, students asked questions by text messaging and/or by phoning their peers that were assigned to their group using the entourage cell phone feature. What follows is a description of the courses that were chosen and the pedagogy and instructional design behind the cell phone activities that were integrated into the learning environment.

One area where mobile technology could enhance the learning environment was through promoting social engagement. Since Montclair State University's student population consisted primarily of commuters, it was our desire to provide an additional venue to provide a sense of community for incoming freshman. In order to accomplish this, the Campus Connect program was integrated into the New Student Experience course in order to promote collaboration and enhance social awareness. Given that this course only met weekly, the use of mobile technology in this venue would allow for students to feel connected outside of class time.

As part of the process, every New Student Experience course was created on the Campus Connect site, which contained an updated course roster. During course orientation, students activated their accounts on the Campus Connect site by registering their cell phones and signing up for alerts. This would provide a communication mechanism to promote social engagement among the group. For example, if students had any questions about an assignment, it was recommended that the students use their cell phones to communicate with each other and the instructor as opposed to using email. Mobile technology provided the means to foster communication among the class and contribute to the sense of community, which is one of the objectives of the New Student Experience course.

In addition to the learners using mobile technology to foster communication, cell phones were used to enhance class discussions that addressed a variety of topics specified by the New Student Experience curriculum. Many of the course discussions centered on topics where students often expressed discomfort sharing their responses in a public setting. For example, topics of discussion include the number of hours allocated to studying, handling peer pressure, alcohol consumption, etc. It was not uncommon for these topics to cause a level of discomfort among students, especially in a classroom setting, specifically when the instructor would ask for a show of hands in response to one of these topics. The mobile technology freed students up to answer normatively threatening questions candidly because they were afforded a level of anonymity they would not otherwise have.

In order to facilitate these class discussions, the instructor would send weekly polls to students prior to the class meeting. Students would receive

Figure 3. Student responses from an open poll

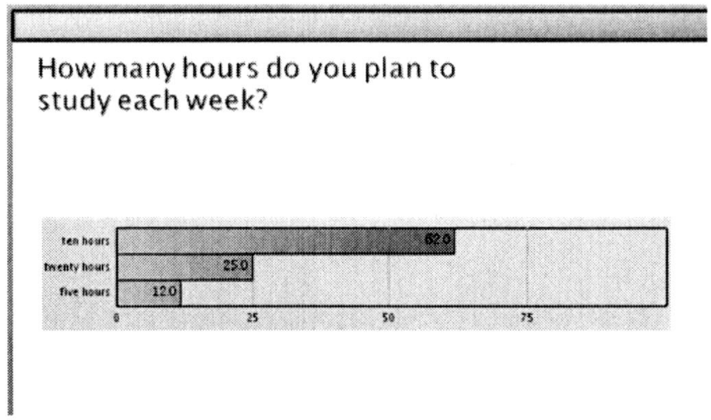

an alert on their cell phone, which contained the survey question. Students would anonymously respond to these polls by texting their responses to survey questions. Anonymous results were immediately calculated and displayed graphically during the next in class meeting, promoting a rich discussion around these topics. An example of students responding anonymously from their cell phone to an open poll question addressing time management can be seen in Figure 3.

One other example of using cell phones to facilitate survey questions was demonstrated in an Introduction to German class, where students were required to respond to questions about famous Germans. Originally, this activity was completed independently using a traditional paper and pen. However, based on the findings of past research as noted earlier noting direct experience as the learning preference for today's learner (Jones, 2002; McGriff, 2001; Oblinger & Hawkins, 2005), this assignment was adapted to include mobile technology. This occurred by creating polls with multiple choice answers that were accessible from the cell phone. The instructor would put together a list of trivia questions (in German) about famous Germans, such as "What did Goethe write?" or "Who is Joseph Ratzinger?" Each question was set up as a poll with multiple choice answers accessed from the cell phone. In groups of three or four, the students were sent out to use all resources available to them (computer lab, library, cell phone with web access, etc.) to find the answers to the questions and reply to the poll. The team with the most correct answers in the least amount of time won. Of course the prize was German chocolates.

In addition to polling activities, the learners in the German class participated in a scavenger hunt as a vocabulary building exercise facilitated by mobile technology. In the chapter on how free time is spent, the students received a list of about 25 verbs. In teams of four to five students, the students had 15 minutes to walk around campus and find people engaging in activities from the list. Using cell phones, the students would take a picture of the activity and send it to the instructor's phone with a full sentence description in German of the activity in the photo. It was also permitted for the students to stage the pictures themselves. The team to send the instructor the most pictures with accurate sentence descriptions won candy.

Cell phones were also adapted in field work activities in a Linguistics course where students were studying the dimensions of gender differences in speech. The class activity required students to go to a public setting and record their observations by texting from their cell phone to their class blog. If permitted, these text entries were reinforced by pictures taken with the cell phone. Students were required to reply to each other's blog posts and share their fieldwork experiences with each other. Previously, this assignment did not include any digital images or peer review; rather, individual observations were submitted only to the instructor for grading. Adapting mobile technology allowed for a more constructivist approach to learning. The learner participated in a more hands-on approach to the topic and was able to construct a deeper meaning of gender differences in speech not only through their experiences but through the immediate feedback received through the blog posts.

One other example of using cell phones to facilitate a field work activity was seen in an Urban History course. Students, studying fires as a historical event, were required to visit a local fire house and interview a fireman/firewoman and post the highlights of the discussion on their blog. Students would text these highlights along with pictures to their blog from their cell phone. As in the Linguistics course, students were able to receive immediate feedback about their experiences from their peers, which was not originally part of the assignment.

The results from the student activities in Phase II were mixed. Out of the 100 students who participated in the pilot, 45 students responded. The quantitative data indicated that students responded favorably to using cell phone functions that provided immediacy to instructor postings. For example, 67% and 64% of the students surveyed responded favorably to viewing Blackboard announcements and University email, respectively.

Despite these favorable results, the students' reactions to using the cell phones to facilitate group activities were less favorable. This was caused by the administrative and operational details that undermined the pedagogical efficacy of the cell phone technology. For example, in those classes where the groups were not automatically created based on a batch enroll process, the group activities were thwarted. For these classes, in order to be part of a class group, students had to accept an email invitation from the instructor. This posed a challenge, because students, who deleted their email or overlooked the invite, were never included in the group. As seen in Figure 4, this challenge had a negative impact on the student's reaction to using the group function. The survey results indicated that the majority of the students either chose not to use the group function or did not have a favorable experience. However, those students who were in the class that had the groups automatically created according to course roster, showed favorable results.

Despite this challenge, survey results indicated that the cell phone activities were relevant to the topics presented, where 60% of the students

Figure 4. Phase II post survey results for group functionality

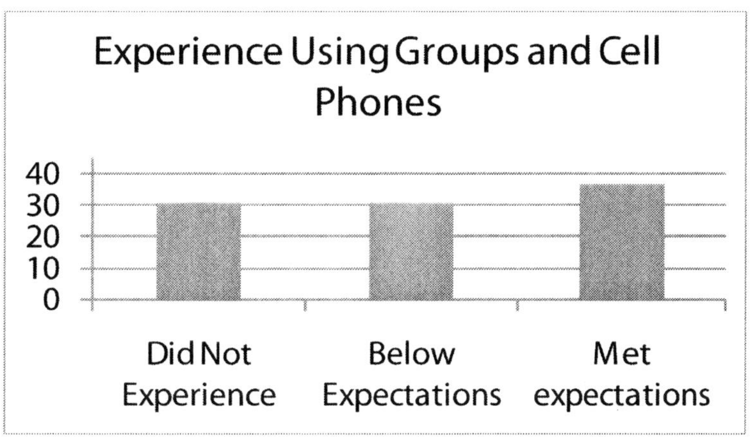

responded favorably. In addition, students were satisfied with the cell phone as a teaching tool, where 58% of students responded favorably. These results indicated that the instructional design strategies suggested in Phase II had a positive impact on the student's learning experience. This also confirmed that the transition from using a passive use of mobile technology (viewing podcasts) to a more active use that involved direct experience activities enhanced the learning process.

Qualitative results for Phase II based on informal instructor inquiries indicated that the students were pleased with the academic use of the cell phone, the activities were engaging, and the activities reinforced the subject matter. These comments corroborated the findings of the quantitative data and reinforced the recommended instructional design strategies. However, despite these positive comments, students described dissatisfaction with administrative policies related to the cell phone plan, model, and required fees. Students did not want to be required to give up their own personal phone or carrier for the University's cell phone.

In addition, to administrative concerns, anecdotal comments expressed from students indicated that many of the learners were not comfortable with some of the required classroom activities due to a lack of adequate support and training. For example, students had difficulty participating in the blogging exercises, which required the mobile phone to be connected to the student's personal blog. It was assumed that because of the students' comfort level of using mobile phones, there would not be any difficulty connecting the phone to their personal blog. This assumption proved to be wrong, where many students not only had problems connecting to the blog, but had difficulties creating a blog account. Once again, classroom instruction had to be provided in order to overcome this obstacle, which interrupted the instructor's lesson plan. This confirms past research where the comfort level of student's use of technology may not be synonymous with the "gadgets" they use in their everyday lives. "But though they may never need to consult an instruction manual for an electronic gadget, their comfort with technology may not be synonymous with competency. Students' underlying understanding of the technology may be shallow (Oblinger & Hawkins, 2005, p. 12).

SOLUTIONS AND RECOMMENDATIONS

Phase III of the Campus Connect program began during the fall 2008 semester. Based on the results from the prior two phases, Montclair made further modifications to the program so that the administrative and pedagogical challenges reported by the faculty and students involved in the first two phases could be addressed.

The first change addressed the student's requirement to purchase University cell phone and service plan. Initially, this mandate was enforced to ensure that all students would have access to the applications provided by the Campus Connect program. Since the change in technology has made cell phone applications more ubiquitous and available through open source and individual providers, this administrative policy has been modified. Currently, students are now allowed to participate in the academic activities and applications regardless of their phone type or carrier. Students however, will still have the option of participating in a Montclair provided cell phone and carrier, if they so desire.

The second change addressed the student's comfort level of using their cell phone. Policy related to the phone delivery has now been established, which requires students to attend an orientation, where they will receive formal instruction on how to activate the cell phones. In addition, students will receive formal training on how to access the applications and required course activities that will leverage the cell phone. In addition, training will involve students experimenting with polling and blogging as well as participating in a variety of group activities that will be integrated into the learning environment. Besides the student orientations, training sessions were scheduled throughout the summer and during the first two weeks of the fall semester, as an outreach to students and faculty who need additional support. In addition to these sessions, the "Campus Connect Doctor" was available at strategic locations throughout the campus for additional one-on-support. Finally, in class sessions were provided throughout the semester, for additional support.

The third change provided standardized lesson plans for the New Student Experience courses. The cell phone activities established in Phase II continued as part of the curriculum. However, instead of offering these activities for an isolated lesson, these activities became an integral component of the course, where students were required to use cell phones for every learning unit. A template within the University's course management system (Blackboard) was created providing curriculum based lessons that matched the learning objectives of the specific unit, fostering the use of mobile technology. This template, along with training and instructional design support, was offered to the instructors slotted to teach the new student experience course.

In order to alleviate the challenges associated with the group functionality as previously described, modifications were made so that the Rave application communicated with the University's student database. This enabled groups to be formed in an automated way for every course involved in the Campus Connect program, matching the course ID and student roster as it conformed to the Registrar's database. This change facilitated group projects and communication among the students, without the instructor having to send an invitation to the student to join the group.

Finally, additional resources were provided to support the Campus connect program. These resources included not only Montclair training and instructional design personnel, but also administrative staff in order to manage the day-to-day operations of the program. Improved communication and collaboration among the various departments provided the necessary commitment in order to forge ahead with the program's goals and objectives.

In addition to the cell phone activities described in Phase II of this program, supplementary activities will be integrated into the learning environment based on the discipline and the instructor's learning objectives. These activities will include accessing content specific literature from the mobile phone through RSS feeds as well as viewing video podcasts to generate class discussion. Experimentation with accessing course documents posted within the Google docs and calendaring function is also being considered.

CONCLUSION

At recent Technology and Learning Conferences (MCCC, 2006; Northeast Connect, 2007), the topic of the personalization of learning was the focus of the keynote speaker. According to Mark Prensky, teaching and learning are not the same, and just because we teach something, does not necessarily mean that the students will learn. Educators need to try new things, experiment, and invent, and in the process generate engagement and form a partnership with their students.

The 21st century learner has transformed the educational process to a new level, where learning can no longer be pushed on students. These digital learners have the desire to get information the way they want it and when they want it, which inspires educators to discover new methods of pedagogy recognizing the benefits of technology (New Media Consortium, 2005). The complexity of the learning environment, the diversity of learners as well as their unique learning styles and the multitude of devices that can be brought to bear on the learning environment suggest that contemporary classrooms are a delicate ecosystem.

Those of us charged to design instruction would do well to heed the wisdom of those Inuit artists described by Renee Dubos and to seek the "god within" these complex learning systems. Anyone involved in educating the 21st century learner needs to have a unique insight about their media, in this instance cell phone technology, in order to fully realize it's potential.

Cell phone technology is ubiquitous among today's learner and is a tool that is easy to use, inexpensive, and can provide content in a variety of formats. The Horizon Report, a project co-sponsored by The New Media Consortium (NMC, 2008) and Educause, identifies emerging technologies that are likely to change the way in which we conduct the business of higher education. For two consecutive years, this report focused on the rich media, social networking and pervasiveness of cell phone technology as a force that will influence higher learning over the next several years (Horizon Report, 2007 & 2008). Part and parcel of the expected influence of the technology is the furious pace with which it is evolving and changing. Although the Campus Connect Program met the criteria for today's learner as described by the NMC, it became apparent in this study that selecting the appropriate technology is not enough in order to have a successful learning environment. This was evident in both Phases of this study, where both instructional design and administrative challenges that the students faced became a deciding factor on whether students were motivated to use the technology.

One would conclude from this study that the pedagogy, which leveraged the technology, is the most important element to be considered when implementing new endeavors. The importance of pedagogical application and instructional design should continue to remain in the forefront of any innovative technological endeavor. The modifications that were applied in Phase II, with a greater emphasis on specific activities with direct learner involvement that complemented the use of the cell phone, resulted in favorable learning experiences. However, it is equally important that administrative policy and assumed technological comfort not be overlooked, in order to have institutional buy in.

Technology has become a staple of the 21st century learning environment, and as technology changes so should the pedagogy providing the framework to empower students to engage in successful learning. Although the use of cell phones accommodated the learner's mind-set of receiving information the way they wanted it and when they wanted it, it became obvious from this study, that if the information were not worth getting, then the learner would not use the technology. A positive learning experience does not rely on the type of technology that has been selected; rather, a positive learning experience relies on the methodologies that are being used in order to integrate the chosen technology appropriately into the learning environment. Finally, despite the integration of pedagogical applications that support mobile technology, it is equally important to have administrative polices and training agendas in order to maintain a learning environment that is not impeded by technological challenges and policy demands.

Today's college student is a different person than five years ago. They are technologically savvy but not always competent with using the tools that represent their primary mode of communication. Universities need to take this into consideration while at the same time harness the power of the mobile device to foster a stronger sense of community and to establish virtual campuses. When deployed effectively, student-focused mobile applications can ensure that students are in constant real-time connection with the University, keeping them safe and secure, academically and socially engaged and, therefore, better able to succeed in their academic pursuits.

Real success takes time, and universities need to start executing the vision of their campus mobility program early. Montclair State University has embraced mobile technology to enrich the living and learning experiences of our students and we have every expectation that this compelling technology will continue to transform the ways in which we service their administrative, academic and social needs for many years to come. The research and strategies that were implemented for the Campus Connect program, should be used as an example for educators who look to explore innovative uses of any technology, not just cell phones, and realize that if used appropriately it can be used as an example to empower teachers and fuel administrators to broaden their horizons to meet the needs of the 21st century learner. However, like The God Within it is important to realize that the campus is a delicate ecosystem and that cell phone technology can refine the campus environment if used and implemented appropriately in a way that all members of the community are able to fully realize its potential.

REFERENCES

Albright, M. J., & Nworie, J. (2008). Rethinking academic technology leadership in an era of change. *EDUCAUSE Quarterly*, (1): 15–23.

Campus Technology. (2007). *Campus technology innovators: Cellular/mobile.* Retrieved July 7, 2008, from http://campustechnology.com/article.aspx?id=49202

Dubos, R. J. (1972). *A God within.* New York, NY: Scribner.

Dunn, R., & Griggs, S. (2000). *Practical approaches to using learning styles in higher education.* Westport, CT: Greenwood Publishing Group, Inc.

Jones, S. (2002). The internet goes to college: How students are living in the future with today's technology. *USDLA Journal, 16*(10).

McGriff, S. (2001). *ISD knowledge base. Pennsylvania State University, Portfolio Web site.* Retrieved June 6, 2006, from http://www.personal.psu.edu/faculty/s/j/sjm256/portfolio/kbase/IDD/ISDModels.html

McLaughlan, R. G., & Kirkpatrick, D. (2004). Online roleplay: Design for active learning. *European Journal of Engineering Education, 29*(4), 477–490. doi:10.1080/03043790410001716293

Montclair State University Board of Trustees. (2002). Retrieved March 31, 2008, from http://www.montclair.edu/provost/strategicplan.pdf

Montclair State University Mission Statement. (2002). Retrieved October 20, 2008, from http://www.montclair.edu/welcome/msumission.html

Montgomery Community County College, MCCC (2006, October 6). Twelfth annual technology and learning conference. *Keynote presented by Marc Prensky on Overcoming Our Digital Immigrant Accents: Understanding the Implications of Emerging Digital Native Behaviors for Education, Training and Learning.* Retrieved October 21, 2006.

New Media Consortium. NMC. (2005). *A global imperative: The report of then21st century literacy summit* (Issue Brief No. 0-9765087-1-0). Stanford, CA

New Media Consortium. NMC (2007). *2007 Horizon report.* Retrieved July 7, 2008, from http://www.nmc.org/horizon/2007/report

New Media Consortium. NMC (2008). *2008 Horizon report.* Retrieved July 7, 2008, from http://www.nmc.org/publications/2008-horizon-report

Northeast Connect. (2007). *2007 annual conference: Broadly exploring new directions in learning.* Retrieved July 7, 2008, from http://www.northeastconnect.org/?page_id=53

Oblinger, D. (2004). Educating the net generation. *Keynote address delivered at Educause 2004*. Denver, CO. available at Oblinger, D. & Hawkins, B. (2005, September/October). The myth about students. *Educause Review.*

Prensky, M. (2001). Digital natives, digital immigrants, part II: Do they really think differently? *On the Horizon, NCB University Press, 9*(6).

Prensky, M. (2004). *The emerging online life of the digital native*. Retrieved October 21, 2008, from http://www.marcprensky.com/writing/Prensky-The_Emerging_Online_Life_of_the_Digital_Native-03.pdf

Rave Wireless. (2008). *Rave Campus.* Retrieved July 7, 2008, from http://ravewireless.com/campus.shtml

Tunstall, D. F. (1995). *Is your school ready for the 21st century? A general overview for planning purposes* (NCES Publication No. 383-443). Washington, DC U.S. Department of Education (2003). *Distance education at degree-granting postsecondary institutions: 2000-2001* (NCES Publication No. 2003-017). Washington, DC

Valdez, G., McNabb, M., Foertsch, M., Anderson, M., Hawkes, M., & Raack, L. (2004). *Computer based technology and learning: Evolving uses and expectations.* Retrieved March 8, 2007, from North Central Regional Educational Laboratory Web site: http://www.ncrel.org/tplan/cbtl/toc.htm

Chapter 19
Contextual Learning and Memory Retention
The use of Near Field Communications, QR Codes, QBIC, and the Spacing Effect in Location Based Learning

David Metcalf
University of Central Florida, USA

David Rogers
University of Central Florida, USA

ABSTRACT

An important part of multiplatform or blended learning is designing learning environments that take full advantage of the relative strengths and weakness of the various platforms employed to meet learning objectives. The desktop has strengths that are conducive to immersive learning environments, whereas mobile devices excel in contextual learning and performance support roles. Blended learning then, is not merely porting the same content from one platform to another, but recognizing the need for unique implementations. This chapter will examine two general applications in which mobile learning takes advantage of the flexibility afforded by the platform. In the first case we will explore the possibilities presented by physical hyperlinks through the use of Near Field Communications, QR codes, and image recognition software. In addition to providing contextually relevant information, the mobile platform is ideal for providing enhanced conceptual retention. The Spacing Effect demonstrates that memory decays according to a well-defined logarithmic curve. Once this curve has been optimized for an individual, it is possible to determine the most productive times to review learning objectives. Mobile devices are the perfect platform to review material initially mastered on a desktop or in a classroom, and these scheduled sessions can boost retention times dramatically. Contextual Learning and Enhanced Retention are two applications that cater to the strengths of mobile devices, and augment a holistic multiplatform approach to learning.

DOI: 10.4018/978-1-60566-703-4.ch019

INTRODUCTION: THE VALUE OF BLENDED LEARNING

Within the field of simulation & training, the value of increasingly immersive environments is readily apparent. The ability to accurately model reality and provide the user with artificial stimuli allows for the simulation of educational experiences that would be either too costly or too dangerous to conduct in physical space. This approach to technology enabled learning represents one distinct end of the spectrum, where the physical world is largely or entirely replaced by a digital recreation. At the other end of the spectrum are educational experiences that also rely on digital information, but the use of this information is intended to augment, rather than replace, physical space.

Mobile technology provides us with the tools necessary to allow for "contextual learning." And this approach, which situates learning objectives firmly in their operational context, can provide results and capacities that are not easily achieved with traditional methods.

However, the use of mobile technology in learning is a distinct tool that achieves its greatest potential when used as one component of a multiplatform approach. The strengths of a mobile device, accessibility and portability, can also be liabilities depending on the nature of the educational objective. But if classroom, desktop, and mobile approaches can be integrated to reinforce each other by capitalizing on the strengths of each, then dramatic increases in efficiency and efficacy are possible.

BACKGROUND: LOCATION BASED AND CONTEXTUAL LEARNING

One of the key advantages of the mobile platform is the ability to access a learning objective from any location and at any time. A recent study published in the Australasian Journal of Educational Technology (Pettit & Kukulska-Hulme, 2007) describes how many users in urban areas found their daily commutes to be one of the best times to study educational content on their mobile device. This unparalleled accessibility lowers the barriers for short learning experiences and encourages frequent interactions. A mobile learning session can be very spontaneous as opposed to traditional classroom instruction that requires a tremendous amount of organization and planning to implement.

Typical classroom teaching sessions are scheduled several months in advance. Likewise, web based seminars rely on rigid scheduling, although by distributing the audience they allow for much wider participation. Both forms of instruction enjoy unique advantages but they incur a high opportunity cost in terms of the required commitments from the instructors and participants.

The mobile learning environment is radically different. The typical user interacts with their mobile device frequently (several times an hour, but only for a few seconds at a time). These brief interactions can be used to provide valuable learning experiences, but they must be structured very differently than those designed for other formats. Specifically there are two considerations that should characterize mobile learning content. The first is a streamlined user interface. Even the simple entry of a URL can be so disruptive to the mobile experience as to present an insurmountable barrier to the typical mobile user. The second consideration is the contextual and temporal relevance of the content. In general, there should be some compelling relationship between the mobile content and the users physical location to create an effective learning experience. One exception to this general principle, which will be discussed later in the chapter, is the temporal relevance of scheduled review sessions.

USER INTERFACE

One of the most important design requirements of a user interface is visual consistency. A theory known as "contextual cueing" asserts "invariant properties of the visual environment such as stable spatial layout... [allow] us to interact more effectively with the visual world." (Chun, 2000). With regards to software applications, a well-designed user interface transparently guides the user through the content, and equips them with the tools that they need to understand and manipulate new material.

Our reliance on these interfaces becomes so ingrained that in some respects, users cease to notice them at all. But even a subtle change can be very disruptive. Anecdotally, there are many who experience these disruptions each time Microsoft upgrades their suite of office tools. Users accustomed to the old interfaces are faced with the fleeting frustrations of learning to navigate the new menus.

An interesting corollary, and one especially pertinent to our present discussion are the reflexive attempts to interact with a user interface in an inappropriate context. When one has trouble hearing a colleague, often times the first instinct is to "turn up the volume", as if reaching for a knob on the radio. The desire to navigate the physical world with the same interactive framework as those used on computers is a recurring fantasy in popular culture.

What gives rise to that impulse is our increasing acculturation to digital spaces and corresponding alienation to the physical. As Chun explains "Sensitivity to such regularities presented in visual context serves to guide visual attention, object recognition and action. Observers are tuned to learning contextual information in an implicit manner, and in humans this learning appears to be mediated by an intact hippocampal system."

The real promise of mobile learning is that the potential exists to develop user interfaces that allow our minds to interact consistently within digital and physical space. Where a browser pane provides consistent visual organization for digital content, a well-designed mobile interface can serve the same function in the physical world. To achieve this, mobile devices must successfully integrate their interfaces with external stimuli, locational awareness, and fluid user interactions.

In the desktop learning environment, entering URLs, navigating from page to page, and writing blocks of text is a natural and comfortable process, but on a mobile device, nothing could be more tedious. This presents a challenge to the effective implementation of mobile content, but this challenge is being answered with a number of innovative strategies.

One approach with unique application to mobile devices is the use of accelerometer data. Instead of relying on passwords to facilitate secure data transactions software developers have found ways to authenticate users through a unique series of motions, almost like a haptic signature. An application for the iPhone developed by Tapulous allows users to share contact details simply by holding two phones together and shaking them. The accelerometer data from each device is uploaded to a server that then delivers the updated contact information though the data network. This transaction occurs without any direct communication between the phones themselves (Marshall, 2008).

Philippe Khan, the inventor of the Camera phone promises to take haptic interfaces even further with his recently announced MotionX software (MarketWatch, 2008). Through careful analysis of accelerometer data it is possible for a mobile device to recognize when a user is sitting or standing, when they are walking or running, and to use movement as input commands. These approaches are in their infancy, and current applications are little more than novelties, but they hold great promise for the future.

But there are other methods of facilitating mobile interactions that are quite mature, and on the cusp of widespread adoption. Optical recognition of QR codes is an input method used extensively in Japan, and recently a partnership between CitySearch and Antenna Audio launched a pilot program in San Francisco to gauge their effectiveness in the United States (Kim, 2008).

A QR (Quick Response) code is very similar to a classic bar code that stores information in a series of thick and thin black bars that can be easily read by a scanner and converted into an alphanumeric sequence. Traditional bar codes are one-dimensional and can only represent a small number of characters. QR codes however, are two-dimensional and are capable of storing much more information.

In fact it's possible to store and retrieve up to 7,089 numeric characters in a single QR code (Denso-Wave, 2003). However, the limited resolution of the typical scanner would be unable to read such a large code. But for storing URL's, the QR code is ideal. The format also incorporates Reed-Solomon Error Correction that adds robustness to the encoding and serves to minimize false readings from image distortion.

During the San Francisco pilot more than 500 restaurants and business participated by displaying QR codes in the windows of their establishments. When scanned by a camera phone equipped with the appropriate software, the mobile device would access a specially formatted website that contained news, reviews, and other information related to the business. The advantage to businesses and consumers is that the codes facilitate an information transaction that is contextually relevant.

These types of interactions serve as a helpful prototype for potential mobile learning applications. While a user is unlikely to search for restaurant reviews as they pass by the establishment, the ability to quickly scan an image facilitates the interaction. In the same way, learning objectives can be associated with physical objects and specific locations.

Semapedia.org is perhaps the most concerted effort to create opportunities for mobile learning based on the use of QR codes. Their site contains an encoding program that allows the user to create their own QR codes linked to specific pages within the Wikipedia. The resultant image can then be printed and posted anywhere to provide users a convenient way to access the page. The QR code featured in figure 1, encodes the URL for the Wikipedia page of the Institute for Simulation & Training.

Also gaining in popularity is the use of Near Field Communications (NFC) and Radio Frequency Identification (RFID). NFC is a communications protocol designed to pass small amounts of information over a very short distance. A device equipped with an NFC radio is capable of creating an electromagnetic field that will activate an RFID tag. These tags, widely used for security and access control, contain an antenna that relies on induction to transmit the stored information. The average RFID tag can only broadcast a small amount of information, around 128 bits, but it can do so relying entirely on the ambient power provided by the reader (ECMA, 2004).

Storage limitations of current generation RFID tags preclude the encoding of full URL's, although IP addresses fit comfortably. The most common use of RFID is to facilitate an information transaction within a pre-established database, and these applications range from tracking inventory to wireless credit card purchases.

NFC can accomplish the same objective of linking information with physical spaces and objects as QR codes, but it has some unique advantages that sets it apart. QR codes require the user to conduct an active scan, and the reliance on the camera requires the device to be precisely oriented. NFC on the other hand is omni-directional and can be either active or passive.

What both these approaches have in common is that they create physical hyperlinks with a high degree of granularity. Not only can a specific area be linked to a data field, but objects within the area can be individually referenced. Swingline, the office supply company, has even suggested that it may be possible to embed RFID tags in

Figure 1. Image of IST QR code

Semapedia.org

Institute for Simulation and Training

Hey! This is a physical hyperlink to Wikipedia. Scan the code or enter the URL in your phone:

http://semapedia.org/w/32833

each staple thereby tagging every document (Hand, 2007). RFID "Post it" notes have already been developed by a team of researchers at MIT (Mistry, Maes, 2008).

RFID tags and QR codes provide a mobile device with a contextual clue that an instructional designer can associate with a specific learning objective. But some software developers are taking the ability to recognize objects and environments one step further, and developing strategies that may eliminate the need for such tags altogether.

Sophisticated pattern recognition algorithms have the potential to recognize any object in much the same way as a scanner reads a barcode. When an object is photographed, recognition software can identify a series of variables, plot the dimensions between those variables, and match their findings with a database. This method has come to be known as QBIC (Queries based on Image Content).

One of the areas where this technique has become quite advanced is biometrics and facial recognition. For law enforcement, private security companies, and intelligence analysts it is tremendously useful to be able to scan recorded images and match the faces against a database.

In 2001, Superbowl XXXV was equipped with a facial recognition system that scanned ticket holders as they entered the stadium, and cross-referenced them with a police database (McCullagh, 2001). More recently, Neven Vision, a company acquired by Google produced a mobile version of this application for the Los Angeles Police Department. (Durst, 2007) Instead of relying upon criminals to helpfully provide law enforcement with a QR code linked to a database of their prior convictions, the mobile phone is able to measure the physical proportions of the face to obtain the match.

If individual faces can be recognized and linked to a database, so can any physical object with a uniform appearance. The most easily identifiable images of all are those that are deliberately designed for recognition, that is to say logos and corporate brands. But even specific products like books and CDs can be readily identified by pattern recognitions software. One such company, Mobot, has an application that will recognize movie posters and use them as a link to provide the user with ticketing and show time information (Durst, 2007).

These approaches lend themselves well to learning applications. The ability to create a hyperlink to any physical object would allow educators to create immersive and interactive experiences. An engineer could obtain specification sheets for a part simply by photographing it with a mobile device. Store clerks could learn about their product lines, and shoppers could make informed choices based on information obtained from a variety of sources. Currently product information, at the point of sale, is provided exclusively by the seller; but the ability to learn about a product from an impartial third party information source will make it possible to consider factors ranging from the environmental impact to the ethics of the supply chain.

Depending on the proposed mobile learning exercise, the use of these physical hyperlinks may be appropriate. At other times however, it may be more advantageous to take a different approach. The most common way to provide information based on the users location is through the use of Location Based Services (LBS).

Instead of relying on an information transaction between the users mobile device and a physical tag, location based services using various positioning technologies to determine the users location and then provide relevant information.

There are essentially three positioning techniques that are currently employed by mobile devices, each of which have their own strengths and weaknesses. The most basic technique is known as "Cell ID." When operating on a wireless network, a mobile device communicates with the nearest cellular antenna. The positions of these antennas are stored in a database, and simply by virtue of the connection, the users position can be narrowed down to a specific cell site. But these sites often cover several square miles and the users position could be anywhere within that range. To narrow the area, some carriers are able to identify the operative element in the antenna array, which indicates a smaller portion of the total coverage area.

More accurate still is the method known as radio triangulation. When a mobile device is in range of at least three towers, it is possible to measure the latency in timing signals, and by approximating the distance to each tower a users position can be specified within a hundred meters. This approach works best in urban areas where the cell sites are numerous and closely packed together. In rural areas where there are fewer towers available it may not be possible to triangulate a position.

The most accurate positioning technology by far is the Global Positioning System. In the much the same way as radio triangulation, a mobile device equipped with a GPS chip is able to receive timing signals from a constellation of satellites in low earth orbit. When a clear signal can be obtained from three or more satellites, the users position can be determined to within a few meters. But this accuracy comes with a price. From a cold start it take a mobile device more than ten minutes to acquire the signals. In order to overcoming this shortcoming, network carriers have implemented what is known as Assisted – GPS. AGPS uses cellular positioning data to establish a rough location, which enables the device to acquire a precise location much more quickly.

Once the position of the user is known it is possible to offer a host of specialized services. These services are generally classified according to the persistence of the positioning technique. Those in which the user requests a discrete piece of information based on his current location are known as "Concierge" services. But those in which the users location is constantly monitored are known as "Tracking" services.

One type of Location Based Service that has many possible applications for mobile learning is known as "Geo-Fencing." Under this model, regions are specified geographically and linked to specific data sources. When the user moves from one region to another, the new information relevant to his current location is uploaded to the mobile device.

CONTEXTUAL RELEVANCE

What all these approaches have in common is that they allow the relationships between pieces of information to be rooted in physical space and specific geographic locations rather than existing as mere semantics.

Much of the knowledge that an individual needs depends in large part upon where they are, and what they are doing at the time. And it is essential that approaches to mobile learning take these elements into consideration. In essence, the best approaches to mobile learning must be contextually aware. By extrapolating from the users position, scheduling information, QR codes, RFID tags, accelerometer data, and any other available information, the mobile device should be able to anticipate the needs of the user.

Consider the arrangement of rooms in a typical home. The bedroom, the bathroom, the kitchen, the study - each is designed to support a specific range of activities. Each contain equipment appropriate to its function, and associated with the purpose of each room is specific field of knowledge and information. Cooking and nutrition are most relevant to the kitchen, whereas health and hygiene are associated with the bathroom.

By extension, the whole range of human experience takes place within zones designed to facilitate a specific purpose. Applications to learning could be as simple linking displays in a museum to a database accessible through mobile devices, or as intricate as a pharmaceutical laboratory where every piece of equipment and associated process can be studied through an intuitive interaction. The same tag could direct the user to certification training, maintenance provisions, or any other pertinent set of instructions.

A mobile device is defined by the fact that a user may take it with him wherever he goes. And if such a device is to be used for learning, it will be most effectively employed when it capitalizes on its strengths and provides learning experiences that are contextually relevant.

For example, Växjö University has been engaged in a series of experiments intended to bring these elements together to create a unique and powerful learning experience. AMULETS, Advanced Mobile and Ubiquitous Learning Environment for Teachers and Students, is the work of Marcelo Milrad and his colleagues. Together they are undertaking innovative approaches to Geography, History, Natural Sciences, and Physical Education. Using mobile phones, PDAs, and GPS devices, they have effectively blended classroom instruction with mobile contextual elements allowing students to master learning objectives while moving seamlessly between different environments (Kurti, Milrad, & Spikol, 2007).

These approaches to learning represent a significant departure from established models of instruction. Presently the idea of a "field trip" is regarded as a somewhat superfluous activity, with little educational value... a diversion from serious academic endeavor. However, with advent of contextual learning, we are beginning to see the value of establishing relationships between knowledge and physical space.

PRINCIPLES OF MEMORY IN CONTEXTUAL LEARNING AND RETENTION

Having considered the importance of location and physical space, it is appropriate that we also consider the significance of learning objectives with respect to time. Just as learning objectives vary with respect to the location of the user, the ideal object of study varies over time. The mind operates according to well-defined rhythms, and learning strategies that anticipate these fluctuations hold tremendous promise.

In order to understand the value of the mobile learning platform with respect to memory, it's important to review some background research on the nature of memory and the methods that can be employed to maximize retention.

For our purposes it's not necessary to delve too deeply into the underlying neuroscience. There is an ongoing debate regarding the biological foundations for certain phenomena that will certainly yield new discoveries in the future. In the meantime however, there is much to occupy the instructional designer looking to capitalize on the fruits of their research.

The use of mobile technology as a performance support aid with regards to short-term memory is a well-established model. Short-term memory is typically defined as the ability to recall a very short sequence of elements (from 5-9) for less than a minute. Mobile devices excel at providing small portions of information at frequent intervals and thus have become a mainstay in performance support applications (Metcalf, 2006). However, there is much evidence to suggest that multiplatform learning may hold the key to long-term retention as well.

One discovery in particular holds great potential for mobile learning, and that is the phenomena known as the "Spacing Effect." First discovered in 1885 by cognitive psychologist, Herman Ebbinghaus, the Spacing Effect is a mnemonic phenomena characterized by the logarithmic decay of recall. The mind loses the ability to recall information according to a defined curve, and the most efficient time to study a specific piece of information is the point along the curve at which it will soon be forgotten. Once the information has been reviewed, the curve flattens and the optimal study interval increases. After a few repetitions the information has been firmly imprinted in long-term memory (Crowder, 1976).

Historically, it has not been feasible for educational programs to take advantage of the Spacing Effect. Not only does memory retention differ to some degree from one individual to another, but it also varies from one piece of information to another. To maximize the impact of a learning system based on the Spacing Effect requires that curves for each learning objective be plotted independently, which simply wasn't practical before personal computers were readily available. And while desktop computers have the processing power to effectively plot learning curves, they are not always conveniently accessible for students and mobile professionals.

However, recent advances in mobile technology have made it possible to implement a learning system with the potential to conform to the users memory retention curve and provide learning objectives when they are most needed.

There are a number of different theories that have been proposed in explanation of the Spacing Effect, and it is likely that some combination of these theories comprise the fundamental mechanism. Perhaps the leading theory is known as Contextual Fluctuation (Raaijmakers, 2003). Contextual Fluctuation asserts that every time a student is exposed to a learning objective the encoded memory is associated with the given context. When the information is spaced in its presentation, there is a greater variation in context and hence less overlap of retrieval cues. With each successive presentation a new association is formed further strengthening the ability the recall the information.

Accessibility theories are not as rigorously defined, but have some support and provide insight into the nature of memory. According to these theories the spacing interval increases the difficulty of recall and that increasing cognitive load creates a more lasting impression. (Pavlik, Anderson, 2005) There could certainly be some overlap between theories of accessibility and contextual fluctuation, given that exposure to a learning objective in a foreign context is likely to increase the difficult of recall.

Habituation to stimulus is another possible explanation (Hintzman, 1974). According to this model the learning objective is regarded as a neural stimulus that decreases in potency with successive exposures. But widely spaced intervals encourage the down regulation to dissipate allowing the mind to regain its sensitivity to the stimulus.

This chapter will focus primarily on Contextual Fluctuation Theory because it enjoys wide support within the research community and has promising implications for multiplatform learning.

PRACTICAL APPLICATION

What Cognitive Psychologists have developed in the laboratory, computer programmers have since implemented in software applications. Despite the ability to nail down the exact mechanisms of memory, researchers have been able to develop very accurate mathematical models that predict the forgetting curves and calculate the recall strength of spaced repetitions.

Drawing upon these models, computer programmers have adapted them for the practical task of allowing students to study and retain learning objectives. One application in particular, known as SuperMemo, has enjoyed widespread popularity and contains a successful memory algorithm refined through years of use and testing (Wozniak, 1990).

A relatively simple program loop, it tests the users ability to recall a learning objective and based on the users response calculates the optimal spacing for the next presentation. If the user is able to readily recall the information the next interval will be increased, and if not, the interval will be shortened.

Having discovered an algorithm that has been empirically shown to enhance retention, it would be foolish not to capitalize upon this discovery to develop a multiplatform learning system. In fact to overlook such a discovery would be a tragic waste, not only of a hundred and twenty years of painstaking research by cognitive psychologists, but also of potentially millions of man-hours of wasted effort for those who acquire learning objectives only to forget them days or weeks later (Dempster, 1988).

Perhaps it is appropriate to pause to consider just how inefficient contemporary educational systems can be. Many courses are taught in such a way that the student is encouraged to prepare for examinations in the least effective way possible. Memory research clearly demonstrates that "cramming", the practice of marathon study sessions just before an exam, lead to low long-term retention rates (APS, 2007). Nevertheless, this is the norm for many students across academic disciplines. And while an all night study session may enable the student to achieve a passing grade, the retrieval strength achieved by such methods are so low that students are unlikely to recall the material weeks after the test.

Reflecting on courses taken as an undergraduate serves as a powerful, though anecdotal confirmation. How many graduates retain the barest vocabulary of the foreign languages they were required to study? And what of the many other courses in which the student briefly masters a body of material in order to pass an exam, and then promptly forgets? Consider just how many millions of hours are spent in the classroom and in private study by students all over the world seeking to gain mastery over a body of knowledge. The fact that a significant portion of this is lost shortly after graduation can be regarded as nothing short of a tragedy.

But fortunately this is a problem to which there is a solution. With a subtle shift in emphasis we can transition our focus away from periodic demonstrations of subject matter mastery to the acquisition and maintenance of information fluency. The study of foreign languages provides an ideal metaphor for demonstrating these contrasting approaches.

Imagine if a student were to study Spanish by taking discrete courses throughout his academic career. Freshman year he may take a class on nouns, sophomore year the course might be verbs, and senior year there could be a semester devoted to other parts of speech. Traditional methods of study would mean that after the satisfactory completion of each course, without the opportunity or incentive to maintain the knowledge, it would largely be forgotten. And upon graduation the student would have failed to acquire any useful command of the Spanish language, and would only retain the meager crumbs of half forgotten words and phrases.

But if we were to approach the instruction from the perspective of information fluency, we could achieve a radically different outcome. Instead of focusing entirely on discrete units of information assessed on a semester basis, learning objectives would be tracked and reviewed at precisely determined intervals throughout the entire course of study. What's more, the process whereby later courses build upon the foundations laid in previous courses can be organized deliberately to correspond to the retention curve of the student.

Using our example of foreign language mastery, this means that the student would continue to review learning objectives from introductory courses all throughout his academic career. But these review sessions need not be mere repetitions of previous subject matter. Rather it's possible to incorporate the more basic elements of review into exercises that are used to teach the more advanced concepts. For example vocabulary words would recur in exercises according to their assigned retention interval. These tightly integrated elements, which exploit the rhythms of memory, would allow a student to complete their course of study with a cohesive body of knowledge rather than a disparate jumble of fragments.

IMPLEMENTATION

Many learning and study strategies can be well implemented by the student himself. Students often develop tricks and habits of the mind that are tailored to their individual needs. But capitalizing on the benefits of the Spacing Effect is not as simple as making a set of flash cards. For this approach to be truly effective, it needs to be implemented on a curricular level.

Course designers need to think holistically about learning programs and plan from the beginning with retention in mind. From an afternoon workshop to a medical program that spans the better part of a decade, there are effective strategies that can be employed to ensure long-term retention of learning objectives.

One of the first questions to be considered is how long should the information should be retained. According to the research, once information has been reviewed for three years it has essentially achieved permanent recall (Wickens, 1999). That is to say, if a medical student periodically reviews a specific drug interaction, after three years they should be able to recall the information indefinitely. But three years worth of review may be a little excessive for a new employee who is completing an office orientation module. In either case, the retention strategy will be the same; only the number of review sessions will be vary.

The next step would be to take the course materials and parse them into discrete learning objectives. This process is very similar to the way in which an examination would be prepared. When preparing a test, the goal is to create a series of questions that will cover the scope of the course content. In the same way, each lessons covers a range of topics, and these topics need to be distilled into a concise reminder.

The reminder need not be a representation of the material; rather it is only necessary that it stimulate the student to recall the learning objective. But in order to track recall, and determine the optimum spacing, it is important that these reminders be structured in the form of a challenge. Failure to recall the information will indicate that it is necessary to decrease the interval, while a successful recall will increase the optimum review interval.

Finally, according to the theory of Contextual Fluctuation, there is value in varying the context to which the memory is associated. This can be accomplished in a number of different ways. One of the easiest ways to vary the context of review is to simply use a mobile device to deliver the recall challenges. Throughout the course of a student's movements, review notifications that occur according to scheduled intervals will naturally arrive in different contexts, providing the student with a broader range of associations.

Additionally the content of these reminders can vary as well to encourage deeper cognation, by forcing the student to review a novel presentation of the learning objective. This could be a simple as displaying vocabulary words in a different font or as intricate as presenting a new variety of symptoms intended to recall a specific diagnosis.

Once the learning objectives are identified and the recall challenges are created, a learning management system can be used to deliver the challenges and track student retention. This is where all the elements of multiplatform learning come together. While the initial instruction may have occurred in a classroom or at a workstation, the recall challenges can be routed through the LMS to the student's mobile device.

Depending on the content, an SMS may be sufficient to remind the student of an objective and allow them to respond with an answer. More intricate reminders may require embedded URL's or images. When the student responds to the reminder, the LMS tracks their performance. If they fail to recall the information, the learning objective can be represented by the LMS and the interval decremented for the next review. If they succeed, the next challenge will occur at the next appropriate interval.

A possible objection to the implementation of such a system is the fact that it could be considered a nuisance to receive many such reminders. No one wants to review learning objectives from their sexual harassment training three years later. But there are a few things to consider.

First, the granularity of the interval is not so specific as to necessitate an immediate response. The student need not excuse himself from dinner to identify the infection vector for African Trypanosomiasis. Instead he can freely review the material at his convenience. In fact, a well-designed recall challenge will allow the student to remember the information through a very brief interaction.

Whatever irritation scheduled reminders may provoke are likely to be overcome by the realization that this system effectively automates the process of studying and that it does so in the most efficient way possible. Far from occasioning resentment, future students may rightly wonder how previous generations were able to remember anything at all.

CONCLUSION

The field of mobile learning is ripe for widespread adoption. Historically there have been a host of barriers that prevented the successful implementation of a multiplatform or blended approach to learning. Previous generations of mobile devices simply were not powerful enough to support dynamic learning applications. However current generations of hardware rival laptops in terms of processor speed and graphics support. Mobile user interfaces formerly possessed the grace of a three-toed sloth, eliciting despair from their unfortunate users. But today's multi-touch interfaces, haptic input, and optimized mobile browsers are a pleasure to use.

Institutional inertia and public perception are the remaining barriers that have yet to be overcome. As educators and instructional designers are increasingly exposed to the unique advantages of each element in a multi-platform approach, it is likely that we will begin to see learning programs that are designed at the curricular level to take full advantage of the tools available. Contextual learning and enhanced retention provide a powerful incentive to those willing to explore the possibilities presented by mobile learning.

Mobile learning allows us to associate learning objectives with their physical manifestations, creating a learning environment grounded in direct experience. And while much attention is currently focused on the ability to create immersive simulations which substitute artificial stimuli for the external world, perhaps there is even more value in engaging with the world directly?

REFERENCES

Association for Psychological Science. (2007). Back To School: Cramming Doesn't Work In The Long Term. *ScienceDaily.* Retrieved August 6, 2008, from http://www.sciencedaily.com– / releases/2007/08/070829122934.htm

Chun, M. M. (2000, May). Contextual Cueing of Visual Attention. *Trends in Cognitive Sciences*, *4*(5), 170–178. doi:10.1016/S1364-6613(00)01476-5

Crowder, R. G. (1976). *Principles of learning and memory*. Oxford, England: Lawrence Erlbaum.

Dempster, F. N. (1988). The spacing effect: A case study in the failure to apply the results of psychological research. *The American Psychologist*, *43*(8), 627–634. doi:10.1037/0003-066X.43.8.627

Denso-Wave. (2003) *About 2D Code*. Denso-Wave. http://www.denso-wave.com/qrcode/aboutqr-e.html

Durst, S. (2007). See It, Snap It, Search It. *Business 2.0*.

ECMA International (2004, December). *Standard ECMA-340 Near Field Communication Interface and Protocol (NFCIP-1)* (2nd ed.).

Hand, K. (2007, February 26). Pimp my cube 2017. *Popular Science*.

Hintzman, D. L. (1974). Theoretical implications of the spacing effect. In R. L. Solso (Ed.), *Theories in cognitive psychology: The Loyola Symposium* (pp. 77-99). Oxford, England: Lawrence Erlbaum Associates, Inc.

Kim, R. (2008) "Bar codes create bridge for window-shoppers." San Francisco Chronicle: March 27th 2008.

Kurti, A., Milrad, M., & Spikol, D. (2007, July 18-20). Designing Innovative Learning Activities Using Ubiquitous Computing. *In Proceedings of the 7th IEEE International Conference on Advanced Learning Technologies*, (pp. 386-390), Niigata, Japan, ISBN 978-07695-2916-5.

MarketWatch. (2008, July 10). Fullpower Announces MotionX-Poker Available on Apple's App Store. *MarketWatch*.

Marshall, M. (2008, July 9). Tapulous wants to be leading application ecosystem for the iPhone. *VentureBeat*.

McCullagh, D. (2001, February 2). Call it Super Bowl Face Scan I. Wired Magazine.

Metcalf, D. (2006). *mLearning: Mobile Learning and Performance Support in the Palm of your Hand*. Amherst, Massachussets: HRD Press.

Mistry, P., & Maes, P. (2008). Quickies: Intelligent Sticky Notes. In *Proceedings of 4th International Conference on Intelligent Environments* (IE08). Seattle, USA.

Pavlik Jr, P. I., & Anderson, J. R. (2005). Practice and Forgetting Effects on Vocabulary Memory: An Activation-Based Model of the Spacing Effect. *Cognitive Science: A Multidisciplinary Journal, 29*(4), 559-586.

Pettit, J., & Kukulska-Hulme, A. (2007). Going with the Grain: Mobile Devices in Practice. *Australasian Journal of Educational Technology, 23*(1), 17–33.

Raaijmakers, J. G. W. (2003). Spacing and repetition effects in human memory: Application of the SAM model. *Cognitive Science*, *27*, 431–452.

Whitten, W. B., & Bjork, R. A. (1977). Learning from tests: Effects of spacing. *Journal of Verbal Learning and Verbal Behavior*, *16*, 465–478. doi:10.1016/S0022-5371(77)80040-6

Wickens, T. D. (1999). Measuring the time course of retention. In C. Izawa (Ed.), *On human memory: Evolution,progress, and reflections on the 30th anniversary of the Atkinson-Shiffrin model* (pp. 245–266). Mahwah, NJ: Lawrence Erlbaum Associates, Inc.

Wozniak, P. A. *(1990)*. Optimization of learning, Master's Thesis, University of Technology in Poznan, 1990 and adapted for publishing as an independent article on the web.

Chapter 20

Development of a Museum Exhibition System Combining Interactional and Transmissional Learning

Shinichi Hisamatsu
The University of Tokyo, Japan

ABSTRACT

"Hands-on" exhibitions, which not only present objects for viewing but also stimulate learning by allowing visitors actually able to touch them, is gaining increasing popularity at museums. By actually handling an exhibited object, the visitor can get a better understanding of the characteristics of the object that cannot be fully grasped by just looking it, such as the object's underlying structure and hidden aspects, as well as tactile information like the object's weight, hardness, and so on. The experience also arouses curiosity and interest and becomes a learning opportunity for the viewer. The author has developed an interactive exhibition system for museums, which combines learning based on the interaction with physical objects and knowledge transmission. In this system, the user handles and looks at an actual physical object, which appears just like the original object and talks directly to the user. This "conversation" with the object as the user "grasps" (in both senses) the object deepens the user's understanding of and interest in the object. This "narrative" feedback to the user is achieved through the active linkage between, in the case presented here, a fossil in real space and three-dimensional computer graphics employing Augmented Reality (AR). The system uses RF-ID technology to determine the level of the user's "grasping" state and to feed back information to the user. In this chapter, the author presents the actual implementation of this interactive system at a museum and a school. The system was tested with elementary and junior high school students and I present results of the trials that show the convenience of the system and its beneficial effect on learning.

DOI: 10.4018/978-1-60566-703-4.ch020

INTRODUCTION

The functions of museums can be broadly divided into the following:

- Introducing
- Exhibiting
- Disseminating knowledge
- Researching
- Storing
- Managing

For some time, museums have been experimenting with a diverse range of exhibiting modes, for example in contrast with the conventional exhibition format.

In contrast with the conventional exhibition format in which exhibited objects cannot be touched, Caulton (1998) presents an interactive exhibition system called "hands on." Hands on is characterized by the interaction with physical objects, and according to Koran et al. (1986), exhibits with which it is possible to interact attract more visitors and draw their attention longer than exhibits without interaction.

However, through the use of the sole interaction, for many exhibits some parts might remain impossible to be seen, and the information might be difficult to read. To solve this problem, one can suggest the effectiveness of knowledge-transmission content, but this type of content has a tendency to present information one-sidedly and might face difficulties attracting visitor attention due to a lack of interaction. Exhibitions are also organized with hands on parts and knowledge-transmission parts in parallel, but physical objects and content are then separated, which might result in a decoupling of the knowledge architecture.

Koshinishi (1996) said that exhibitions in museums should not rely on single senses such as vision or audition to appeal to visitors, but should also assist their intuitive understanding by appealing in a composite way to many sensations.

For that purpose, Miles (1986) delineates several necessary roles for an exhibition medium:

1. Attract the visitor
2. Maintain the visitor's attention
3. Revive the visitor's knowledge
4. Provide information to the visitor
5. Stimulate the visitor's reaction
6. Give feedback to the visitor

"Participatory devices" have been introduced as media achieving such requirements. A "participatory device" plays speech or video content when an action is performed by the visitor, such holding a button down or manipulating a pointing device.

For exhibits to which the visitor is attracted, this kind of system indeed appeals to his/her senses as it combines with the exhibited item at this particular place and time to provide additional information. However, exhibited items are usually set in glass cases, making it impossible for the visitor to touch them. Interaction between the visitor and the exhibited item thus becomes limited, and the acquisition of knowledge from the exhibit is inhibited.

Taking such problems into account, hands on exhibitions, where visitors can actually take the items in their hands and observe them, are attracting attention. Hands on displays are headed for a learning model where the learner "discovers by himself through trial-and-error", and follows the constructivist idea which considers the learner as a being reconstructing his/her interpretation of the world and the information he/she holds through interaction with the surrounding world. Through the experience of

1. Touching the item
2. Manipulating
3. Trying one thing or another
4. Sensing something through manipulation
5. Searching until one fully understands what one has sensed

hands on displays arouse the spontaneous searching activity of the visitors, thus providing an environment that stimulates learning and the interest for the exhibition (Nezu, 2003).

To explore this concept, numerous hands-on exhibitions have been created up to now. For example, Japan's National Museum of Emerging Science and Innovation has developed a physical model of the system that allows information "packets" to be exchanged on the Internet. By making their own packets (in this case, groups of white or black balls) with actual designated address nodes and "inputting" them into the physical model, users can confirm with their own eyes how these information packets arrive at the destination address nodes.

Another example is the Ann Arbor Hands-On Museum in the United States, which, as its name makes clear, is a museum that specializes in hands-on exhibitions. The museum has numerous hands-on exhibitions that allow users to understand clearly, through an interactive experience, basic principles of physics, chemistry, and other sciences.

Even though hands on displays are able to reduce the distance between the exhibited items and the visitor and stimulate an autonomous behavior, in such conventional methods the additional information is still provided in one modality, for example by the exhibition guideboards, and the items and the additional information are still decoupled, making these approaches insufficient to assist the viewers' intuitive understanding. It is necessary to provide the items and the additional information in a more integrated and mutually complementary manner, more tightly linked to the context of each visitor. Engaging an interaction involving his/her five senses by a mutual feedback between the visitor and the exhibits on their reciprocal states, such a goal becomes realizable.

A related research project called Tangible Bits has been conducted by Ishii (1997) at the MIT media lab. Tangible Bits attempt to develop interactive media and user interfaces involving tactile sensation with a wide variety of sensors. Ishii explains the concept of Tangible Bits as follows:

"Tangible Bits is our vision of Human Computer Interaction (HCI) which guides our research in the Tangible Media Group. People have developed sophisticated skills for sensing and manipulating our physical environments. However, most of these skills are not employed by traditional GUI (Graphical User Interface). Tangible Bits seeks to build upon these skills by giving physical form to digital information, seamlessly coupling the dual worlds of bits and atoms.

Guided by the Tangible Bits vision, we are designing 'tangible user interfaces' which employ physical objects, surfaces, and spaces as tangible embodiments of digital information. These include foreground interactions with grip-able physical objects and augmented surfaces, exploiting the human senses of touch and kinesthesia. We are also exploring background information displays which use 'ambient media' -- ambient light, sound, airflow, and water movement. Here, we seek to communicate digitally-mediated senses of activity and presence at the periphery of human awareness.

The goal is to change the 'painted bits' of GUIs (Graphical User Interfaces) to 'tangible bits,' taking advantage of the richness of multimodal human senses and skills developed through our lifetime of interaction with the physical world."

In interactive media, most emphasis had so far been put on vision and audition, but the use of tactile perception broadens the expression range. To illustrate this idea, Ishii et al. have developed several systems such as "Music Bottle", which plays different music or news when the cork of a bottle is opened, depending on the bottle, or "I/O bulb", which can be used as an intuitive support system for architecture and design or a visualization tool for physical simulation.

Tangible Bits facilitates the intuitive manipulation of computers, and constitutes a natural link between the user and the physical object. One can thus expect that introducing such a technology in hands-on displays should enable us to provide a deeper experience.

In order to realize this transmissional learning utilizing tangible bits, we have developed for this study an interactive display system called "Monogatari" (in English, "narrative exhibition"), which employs ubiquitous technology to combine interaction and knowledge-transmission by embedding the media content from the transmissional side directly into the interaction with the exhibited item.

PREVIOUS WORK (LITERATURE REVIEW)

Ayres et al. (1998) showed through a study of primary school students at the Discovery center in East Tennessee USA, that exhibitions based on multimedia stimulated the understanding of scientific concepts more than the usual hands on displays. This effect is thought to result from the fact that multimedia content delivers the information by appealing to several senses and making full use of text and video.

Nakasugi et al. (2002) developed a system based on a wearable computer called "Past Viewer", that overlays videos onto present scenes such as historical buildings, etc. By enabling an embodiment of the knowledge about the object, this system deepens the user's interest.

Kondo et al. (2006) also performed user-behavior-analysis in a science museum exhibition where they visually combined, using Mixed Reality technology, the displayed skeletal preparations of dinosaurs with a 3D model reconstruction of their bodies.

Ohashi et al. (2000) developed a Voice Track-back system, which allows learners to leave voice message using cellular phones, PDAs, etc. whenever they like and then to access these messages on the Internet. Onishi and his team conducted an empirical experiment with this system at a zoo. In this experiment, each subject was given one area of responsibility, and the subjects collected and shared information using the Voice Track-back system. It was observed that by organizing their impressions into a report, the subjects were able to effectively reflect on their impressions.

Kato et al. (2000) have developed technology that combines computer graphics with actual images to use printed two-dimensional markers for virtual object interaction. Markers labeled with commands are picked up by webcams, and when they are detected by computer processing of the image, virtual images are superimposed on the original object.

Papagiannakis et al. (2008) have produced a survey of various technologies for mobile mixed reality/augmented reality systems. This survey also refers to the use of RF-ID technology, which the research presented in this paper uses for exploring the interaction between real and virtual space. One difference between the utilization of RF-ID is that while Papagiannki et al. (2008) use a device for specifying RF-ID tags for a user's position, we use RF-ID tags for detecting the physical state of the user's grasping of an object.

Monogatari System

In this study, we focused on having the user "pick up and see" an object as an interface for allowing the user to become engaged with the exhibited object in a more natural way and to appreciate it. Our objective was to aid the user's deeper understanding of the exhibited object by letting the user actually hold the object and feel it — its weight, surface features, and so on — and providing the user with overlapping additional information about the object. The Monogatari system also

Figure 1. System diagram

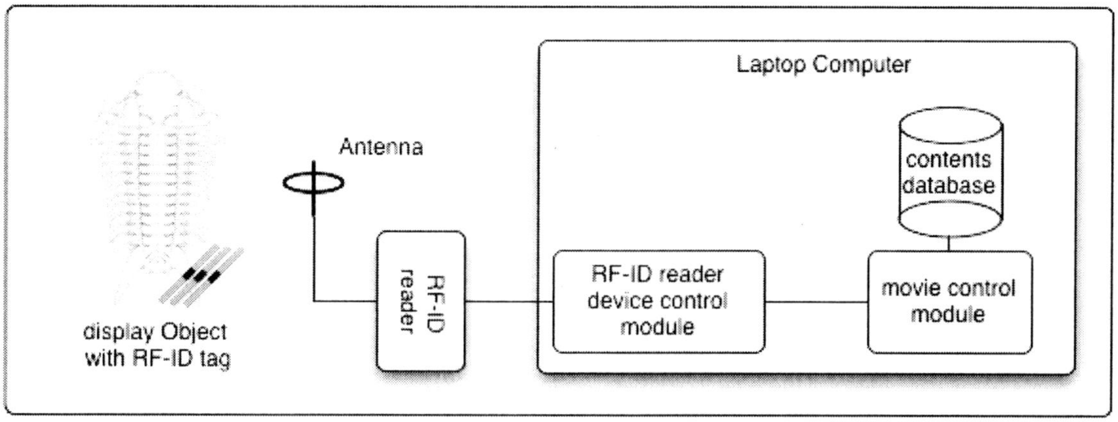

enhances the user's understanding by changing the additional information provided, depending on how the user is holding the object and how (where) the user is looking at it. For example, when the user is holding, in the case of this study, the fossil of a prehistoric animal and begins to focus on the eye of the animal, explanation is provided about the animal's eye. Since the system allows the object to talk directly to the user to provide the explanation, the user is aware of the presence of only two parties—the object and the user—making the user more engrossed in the interaction and heightening the user's concentration on the experience of the Monogatari.

The Monogatari system presented in this study is an exhibition system for museums that combine knowledge-transmission based learning and interaction with physical objects by using RF-ID technology and Augmented Reality.

Monogatari consists of software that manages the exhibited item and the installation and the presentation of additional information.

1. The user wears a ring-shaped antenna device, takes the exhibited item in his/her hands and observes it.
2. By sensing the quality of the user's grip of the user, the system determines which part of the item the user's attention is focused.
3. Depending on the place where the user's attention is focused, video content is played, superimposed onto the item.
4. When the user changes the way he/she is grasping the physical object and a variation of the quality of the user's grip is detected, the content is switched.
5. When all the content has been seen, an item of concluding content is played to signify to the user that he/she completed the viewing of all the content.

By enabling the user observing the item through such a sequence, the additional information is provided as if the item itself was talking to the user.

RF-ID Technology

The system uses RF-ID technology in order to sense how the user is griping the item. RF-ID is a technology that reads and discriminates the IDs of so-called "tags". These tags are composed of an antenna and a chip with a unique ID. The use of these tags, originally found in supply chain management (SCM), logistics, is now spreading. Here, several tags are embedded in a fixed pat-

Table1. μ-Chip specifications

Size	0.4×0.4mm
Frequency	2.45GHz
Memory size	128bit
Writing	Disabled

Figure 2. μ-chip RF-ID tag

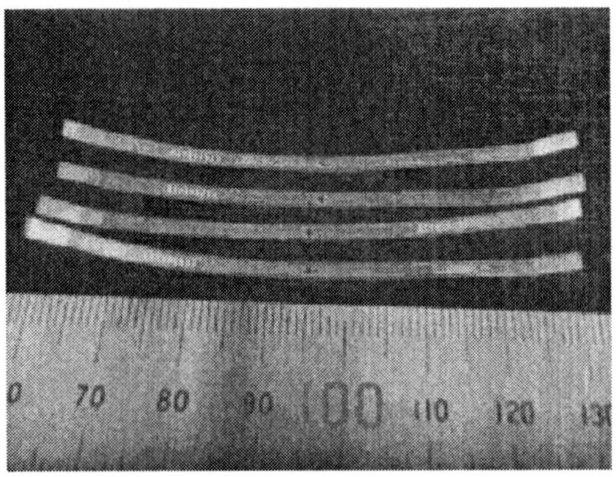

tern in the exhibited item, and how the user grips the item is sensed by reading the tag ID with the ring-shaped antenna worn on the finger.

As the RF–ID tags are to be pasted on the item or embedded into it, the size with the antenna included should be small. For the same reason, active-type tags with built-in batteries must be discarded, and one must use passive-type tags, for which the power is supplied by the reader and no built-in power source is thus required.

Table 2. MRJ200A specifications

Telecommunication standard	ARIB STD-T81
Frequency	2.407-2.426GHz
Stable reagin distance	60-100mm (with patch antenna)
Transmission power	10mW/MHz
Modulation system	A1D(ASK)
Interface	RS-232C
Power consumption	In use: 2.3W In standby: 0.3W

After performing experiments using several types of tags, we selected Hitachi's μ-Chip. The μ-Chip is an ultra-small (0.4mm square size) passive RF-ID chip performing communication in the 2.4GHz band, which can be used legally in Japan. The size of the inlet including the chip and the antenna is 51 x 1.5 x 0.25 (mm), and the tag can be used even when bent, making it suitable for a pasted or embedded use on exhibition items.

We used as RF-ID reader device YAGI Antenna's MRJ200A, whose specifications are given in Table 2.

As shown in Table 2, the transmission power is 10mW/MHz, and the device can thus be used legally in Japan without specific authorization.

We note that the MRJ200A cannot read several tags simultaneously, but this is not an issue for the application considered in this study.

We used the RS-232C as an interface to connect the device to a PC and to control it.

In order to use the antenna as a ring, we built a small loop antenna. This small loop antenna does not detect the electric field but the magnetic field, and can be made extremely small compared to usual antennas.

The diameter of the antenna was set to 15mm so that primary school and junior high school students can wear it without discomfort. Moreover, a small loop antenna can, by its nature, easily lose its gain through bending, so we secured its intensity by using steel wire.

As the adverse effect of the feeder cable coming from the RF-ID reader device acting as an antenna cannot be neglected, we used a balun device with the small loop antenna.

The antenna was fixed with Velcro tape on the inside of the first joint of the thumb.

Projection Equipment

The system uses a tablet-type laptop PC set at the back of the display box to show information in the real world by projecting it on a half-mirror. This leads to an AR (Augmented Reality) system in which fossil replicas and video content such as

Figure 3. Ring-shaped antenna

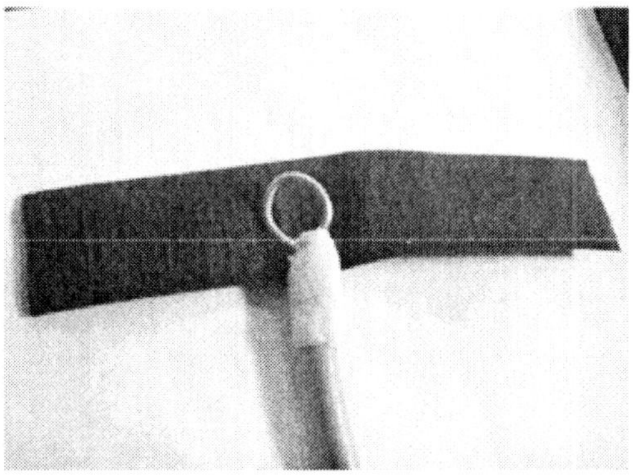

3D animations, texts or photographs, are superimposed and can be seen simultaneously. By allowing the user to see video content while touching the physical object (for example, when looking at the head of a trilobite, showing video content about the trilobite's high-resolution eyes), one can display video content related to the interaction with the object and thus stimulate the user's interest and knowledge building.

Software

The control of the reader and the video projection is performed by the laptop PC installed in the display box.

Figure 4. Display box structure

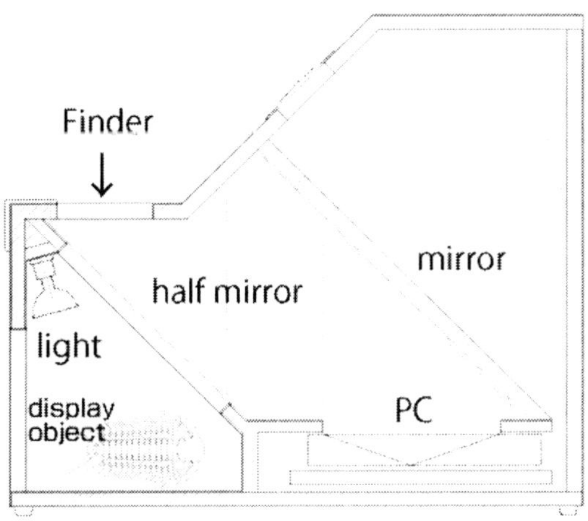

The software is composed of an RF-ID reader control module and a movie control module.

RF-ID Reader Device Control Module

This module performs the initialization of the RF-ID reader device, the reading control and the code analysis. At regular time intervals (usually every second), the ID is read by the RF-ID reader device; if an ID was read, after normalization and check for duplicate, if the ID is valid, it is handed to the movie control module. The module is a Microsoft Windows application created with C# and .NET Framework. User session administration is maintained such that if content has already been displayed to a user and the same signal is received again this content is not re-displayed. If the user takes off the ring-shaped antenna and puts it on the display box, a specific RF-ID tag is sensed and the session is reset.

Movie Control Mmodule

This module receives a signal from the RF-ID reader device control module and, after comparing it with a database, displays the appropriate content. It is implemented using Macromedia Flash Player 8.

A standby display is shown until the RF-ID reader device control module receives a signal. The corresponding movie is then played, and when it is finished the system goes back to the standby display. When all the contents have been played, the user is informed of the end of the session.

When a movie is being played, signals coming from the RF-ID reader device control module are ignored, and priority is given to the movie.

```
show_waiting_screen();
if(detect_grip){ show prologue
movie};
while(detect_grip_of_new_pat-
tern){
```

Figure 5. Display box

Figure 6. Example of projected video

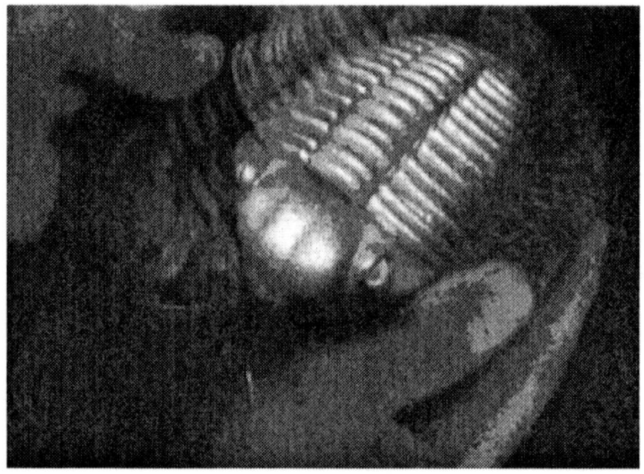

```
show_movie(pattern);
   //ignore pattern recognition
   show_waiting_screen();
   if(all_movie_are_shown)
      break;
show epilogue movie();
```

Database

We implemented two databases. One database is part of the pattern recognition system, and is embedded in the movie control module. It stores elements used in the estimation of the quality of the user's grip (combination of RF-ID and machine learning materials).

The other database stores pattern-content relations. The movie control module asks this database what movie it should play.

Content

In this study, we selected the trilobite as a first subject. These days it is possible to take a trilobite fossil in one's hands and observe it, but this interaction is not enough to learn about the original aspects of the fossil. Our system assists

Figure 7. Movie diagram

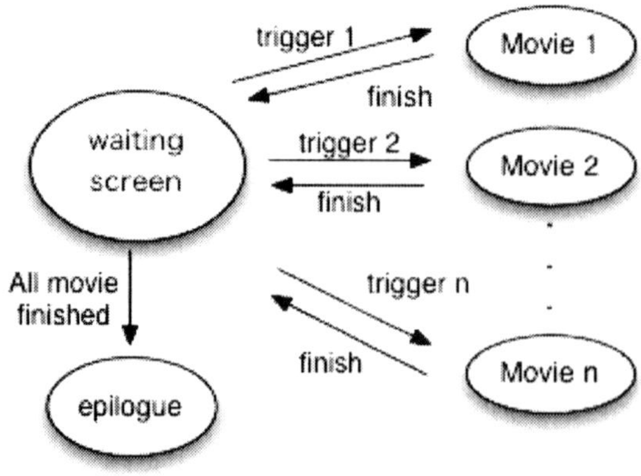

Figure 8. Movie of the trilobite's defensive posture

knowledge building about the trilobite by linking the fossil with its original aspect.

A trilobite fossil replica was built for exhibition. RF-ID tags were embedded about 2 - 3mm inside the surface of the replica. To avoid interference between tags, they were dispersed at intervals of 10mm. A total of 20 tags were embedded on the front and back of the replica.

The video content currently features the following movies, but additional content can be added as needed:

1. **Standby display:** This video invites the visitor to use Monogatari and encourages him/her to try different ways of holding the item
2. **Overview of trilobite:** This video explains basic information about trilobite and fossils. What is a trilobite? When and where did they live? How the fossil was made and so on.
3. **The trilobite and its congeners:** This video presents the different types of trilobite and explains how it evolved.
4. **Background to the age of the trilobite:** This video explains the condition of the Earth and other living organisms during the Cambrian and Permian periods, when the trilobite lived.

Figure 9. Compound eye of the trilobite

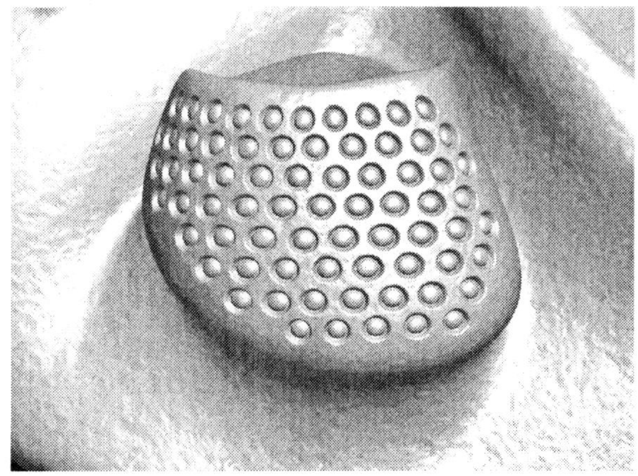

5. **Life form:** This video explains the characteristics of the trilobite's body and the characteristic means of defense it is thought to have used against the attacks by its enemies.

6. **Presentation of the compound eye:** This video presents the principle, characteristics and advantage (the high sensibility to movement within eyesight) of the trilobite's compound eye.

When users first hold the fossil, the second video is always shown. After that, the videos change, depending on the way the user holds the trilobite, and there is no sequence to the video themselves. This way of presenting information both provides basic knowledge to the user, who has no prior knowledge of the trilobite and fossils—the objects of study—by showing a basic overview of the trilobite at first, and gives the user a feeling of intimacy (ice-breaking) with the object, thus motivating the user to learn more.

In order to make the user feel the presence of only two parties—the object and the user—and, by doing so, immerse the user more deeply into the experience, we designed the narration as if the trilobite were talking directly to the user.

Apart from the standby display, we kept the content length within 60-90 seconds to maintain peak viewer/listener interest.

System Design Guideline

In the course of the development of our system, we found some principles for its implementation and deployment especially with regard to children.

Learning Guidelines

- **Limit the age range of the targeted user:** When aiming at children, as the size of their hands and the way they grip physical objects differs with their stage of growth, the design of the antenna and the griping patterns will differ accordingly. It is thus necessary to fix the targeted age range. It is still possible however to deal with several age ranges by preparing several antennas and switching between pattern recognition algorithms.
- **Preferably provide both speakers and headphones for the speech output:** As Monogatari is a system designed to be used by one person at a time, and as in exhibitions in museums, etc., noise often comes from other exhibits, headphones are usually more suitable. However, when it is possible to be physically isolated from other noise sources, the use of speakers is less burdensome to the user.
- **Make a large number of short videos:** If a video is too long, the user's concentration drops and he/she gets tired. The user might then quit in the middle of the video, or give up the interaction with the object to focus on just watching the video. It is thus important to make video content with an appropriate short length according to the part of the object which was touched. We heuristically used a 60-90 second length as a standard.
- **Videos coexisting with the object:** A sense of reality can be provided to the user by creating video content based on the fact that the user watches it as overlapped onto the object, and that the video is not complete by itself but should feature the object as one of its components. For example, in our research, when explaining the structure of the eye of the trilobite, we did not use a Computer Graphics model of the eye or an enlarged view of a picture, but created the video such that it attracts attention to eye part of the object.
- **Always play an epilogue video:** When all the available videos have been watched, we inform the user of the end of the experiment by playing an epilogue video saying "Thank you" or "Good bye". Without such a video, the user does not know when to stop.
- **Make the narration in the videos talk directly to the user:** As in Monogatari the immersive feeling is provided to the user by staging a conversation between the physical object and the user, if the user can feel through the narration another presence which is neither the object nor the user, he/she may lose his/her interest. The narration is thus made not as a third person but directly as if the object is speaking to the user. Sound effects can be useful, but background music should be avoided.

Technical Guidelines

RF-ID Tag Embedding

Many RF-ID tags are directional, and their detection accuracy degrades as the antenna deviates from the directionality region. We should thus embed the tag such that the directionality is perpendicular to the surface of the physical object. Here, we embedded the μ-Chip tags about 2 to 3mm inside the surface of the object, and such that the tags were parallel to the surface, as the directionality of μ-Chip tag lies in the perpendicular direction.

Disposition of the RF-ID Tags

The disposition of the RF-ID tags must be designed according to the envisioned griping patterns. To avoid interference between tags, one must ensure that they are dispersed with a certain interval between them. Here, we arranged the tags such that they were at least 12mm apart.

Reading Interval of the Tags

The interval at which information is read from the tags influences the reaction of the system. If it is too long, changes in the grasping state cannot be sensed, while if it is made too short, due to the characteristics of passive tags, false detections may occur. Here, by using a reading interval of 1s, we were able to combine accurate reading and reaction speed.

Prepare Half-Mirrors with Different Degrees of Transparency

Making the room where the system is used dimmer facilitates the visualization of the video projected on the object. We made the brightness of the lamp inside the display box variable, such that light can be adjusted before use to make the object and the video easily viewable. Finally, by preparing several half-mirrors with different degrees of transparency and switching between them, it is possible to adapt flexibly to the brightness of the surrounding area.

EVALUATION

Preliminary Experiment

We conducted a preliminary experiment at first in order to confirm the working of the Monogatari and test the usability of the system in an actual museum environment.

A preliminary test was performed in August 2007 at the Kanagawa Prefecture Museum of Natural History.

The Monogatari system was installed in a corner of the exhibition space. A survey based on a questionnaire and an interview was conducted on primary school and junior high-school students (6 to 13 year old) who visited the exhibition, after they used the system.

As a result of this experiment, the following problems emerged:

- **Insufficient sensitivity of the RF-ID:** With the antenna on and while holding the item, sometimes the RF-ID tag could not be read and the grasping state could not be sensed, or the response was bad. The grasping state could then be sensed after slightly changing the grip, but it appeared that it was necessary to raise the sensitivity of the RF-ID to display movies to the user in a more natural way. We have thus re-designed the antenna and the cable between the antenna and the RF-ID reader device in order to raise the antenna gain and hold down the capacity of the cable. We also investigated several patterns for the embedding of the RF-ID tags to diminish the interferences among tags.
- **Simultaneous viewing:** As the exhibition is performed using a display box into which the user has to look, the experience can only be had by one user for one machine at a time. However, in the target age-group we considered this time, visitors have difficulties waiting for their turn and try to have a look at what the preceding user is doing. We should thus consider a mechanism which enables persons other than the user to attend too, such as external displays for example.
- **Usability:** The software used this time could be easily manipulated by the junior-

high school students, but primary school students sometimes encountered difficulties and in some situations were puzzled by how to use it. Modifications of the system and the content, such as the addition of navigation to the software, and the use of plain words in the explanations are necessary. We added guidance and navigation animation to the video content to fix the problem.

Through the results of the preliminary experiment, we made the above-mentioned improvements and added refinements to the system. After that, in order to confirm the efficaciousness of the improvements and evaluate the system itself, we conducted the following evaluation.

Evaluation at a Junior High School

In July 2008, experimental evaluation was performed at the Takehaya Junior High School attached to Tokyo Gakugei University. By this evaluation we aimed at confirming the pedagogical effect as well as the stability and reliability of our system from a different angle than the experiment described above.

In this evaluation, the Monogatari system was used by 20 students from first year students to third year students, with half of them in an experimental group and the other half in a control group. The system used for the experimental group was the full-feature Monogatari system presented above, while for the control group the system consisted of the object (trilobite fossil) and a laptop PC. Students in the control group could look at the trilobite fossil but could not touch it, and watched the same video content as the experimental group on the laptop PC, starting each video by a key press. For both groups, an immediate recall test on the amount of knowledge acquired was performed after use of the system. To confirm the modification in their motivation and more generally in their perception towards learning and knowledge acquisition, the students were asked to fill out a questionnaire right after they used the system. (The questionnaire items are reported in the Appendix.)

As mentioned above, prior to this test, we responded to various problems that became apparent in the preliminary experiment, such by as increasing the sensitivity of the RF-ID reader antenna, and made other refinements to further stabilize the movements. As a result, we could confirm, through observations and questions, that the students in the experimental group had considerably less difficulty using the system than users in the preliminary experiment.

For example, in a sketching task asking students to make a drawing of the trilobite, where

Figure 10. Illustrations drawn by the students in the experimental group

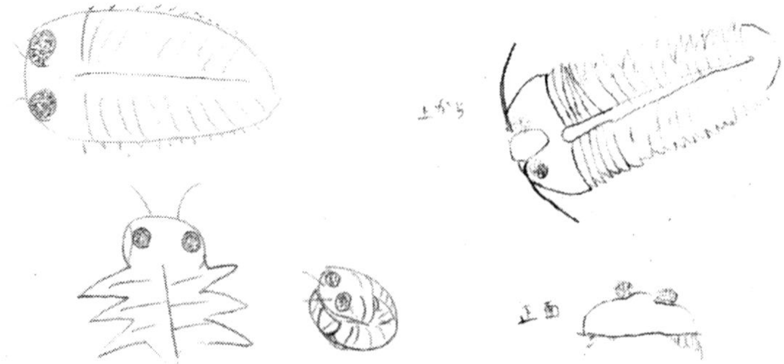

most drawings from the control group represented the trilobite with the back in front, while in the experimental group the trilobite was represented generally correctly in various angles and positions, results show that a better understanding and knowledge of the trilobite could be acquired by the students from the experimental group.

FUTURE RESEARCH

We are currently improving the system according to the results of the evaluation, and plan to perform a temporary exhibition in a natural history museum and experimentally evaluate the effects on learning and interest among children and students of various ages.

Expecting this project to grow as a full system, we also plan to enrich the movie content and to expand the concept to other topics than the trilobite.

Other examples of applications can be considered, such as training for surgical operations by combining the system with medical mannequins. In such cases, the necessity for a wide field of vision is expected to require large-scale equipment.

On the other hand, we are also considering the use of Monogatari in a mobile environment. Although we discarded this option here due to the unknown influence on children's health, it is possible to replace the large display box by a wearable system using a Head Mount Display (HMD). Using portable devices such as the iPhone and combining information obtained from the grip interface with position information acquired through GPS, we can develop context-aware content able to interact with the user according to "what" was touched, and to "when", "where" and "how" it was touched.

For example, in a situation where the same vegetable is eaten differently according to the region, we can imagine explaining to the user the way to eat it in the region where he/she is.

Content with a game element can also be created, such as a quest where the indication for the next thing to look for is written on the back of a stone in the street.

We are also considering the introduction of the grip interface by itself into Factory Automation (FA). For the assembly or inspection of large-scale industrial products, we can imagine an application where, by wearing a glove with a built-in antenna, one can check that all the inspection items in the manual have indeed been inspected, thus avoiding omissions in the process.

REFERENCES

Akiko, N., & Hirohisa, O. (1997). *On the idea of hands on in children's museums: Formative function of children's museums as an environment of children's play*. Bulletin of Tokyo Gakugei University, Series I, science of education, 48, (pp. 345-354).

Ayres, R., Papagiannakis, G., Kim, H. S., & Magnenat-Thalmann, N. (2005, February). Believability and presence in mobile mixed reality environments. In *Proceedings of the IEEE workshop on Virtuality Structures (VR2005)*. IEEE.

Caulton, T. (1998). *Hands-on exhibitions: Managing interactive museums and science centre*. Routledge.

George, P, Gurminder, S., & Magnenat-Thalmann, N. (2008). A survey of mobile and wireless technologies for augmented reality systems. *Journal of Computer Animation and Virtual Worlds*.

Hida, N., & Yamauchi, Y. (2005). Development of interactive exhibition systems for museum. *Japan Society for Educational Technology 21st conference* (pp. 567-568).

Ishii, H., & Ullmer, B. (1997). Tangible bits: Towards seamless interfaces between people, bits and atoms. In *Proceedings of the Conference on Human Factors in Computing Systems (CHI '97)*. ACM Press.

Kato, H., Billinghurst, M., & Poupyrev, I. Imamoto1, K., & Tachibana, K. (2000). Virtual object manipulation on a table-top AR environment. In *Proceedings. of the IEEE and ACM International Symposium on Augmented Reality 2000* (pp. 111-119).

Kondo, T. (2006). Augmented learning environment using mixed reality technology. *Japan Journal of Educational Technology, 29*(Suppl), 145–148.

Kondo, T., Shibasaki, J., & Arita, H. (2006). A proposal for a mixed reality museum display system. *Japan Journal of Educational Technology, 30*(Suppl.), 45–48.

Koran, J. J. Jr, Koran, M. L., & Longino, S. J. (1986). The relationship of age, sex, attention, and holding power with two types of science exhibits. *Curator, 29*, 227–235.

Koshinishi, M. (1996). Seeing to watching, and doing. *Kokugakuin Journal, 11*, 96–100.

Melear, C. T. (1998). *Increased learning of physical science concepts via multimedia exhibits compared to hands-on exhibits in a science museum.* Paper presented at the Annual Meeting of the National Association for Research in Science Teaching, San Diego, CA.

Miles, R. S. (1986). *The design of educational exhibits.* Routledge.

Nakasugi, H., & Yamauchi, Y. (2002). Past viewer: The development of a wearable learning system for history education, Japan. *Journal of Educational Technology, 26*, 167–172.

Nartin, D. (1998a). Multimedia case study: Manchester United Museum. *Museum Practice, 9*, 65–68.

Nartin, D. (1998b). Multimedia case study: Museum of Science and Industry in Manchester. *Museum Practice, 9*, 69–71.

Pangaro, G., Maynes-Aminzade, D., & Ishii, H. (2002, October 27 - 30). The actuated workbench: Computer-controlled actuation in tabletop tangible interfaces. In *Proceedings of Symposium on User Interface Software and Technology (UIST '02)*, Paris, France.

Ratti, C., Wang, Y., Ishii, H., Piper, B., & Frenchman, D. (2004). Tangible user interfaces (TUIs): A novel paradigm for GIS. *Transactions in GIS, 8*(4), 407–421. doi:10.1111/j.1467-9671.2004.00193.x

Reznik, M. (2000). *Learning with digital manipulatives: New frameworks to help elementary-school students explore "advanced" mathematical and scientific concepts.* Proposal to the National Science Foundation. Brave, S. (1997). InTouch: A medium for haptic interpersonal communication. In *Proceedings of the Conference on Human Factors in Computing Systems (CHI '97)*, ACM Press.

Simpson, R. (1998). Multimedia special: Effective audio-visual. *Museum Practice, 9*, 44–46.

Ullmer, B., & Ishii, H. (2005). Token+constraint systems for tangible interaction with digital information. ACM Trans. *Computer-Human Interact, 12*(1), 81–118. doi:10.1145/1057237.1057242

APPENDIX

During the experimental evaluation, a survey was conducted based on the following questionnaire.

Please answer the following questions:

1. Make a drawing of the trilobite.
2. Please answer the following questions (2-1 to 2-6)
   - 2-1. Explain what you know about the period where the trilobite lived.
   - 2-2. Explain what you know about the different types of trilobite.
   - 2-3. Explain what you know on the means of defense the trilobite used when attacked by its enemies.
   - 2-4. Explain what you know about the trilobite's eye.
   - 2-5. Explain what you know about the trilobite's legs.
   - 2-6. If you know things about the trilobite other than what you answered in questions 2-1 to 2-5, please explain them.
3. Read the sentences 3-1 to 3-18, select for each of them from 1 to 5 the alternative which corresponds best to your situation, and circle the corresponding number. 1. Totally disagree 2. Somewhat disagree 3. No opinion 4. Somewhat agree 5. Totally agree
   - 3-1. I want to know more about the trilobite
   - 3-2. I want to learn about fossils other than the trilobite.
   - 3-3. I would like to go to a museum.
   - 3-4. I would like to go to the library and read books on the trilobite.
   - 3-5. I would like to go to the library and read books on fossils other than the trilobite.
   - 3-6. I would like to ask my teacher at school about the trilobite.
   - 3-7. I would like to ask my teacher at school about fossils other than the trilobite.
   - 3-8. I would like to watch a TV program talking about the trilobite.
   - 3-9. I would like to watch a TV program talking about fossils other than the trilobite.
   - 3-10. I would like to dig trilobite fossils myself.
   - 3-11. I would like to dig fossils other than the trilobite.
   - 3-12. I would like to search for information on the trilobite on the internet.
   - 3-13. I would like to search for information on fossils other than the trilobite on the Internet.
   - 3-14. I would like to talk with my family and friends about the trilobite.
   - 3-15. I would like to talk with my family and friends about fossils other than the trilobite.
   - 3-16. I would like to play a game concerning fossils.
   - 3-17. I would like to touch a trilobite fossil.
   - 3-18. I would like to touch a fossil other than the trilobite.

Compilation of References

3GPP. (2003). 3GPP TS 23.140: MMS Functional Description. Retrieved 24 July, 2008, from http://www.3gpp.org

Abbas, N., Gravell, A., & Wills, G. (2008, June 10-14). Historical Roots of Agile Methods: Where did "Agile Thinking" Come from? In *Agile processes and eXtreme programming in Software Engineering*, Limerick, Ireland.

Abowd, G. D., Atkeson, C. G., Hong, J., Long, S., Kooper, R., & Pinkerton, M. (1997). Cyberguide: A mobile context-aware tour guide. *Wireless Networks*, *3*(5), 421–433. doi:10.1023/A:1019194325861

Advanced Distributed Learning Initiative. (2006). *SCORM® Shareable Content Object Reference Model SCORM 2004 3rd Edition.*

Ahern, S., Davis, M., Eckles, D., King, S., Naaman, M., Nair, R., et al. (2006, September). ZoneTag: Designing context-aware mobile media Capture to increase participation. Paper presented at *Workshop on Pervasive Image Capture and Sharing at the Eight International Conference on Ubiquitous Computing*, Orange County, CA.

Ajzen, I. (1991). The theory of planned behaviour . *Organizational Behavior and Human Decision Processes*, *50*, 179–211. doi:10.1016/0749-5978(91)90020-T

Akhras, F. N., & Self, J. A. (2000). Modeling the process, not the product, of learning. In S. P. Lajoie (Ed.), *Computers as cognitive tools, volume two: No more walls* (pp. 3-28). Mahwah, NJ: Lawrence Erlbaum Associates.

Akiko, N., & Hirohisa, O. (1997). *On the idea of hands on in children's museums: Formative function of children's museums as an environment of children's play.* Bulletin of Tokyo Gakugei University, Series I, science of education, 48, (pp. 345-354).

Albright, M. J., & Nworie, J. (2008). Rethinking academic technology leadership in an era of change. *EDUCAUSE Quarterly*, (1): 15–23.

Aleven, V., Stahl, E., Schworm, S., Fischer, F., & Wallace, R. M. (2003). Help seeking and help design in interactive learning environments. *Review of Educational Research*, *73*(3), 277–320. doi:10.3102/00346543073003277

Alier, M. (2007). A social constructionist approach to learning communities: Moodle. In M. D. Lytras (Ed.), *Open Source for Knowledge Management: Strategies beyond tools* (pp. 369-381). Hersley: Idea Group Publishing.

Alier, M., & Casany, M. (2008). Moodbile: Extending Moodle to the Mobile on/offline Scenario. In I. Arnedillo & P. Isaías (Ed.), *IADIS International Conference Mobile Learning* (pp.11-18). IADIS PRESS.

Alier, M., Casany, M., & Casado, P. (2007). A Mobile extension to a Web based Moodle virtual classroom. In P. Cunningham & M. Cunningham (Ed.), *Expanding the Knowledge Economy: Issues, Applications, Case Studies, 4*, 1169-1176. IOS Press.

Al-khamayseh, S., Zmijewska, A., Lawrence, E., & Culjak, G. (2007). Mobile learning systems for digital natives. In *Proceedings of the Sixth IASTED International Conference on Web Based Education* (pp. 252–257). Chamonix, France.

Allen, M., Witt, P. L., & Wheeless, L. R. (2006). The role of teacher immediacy as a motivational factor in student learning: Using meta-analysis to test a causal model. *Communication Education*, *55*(1), 21–31.

Allwright, R. L. (1984). The importance of interaction in classroom language learning. *Applied Linguistics*, *5*(2), 156–171. doi:10.1093/applin/5.2.156

Amazon Askville. (2006). Retrieved 24 July, 2008, from http://askville.amazon.com

Andersen, J. F. (1979). Teacher immediacy as a predictor of teaching effectiveness. In D. Nimmo (Ed.), *Communication Yearbook 3* (pp. 543-559). New Brunswick, NJ: Transaction Books.

Andersen, J. F. (1981). Three investigations exploring relationships between perceived teacher communication behaviors and student learning. *Communication Education, 30*(4), 377–392.

Anderson, K., & Witfelt, C. (2003). Educational game design: bridging the gab between computer based learning and experimental learning environments. *Conference proceedings of the Sixth International Conference on Computer-based learning in Science, 1*, 402-413.

Anderson, R. J., Anderson, R., Vandergrift, T., Wolfman, S., & Yasuhara, K. (2003). *Promoting Interaction in Large Classes with Computer Mediated Feedback.* Paper presented at 5th International Conference on Computer Support for Collaborative Learning (CSCL), Bergen, Norway.

Anderson, T. (2004). Towards a theory of online learning. In T. Anderson & F. Elloumi (Eds.), *Theory and practice of online learning* (pp. 33-60). Canada: Athabasca University Press.

Andronico, A., Carbonaro, A., Colazzo, L., & Molinari, A. (2004). Personalisation services for learning management systems in mobile settings. *International Journal of Continuing Engineering Education and Lifelong Learning, 14*(4/5), 353–369. doi:10.1504/IJCEELL.2004.005726

Arcavi, A. (2003). The role of visual representations in the learning of mathematics. *Educational Studies in Mathematics, 52*(3), 215–241. doi:10.1023/A:1024312321077

Argyle, M. (1992). *The social psychology of everyday life*. London: Routledge.

Argyle, M., & Dean, J. (1965). Eye-contact, distance and affiliation. *Sociometry, 28*(3), 289–304. doi:10.2307/2786027

Arina, T. (2008). Learning Zeitgeist: The Future of Education is Just-in-Time, Multidisciplinary, Experimental. In *Masternewmedia*. Retreived July, 20, 2008 from http://www.masternewmedia.org/news/2008/02/13/learning_zeitgeist_the_future_of.htm

Arnedillo-Sánchez, I. (2008). The mobile digital narrative tool. In [Lisbon, Portugal.]. *Proceedings of IADIS Mobile Learning, 2008*, 77–83.

Arrigo, M., Di Giuseppe, O., Fulantelli, G., Gentile, M., Novara, G., Seta, L., et al. (2007, October 16-19). A Collaborative mlearning environment. In *Proceedings of mLearn* (pp. 13-21). Melbourne, Australia.

Arvind, A., & Hicks, J. (2006). A Mobile phone Ecosystem: MIT and Nokia's Joint Research Venture . *IEEE Intelligent Systems, 21*(5), 78–79. doi:10.1109/MIS.2006.84

Association for Psychological Science. (2007). Back To School: Cramming Doesn't Work In The Long Term. *ScienceDaily*. Retrieved August 6, 2008, from http://www.sciencedaily.com– /releases/2007/08/070829122934.htm

Attewell, J., & Webster, T. (2005). Engaging and supporting mobile learners. In J. Attewell & C. Savill-Smith (Eds.), *Mobile learning anytime everywhere* (pp. 15). Learning and Skills Development Agency.

Atun, R., & Sittampalam, S. (2006, March). *The upward trend in healthcare spend, The role of mobile phones in increasing accessibility and efficiency in healthcare* (Paper No. 4), The Vodafone Policy Paper Series.

Atun, R., Greene, S., Morris, S., Franklin, V., & Samyshkin, Y. (2006, March). *Use of mobile technologies to enhance control of type 1 diabetes in young people: economic evaluation*, (Paper No. 4), The Vodafone Policy Paper Series.

Ayres, R., Papagiannakis, G., Kim, H. S., & Magnenat-Thalmann, N. (2005, February). Believability and presence in mobile mixed reality environments. In *Proceedings of the IEEE workshop on Virtuality Structures (VR2005)*. IEEE.

Baader, F., Calvanese, D., Mc Guinness, D., Nardi, D., & Patel-Schneider, P. (2002). *The Description Logic Handbook*. Cambridge University Press.

Bachfischer, A., Lawrence, E., Litchfield, A., Dyson, L., & Raban, R. (2008). Student Perspectives about Using Mobile Devices in Their Studies, *IADIS International Conference Mobile Learning* (pp. 8-18). Algarve, Portugal.

Bakhtin, M. (1986). The problem of speech genres (V. McGee, Trans.). In C. Emerson & M. Holquist (Eds.), *Speech genres and other late essays* (pp. 60-102). Austin: University of Texas Press.

Balint, L. (1995). Adaptive interfaces for human-computer interaction: a colourful spectrum of present and future options. *Systems, Man and Cybernetics, Intelligent Systems for the 21st Century, IEEE International Conference, 1*(22), 292 - 297.

Balme, L. (2008). *Interfaces homme-machine plastiques: Une approche par composants dynamiques.* Unpublished doctoral dissertation, University Joseph Fourier, Grenoble, France.

Bär, H., Häussge, G., & Rößling, G. (2007). An integrated system for interaction support in lectures. *In Proceedings of the 12th Annual SIGCSE Conference on innovation and Technology in Computer Science Education ITiCSE '07* (pp. 281- 285). New York: ACM.

Barajas, M., Chrysos, A., Bosco, A., Fonollosa, M., Alvarez, I., & Sancho, J. M. (1998). *Virtual Classrooms in Traditional Universities: Changing Teaching Cultures Through Telematics.* Paper presented at the World Conference on Educational Multimedia and Hypermedia, Freiburg, Germany.

Barbosa, D., Geyer, C., & Barbosa, J. (2005). Globaledu: An architecture to support learning in a pervasive computing environment. In *Proceedings of the IFIP Workshop on Educational Technology.* Perth, Australia. Springer-Verlag.

Bardram, J. E., & Hansen, T. R. (2004). The AWARE architecture: supporting context-mediated social awareness in mobile cooperation. *Proceedings of ACM conference on Computer supported cooperative work* (pp. 192-201).

Battistich, V., Solomon, D., & Delucchi, K. (1993). Interaction processes and student outcomes in cooperative learning groups. *The Elementary School Journal, 94*(1), 19. doi:10.1086/461748

Beaudouin-Lafon, M. (2000). Ceci n'est pas un ordinateur. Perspectives sur l'Interaction Homme-Machine. *Technique et Science Informatique, 19*(1-2-3), 69-74.

Bechhofer, S., Möller, R., & Crowther, P. (2003). The DIG Description Logic Interface. In *Proceedings of the 16th International Workshop on Description Logics (DL'03), volume 81 of CEUR Workshop Proceedings.* Rome, Italy.

BECTA (British Educational and Technology Agency). (2003). *Becta ICT Research Online.* Retrieved July 30, 2007, from http://www.becta.org.uk.

Bederson, B., Meyer, J., & Lance, G. (2000) Jazz: An extensible zoomable user interface graphics toolkit in Java. In *Proceedings of the 13th Annual ACM Symposium on User Interface Software and Technology* (pp. 171-180). ACM press.

Birchfield, D., Savenye, W., & Thornburg, H. (2008). *AMEEd Edulink.* Retrieved July 10, 2008. from http://ame4.hc.asu.edu/edulink

Bishop, J. (2002). *Development and evaluation of a virtual community.* Unpublished BSc(Hons), University of Glamorgan, Pontypridd.

Bishop, J. (2003). The internet for educating individuals with social impairments. *Journal of Computer Assisted Learning, 19*(4), 546–556. doi:10.1046/j.0266-4909.2003.00057.x

Bishop, J. (2004). *The potential of persuasive technology for educating heterogeneous user groups.* Unpublished MSc, Univeristy of Glamorgan, Pontypridd.

Bishop, J. (2007a). Evaluation-centred design of E-learning communities: A case study and review. In V. Grout, D. Oram, & R. Picking (Eds.), *Proceedings of the second international conference on internet technologies and applications* (ITA07 ed.) (pp. 1-9). Wrexham: University of Wales Press.

Bishop, J. (2007b). Increasing participation in online communities: A framework for human–computer interaction. *Computers in Human Behavior, 23*(4), 1881–1893. doi:10.1016/j.chb.2005.11.004

Bishop, J. (2007c). Ecological cognition: A new dynamic for human-computer interaction. In B. Wallace, A. Ross, J. Davies, & T. Anderson (Eds.), *The mind, the body and the world: Psychology after cognitivism* (pp. 327-345). Exeter: Imprint Academic.

Bishop, J. (2007d). *The psychology of how christ created faith and social change: Implications for the design of E-learning systems.* University of Winchester.

Blackboard. (2008). Retrieved July 10, 2008, from http://www.blackboard.com/us/index.bb

Bloom, B. S. (1956). *Taxonomy of educational objectives: The classification of educational goals.* New York: McKay.

BlueGrid. (2008). *Grid computing for J2ME Bluetooth devices.* Retrieved July 20, 2008, from https://bluegrid.dev.java.net

BlueMGF. (2007). *Bluetooth multiplayer games framework*. Retrieved July 20, 2008, from http://sourceforge.net/projects/bluemgf

Bluetooth. Core System Architecture. *Retrieved from:* http://www.bluetooth.com/Bluetooth/Technology/Works/Core_System_Architecture.htm#10, July 26, 2007.

Bondendorf, E., Schertler, M., & Cohen, E. (2006) Producing reusable web based multimedia presentations. In Alex Koohang (Eds.), *Interdisciplinary Journal of knowledge and Learning Objects, 2*, 127-142.

Borba, M. C., & Villarreal, M. E. (2005). *Humans-with-media and the reorganization of mathematical thinking: Information and communication technologies, modeling, visualization and experimentation*. New York: Springer.

Bosch, J., Friedrichs, S., Jung, S., Helbig, J., & Scherdin, A. (2007). Service orientation in the enterprise. *Computer, 40*(11), 51–56. doi:10.1109/MC.2007.399

Boyd, C. (2008). *Text Messages and Disaster Recovery, World a BBC Service and WGBH-Boston co-production*. Retrieved 23 March, 2008, from http://newsbbc.co.uk/2/hi/technology/4149977.htm.

Bradley, C., Haynes, R., & Boyle, T. (2005). Adult multimedia learning with PDAs - the user experience. *Proceedings of the 4th World Conference on mobile learning (mLearn)*. Retrieved June 3, 2008, from http://www.mlearn.org.za/CD/papers/bradley-1.pdf

Brandt, E., Björgvinsson, E., Hillgren, P.-A., Bergqvist, V., & Emilson, M. (2002, October 19-23). PDA's, Barcodes and Video-films for Continuous Learning at an Intensive Care Unit. *Proceedings of NordiCHI*.

Brickell, G., Herrington, J., & Harper, B. (2005). Technology-supported collaboration on field-based authentic tasks. *Proceedings of EdMedia 2005 World Conference on Educational Multimedia, Hypermedia and Telecommunications* (pp. 1961-1968), Norfolk, VA AACE.

Brockett, R. G., & Hiemstra, R. (1991). *Self-direction in adult learning: perspectives on theory, research and practice*. London: Routledge.

Brown, B., Lundin, J., Rost, M., Lymer, G., & Holmquist, L. E. (2007) Seeing ethnographically: Teaching ethnography as part of CSCW. In L. J. Bannon, I. Wagner, C. Gutwin, R. Harper, & K. Schmidt (Ed.), *Proceedings of the 10th European Conference on Computer-Supported Cooperative Work* (pp. 411-430). Springer.

Brown, J. S., & Burton, R. R. (1978). Diagnostic models for procedural bugs in basic mathematical skills. *Cognitive Science, 2*, 155–191.

Brown, J. S., & Duguid, P. (2002). *The social life of information*. Cambridge, MA: Harvard Business School Press.

Brown, J. S., Collins, A., & Duguid, P. (1989). Situated cognition and culture of learning. *Educational Researcher, 18*(1), 32–42.

Brown, R. E. (2001). The process of community-building in distance learning classes. *Journal of Asynchronous Learning Networks, 5*(2), 18–35.

Brown, S. A., Chervany, N. L., & Reinicke, B. A. (2007, September). What matters when introducing new information technology. *Communications of the ACM, 50*(9), 91–96. doi:10.1145/1284621.1284625

Browne, D., Totterdell, P., & Norman, M. (1990). *Adaptive User Interfaces*. Computer and People Series. Academic Press, ISBN 0-12-137755-5.

Brown-Martin, G. (2008). It's the learning, stupid! In I. Arnedillo & P. Isaias (Ed.), *IADIS International Conference Mobile Learning* (pp. XVII-XIX). IADIS PRESS.

Bruner, J. (1966). *Toward a Theory of Instruction*. Cambridge, MA: Harvard University Press.

Bruner, J. (1996). *The Culture of Education*. Cambridge, MA: Harvard University Press.

Brusilovsky, B. (2003). Developing adaptive educational hypermedia system. In T. Murray, S. Blessing, & S. Ainsworth (Eds.), *Authoring Tools for Advanced Technology Learning Environments* (pp. 337–409), Dordrecht, Kluwer Academic Publishers.

Brusilovsky, P. (1996). Methods and techniques of adaptive hypermedia. *User Modeling and User-Adapted Interaction, 6*(2-3), 87–129. doi:10.1007/BF00143964

Brusilovsky, P. (2001). Adaptive hypermedia. In Alfred Kobsa (Ed.), *User Modeling and User Adapted Interaction, Ten Year Anniversary Issue, 11*(1/2), 87-110.

Buckingham, D. (2003). *Media education: Literacy, learning and contemporary culture*. Cambridge, UK: Polity Press.

Buckingham, D., Harvey, I., & Sefton-Green, J. (1999). The difference is digital: Digital technology and student media production. *Convergence: The Journal of Research into New Media Technologies, 5*(4), 10–21. doi:10.1177/135485659900500402

Budde, P. (2008). *Australian Mobile Market Statistics,* [REMOVED HYPERLINK FIELD]*Mobile Data Association.* Retrieved 12 Oct 2008 from http://www.themda.org/Page._

Bullen, M. (1998). Participation and critical thinking in online university distance education. *Journal of Distance Education, 13*(2), 1–32.

Burden, K., & Kuechel, T. (2004). *Evaluation report of the Teaching and Learning with Digital Video Assets Pilot 2003-2004: Becta.* Retrieved from http://www.becta.org.uk/page_documents/research/evaluation_dv_assets03.pdf.

Bybee, R. W. (1997). *Achieving Scientific Literacy: From Purposes to Practices.* New York: Heinemann.

Byrne, M. D., Catrambone, R., & Stasko, J. T. (1999). Evaluating animations as student aids in learning computer algorithms. *Computers & Education, 33*(4), 253–278. doi:10.1016/S0360-1315(99)00023-8

Byrne, P., Arnedillo Sánchez, I., & Tangney, B. (2008, October). A mobile computer supported collaborative learning tool for digital narrative pProduction. In *Proceedings of the MLearn2008 Conference: The Bridge From Text To Context* (pp. 66-73). UK: University of Wolverhampton.

Calvary, G., Coutaz, J., Thevenin, D., Limbourg, Q., Bouillon, L., & Vanderdonckt, J. (2003). A unifying reference framework for multi-target user interfaces. *Interacting with Computers, 15*(3), 289–308. doi:10.1016/S0953-5438(03)00010-9

Campus Technology. (2007). *Campus technology innovators: Cellular/mobile.* Retrieved July 7, 2008, from http://campustechnology.com/article.aspx?id=49202

Cao, Y., Tin, T., McGreal, R., Ally, M., & Coffey, S. (2006). The Athabasca University Mobile Library Project: Increasing the Boundaries of Anytime and Anywhere Learning for Students. *International conference on Wireless Communications and Mobile Computing (IWCMC '06)* (pp. 1289-1294).

Carboni, D. (2004). *Services in Pervasive Computing Environments: from Design to Delivery,* Unpublished doctoral dissertation, CRS4 – Center for Advanced Studies, Research and Development in Sardinia, Italy.

Carr, B., & Goldstein, I. P. (1977). Overlays: a theory of modeling for computer-aided instruction. *AI Lab Memo 406 (Logo Memo 40).* Massachusetts Institute of Technology, Cambridge, MA.

Casany, M., Alier, M., & Casado, P. (2007). JDME-MicroDB: A new Open Source Lightweight Database Engine for J2ME Mobile Devices. In S. Kim et al. (Ed.), *Multimedia and Ubiquitous Engineering* (pp. 247-252). IEEE Computer Society.

Cascio, J. (2008). *World Changing.* Retrieved 23 March, 2008, from, http://www.worldchanging.com/archives/00128.htm.

Casey, D. (2005, October 24-28). u-Learning = e-Learning + m-Learning. *Proceedings of e-Learn 2005:World Conference on e-Learning in Corporate, Government, Healthcare & Higher Education,* Vancouver, Canada. Retrieved June 8, 2008 from http://neptune.netcomp.monash.edu.au/staff/dcasey/papers/2005/E-Learn2005/paper6792.doc.

Castells, M., Fernández-Ardèvol, M., Linchuan Qiu, J., & Sey, A. (2007). *Mobile Communication and Society: A Global Perspective.* Cambridge, Massachussets: MIT Press.

Castillo, S., & Ayala, G. (2008). ARMOLEO: An Architecture for Mobile Learning Objects. In *Proceedings of 18th International Conference on Electronics, Communications and Computers* (pp. 53-58). Cholula, Puebla, Mexico. IEEE Computer Society.

Caulton, T. (1998). *Hands-on exhibitions: Managing interactive museums and science centre.* Routledge.

Ceccucci, W., & Tamarkin, D. (2006). *Ice breakers for the traditional or online classroom.* Palm Springs, CA.

Cesarini, M., Monga, M., & Tedesco, R. (2004, March 14-17). Carrying on the e-learning process with a workflow management engine. In *Proceedings of the 2004 ACM Symposium on Applied Computing* (pp. 940-945), Nicosia, Cyprus. SAC '04. New York: ACM.

Chakraborty, D., Perich, F., Avancha, S., & Joshi, A. (2001). DReggie: Semantic Service Discovery for M-Commerce Applications. In *Proceedings of Workshop on Reliable and Secure Applications in Mobile Environment.*

Chan, T. W., & Chou, C. Y. (1997). Exploring the design of computer supports for reciprocal tutoring. *International Journal of Artificial Intelligence in Education, 8,* 1–29.

Chan, T., Sharples, M., Vavoula, G., & Lonsdale, P. (2004). Educational metadata for mobile learning. In *Proceedings of the 2nd IEEE International Workshop on Wireless and Mobile Technologies in Education* (pp. 197-198). Washington, DC, USA. IEEE Computer Society.

Chang, C. T., Chang, C. Y., & Sheu, J. P. (2003). BlueCube: Constructing a hypercube parallel computing and communication environment over Bluetooth radio system. In *International Conference on Parallel Processing* (pp. 447–454).

Chauhan, D., Pritchard, C., & Sussex, J. (2008, March). *The upward trend in healthcare spend, the role of mobile phones in increasing accessibility and efficiency in healthcare*, (Paper No. 4), The Vodafone Policy Paper Series.

Chen, T.-S., Chang, C.-S., Lin, H.-S., & Yu, H.-L. (2008). Context-Aware Writing in Ubiquitous Learning Environments. In *Proceedings of Fifth IEEE International Conference on Wireless, Mobile, and Ubiquitous Technology in Education* (pp. 67-73). Beijing, China.

Chen, Y. S., Kao, T. C., Sheu, J. P., & Chiang, C. A. (2002). A mobile scaffolding-aid-based bird-watching learning system. In *Proc. of the IEEE International Workshop on Wireless and Mobile Technologies in Education 2002* (pp. 15–22), IEEE Computer Society.

Choi, Y. B., Abbott, T. A., Arthur, M. A., & Hill, D. N. (2007). Toward a future wireless classroom paradigm. *International Journal of Innovation and Learning*, *4*(1), 14–25. doi:10.1504/IJIL.2007.011472

Chou, C. Y., Chan, T. W., & Lin, C. J. (2003). Redefining the learning companion: The past, present, and future of educational agents. *Computers & Education*, *40*(3), 255–269. doi:10.1016/S0360-1315(02)00130-6

Christensen, L. J., & Menzel, K. E. (1998). The linear relationship between student reports of teacher Immediacy behaviors and perceptions of state motivation, and of cognitive, affective, and behavioral learning. *Communication Education*, *47*(1), 82–90.

Christophel, D. M. (1990). The relationships among teacher immediacy behaviors, student Motivation, and learning. *Communication Education*, *39*(4), 323–340.

Chun, M. M. (2000, May). Contextual Cueing of Visual Attention. *Trends in Cognitive Sciences*, *4*(5), 170–178. doi:10.1016/S1364-6613(00)01476-5

Cingil, I., Dogac, A., & Azgin, A. (2000). A broader approach to personalization. *Communications of the ACM*, *43*(8), 136–141. doi:10.1145/345124.345168

Clark, A. (1997). *Being there: Putting brain and world together again*. Cambridge, MA: MIT press.

Claroline (2008). Retrieved July 10, 2008, from http://www.claroline.net

Cockton, G. (1987). Some Critical Remarks on Abstractions for Adaptable Dialogue Managers. In D. Diaper & R. Winder (Eds.), People *and Computers III. Proceedings of the Third Browne Conference of the British Computer Society, Human Computer Interaction Specialist Group* (pp. 325343). University of Exeter.

Collins, A., Brown, J. S., & Newman, S. E. (1989). Cognitive Apprenticeship: Teaching the Craft of Reading, Writing, and Mathematics. In L. B. Resnick (Ed.), *Knowing, Learning and Instruction* (pp. 453 - 494). Hillsdale N.J: Lawrence Erlbaum Associates.

Collins, A., Neville, P., & Bielaczyc, K. (2000). The role of different mMedia in designing learning environments. *International Journal of Artificial Intelligence in Education, 11*, 144–162.

Collins, T. G. (2005). English Class on the Air: Mobile Language Learning with Cell Phones. In *Fifth IEEE International Conference on Advanced Learning Technologies (ICALT'05)*, pp. 402-403.

Colucci, S., Di Noia, T., Di Sciascio, E., Donini, F. M., & Mongiello, M. (2003). Concept Abduction and Contraction in Description Logics. In *Proceedings of the 16th International Workshop on Description Logics (DL'03)*. Volume 81 of CEUR Workshop Proceedings, Rome, Italy.

Compeau, D., Higgins, C., & Huff, S. (1999). Social Cognitive Theory and Individual Reactions to Computing Technology: A Longitudinal Study. *MIS Quarterly*, *23*(2), 146–158. doi:10.2307/249749

Coop, R. H., & White, K. (1974). *Psychological concepts in the classroom*. London: Harper & Row.

Cooper, A., & Reimann, R. (2003). *About Face 2.0: The Essentials of Interaction Design*. John Wiley & Sons.

Cooper, M., Foote, J., Girgensohn, A., & Wilcox, L. (2005) Temporal event clustering for digital photo collections. In *ACM Transactions on Multimedia Computing, Communications, and Applications* (pp. 269-288). ACM press.

Corder, N. (2002). *Learning to teach adults: An introduction*. London, UK: Routledge Falmer.

Corsini, L. S., Crittenden, V. L., Keeley, R. C., Trompter, G. M., & Viechnicki, B. (2000). Integrating cross-functional undergraduate teaching and learning: A work in progress. *Marketing Education Review*, *10*(3), 1–17.

Cortez, C., Nussbaum, M., Santelices, R., Rodrigues, P., & Zurita, G. (2004). Teaching Science with Mobile Computer Supported Collaborative Learning (MCSCL). In *Proc. of the IEEE International Workshop on Wireless and Mobile Technologies in Education 2004* (pp. 67–74), IEEE Computer Society.

Costagliola, G., Ferrucci, F., & Fuccella, V. (2006, July 11-14). SCORM run-time environment as a service. In *Proceedings of the 6th international Conference on Web Engineering* (ICWE '06), Palo Alto, CA, USA, (vol. 263, pp. 103-110). New York: ACM.

Crowder, R. G. (1976). *Principles of learning and memory*. Oxford, England: Lawrence Erlbaum.

Cuppens, E., Raymaekers, C., & Coninx, C. (2006). A model-based design process for interactive virtual environments. In S.W. Gilroy & M.D. Harrison (Eds.), *Interactive Systems. (LNCS, 3941*, pp. 225-236.) Springer, Heidelberg.

Curtis, M., Luchini, K., Bobrowsky, W., Quintana, C., & Soloway, E. (2002). Handheld Use in K-12: A Descriptive Account. In U. Hoppe, M. Milrad & Kinshuk (Eds.), *Proceedings of IEEE International Workshop on Wireless and Mobile Technologies in Education (WMTE'02)* (pp.23-30). New York: IEEE Computer Society Press.

Danesi, M. (2007). *The quest for meaning: A guide to semiotic theory and practice*. Toronto Studies in Semiotics and Communication, University of Toronto Press, Toronto,Canada.

Darling, L. (1984, October). What do nurses want in a mentor? *The Journal of Nursing Administration*, *14*(10), 42–44.

Davis, F. D. (1989). Perceived usefulness, perceived ease of use, and user acceptance of information technology. *MIS Quarterly*, *13*(3), 319–340. doi:10.2307/249008

Davis, F. D. (1993). User acceptance of information technology: system characteristics, user perceptions and behavioural impacts. *International Journal of Man-Machine Studies*, *38*, 475–487. doi:10.1006/imms.1993.1022

de Aguilera, M., & Mendiz, A. (2003). Video games and education: (education in the face of a "parallel school"). [CIE]. *Computers in Entertainment*, *1*, 1–10. doi:10.1145/950566.950583

De Bra, P., & Ruiter, J.-P. (2001). AHA! Adaptive hypermedia for all. In *Proc. of the World Conference of the WWW and Internet 2001* (pp. 262-268), AACE.

de Crom, N. (2005, October). The "ME"-Learning Experience: PDA Technology and eLearning in Ecotourism at TUT. *Fourth World Conference on Mobile Learning (mLearn).*

De Jong, T., Specht, M., & Koper, R. (2007, October 17-19). *ContextBlogger: learning by blogging in the real world.* Paper presented at the conference ePortfolio 2007., Maastricht, The Netherlands.

De Jong, T., Specht, M., & Koper, R. (2008a). Contextualised media for learning. *Educational Technology & Society*, *11*(2), 41–53.

De Jong, T., Specht, M., & Koper, R. (2008b). A reference model for mobile social software for learning. [IJCEELL]. *International Journal of Continuing Engineering Education and Lifelong Learning*, *18*(1), 118–138. doi:10.1504/IJCEELL.2008.016079

Decker, K. (1998). Coordinating human and computer agents. *Coordination Technology for Collaborative Applications-Organizations, Processes, and Agents. LNCS*, *1364*, 77–98.

Dempster, F. N. (1988). The spacing effect: A case study in the failure to apply the results of psychological research. *The American Psychologist*, *43*(8), 627–634. doi:10.1037/0003-066X.43.8.627

Denso-Wave. (2003) *About 2D Code.* Denso-Wave. http://www.denso-wave.com/qrcode/aboutqr-e.html

Desouza, K. C., & Hensgen, T. (2003). 9 -11 information failures: A semiotic approach. *IT Professional*, *5*(2), 61–63. doi:10.1109/MITP.2003.1191793

Dey, A. K. (2001). Understanding and Using Context. *Personal and Ubiquitous Computing, 5*(1).

Di Noia, T., Di Sciascio, E., & Donini, F. M. (2004). Extending Semantic-Based Matchmaking via Concept Abduction and Contraction. In *Proceedings of the 14th International Conference on Knowledge Engineering and Knowledge Management (EKAW2004).* Whittlebury Hall, UK. Springer.

Di Noia, T., Di Sciascio, E., & Donini, F. M. (2007). Semantic matchmaking as non-monotonic reasoning: A description logic approach. [JAIR]. *Journal of Artificial Intelligence Research*, *29*, 269–307.

Di Noia, T., Di Sciascio, E., Donini, F. M., & Mongiello, M. (2003a). A system for principled Matchmaking in an electronic marketplace. In *Proceedings of International World Wide Web Conference (WWW '03)* (pp. 321-330). Budapest, Hungary. ACM Press.

Di Noia, T., Di Sciascio, E., Donini, F. M., & Mongiello, M. (2003b). Abductive matchmaking using description logics. In *Proceedings of Eighteenth International Joint Conference on Artificial Intelligence (IJCAI-03)* (pp. 337-342), Acapulco, Mexico. MK.

Dieterich, H., Malinowski, U., Kühme, T., & Schneider-Hufschmidt, M. (1993). State of the Art in Adaptive User Interfaces. In M. Schneider-Hufschmidt, T. Kühme, & U. Mallinowski (Eds.), *Adaptive User Interfaces, Principles and Practice* (pp. 13-48). Included in series Human Factors in Information Technology, 10.

Dillenbourg, P. (2000). Virtual learning environments. In *Proceedings of Workshop on Virtual Learning Environments during EUN Conf. 2000, Learning in the new millennium: Building new education strategies for schools*.

Dimitracopoulou, A. (2005). Designing collaborative learning systems: Current trends & future research agenda. In *Proceedings of Computer Support for Collaborative Learning: Learning 2005: The Next 10 Years*! (pp. 115-124). Taipei, Taiwan.

Dinkmeyer, D., & Dreikurs, R. (1963). *Encouraging children to learn: The encouragement process*. Englewood Cliffs, NJ: Prentice-Hall.

Dix, A., & Finlay, J. Abowd, G., & Beale, R. (1993). *Human-Computer Interaction*. Prentice-Hall

Dochev, D., & Hristov, I. (2006). Mobile learning applications: ubiquitous characteristics and technological solutions. *Cybernetics and Information Technologies*, *6*(3), 63–74.

Docq, F., Lebrun, M., & Smidts, D. (2007). Claroline, une plate-forme d'enseignement/apprentissage sur Internet: pour propulser la pédagogie active et l'innovation? In M. Frenay, B. Raucent, & P. Wouters (Eds.), *Actes du 4ième colloque Questions de pédagogies dans lenseignement supérieur* (pp. 99–109). Presses universitaires de Louvain, Louvain-la-Neuve.

Dokeos, (2008). Retrieved July 10, 2008, from http://www.dokeos.com/

Dominick, J. (2002). Ready or Not – PDAs in the Classroom, *Campus Technology Magazine*, Retrieved 24 July, 2008 from http://www.campustechnology.com/article.asp?id=6705.

Doolan, D. C., Tabirca, S., & Yang, L. T. (2006). Mobile parallel computing. In *5th International Symposium on Parallel and Distributed Computing* (pp. 161-167).

Dougiamas, M., & Taylor, P. C. (2003, June 23-28). Moodle: Using learning communities to create an open source course management system. In *Proceedings of the World Conference on Educational Multimedia Hypermedia & Telecommunication (ED-MEDIA'2003)*, Honolulu (pp. 171–178). Charlottesville: AACE Press.

Dourish, P., & Bellotti, V. (1992) Awareness and Coordination in Shared Workspaces. In *Proceedings of the 1992 ACM conference on Computer-supported cooperative work* (pp. 107-114). ACM press.

Drachsler, H., Hummel, H., & Koper, R. (2008). Personal recommender systems for learners in lifelong learning: requirements, techniques and model. *International Journal of Learning Technology*, *3*(4), 404–423. doi:10.1504/IJLT.2008.019376

Druin, A. (1999). Cooperative Inquiry: Developing New Technologies for Children with Children. *Proc. CHI '99* (pp. 592-599). Pittsburgh, PA, USA: ACM

Dubos, R. J. (1972). *A God within*. New York, NY: Scribner.

Duffy, A. (2004). *Failing students - a qualitative study of factors that influence the decisions regarding assessment of students, article for Nursing and Midwifery Council*. http://www.nmc-uk.org/aArticle.aspx?ArticleID=1675 (last accessed Dec, 2007)

Dunn, R., & Griggs, S. (2000). *Practical approaches to using learning styles in higher education*. Westport, CT: Greenwood Publishing Group, Inc.

Durst, S. (2007). See It, Snap It, Search It. *Business 2.0*.

Eagle, N., & Pentland, A. (2005). Social serendipity: mobilizing social software. *Proceedings of Pervasive Computing, IEEE* (pp. 28-34). April.

ECMA International (2004, December). *Standard ECMA-340 Near Field Communication Interface and Protocol (NFCIP-1)* (2nd ed.).

Edelson, D. C., Pea, R. D., & Gomez, L. (1995). Constructivism in the Collaboratory. In B.G. Wilson (Ed.), *Constructivist learning environments: Case studies in instructional design.* Englewood Cliffs, NJ: Educational Technology Publications.

Efimova, L., & Fiedler, S. (2004, March 24-26). Learning Webs: Learning In Weblog Networks. *Proceedings of Web-based communities*, Lisbon, Portugal.

Ekman, P., & Friesen, W. V. (1975). *Unmasking the face: a guide to recognizing emotions from facial clues.* Prentice-Hall, Englewood Cliffs, NJ.

El-Bishouty, M. M., Ogata, H., & Yano, Y. (2007). PERKAM: Personalized knowledge awareness map for computer supported ubiquitous learning. *Journal of Educational Technology & Society, 10*(3), 122–134.

Ellis, K. (2004). The impact of perceived teacher confirmation on receiver apprehension, motivation, and learning. *Communication Education, 53*(1), 1. doi:10.1080/0363452032000135742

Engeström, Y. (1993). Developmental studies of work as a test bench of activity theory: The case of primary care medical practice. In J. Lave, & S. Chaiklin (Eds.), *Understanding practice: Perspectives on activity and context* (pp. 63-104). Cambridge: Cambridge University Press.

English, R., Brookes, S. T., Avery, K., Blazeby, J. M., & Ben-Shlomo, Y. (2006). The effectiveness and reliability of peer-marking in first-year medical students. *Medical Education, 40*(10), 965–972. doi:10.1111/j.1365-2929.2006.02565.x

Ericsson, (2005). MMS Services Moving Forward. *Ericsson News*, Retrieved 24 July, 2008 from http://www.ericsson.com/solutions/news/2005/q3/20050725_mms.shtml

Fallon, C., & Brown, S. (2003). *E-Learning Standards.* Boca Raton, St. Lucie Press.

Fasel, B., & Luettin, J. (2003). Automatic facial expression analysis: A survey. *Pattern Recognition, 36*(1), 259–275. doi:10.1016/S0031-3203(02)00052-3

Faulkner, X., & Fintan, C. (2005). When Fingers do the talking: a study of text messaging. *Interacting with Computers, 17*, 167–185. doi:10.1016/j.intcom.2004.11.002

Fensham, P. J., Gunstone, R. F., & White, R. T. (1994). Science content and constructivist views of learning and teaching. In P. J. Fensham, R. F. Gunstone, & R. T. White (Eds.), *The content of science* (pp. 1-8). London: The Falmer Press.

Fischer, G. (2000). User Modeling in Human Computer Interaction [UMUAI]. *User Modeling and User-Adapted Interaction, 11*, 65–86. doi:10.1023/A:1011145532042

Fletcher, J. D. (1975). Modeling of learner in computer-based instruction. *Journal of Computer-based Instruction, 1*, 118–126.

Florins, M., Montero, F., Vanderdonckt, J., & Michotte, B. (2006, May 23-26). Splitting Rules for Graceful Degradation of User Interfaces. In *Proceedings of the 8th International Working Conference on Advanced Visual Interfaces (AVI'2006)* Venezia (pp. 59-66). New York: ACM Press.

Flynn, M., Pendlebury, D., Jones, C., Eldridge, M., & Lamming, M. (2000). The satchel system architecture: Mobile access to documents and services. *ACM Journal of Mobile Networks and Applications, 5*(4), 243–258. doi:10.1023/A:1019172931873

Fogg, B. (1999). Persuasive technologies: Now is your chance to decide what they will persuade us to do- and how they'll do it. *Communications of the ACM, 42*(5), 26–29. doi:10.1145/301353.301396

Fontana, J. (2003). *Collaborative software ages slowly.* Retrieved July 30, 2008, from http://www.networkworld.com

Foroughi, R., & Rieger, T. (2005). A virtual character for face to face interaction with children in E-learning. In D. Shepherd (Ed.), *Creative engagements* (pp. 107-111). Oxford: Inter-Disciplinary Press.

Fragiacomo, L. (2005). Capturing the Y, MIS Managing Information Strategies (www.misweb.com), *Communications of the ACM, 42(5*), 54-59 Australia, Summer 2004/2005.

Freud, S. (1933). *New introductory lectures on psychoanalysis.* New York: W.W. Norton & Company, Inc.

Fung, Y. H. (2004). Collaborative online learning: interaction patterns and limiting factors. *Open Learning, 19*(2), 135–149. doi:10.1080/0268051042000224743

Gäardenfors, P. (1988). *Knowledge in Flux: Modeling the Dynamics of Epistemic States.* Bradford Books, MIT Press, Cambridge, MA.

Garett, J. J. (2005). *Ajax: A new approach to Web applications*. Retrieved on July 30, 2008, from http://www.adaptivepath.com/ideas/essays/archives/000385.php

Garrison, D. R., Anderson, T., & Archer, W. (1999). Critical inquiry in a text-based environment: Computer conferencing in higher education. *The Internet and Higher Education, 2*(2-3), 87–105. doi:10.1016/S1096-7516(00)00016-6

George, P, Gurminder, S., & Magnenat-Thalmann, N. (2008). A survey of mobile and wireless technologies for augmented reality systems. *Journal of Computer Animation and Virtual Worlds.*

Georgiev, T., Georgieva, E., & Smrikarov, A. (2004, June 17-18). M-learning: a new stage of e-learning, *Proceedings International Conference on Computer Systems and Technologies - CompSysTech'2004,* Rousse, Bulgaria. Retrieved June 8, 2008 from http://ecet.ecs.ru.acad.bg/cst04/Docs/sIV/428.pdf

Germán Sánchez, V., Téllez Mora, F., & Morales Gamboa, R. (2007). Collaborative environment for the composition of learning scenarios. In Gelbukh & Zechinelli (Eds.), *IEEE Editorial. ENC07 Conference,* Mexico.

Geser, H. (2004). *Towards a sociological theory of the mobile phone.* (University of Zurich Release 3.0).

Geske, J. (1992). Overcoming the drawbacks of the large lecture class. *College Teaching, 40*(4), 151–154.

Gibson, J. J. (1979). *The ecological approach to visual perception.* Boston: Houghton Mifflin.

Gleason, M. (1986). Better communication in large courses. *College Teaching, 34*(1), 20–24.

Gokhale, A., Natarjan, B., Schmidt, D. C., Nechypurenko, A., Wang, N., Gray, J., et al. (2002). CoSMIC: An MDA Generative Tool for Distributed Real-time and Embdedded Component Middleware and Applications. *Proceedings of the OOPSLA 2002 Workshop on Generative Techniques in the Context of Model Driven Architecture*, Seattle, WA.

González Calleros, J. M., Vanderdonckt, J., & Arteaga, J. M. (2007, June 6-8). A method for developing 3D user interfaces of information systems. In *Proceedings of the 6th International Conference on Computer-Aided Design of User Interfaces (CA-DUI'2006, 5*(7), pp. 85-100). Springer-Verlag, Berlin. Guerrero, J., Vanderdonckt, J., & Gonzalez, J.M. (2008). FlowiXML: A step towards designing workflow management systems. *Journal of Web Engineering, 4*(2), 163–182.

Goodman, B., Hitzeman, J., Linton, F., & Ross, H. (2003). Towards intelligent agents for collaborative learning: Recognizing the role of dialogue participants. *Proceedings of the International Conference on AI in Education,* Honey, P., & Mumford, A. (1986). *Manual of learning styles.* Maidenhead, UK: Peter Honey Publications.

Gordijn, J., & Akkermans, J. M. (2001). A Conceptual Value Modeling Approach for e-Business Development. In *Proceedings of the First International Conference on Knowledge Capture, Workshop Knowledge in e-Business (K-CAP 2001),* (pp. 29-36) http://www.cs.vu.nl/~gordijn/er2001.pdf

Gordin, D. N., & Pea, R. D. (1995). Prospects for scientific visualization as an educational technology. *Journal of the Learning Sciences, 4*(3), 249–279. doi:10.1207/s15327809jls0403_1

Gordin, D. N., Edelson, D., & Pea, R. D. (1995). The Greenhouse Effect Visualizer: A tool for the science classroom. *Proceedings of the Fourth American Meteorological Society Education Symposium.*

Gordin, D., Polman, J., & Pea, R. D. (1994). The Climate Visualizer: Sense-making through scientific visualization. *Journal of Science Education and Technology, 3,* 203–226. doi:10.1007/BF01575897

Gorham, J. (1988). The relationship between verbal teacher immediacy behaviors and student learning. *Communication Education, 37*(1), 40–53.

Gorham, J., & Zakahi, W. R. (1990). A comparison of teacher and student perceptions of immediacy and learning: Monitoring process and product. *Communication Education, 39*(4), 354.

Gough, N., & McCulloch, B. (2006). *The role of mobile phones in increasing accessibility and efficiency in healthcare, Moving the debate forward* (Paper No. 4), The Vodafone Policy Paper Series, March 2006.

Grasso, A., & Roselli, T. (2005). Guidelines for designing and developing contents for mobile learning. In *Proceedings of the 3rd IEEE International Workshop on Wireless and Mobile Technologies in Education (WMTE 2005)* (pp. 123–127). Tokushima, Japan. IEEE Computer Society.

Gray, M., & Smith, L. (2000). The qualities of an effective mentor from the student nurse's perspective: Findings from a longitudinal qualitative study. *Journal of Advanced Nursing,* 32(6), 1542–1549. doi:10.1046/j.1365-2648.2000.01606.x

Green, H., & Hannon, C. (2007). *Young people are spending their time in a space which adults find difficult to supervise or understand.* London: Demos.

Greenfield, A. (2006). Everyware: The Dawning Age of Ubiquitous Computing. *New Riders Publishing., ISBN-13*, 978–0321384010.

Gross, T., Wirsam, W., & Graether, W. (2003). AwarenessMaps: Visualizing awareness in shared workspaces. In *CHI '03 Extended abstracts on Human factors in computing systems* (pp. 784-785). ACM press.

Grudin, J., & Pruitt, J. (2002) Personas, Participatory Design and Product Development: An Infrastructure for Engagement. *Proceedings Participatory Design Conference (PDC)* 2002, 144-161.

Guerrero, J., Vanderdonckt, J., González, J. M., & Winckler, M. (2008). Modeling user interfaces to workflow information systems. Paper presented in *4th International Conference on Autonomic and Autonomous Systems* (ICAS'2008). Springer-Verlag.

Guha, M. L., Druin, A., Chipman, G., Fails, J. A., Simms, S., & Farber, A. (2005, January). Working with Young Children as Technology Design Partners. *Communications of the ACM*, *48*(1), 39–42. doi:10.1145/1039539.1039567

Hacker, R. J., & Rowe, M. J. (1997). The impact of National Curriculum development on teaching and learning behaviours. *International Journal of Science Education*, *19*(9), 997–1004. doi:10.1080/0950069970190901

Hämäläinen, P., Lindholm, M., Nykänen, A., & Höysniemi, J. (2004, April). Animaatiokone: An installation for creating clay animation. In *Proceedings of CHI* (pp. 24-29). Vienna, Austria.

Hand, K. (2007, February 26). Pimp my cube 2017. *Popular Science.*

Hargreaves, A. (2003). *Teaching in the Knowledge Society.* Maidenhead: Open University Press.

Hartenian, S., Schelienger, M., & Frederickson, P. (2001). Creation and Assessment of an Integrated Business Course: One College's Experience. *Journal of Education for Business*, *76*(3), 149–159.

Hartnell-Young, E. (2007). Making the Connections: Theory and Practice of Mobile Learning in Schools. In A. Norman & J.Pearce (Eds.), *Proceedings of MLearn Melbourne 2007: Making the Connections* (pp. 87-95). Melbourne: University of Melbourne Press.

Hauser, C., & O'Connor, A. (2007). Virginia Tech Shooting Leaves 33 Dead. *The New York Times Online*. Retrieved 20 Feb 2008 from http://www.nytimes.com/2007/04/ 16/us/16cnd-shooting.html.

Head, B. (2008, September 1-3). Who's Teaching who? How the gap between students – the digital natives – and teachers – the digital immigrants – can be closed. *Education Review Technology Guide*, (www.campusreview.com.au).

Hida, N., & Yamauchi, Y. (2005). Development of interactive exhibition systems for museum. *Japan Society for Educational Technology 21st conference* (pp. 567-568).

Highsmith, J. Beck k., Cockburn A., & Jeffries, R. (2001). *Agile Manifesto.* from www.agilemanifesto.org (last accessed July 2008).

Hine, N., Rentoul, R., & Specht, M. (2003, May 19-20). Collaboration and roles in remote field trips. In Proceedings of MLEARN (London, United Kingdom: Learning and Skills Development Agency) 2003, 69–72.

Hintzman, D. L. (1974). Theoretical implications of the spacing effect. In R. L. Solso (Ed.), *Theories in cognitive psychology: The Loyola Symposium* (pp. 77-99). Oxford, England: Lawrence Erlbaum Associates, Inc.

Hofer, M., & Swan, K. O. (2005). Digital moviemaking—The harmonization of technology, pedagogy and content. *International Journal of Technology in Teaching and Learning & Leading with Technology*, *1*(2), 102–110.

Holmes, B., & Gardner, J. (2006). *E-learning: Concepts and practice.* London: Sage.

Holmevik, J. R., & Haynes, C. (1998). *High wired. On the design, use and theory of educational MOOs.* University of Michigan Press.

Holzinger, A., Nischelwitzer, A., & Kickmeier-Rust, M. D. (2006). Pervasive e-education supports life long learning: Some examples of x-media learning objects (xlo). In *Proceedings of 10th IACEE World Conference on Continuing Engineering Education.* Vienna, Austria.

Holzinger, A., Nischelwitzer, A., & Meisenberger, M. (2005). Mobile phones as a challenge for m-learning: Examples for mobile interactive learning objects (milos). In *Proceedings of the Third IEEE International Conference on Pervasive Computing and Communications Workshops* (pp. 307–311). IEEE Computer Society.

Hoppe, H., Joiner, R., Milrad, M., & Sharples, M. (2003). Guest editorial: Wireless and mobile technologies in education. *Journal of Computer Assisted Learning, 19*(3), 255–259. doi:10.1046/j.0266-4909.2003.00027.x

Horstmanshof, L. (2004, 5-8 December). *Using SMS as a way of providing connection and community for first year students.* Paper presented at the Beyond the Comfort Zone: Proceedings of the 21st annual conference of the Australasian Society for Computers in Learning in Tertiary Education, Perth.

Houser, C., Thornton, P., & Kluge, D. (2002). Mobile learning: cell phones and PDAs for education. *International Conference Proceedings on Computers in Education* (pp. 1149-1150).

Hsi, S., & Fait, H. (2005). RFID enhances visitors' museum experience at the Exploratorium. *Communications of the ACM, 48*(9), 60–65. doi:10.1145/1081992.1082021

Huang, A., Pulli, K., & Rudolph, L. (2005). *Kimono: Kiosk-Mobile Phone Knowledge Sharing System*, Massachusetts Institute of Technology, Nokia Research Centre, HIT Lab NZ, University of Canterbury, Christchurch, New Zealand.

Hupfer, S., Cheng, L., Ross, S., & Patterson, S. (2004) Introducing collaboration into an application development environment. In *Proceedings of the 2004 ACM conference on Computer supported cooperative work* (pp. 21-24). ACM press.

Hwang, G.-J., Tsai, C.-C., & Yang, S. J. H. (2008). Criteria, strategies and research issues of context-aware ubiquitous learning. *Journal of Educational Technology & Society, 11*(2), 81–91.

Hwang, W.-Y., Hsu, J.-L., & Huang, H.-J. (2007, October 16-19). A study on ubiquitous computer supported collaborative learning with hybrid mobile discussion forum. In *Proceedings of mLearn* (pp. 101-109), Melbourne, Australia.

IEEE (2002). *IEEE Standard for Learning Object Metadata*, std 1484.12.1-2002 edition.

IMS Global Learning Consortium. IMS Abstract Framework: White Paper. (2003). Retrieved June 10, 2008 from http://www.imsglobal.org/af/index.html

IMS Global Learning Consortium. IMS Tools Interoperability Guidelines. (2006). Retrieved June 10, 2008 from http://www.imsglobal.org/ti/index.html

Ishii, H., & Ullmer, B. (1997). Tangible bits: Towards seamless interfaces between people, bits and atoms. In *Proceedings of the Conference on Human Factors in Computing Systems (CHI '97)*. ACM Press.

Jackson, P. Z. (2001). *The inspirational trainer: Making your training flexible, spontaneous & creative.* Kogan Page.

Jameson, A. (2001). *Systems That Adapt to Their Users: An Integrative Perspective.* Unpublished doctoral habilitation, Saarbrücken: Saarland University.

Jonassen, D. J., & Rohrer-Murphy, L. (1999). Activity theory as a framework for designing constructivist learning environments. *Educational Technology Research and Development, 47*(1), 61–79. doi:10.1007/BF02299477

Jones, S. (2002). The internet goes to college: How students are living in the future with today's technology. *USDLA Journal, 16*(10).

June 8, 2008 from http://www.mlearn.org.za/CD/papers/keegan1.pdf

Kalz, M., Van Bruggen, J., Rusmann, E., Giesbers, B., & Koper, R. (2007). Positioning of learners in learning networks with content-analysis, Metadata and Ontologies. *Interactive Learning Environments, 15*(2), 191–200. doi:10.1080/10494820701424585

Kanjo, E., Benford, S., Paxton, M., Chamberlain, A., Woodgate, D., & Stanton Fraser, D. (2007). MobGeoSen: Facilitating Personal GeoSensor Data Collection and Visualization using Mobile Phones. *Personal and Ubiquitous Computing Journal.* Springer ISSN 1617-4909 (Print) 1617-4917.

Kappel, G., Retschitzegger, W., & Schwinger, W. (2000). Modeling Customizable Web Applications – A Requirement's Perspective. In *Proceedings of the International Conference on Digital Libraries, (ICDL 2000),* Kyoto, Japan.

Kaptelinin, V. (1996). Activity theory: Implications for human computer interaction. In B. Nardi (Ed.), *Context and consciousness: Activity theory and human-computer interaction* (pp. 103-116). Cambridge, MA: The MIT Press.

Karagiannidis, C., Koumpis, A., & Stephanidis, C. (1996). Deciding 'What', 'When', 'Why', and 'How' to Adapt in Intelligent Multimedia Presentation systems. ECAI'96. *Workshop on Towards a standard reference model for intelligent multimedia presentation systems,* Budapest, Hungary.

Kato, H., Billinghurst, M., & Poupyrev, I. Imamotol, K., & Tachibana, K. (2000). Virtual object manipulation on a table-top AR environment. In *Proceedings. of the IEEE and ACM International Symposium on Augmented Reality 2000* (pp. 111-119).

Kazi, S. (2004). A conceptual framework for Web-based intelligent learning environments using SCORM 2004. In *Proc. of the IEEE International Conference on Advanced Learning Technologies 2004* (pp. 12–15), IEEE Computer Society.

Kearney, M., & Schuck, S. (2005). Students in the director's seat: Teaching and learning with student-generated video. In *Proceedings of World Conference on Educational Multimedia, Hypermedia and Telecommunications* (pp. 2864-2871), Montreal, Canada.

Kearney, P., Plax, T. G., & Wendt-Wasco, N. J. (1985). Teacher immediacy for affective learning in divergent college classes. *Communication Quarterly, 33*(1), 61.

Keegan, D. (2005, October). The incorporation of mobile learning into mainstream education and training. *MLearn 2005 conference*, Cape Town. Retrieved

Keil-Slawik, R., Hampel, T., & Esmann, B. (2005). Reconceptualizing learning environments: A framework for pervasive e-learning. In *Proceedings of the Third IEEE International Conference on Pervasive Computing and Communications Workshops* (pp. 322–326). IEEE Computer Society.

Kendon, A. (1967). Some functions of gaze-direction in social interaction. *Acta Psychologica, 26*(1), 22–63. doi:10.1016/0001-6918(67)90005-4

Kennedy, G., Krause, K., Judd, T., Churchward, A., & Gray, K. (2006). *First Year Students Experiences with Technology: Are they really Digital Natives*? Retrieved May 5, 2008, from http://www.bmu.unimelb.edu.au/research/munatives/natives_report2006.pdf

Khalil, C. I. J. (2001). *The Design of a Multimedia Adaptive Interface for Process Control Using a Multi-Agent Approach.* Unpublished doctoral dissertation, Computer Science, Loughborough University, UK.

Kiczales, G., Lamping, J., Mendhekar, A., Maeda, C., Lopes, C., Loingtier, J.-C., & Irwin, J. (1997). Aspect-Oriented Programming, Proceedings of the *European Conference on Object-Oriented Programming*, 1241, 220–242.

Kim, R. (2008) "Bar codes create bridge for window-shoppers." San Francisco Chronicle: March 27th 2008.

Kling, R. (1996) *Computerization and Controversy: Value Conflicts and Social Choices.*, San Diego, CA: Academic Press.

Klopfer, E., Squire, K., & Jenkins, H. (2002, August). Environmental Detectives: PDAs as a window into a virtual simulated world. *Proceedings of IEEE International Workshop on Wireless and Mobile Technologies in Education (WMTE'02)* (pp. 95-98).

Kobsa, A., Koenemann, J., & Pohl, W. (2001). Personalized Hypermedia Presentation Techniques for Improving Online Customer Relationships. *The Knowledge Engineering Review, 16*(2), 111–155. doi:10.1017/S0269888901000108

Kolb, D. A. (1984). *Experiential learning: Experience as the source of learning and development.* London: Prentice Hall.

Kondo, T. (2006). Augmented learning environment using mixed reality technology. *Japan Journal of Educational Technology, 29*(Suppl), 145–148.

Kondo, T., Shibasaki, J., & Arita, H. (2006). A proposal for a mixed reality museum display system. *Japan Journal of Educational Technology, 30*(Suppl.), 45–48.

Kong, S. C. (2008a). The development of a cognitive tool for teaching and learning fractions in the mathematics classroom: a design-based study. *Computers & Education, 51*(2), 886–899. doi:10.1016/j.compedu.2007.09.007

Kong, S. C. (2008b). Collaborative learning in a mobile technology supported classroom. In J. Lumsden (Ed.), *Handbook of research on user interface design and evaluation for mobile technology* (pp. 269-284). Hershey, PA: Information Science Reference.

Kong, S. C., & Kwok, L. F. (2002). Modeling a cognitive tool for teaching the addition/subtraction of common fractions. *International Journal of Cognitive Technology, 1*(2), 327–352. doi:10.1075/ijct.1.2.07kon

Kong, S. C., & Kwok, L. F. (2003). A graphical partitioning model for learning common fractions: Designing affordances on a web-supported learning environment. *Computers & Education, 40*(2), 137–155. doi:10.1016/S0360-1315(02)00118-5

Kong, S.C., & Kwok, L. F. (2005). A cognitive tool for teaching the addition/subtraction of common fractions: A model of affordances. *Computers & Education, 45*(2), 245–265. doi:10.1016/j.compedu.2004.12.002

Koper, R. (2005). *TENCompetence: Building the European Network for Lifelong Competence Development* (Project Draft). Heerlen: Open Universiteit Nederland.

Koper, R., & Tattersall, C. (2004). New directions for lifelong learning using network technologies. *British Journal of Educational Technology*, *35*(6), 689–700. doi:10.1111/j.1467-8535.2004.00427.x

Koper, R., & Tattersall, C. (2005). *Learning Design: A Handbook on Modelling and Delivering Networked Education and Training* (1st ed.). Heerlen: Springer-Verlag.

Koper, R., Olivier, B., & Anderson, T. (2003). *IMS Learning Design Information Model: IMS Global Learning Consortium.*

Koran, J. J. Jr, Koran, M. L., & Longino, S. J. (1986). The relationship of age, sex, attention, and holding power with two types of science exhibits. *Curator*, *29*, 227–235.

Koshinishi, M. (1996). Seeing to watching, and doing. *Kokugakuin Journal*, *11*, 96–100.

Kramer, B. J., & Strohlein, G. (2006). *Exploring the Use of Cellular Phones for Pervasive eLearning*. Paper present at 4th IEEE International Conference on Pervasive Computing and Communications Workshops, Pisa, Italy.

Kraus, S., Sycara, K., & Evenchik, A. (1998). Reaching agreements through argumentation: A logical model and implementation. *Artificial Intelligence*, *104*(1-2), 1–69. doi:10.1016/S0004-3702(98)00078-2

Kravcik, M., Kaibel, A., Specht, M., & Terrenghi, L. (2004). Mobile collector for field trips. *Educational Technology & Society*, *7*(2), 25–33.

Kurti, A., Milrad, M., & Spikol, D. (2007, July 18-20). Designing Innovative Learning Activities Using Ubiquitous Computing. *In Proceedings of the 7th IEEE International Conference on Advanced Learning Technologies*, (pp. 386-390), Niigata, Japan, ISBN 978-07695-2916-5.

Kutnick, P., Ota, C., & Berdondini, L. (2008). Improving the effects of group working in classrooms with young school-aged children: Facilitating attainment, interaction and classroom activity. *Learning and Instruction*, *18*(1), 83–95. doi:10.1016/j.learninstruc.2006.12.002

Kyttä, M. (2003). *Children in outdoor contexts: Affordances and independent mobility in the assessment of environmental childfriendliness.* (Doctor of Philosophy, Helsinki University of Technology).

Lange, D. B., & Oshima, M. (1999). Seven good reasons for mobile agents. *Communications of the ACM*, *42*(3), 88–89. doi:10.1145/295685.298136

Larman, C. (2004). *Agile and Iterative Development: A manager's guide*. Pearson Education.

Larson, H. P., Menke, R., Tollefson, S. J., Harrison, E., & Berman, E. (2001). The University of Arizona Teaching Teams Program: A "Just-in-Time" Model for Peer Assistance. In J. E. Miller, J. E. Grocia, & M. S. Miller (Eds.), *Student-Assisted Teaching: A Guide to Faculty-Student Teamwork* (pp. 27-33). Bolton, MA: Anker Publishing.

Lave, J., & Wenger, E. (1991). *Situated learning legitimate peripheral participation.* Cambridge, MA: Cambridge University Press.

Lehman, M. M. (1980). Programs, life cycles and laws of program evolution. *Proceedings of the IEEE*, *68*, 1060–1076. doi:10.1109/PROC.1980.11805

Lei, J., Conway, P. F., & Zhao, Y. (2007). *The digital pencil: One-to-one computing for children.* New York: Lawrence Erlbaum Associates.

Lemlouma, T., & Layaida, N. (2004). Context-aware adaptation for mobile devices. In *Proc. of the IEEE International Conference on Mobile Data Management* (pp. 106–111), IEEE Computer Society.

Leong, K. C., Chen, W. S., Leong, K. W., Mastura, I., Mim, O., & Sheikh, M. A. (2006). The use of text messaging to improve attendance in primary care: a randomized controlled trial. *Family Practice*, *23*(6), 699–705. doi:10.1093/fampra/cml044

Lester, J. C., Converse, S. A., Stone, B. A., Kahler, S. E., & Barlow, S. T. (1997). Animated pedagogical agents and problem-solving effectiveness: A large-scale empirical evaluation, *2330*.

Leuhn, C., & Chan, Y. (2003). Mobile learning: a new paradigm in electronic learning. In *Proc. of the IEEE International Conference on Advanced Learning Technologies 2003* (pp. 76–80), IEEE Computer Society.

LeWinter, B. Y. (2003). *Inspire.* In A. Mack-Kirschner (Ed.), *Powerful classroom stories from accomplished teachers* (pp. 137-139). Thousand Oaks, CA: Corwin Press.

Li, Q., Lau, R. W., Shih, T. K., & Li, F. W. (2008). Technology supports for distributed and collaborative learning over the internet. ACM Tranactions on Inter. Technology, 8(2).

Lickona, T. (1983). *Raising good children: Helping your child through the stages of moral development.* London: Bantam Books.

Limbourg, Q., & Vanderdonckt, J. (2004). UsiXML: A user interface description language supporting multiple levels of independence. In M. Matera & S. Comai (Eds.), *Engineering Advanced Web Applications* (pp. 325-338). Paramus: Rinton Press.

Ling, Z., Ying, Y., Qi, L., & Jia, Y. (2006). Research on personalized e-learning system by using mobile agents. In *Proceedings of the 1st International Symposium on Pervasive Computing and Applications* (pp. 311–315). IEEE Computer Society.

LIP(2001). IMS Learner Information Packaging Information Model Specification.

Lipman, M. (1991). *Thinking in education.* Cambridge: Cambridge University Press.

Lipponen, L. (2002). *Exploring foundations for computer-supported collaborative learning.* Retrieved October 12, 2007 from http://newmedia.colorado.edu/cscl/31.html

Littlejohn, A. (2003). Issues in reusing online resources. In A. Littlejohn (Ed.), *Reusing online resources: A sustainable approach to E-learning* (pp. 1-6)

Loidl, S. (2006). Towards pervasive learning: WeLearn. Mobile. A CPS package viewer for handhelds. *Journal of Network and Computer Applications, 29,* 277–293. doi:10.1016/j.jnca.2005.03.002

LOM. (2002). *Institute of Electrical and Electronics Engineers Learning Technology Standards Committee: IEEE standard for learning object metadata.* IEEE Standard 1484.12.1.

Lonchamp, J. (2006). Supporting synchronous collaborative learning: A generic, multi-dimensional model. *International Journal of Computer-Supported Collaborative Learning, 1*(2), 247–276. doi:10.1007/s11412-006-8996-7

Lopez, J., & Szekely, P. (2001). Automatic web page adaptation. *CHI2001-Workshop,* Transforming the UI for anyone anywhere, Seattle.

Lubega, J., McCrindle, R., Williams, S., Armitage, U., & Clements, I. (2004). Uses of mobile phones in higher education. Cantoni & McLaughlin (Eds.), In *Proceedings of EDMEDIA 2004* (pp. 3951-3956). Lugano, Switzerland.

Lymer, G., Lundin, J., Brown, B., Rost, M., & Holmquist, L. E. (2007). Web based platforms in co-located practice: The use of a Wiki as support for learning and instruction. In *Proceedings of Computer Supported Collaborative Learning 2007.* Lawrence Erlbaum Associates, p.

MacDonald, C.J., Stodel, E.J., Thompson, T-L., Muirhead, W., Hinton, C., Carson, B., & Banit, E. (2005). Addressing the eLearning contradiction: A collaborative approach for developing a conceptual framework learning object. Interdisciplinary Journal of knowledge and Learning Objects, 2, 79-98.

Madden, M., Chung, P. W. H., & Dawson, C. W. (2008). The effect of a computer-based cartooning tool on children's cartoons and written stories. *Computers & Education, 51*(2), 900–925. doi:10.1016/j.compedu.2007.09.008

Malinowski, U., Kuhme, T., Dietrich, H., & Schneider-Hufschmidt, M. (1992). A taxonomy of adaptive user interfaces. In Monk, Diaper, Harrison (Eds.), *People and Computers VII,* (pp. 391-414). Cambridge University Press.

Mandl, H., Gruber, H., & Renkl, A. (1995). Situiertes Lernen in multimedialen Lernumgebungen. In L. J. Issing & P. Klimsa (Eds.), *Information und Lernen mit Multimedia* (pp. 167-178). Weinheim: Psychologie Verlags Union.

Mann, B. L. (2006). *Selected styles in web-based educational research.* New York: Information Science Pub.

Mann, C., & Stewart, F. (2000). *Internet communication and qualitative research: A handbook for research online.* London: Sage Publications.

Mantovani, G. (1996). *New communication environments: From everyday to virtual.* London: Taylor & Francis.

Margain, L., Muñoz, J., & Álvarez, F.J. (2008). A *Methodology for Design Collaborative Learning Objects. ICALT 2008,* Santander, Spain.

Markett, C., Sanchez, A., Weber, S., & Tangney, B. (2006). Using Short Message Service to Encourage Interactivity in the Classroom. *Computers & Education, 46,* 280–293. doi:10.1016/j.compedu.2005.11.014

Markett, C., Sanchez, I. A., Weber, S., & Tangney, B. (2006). Using short message service to encourage interactivity in the classroom. *Computers & Education, 46*(3), 280–293. doi:10.1016/j.compedu.2005.11.014

MarketWatch. (2008, July 10). Fullpower Announces MotionX-Poker Available on Apple's App Store. *MarketWatch.*

Marsh, J., & Thompson, P. (2001). Parental involvement in literacy development: using media texts. *Journal of Research in Reading, 24*(3), 266–278. doi:10.1111/1467-9817.00148

Marshall, M. (2008, July 9). Tapulous wants to be leading application ecosystem for the iPhone. *VentureBeat.*

Martínez Ruiz, F. J., González Calleros, J. M., Guerrero García, J., Muñoz Arteaga, J., & Vanderdonckt, J. (2008, June 9-11). ECOOL: Genaration of Collaborative Multiplatform Scenarios with Interactive Learning Objects. In M. Lozano & J.A. Gallaud (Eds.), *Servicio de Publicaciones de la Universidad de Castilla - La Mancha* (pp.361-370), Albacete.

Massimi, M., Baecker, R. M., & Wu, M. (2007, October 15-17). Using participatory activities with seniors to critique, build, and evaluate mobile phones. In *Proceedings of the 9th international ACM SIGACCESS Conference on Computers and Accessibility* (pp. 155-162), Tempe, Arizona, USA. Assets '07. New York: ACM

Mattern, F. (2004, November). Wireless Future: Ubiquitous Computing. *Proceedings of Wireless Congress*, Munich, Germany.

Matthews, M. R. (1993). Constructivism and science education: Some epistemological problems. *Journal of Science Education and Technology, 2*(1), 359–370. doi:10.1007/BF00694598

Matthews, M. R. (Ed.). (1998). *Constructivism in science education: A philosophical examination.* Dordrecht: Kluwer.

Maurer, H. (1998). Web-based knowledge management. *Computer, 31*(3), 122–123. doi:10.1109/2.660197

McAdams, U. (2006). *SMS does SOS.* Retrieved 23 March 2008. from http://www.fcw.com/12_11/news/92790-1.htm, FLW staff.

McClatchey, S. (2006). The Consumption of Mobile Services by Australian University Students . *International Journal of Mobile Marketing, 1*, 35–42.

McClure, J. (2003). *How to find your dream job and make it a reality: Solutions for a meaningful and rewarding career.* Oxford: Trafford Publishing.

McCombs, B. L. (1991). Motivation and lifelong learning. *Educational Psychologist, 26*(2), 117–127. doi:10.1207/s15326985ep2602_4

McConnell, D. (2000). *Implementing computer supported cooperative learning* (2nd ed). London: Kogan Page.

McCroskey, J. C. (1994). Assessment of affect toward communication and affect toward instruction in communication. In M. Morreale & S. Brooks (Ed.), *1994 SCA Summer Conference Proceedings and Prepare Remarks* (pp. 55-65).

McCroskey, J. C., & Richmond, V. P. (1992). Increasing teacher influence through immediacy. *Power in the Classroom: Communication, Control, and Concern, 45*, 200–211.

McCroskey, J. C., Fayer, J. M., Richmond, V. P., Sallinen, A., & Barraclough, R. A. (1996). A multi-cultural examination of the relationship between nonverbal immediacy and affective learning. *Communication Quarterly, 44*(3), 297.

McCullagh, D. (2001, February 2). Call it Super Bowl Face Scan I. Wired Magazine.

McGriff, S. (2001). *ISD knowledge base. Pennsylvania State University, Portfolio Web site.* Retrieved June 6, 2006, from http://www.personal.psu.edu/faculty/s/j/sjm256/portfolio/kbase/IDD/ISDModels.html

McLaughlan, R. G., & Kirkpatrick, D. (2004). Online roleplay: Design for active learning. *European Journal of Engineering Education, 29*(4), 477–490. doi:10.1080/03043790410001716293

Mehrabian, A. (1969). Some referents and measures of nonverbal behaviour. *Behavior Research Methods and Instrumentation, 1*(6), 205–207.

Mehrabian, A. (1971). *Silent Messages.* Belmont, CA: Wadsworth.

Mehrabian, A. (1981). *Silent Messages* (2nd ed.). Belmont, CA: Wadsworth.

Mehrabian, A., & Williams, M. (1969). Nonverval concomitants of perceived and intended persuasiveness. *Journal of Personality and Social Psychology, 13*(1), 37–58. doi:10.1037/h0027993

Melear, C. T. (1998). *Increased learning of physical science concepts via multimedia exhibits compared to hands-on exhibits in a science museum.* Paper presented at the Annual Meeting of the National Association for Research in Science Teaching, San Diego, CA.

Meneses, F., & Moriera, A. (2006) Using GSM CellID positioning for place discovering. Presented at *First Workshop on Location Based Services for Health Care in the International Conference on Pervasive Computing Technologies for Healthcare, (Locare'06)* Innsbruck, Austria.

Merriman, J. (2008). Redefining interoprability. *The Open Knowledge Initiative (OKI)*. Retrieved May 5, 2008, from http://www.okiproject.org/view/html/node/2916

Metcalf, D. (2006). *mLearning: Mobile Learning and Performance Support in the Palm of your Hand*. Amherst, Massachussets: HRD Press.

Michotte, B., & Vanderdonckt, J. (2008, March 16-21). GrafiXML, A Multi-Target User Interface Builder based on UsiXML. In *Proceedings of 4th International Conference on Autonomic and Autonomous Systems (ICAS'2008)*, Los Alamitos: IEEE Computer Society Press

Miles, R. S. (1986). *The design of educational exhibits*. Routledge.

Miner, R. (1992). Reflections on teaching a large class. *Journal of Management Education, 16*(3), 290–302. doi:10.1177/105256299201600302

Mistry, P., & Maes, P. (2008). Quickies: Intelligent Sticky Notes. In *Proceedings of 4th International Conference on Intelligent Environments* (IE08). Seattle, USA.

Mobasher, B., Cooley, R., & Srivastava, J. (2000). Automatic Personnalization Based on Web Usage Mining. *Communications of the ACM, 43*(8), 142–151. doi:10.1145/345124.345169

Montclair State University Board of Trustees. (2002). Retrieved March 31, 2008, from http://www.montclair.edu/provost/strategicplan.pdf

Montgomery Community County College, MCCC (2006, October 6). Twelfth annual technology and learning conference. *Keynote presented by Marc Prensky on Overcoming Our Digital Immigrant Accents: Understanding the Implications of Emerging Digital Native Behaviors for Education, Training and Learning*. Retrieved October 21, 2006.

Morch, A. (1995). Three levels of end-user tailoring: customization, integration, and extansion. *Third Decennial Aarhus Conference* (pp. 41-51), Aarhus, Denmark.

Morch, A., & Mehandjiev, N. D. (2000). Tailoring as Collaboration: The Mediating Role of Multiple Representation and Application Units. *Computer Supported Cooperative Work (CSCW'00)*, (pp. 75-100), Philadelphia, Pennsylvania, USA.

Mortera-Gutiérrez, F. (2002). Instructor interactions in distance education environments. *Journal of Interactive Learning Research, 13*(3), 191–209.

Motiwalla, L. F. (2007). Mobile learning: A framework and evaluation. *Computers & Education, 49*(3), 581–596. doi:10.1016/j.compedu.2005.10.011

Mukherjee, D., Delfosse, E., Kim, J.-G., & Wang, Y. (2005). Optimal adaptation decision-taking for terminal and network quality-of-service. *IEEE Transactions on Multimedia, 7*(3), 454–462. doi:10.1109/TMM.2005.846798

Murphy (2003). *Report 7: Literature Review in Primary Science and ICT*. Bristol.

Murphy, C., Beggs, J., Hickey, I., O'Meara, J., & Sweeney, J. (2001). National Curriculum: compulsory school science – is it improving scientific literacy? *Educational Research, 43*(2), 189–199. doi:10.1080/00131880010021294

Murray, T., Blessing, S., & Ainsworth, S. (Eds.). (2003). *Authoring tools for advanced technology learning environments*. Dordrecht, Kluwer Academic Publishers.

Naismith, L., Lonsdale, P., Vavoula, G., & Sharples, M. (2004). *Literature Review in Mobile Technologies and Learning* (Literature Review No. 11). University of Birmingham.

Naismith, L., Lonsdale, P., Vavoula, G., & Sharples, M. (2004). *Report 11: Literature review in mobile technologies and learning, (No. 11): Nesta Futurelab*. Retrieved from http://www.nestafuturelab.org/download/pdfs/research/lit_reviews/futurelab_review_11.pdf.

Nakabayashi, K. (2004). E-Learning technology standardization – make it converge!! In *Proc. of the International Conference on Computers in Education 2004* (pp. 33–39), Asia-Pacific Society for Computers in Education.

Nakabayashi, K., Hoshide, T., Hosokawa, M., Kawakami, T., & Sato, K. (2007a). Design and implementation of a mobile learning environment as an extension of SCORM 2004 specifications. In *Proc. of the IEEE International Conference on Advanced Learning Technologies 2007* (pp. 369–373), IEEE Computer Society.

Nakabayashi, K., Hosokawa, M., Kawakami, T., & Sato, K. (2007b). Mobile learning environment with content download capability as an extension of SCORM 2004 specification, In *Suppl. Proc. of the 15th Intentional Conference on Computers in Education, Vol.1.* (pp. 45–47), Asia-Pacific Society for Computers in Education.

Nakabayashi, K., Koike, Y., Maruyama, M., Touhei, H., Ishiuchi, S., & Fukuhara, Y. (1995). A distributed intelligent-CAI system on the World-Wide Web. In *Proc. of the International Conference on Computers in Education 1995* (pp. 214–221), Asia-Pacific Society for Computers in Education.

Nakabayashi, K., Moroimoto, Y., & Hada, Y. (2008). Investigation into object oriented architecture for extensible learner-adaptive environment. In *Suppl. Proc. of the 16th Intentional Conference on Computers in Education,* (pp. 161–165), Asia-Pacific Society for Computers in Education.

Nakabayashi, K., Nakamura, A., Kosaka, Y., & Nagaoka, K. (2006). Design and implementation of SCORM 2004 execution engine and its performance evaluation. In *Proc. of the 2006 International Conference on SCORM 2004* (pp. 31–35).

Nakasugi, H., & Yamauchi, Y. (2002). Past viewer: The development of a wearable learning system for history education, Japan. *Journal of Educational Technology, 26*, 167–172.

Nardi, B. A. (1996a). *Context and consciousnesses: Activity theory and human-computer interaction.* Cambridge, Ma: The MIT Press.

Nardi, B. A. (1996b). Studying context: A comparison of activity theory, situated action models and distributed cognition. In B. A. Nardi (Ed.), *Context and consciousnesses: Activity theory and human-computer interaction.* (pp. 35-51). Cambridge, MA: MIT Press.

Nartin, D. (1998a). Multimedia case study: Manchester United Museum. *Museum Practice, 9*, 65–68.

Nartin, D. (1998b). Multimedia case study: Museum of Science and Industry in Manchester. *Museum Practice, 9*, 69–71.

New Media Consortium. NMC (2007). *2007 Horizon report.* Retrieved July 7, 2008, from http://www.nmc.org/horizon/2007/report

New Media Consortium. NMC (2008). *2008 Horizon report.* Retrieved July 7, 2008, from http://www.nmc.org/publications/2008-horizon-report

New Media Consortium. NMC. (2005). *A global imperative: The report of then21st century literacy summit* (Issue Brief No. 0-9765087-1-0). Stanford, CA

New York Times. (2007). Yahoo Introduces Mobile Service Software, late Edition New York Times, New York, N. Y. (9/01/2007)

Ng, W. (2008). Virtual teamwork: students learning about ethics in an online environment. *Journal of Research in Science & Technological Education, 26*(1), 13–29. doi:10.1080/02635140701847421

Ng, W., & Nicholas, H. (2007). Conceptualising the use of online technologies for gifted secondary students. *Roeper Review, 29*(3), 190–196.

Ng, W., & Nicholas, H. (2007). Ubiquitous Learning with Handheld Computers in Schools. In A. Norman & J. Pearce (Eds.), *Proceedings of MLearn Melbourne 2007: Making the Connections* (pp.186-193). Melbourne: University of Melbourne Press.

Northeast Connect. (2007). *2007 annual conference: Broadly exploring new directions in learning.* Retrieved July 7, 2008, from http://www.northeastconnect.org/?page_id=53

O'Hara, K., Kindberg, T., Glancy, M., Baptista, L., Sukumaran, B., Kahana, G., & Rowbotham, J. (2007). Collecting and Sharing Location-based Content on Mobile Phones in a Zoo Visitor Experience. *Computer Supported Cooperative Work (CSCW), 16*(1-2,34), 11-44. Springer

O'Malley, C., & Fraser, D. S. (2006). *Literature review in learning with tangible technologies.* Bristol: Futurelab. Retrieved June, 6, 2008 from http://palmsource.palmgear.com/index.cfm?fuseaction=software.category&catid=209&PartnerREF=&siteid=7

O'Malley, C., Vavoula, G., Glew, J. P., Taylor, J., Sharples, M., & Lefrere, P. (2003). *MOBIlearn project: Report of WP 4 "Guidelines for Learning/Teaching/Tutoring in a Mobile Environment.* Retrieved June 8, 2008 from http://www.mobilearn.org

Oblinger, D. (2004). Educating the net generation. *Keynote address delivered at Educause 2004.* Denver, CO. available at Oblinger, D. & Hawkins, B. (2005, September/October). The myth about students. *Educause Review.*

Ogata, H. (2008). Computer Supported Ubiquitous Learning: Augmenting Learning Experiences in the Real World. In *Proceedings Fifth IEEE International Confer-*

ence on *Wireless, Mobile, and Ubiquitous Technology in Education* (pp. 3-10). IEEE Computer Society.

Ogata, H., & Yano, Y. (2004). Context-aware support for computer supported ubiquitous learning. In J. Roschelle, T.W. Chan, Kinshuk, & S.Yang (Eds.), *Proceedings of the 2nd IEEE (WMTE2004)* (pp. 27-34). New York: IEEE Press.

Ohlsson, S. (1991). Knowledge requirements for teaching: The case of fractions. In P. Goodyear (Ed.), *Teaching knowledge and intelligent tutoring* (pp. 25-59). Norwood, NJ: Ablex Publishing Corporation.

Open Mobile Alliance. (2006). *Wireless Application Protocol*. Retrieved 24 July, 2008, from http://www.openmobilealliance.org/Technical/wapindex.aspx

Oppermann, R. (1994). Adaptively supported adaptability [Duluth, MN, USA: Academic Press Inc.]. *International Journal of Human-Computer Studies, 40*(3), 455–472. doi:10.1006/ijhc.1994.1021

Oreizy, P., Gorlick, M.M., & Taylor, R.N., Heimbigner, D., Johnson, G., Medvidovic, N., et al. (1999). An Architecture-Based Approach to Self-Adaptive Software. *IEEE Intelligent Systems, 14*(3), 54–62. doi:10.1109/5254.769885

Osborne, J., & Hennessy, S. (2003). *Report 6: Literature Review in Science Education and the Role of ICT: Promise, Problems and Future Directions*. Bristol: Nesta Futurelabs. www.nestafuturelab.org

Ozok, A. A., & Salvendy, G. (2004). Twenty guidelines for the design of web-based interfaces with consistent language. *Computers in Human Behavior, 20*(2), 149–161. doi:10.1016/j.chb.2003.10.012

Paiget, J. (1970). *The Science of Education and the Psychology of the Child*. New York: Grossman.

Pangaro, G., Maynes-Aminzade, D., & Ishii, H. (2002, October 27 - 30). The actuated workbench: Computer-controlled actuation in tabletop tangible interfaces. In *Proceedings of Symposium on User Interface Software and Technology (UIST '02)*, Paris, France.

Papazoglou, M. P., Traverso, P., Dustdar, S., & Leymann, F. (2007). Service-oriented computing: State of the art and research challenges. *Computer, 40*(11), 38–45. doi:10.1109/MC.2007.400

Paramythis, A. (2004). Towards Self-Regulating Adaptive Systems. In S. Weibelzahl & N. Henze (Eds.), Proceedings of the Annual *Workshop of the SIG Adaptivity and User Modeling in Interactive Systems of the German Informatics Society* (ABIS04), (pp. 57-63), Berlin.

Paredes, R. G. J., Ogata, H., Saito, N. A., Yin, C., Yano, Y., Oishi, Y., & Ueda, T. (2005). LOCH: Supporting Informal Language Learning Outside the Classroom with Handhelds. *Proceedings of IEEE International Workshop on Wireless and Mobile Technologies in Education (WMTE '05)* (pp. 182-186).

Parsons, D., Ryu, H., & Cranshaw, M. (2006). A Study of Design Requirements for Mobile Learning Environments. In *Proceedings of the Sixth IEEE International Conference on Advanced Learning Technologies* (ICALT'06) (pp. 96-100).

Pask, G. (1975). *Conversation, Cognition, and Learning*. New York Elsevier.

Pask, G. (1976). *Conversation Theory: Applications in Education and Epistemology*. Amsterdam and New York: Elsevier.

Paternò, F. (1999). Model-based design and evaluation of interactive applications. *Applied Computing*. Springer: Heidelberg.

Paternò, F., & Santoro, C. (2002). One Model, Many Interfaces. *International conference Computer-Aided Design of User Interfaces, CADUI'02*, Valenciennes, France.

Patten, B., Arnedillo Sánchez, I., & Tangney, B. (2006). Designing collaborative, constructionist and contextual applications for handheld devices. *Computers & Education, 46*(3), 294–308. doi:10.1016/j.compedu.2005.11.011

Pavlik Jr, P. I., & Anderson, J. R. (2005). Practice and Forgetting Effects on Vocabulary Memory: An Activation-Based Model of the Spacing Effect. *Cognitive Science: A Multidisciplinary Journal, 29*(4), 559-586.

Pavlov, I. P. (1927). *Conditioned reflexes*. London: Routledge and Kegan Paul.

Pea, R. D. (2002). Learning Science through Collaborative Visualization over the Internet. Nobel Symposium (NS 120), 'Virtual Museums and Public Understanding of Science and Culture. May 26-29 2002, Stockholm, Sweden. http://nobelprize.org/nobel/nobel-foundation/symposia/interdisciplinary/ns120/lectures/pea.pdf

Petersen, S. A., & Divitini, M. (2005). Language Learning: From Individual Learners to Communities. *Proceedings of IEEE International Workshop on Wireless and Mobile Technologies in Education (WMTE'05)* (pp. 169-173).

Petropoulakis, L., & Flood, F. (2008). Interactive student engagement using wireless handheld devices. *International Journal of Engineering Education, 24*(1), 46–55.

Petrushin, V. A. (2000). Emotion recognition agents in real world. *AAAI Fall Symposium on Socially Intelligent Agents: Human in the Loop,* Rasmussen, J., Duncan, K., & Leplat, J. (1987). *New technology and human error* John Wiley & Sons Inc.

Pettit, J., & Kukulska-Hulme, A. (2007). Going with the Grain: Mobile Devices in Practice. *Australasian Journal of Educational Technology, 23*(1), 17–33.

Phillips, D. (Ed.). (2000). *Constructivism in education: Opinions and second opinions on controversial issues.* Chicago: University of Chicago Press.

Phillips, D. C. (1995). The good, the bad, and the ugly: The many faces of constructivism. *Educational Researcher, 24*(7), 5–12.

Piaget, J. (1955). *The construction of reality in the child.* London: Routledge & Keegan Paul.

Piaget, J. (1972). *Psychology and epistemology: Towards a theory of knowledge.* London: Penguin

Picard, R. W., Strohecker, C., Papert, S., Bender, W., Blumberg, B., & Breazeal, C. (2004). Affective learning - A manifesto. *BT Technology Journal, 22*(4), 253–269. doi:10.1023/B:BTTJ.0000047603.37042.33

Platts, B. (2004). *ATN Online Learning Surveys Project 2004.* Online Learning Student Survey Report, Combined ATN Universities: Institute for Interactive Media and Learning, University of Technology Sydney. Lawrence, E., Bachfischer, A., Dyson, L. E., & Litchfield, A. (2008). Mobile Learning and Student Perspectives: An mReality Check! *7th International Conference on Mobile Business,* (pp. 287-295). Barcelona, Spain.

Plax, T. G., Kearney, P., McCroskey, J. C., & Richmond, V. P. (1986). Power in the classroom VI: Verbal control strategies, nonverbal immediacy and affective learning. *Communication Education, 35*(1), 43.

P-LearNet project. Pervasive Learning Network, Retrieved November 17, 2008, from http://p-learnet.univ-lille1.fr/

Pogue, L. L., & Ahyun, K. (2006). The effect of teacher nonverbal immediacy and credibility on student motivation and affective learning. *Communication Education, 55*(3), 331–344.doi:doi:10.1080/03634520600748623

Posner, I., Baecker, R., & Homer, B. (1997). Children learning filmmaking using multimedia tools. In *Proceedings of ED-MEDIA/ED-TELECOM,* Calgary, Canada.

Preece, J. (2000). Online communities: Designing usability and supporting socialability. John Wiley & Sons, Inc.

Prensky, M. (2001). *Digital Game-Based Learning.* New York, NY: McGraw-Hill.

Prensky, M. (2001). Digital natives, Digital immigrants. *On the Horizon, 9*(5). MCB University Press.

Prensky, M. (2004). *The emerging online life of the digital native.* Retrieved October 21, 2008, from http://www.marcprensky.com/writing/Prensky-The_Emerging_Online_Life_of_the_Digital_Native-03.pdf

Prensky, M. (2005, September/October). Engage me or enrage me: What today's learners demand. EDUCAUSE Review, 40(5), 60–64.

Prensky, M. (2008). *Mark Prensky writing.* Retrieved Juny 3, 2008 from http://www.marcprensky.com/writing/default.asp

Raaijmakers, J. G. W. (2003). Spacing and repetition effects in human memory: Application of the SAM model. *Cognitive Science, 27,* 431–452.

Ratner, R. K., & Fitzsimons, G. J. (2002). When will consumers act on their attitudes?: New directions in attitude-behavior consistency: Special session summary. *Advances in Consumer Research. Association for Consumer Research (U. S.), 29*(1), 139–142.

Ratti, C., Wang, Y., Ishii, H., Piper, B., & Frenchman, D. (2004). Tangible user interfaces (TUIs): A novel paradigm for GIS. *Transactions in GIS, 8*(4), 407–421. doi:10.1111/j.1467-9671.2004.00193.x

Rave Wireless. (2008). *Rave Campus.* Retrieved July 7, 2008, from http://ravewireless.com/campus.shtml

Reddy, M., & Lewis, S. F. (2002). Building aunt hillary: Creating artificial minds with 'Neural nests'. *Proceed-*

ings of the 3rd International Workshop on Ant Colony Optimization, Ants 2002, Brussels, Belgium.

Reid, M., Burn, A., & Parker, D. (2002). *Evaluation report of the Becta Digital Video Pilot Project: Becta.* Retrieved, from http://www.becta.org.uk/research/reports/digitalvideo/

Reisman, S. (2006, January/February). The Ivory Tower–Failing to Communicate in an Educational Environment. New York: IEEE Computer Society. *IT Professional, 8*(1), 62–66.doi:doi:10.1109/MITP.2006.53

Resta, P., & Laferrière, T. (2007). Technology in support of collaborative learning. *Educational Psychology Review, 19*(1), 65–83. doi:10.1007/s10648-007-9042-7

Reznik, M. (2000). *Learning with digital manipulatives: New frameworks to help elementary-school students explore "advanced" mathematical and scientific concepts.* Proposal to the National Science Foundation. Brave, S. (1997). InTouch: A medium for haptic interpersonal communication. In *Proceedings of the Conference on Human Factors in Computing Systems (CHI '97)*, ACM Press.

Rheingold, H. (2002). *Smart Mobs: The Next Social Revolution*: Perseus Books Group.

Richardson, J., & Lenarcic, J. (2007a, December 5-7). Trigger iz gr8 4 Gen Y: mobile Student Administration via a Text Messaging System. *Australasian Conference on Information Systems (ACIS)*, (pp. 991-998), Toowomba, Australia Richardson, J, & Lenarcic, J. (2007, June 4-6). E-inclusion through text messaging: The emergence of an admisnistrative ecology within a university student population via the use of a mobile academic information delivery system. *20th Bled eConference*, Bled, Slovenia.

Richardson, J., Lenarcic, J., McKay, E., & Craig, C. (2008, March 17-19). Pull with Pearson's Trigger Word Tool to Enable Simplexity: Student Empowerment via a Text Messaging System for Mobile Administration. *IASTED International Conference on Web-Based Education*, Innsbruck, Austria

Rocca, K. A. (2004). College student attendance: impact of instructor immediacy and verbal aggression. *Communication Education, 53*(2), 185–195.doi:doi:10.1080/03634520410001682447

Rodden, K. (2001). *Evaluating similarity-based visualizations as interfaces for image browsing.* Unpublished doctorial dissertation, University of Cambridge Computer Laboratory.

Rodden, K., & Wood, K. (2003). How do people manage their digital photographs? In *Proceedings of the SIGCHI conference on Human factors in computing systems* (pp. 409-416). ACM press.

Rodriguez, J. I. (1996). Clarifying the relationship between teacher nonverbal immediacy and student cognitive learning: Affective learning as the central causal mediator. *Communication Education, 45*(4), 293–305.

Roschelle, J. (2003). Keynote paper: Unlocking the learning value of wireless mobile devices. *Journal of Computer Assisted Learning, 19*(3), 260–272. doi:10.1046/j.0266-4909.2003.00028.x

Roschelle, J. (2003). Unlocking the learning value of wireless mobile devices. Journal of Computer Assisted Learning, 19(3), 260–272. doi:10.1046/j.0266-4909.2003.00028.xdoi:10.1046/j.0266-4909.2003.00028.x

Roschelle, J., & Pea, R. (2002). A walk on the WILD side: How wireless handhelds may change computer-supported collaborative learning. *International Journal of Cognitive Technology, 1*(1), 145–168. doi:10.1075/ijct.1.1.09ros

Roschelle, J., & Teasley, S. D. (1995). The construction of shared knowledge in collaborative problem solving. In C. O'Malley (Ed.), *Computer-supported collaborative learning* (pp. 145-168). Berlin: Springer-Verlag.

Roschelle, J., Sharples, M., & Chan, T. W. (2005). Introduction to the Special Issue on Wireless and Mobile Technologies in Education. [Blackwell Publishing.]. *Journal of Computer Assisted Learning, 21*, 159–161. doi:10.1111/j.1365-2729.2005.00123.x

Rosenberg, M. (2001). *The personalization story, ITworld.com.* Retrieved November 17, 2008, from http://www.itworld.com/Man/2676/ITW010511rosenberg/

Rossi, G., Schwabe, D., & Guimarães, R. (2001). Designing Personalized Web Applications. *10th International World Wide Web Conference (WWW10)*, Hong Kong.

Rouillard, J. (2003). HCII 2003. *Plastic ML and its toolkit*, Heraklion, Crete, Greece.

Rouillard, J. (2008). Contextual QR Codes. *Third International Multi-Conference on Computing in the Global Information Technology, ICCGI 2008*, Athens, Greece.

Rushby, N. J. (1979a). *Computers in the teaching process.* New York: John Wiley & Sons, Inc.

Rushby, N. J. (1979b). *An introduction to educational computing.* Towbridge: Redwood Burn Limited.

Russell, G., & Pitt, I. (2004). Visions of a wireless future in educational technology. In *Proceedings of Informing Science and Information Technology Education Joint Conference* (pp. 747-752). Rockhampton QLD, Australia.

Ruta, M., Di Noia, T., Di Sciascio, E., & Donini, F. M. (2006). Semantic-enhanced Bluetooth discovery protocol for m-commerce applications. *International Journal of Web and Grid Services, 2*(4), 424–452. doi:10.1504/IJWGS.2006.011713

Ruta, M., Di Noia, T., Di Sciascio, E., & Donini, F. M. (2007b). Semantic based collaborative p2p in ubiquitous computing. *Web Intelligence and Agent Systems, 5*(5), 375–391.

Ruta, M., Di Noia, T., Di Sciascio, E., Scioscia, F., & Piscitelli, G. (2007a). If Objects Could Talk: A novel resource discovery approach for pervasive environments. *International Journal of Internet and Protocol Technology, 2*(3/4), 199–217. doi:10.1504/IJIPT.2007.016221

Ryans, D. G. (1964). Research on teacher behavior in the context of the teacher characteristics study. *Contemporary Research on Teacher Effectiveness.*

Scaife, M., Rogers, Y., Aldrich, F., & Davies, M. (1997, March 22-27). Designing For or Designing With? Informant Design for Interactive Learning Environments. *Proc. CHI '97* (pp. 343-350), Atlanta, GA, USA: ACM.

Scanlon, E., Jones, A., & Waycott, J. (2005). Mobile Technologies: prospects for their use in learning in informal science settings. In A.Jones, A. Kukulska-Hulme, & D. Mwanza (Eds.), *Journal of Interactive Media in Education, special issue on Portable Learning: Experiences with Mobile Devices.* [jime.open.ac.uk/2005/25]

Scapin, D. L., & Bastien, J. M. C. (1997). Ergonomic criteria for evaluating the ergonomic quality of interactive systems. *Behaviour & Information Technology, 6*(4-5), 220–231.

Schenker, J., Kratcoski, A., Lin, Y., Swan, K., & van 't Hooft, M. (2007). In M. van 't Hooft & M. Swan (Eds.), Researching ubiquity: ways to capture it all. In *Ubiquitous computing in education: invisible technology, visible impact* (pp. 167-186), Mahwah, NJ: Lawrence Erlbaum Associates.

Schilit, B. N., Adams, N. I., & Want, R. (1994). Context-aware computing applications, In *Proc. of the IEEE Workshop on Mobile Computing Systems and Applications* (pp. 85–90), IEEE Computer Society.

Schon, D. (1983). *Reflective Practitioner.* Basic Books.

Schon, D. (1987). *Educating the Reflective Practitioner: Towards a New Design for Teaching and Learning in the Professions.* Jossey-Bass Inc.

Schön, D. A. (1983). *The Reflective Practitioner: How Professionals think in Action.* London: Maurice Temple Smith.

Schön, D. A. (1987). *Educating the Reflective Practitioner.* San Francisco: Jossey-Bass.

SCORM. (2004). Advanced Distributed Systems (ADL) lab, Sharable Content Object Reference Model.

Scott, S. D., Mandryk, R. L., & Inkpen, K. M. (2003). Understanding children's collaborative interactions in shared environments. *Journal of Computer Assisted Learning, 19*(2), 220–228. doi:10.1046/j.0266-4909.2003.00022.x

Sebeok, T. A. (2001). Signs an introduction to semiotics, *Toronto studies in semiotics and communication.* University of Toronto Press, Toronto, Canada.

Sedig, K., & Liang, H. N. (2006). Interactivity of visual mathematical representations: factors affecting learning and cognitive processes. *Journal of Interactive Learning Research, 17*(2), 179–212.

Sedig, K., & Sumner, M. (2006). Characterizing interaction with visual mathematical representations. *International Journal of Computers for Mathematical Learning, 11*(1), 1–55. doi:10.1007/s10758-006-0001-z

Sefton-Green, J. (2004). *Report 7: Literature Review in Informal Learning with Technology Outside School.* Bristol: Nesta Futurelabs. www.nestafuturelab.org

Severance, C., Hardin, J., & Whyte, A. (2008). The coming functionality mashup in personal learning environments. *Interactive Learning Environments (Journal), 16*(1), 1–15. doi:10.1080/10494820701772645

Sharples, M. (2002). Disruptive devices: mobile technology for conversational learning. *International Journal of Continuing Engineering Education and Lifelong Learning, 12*(5/6), 504–520. doi:10.1504/IJCEELL.2002.002148

Sharples, M. (2006). *Big Issues in Mobile Learning: Report of a Workshop by the Kaleidoscope Network of Excellence Mobile Learning Initiative.* University of Nottingham 2006.

Sharples, M. (Ed.). (2007). *Big issues in mobile learning. Report of a workshop by the Kaleidoscope Network of Excellence Mobile Learning Initiative.* Nottingham, UK: University of Nottingham, Learning Sciences Research Institute.

Sharples, M., Jeffery, N., du Boulay, J. B. H., Teather, D., Teather, B., & du Boulay, G. H. (2002, January 16). Socio-cognitive engineering: A methodology for the design of human-centred technology. *European Journal of Operational Research, 136*(2), 310-323. Retrieved from http://dx.doi.org/10.1016/S0377-2217(01)00118-7

Sharples, M., Lonsdale, P., Meek, J., Rudman, P. D., & Vavoula, G. N. (2007). *An Evaluation of MyArtSpace: a Mobile Learning Service for School Museum Trips.* Paper presented at the conference mLearn 2007 conference, Melbourne.

Sharples, M., Lonsdale, P., Meek, J., Rudman, P., & Vavoula, G. (2007, October 16-19). An evaluation of MyArtSpace: A mobile learning service for school museum trips.In *Proceedings of mLearn* (pp. 238-244), Melbourne, Australia.

Sharples, M., Taylor, J., & Vavoula, G. (2005). Towards a Theory of Mobile Learning. In *Proceedings of mLearn 2005 Conference, 4th World conference on mLearning: Mobile technology: The future of learning in your hands,* Capetown.

Sharples, M., Taylor, J., & Vavoula, G. N. (2007). A Theory of Learning for the Mobile Age. In R. Andrews & C. Haythornthwaite (Eds.), *The SAGE Handbook of E-learning Research* (pp. 221-247). London: Sage.

Shenk, D. (1997). *Data smog: surviving the information glut.* San Francisco, CA: Harper Edge.

Shepherd, R., Story, J., & Mansoor, S. (2004). Parallel computation in mobile systems using Bluetooth scatternets and Java. In *Proceedings of the International Conference on Parallel and Distributed Computing and Networks.* Retrieved July 20, 2008, from http://www.informatics.bangor.ac.uk/~robshep/publications/DynaMP.pdf

Shih, T. K., Lin, N. H., Chang, W., Wang, T., Wen, H., & Yang, J. (2005). The hard SCORM: reading SCORM courseware on hardcopy textbooks. In *Proc. of the IEEE International Conference on Advanced Learning Technologies 2005* (pp. 812–816), IEEE Computer Society.

Simpson, R. (1998). Multimedia special: Effective audio-visual. *Museum Practice, 9,* 44–46.

Singh, D., & Baker, Z. A. (2007). Wireless implementation of a mobile learning environment based on students' expectations. *International Journal of Mobile Learning and Organisation, 1*(2), 198–215. doi:10.1504/IJMLO.2007.012678

Singh, H. (2003). Building effective blended learning programs. Educational Technology, 43(6), 51–54.

Sklar, E., Parsons, S., & Davies, M. (2004). When is it okay to lie? A simple model of contraditcion in agent-based dialogues. *Proceedings of the First Workshop on Argumentation in Muliagent Systems,* Suchman, L. A. (1987). *Plans and situated actions: The problem of human-machine communication* Cambridge University Press.

Slack, F., & Rowley, J. (2002). Online kiosks: the alternative to mobile technologies for mobile users . *Internet Research: Electronic Networking Applications and Policy, 12*(3), 248–257. doi:10.1108/10662240210430928

Slavin, R. E. (1996). Research on cooperative learning and achievement: What we know, what we need to know. *Contemporary Educational Psychology, 21*(1), 43–69. doi:10.1006/ceps.1996.0004

Sloan, H. L., & Delahoussaye, C. P. (2003). Clinical application of the Omaha system with the Nightingale Tracker: a community health nursing student home visit program. *Nurse Educator, 28*(1), 15–17. doi:10.1097/00006223-200301000-00008

Sloodle. (2008). Retrieved May 10, 2008, from http://www.sloodle.org van der Aalst, W., & van Hee, K. (2002): *Workflow Management: Models, Methods, and Systems.* Cambridge: The MIT Press.

Soller, A., Lesgold, A., Linton, F., & Goodman, B. (1999). *What Makes Peer Interaction Effective? Modeling Effective Communication in an Intelligent CSCL.* Paper presented at the AAAI Fall Symposium: Psychological Models of Communication in Collaborative Systems, North Falmouth, MA, USA.

Solomon, J. (1994). The rise and fall of constructivism. *Studies in Science Education, 23,* 1–19. doi:10.1080/03057269408560027

Soloway, E., Norris, C., Blumenfeld, P., Fishman, B., Krajcik, J., & Max, R. (2001). Log on education: Handheld devices are ready-at-hand. *Communications of the ACM*, *44*(6), 15–20. doi:10.1145/376134.376140

Sosnovsky, S., Dolog, P., Henze, N., Brusilovsky, P., & Nejdl, W. (2007). Translation of overlay models of student knowledge for relative domains based on domain ontology mapping. In *Proc. of the 13th Int. Conf. Artificial Intelligence in Education* (pp. 289–296), IOS Press.

Specht, M., & Zimmermann, A. (2006). An Architecture for Contextualized Learning Experiences. Proceedings of the *Sixth IEEE International Conference on Advanced Learning Technologies (ICALT)*, (pp. 169-173). ISBN:0-7695-2632-2.

Spikol, D. (2007, December). Designing mobile games that explore novel learning practices with co-design. In *Proceedings of Research Methods in Informal and Mobile Learning (MIL-RM)* (pp. 41-47), Institute of Education, University of London, UK.

Spiro, R. J., & Jehng, J. (1990). Cognitive flexibility and hypertext: Theory and technology for the non-linear and multidimensional traversal of complex subject matter. In D. Nix & R. Spiro (Eds.), *Cognition, Education, and Multimedia*. Hillsdale, NJ: Erlbaum.

Spiro, R. J., Feltovich, P. J., Jacobson, M. J., & Coulson, R. L. (1992). Cognitive flexibility, constructivism and hypertext: Random access instruction for advanced knowledge acquisition in ill-structured domains. In T. Duffy & D. Jonassen (Eds.), *Constructivism and the Technology of Instruction*. Hillsdale, NJ: Erlbaum.

Stacey, E. (1999). Collaborative learning in an online environment. *Journal of Distance Education*, *14*(2), 14–33.

Stanton Fraser, D., Smith, H., Tallyn, E., Kirk, D., & Benford, S. Rowland, D., Paxton, M., Price, S., & Fitzpatrick, G. (2005). *The Sense Project: a context-inclusive approach to studying environmental science within and across schools*. www.equator.ac.uk/index.php?module=upload&func-download&fileID-227

Steffe, L. P., & Olive, J. (1996). Symbolizing as a constructive activity in a computer micro-world. *Journal of Educational Computing Research*, *14*(2), 113–138.

Steinkuelhr, C. A., Derry, S. J., Woods, D. K., & Hmelo-silver, C. E. (2002). The STEP environment for distributed problem based learning on the World Wide Web. In *Proceedings of Computer Support for Collaborative Learning (CSCL 2002)*.

Stephanidis, C., Paramythis, A., Sfyrakis, M., Stergiou, A., Maou, N., Leventis, A., et al. (1998). Adaptable and adaptive user interfaces for disabled users in AVANTI Project. In S. Triglia, A. Mullery, M. Campolargo, H. Vanderstraeten, & M. Mampaey, (Eds.), *Intelligence in Services and Networks: Technology for Ubiquitous Telecom Services* (pp. 153–166). Springer.

Sticht, T. G. (1975). Applications of the audread model to reading evaluation and instruction. In L. Resnick & P. Weaver (Eds.), *Theory and Practice of Early Reading* (Vol. 1). Hillsdale, NJ: Erlbaum.

Stone, A. (2004). *Mobile scaffolding: An experiment in using SMS text messaging to support first year university students*. Paper presented at the IEEE International Conference on Advanced Learning Technologies, ICALT 2004, Joensuu, Finland.

Storey, M.-A., Čubranić, D., & German, D. (2005). On the use of visualization to support awareness of human activities in software development: A survey and a framework. In *Proceedings of the 2005 ACM symposium on Software visualization* (pp. 193-202). ACM press.

Svanaes, D., & Seland, G. (2004, April 24-29). Putting the users center stage: role playing and low-fi prototyping enable end users to design mobile systems. In *Proceedings of the SIGCHI Conference on Human Factors in Computing Systems* CHI '04 (pp. 479-486), Vienna, Austria. New York: ACM.

Swain, C., Sharpe, R., & Dawson, K. (2003). Using digital video to study history. *Social Education*, *67*(3), 154–158.

Tatar, D., Roschelle, J., Vahey, P., & Penuel, R. W. (2003). Handhelds go to school:Lessons learned. *SRI International* (pp. 30-37).

Taylor, J., Sharples, M., O'Malley, C., Vavoula, G., & Waycott, J. (2006). Towards a Task Model for Mobile Learning: a Dialectical Approach. *International Journal of Learning Technology*, *2*(2/3), 138–158. doi:10.1504/IJLT.2006.010616

Terrenghi, L., Specht, M., & Moritz, S. (2004). RAFT. A role-based appliance for collaborative learning. *Proceedings of 2AD:Second International Conference on Appliance Design*, Bristol, UK: 2AD.

The Age (2008). *Life & Times*. 6/04/2008.

The New Media Consortium & EDUCAUSE Learning Initiative, The Horizon Project 2006.

The New Media Consortium & EDUCAUSE Learning Initiative, The Horizon Project 2007.

Thevenin, D. (2001), *Adaptation en Interaction Homme-Machine: Cas de la plasticité*, Unpublished doctoral dissertation, Université Joseph Fourier, Grenoble, France.

Thevenin, D., & Coutaz, J. (1999). Plasticity of user-interfaces: framework and research agenda. In *Proceedings of 7th IFIP conference on human-computer interaction, INTERACT 1999*, (pp. 110-117), Edinburgh, Scotland.

Thornton, P., & Houser, C. (2004). Using mobile phones in education. In *Proc. of the International Workshop on Wireless and Mobile Technologies in Education 2004* (pp. 3–9), IEEE Computer Society.

Totterdell, P., & Rautenbach, P. (1990). Adaptation as a Problem of Design. In: Browne, D., Totterdell, P., Norman, M., (Eds.): *Adaptive User Interfaces*. Academic say, Press, London, (pp. 5984).

Traxler, J. (2005). Mobile Learning: it's here, but what is it? *Interactions, 9*(1). University of Warwick.

Trigg, R., Moran, T., & Halasz, F. (1987). Adaptability and Tailorability in NoteCards. In Bullinger & Shackel (Eds.), *Proc. INTERACT `87*, North-Holland.

Tunstall, D. F. (1995). *Is your school ready for the 21st century? A general overview for planning purposes* (NCES Publication No. 383-443). Washington, DC U.S. Department of Education (2003). *Distance education at degree-granting postsecondary institutions: 2000-2001* (NCES Publication No. 2003-017). Washington, DC

Uden, L. (2007). Activity theory for designing mobile learning. *International Journal of Mobile Learning and Organisation, 1*(1), 81–102. doi:10.1504/IJMLO.2007.011190

Ullmer, B., & Ishii, H. (2005). Token+constraint systems for tangible interaction with digital information. ACM Trans. *Computer-Human Interact, 12*(1), 81–118. doi:10.1145/1057237.1057242

University Books Roschelle. J., Patton, C., & Tatar, D. (2007). Designing networked handheld devices to enhance school learning. In *Advances in Computers, 70*, 1-60.

Valdez, G., McNabb, M., Foertsch, M., Anderson, M., Hawkes, M., & Raack, L. (2004). *Computer based technology and learning: Evolving uses and expectations.* Retrieved March 8, 2007, from North Central Regional Educational Laboratory Web site: http://www.ncrel.org/tplan/cbtl/toc.htm

Vanderdonckt, J. (2005, June 13-17), A MDA-compliant environment for developing user interfaces of information systems. *Proc. of 17th Conf. on Advanced Information Systems Engineering (CAiSE'05)*, Porto (LNCS, 3520, 16-31). Berlin: Springer-Verlag.

Vanderdonckt, J., Grolaux, D., & Van Roy, P. Limbourg, Macq, Q., & Michel, B. (2005). A Design Space For Context-Sensitive User Interfaces. *International Conference on Intelligent and Adaptive Systems and Software Engineering (IASSE'2005)*, (pp. 207-214). Toronto, Canada.

Vavoula, G. (2007). *Evaluating mobile learning.*

Vavoula, G. N., & Sharples, M. (2002). KLeOS: a personal, mobile, knowledge and learning organisation system. *Proceedings of IEEE International Workshop on Wireless and Mobile Technologies (WMTE '02)* (pp. 152-156).

Vavoula, G. N., & Sharples, M. (2002). KLeOS: A personal, mobile, knowledge and learning organisation system. In M. Mildrad, U. Hoppe, & Kinshuk (Eds.), *Proceedings of the IEEE International Workshop on Mobile and Wireless Technologies in Education (WMTE2002)*, (pp.152-156), Vaxjo, Sweeden.

Vavoula, G., & Sharples, M. (2008, October 7-10). Challenges in evaluating mobile learning. In *Proceedings of the MLearn2008 Conference: The Bridge From Text To Context* (pp. 296-303), University of Wolverhampton, UK.

Vavoula, G., Meek, J., Sharples, M., Lonsdale, P., & Rudman, P. (2006). *A lifecycle approach to evaluating MyArtSpace*.

Vetro, A., & Timmerer, C. (2005). Digital item adaptation: overview of standardization and research activities. *IEEE Transactions on Multimedia, 7*(3), 418–426. doi:10.1109/TMM.2005.846795

Viégas, F., Wattenberg, M., & Dave, K. (2004). Studying cooperation and conflict between author with history flow visualizations. In *Proceedings of the SIGCHI conference on Human factors in computing systems* (pp. 575-582). ACM press.

Voice, X. M. L. Forum. Retrieved November 17, 2008, from http://www.voicexml.org

von Glasersfeld, E. (1995). *Radical constructivism: A way of knowing and learning.* London: Falmer Press.

von Hessling, A., Kleemann, T., & Sinner, A. (2004). Semantic User Profiles and their Applications in a Mobile Environment. In *Proceeding of Artificial Intelligence in Mobile Systems'04* (UbiComp'04).

Vygotsky, L. (1978). *Mind in society: The development of higher psychological processes.* Cambridge, M: Harvard University Press.

Vygotsky, L. S. (1930). *Mind in society.* Cambridge, MA: Webb, N. M. (1982). Student interaction and learning in small groups. *Review of Educational Research, 52*(1982), 421-445.

Vygotsky, L. S. (1962). *Thought and language.* Cambridge, MA: MIT Press.

W3C Recommendation. *OWL Web Ontology Language. Retrieved from: http://*www.w3.org/TR/owl-features/, February 10, 2004.

Wadsworth, P. (2000). What's happening to science? *Primary Science Review, 61*, 31–32.

Wagner, M., & Gansemer-Topf, A. (2005). Learning by teaching others: A qualitative study exploring the benefits of peer teaching . *Landscape Journal, 24*(2), 198–208. doi:10.3368/lj.24.2.198

Wainwright, K., Osterman, M., Finnerman, C., & Hill, B. (2007, October 7-10). Traversing the LMS terrain. In *Proceedings of the 35th Annual ACM Conference on User Services SIGUCCS'2007* (pp. 355–359), Orlando. New York: ACM Press

Wang, A. I., Bjornsgard, T., & Saxlund, K. (2007). Peer2Me – Rapid application framework for mobile peer-to-peer applications. In *International Symposium on Collaborative Technologies and Systems.* Retrieved July 20, 2008, from http://www.idi.ntnu.no/grupper/su/publ/alfw/peer2me-cts2007.pdf

Wang, C., Liu, B., Chang, K., Horng, J., & Chen, G. (2003). Using mobile techniques in improving awareness to promote learning performance. In *Proc. of the IEEE International Conference on Advanced Learning Technologies 2003*, IEEE Computer Society.

Weal, M. J., Cruickshank, D. G., Michaelides, D. T., Millard, D. E., De Roure, D. C., Halloran, J., et al. (2006). A persistent infrastructure for augmented field trips. In *Proceedings of World Conference on Educational Multimedia, Hypermedia and Telecommunications, Ed-Media 06* (in press), Orlando, Florida, USA.

Wei, P. C., Chen, C. H., & Chen, C. W. (2002). Support and optimisation of Java RMI over a Bluetooth environment. In *Concurrency and computation: Practice and experience* (pp 1-21).

Weibelzahl, S. (2002). *Evaluation of Adaptive Systems.* Graduate Programme Human and Machine Intelligence, Kunzenweg 21, University of Education Freiburg, Germany.

Weiser, M. (1991). The computer for the twenty-first century. *Scientific American*, 265(3):94-104. September 1991.

Weiser, M. (1996, March). *Open house. ITP Review 2.0.* Retrieved March 15, 2008 from http://www.itp.tsoa.nyu.edu/~review.

Weiser, M., & Brown, J. S. (1996). *The coming of age of calm technology.* Retrieved March 15, 2008 from http://www.ubiq.com/hypertext/weiser/acmfuture2endnote.htm.

Weng, M. (2007). *A multimedia social networking community for mobile devices.* Unpublished Interactive Telecommunications Program, Tisch School of the Arts / New York University.

Wenger, E. (1987). *Artificial intelligence and tutoring systems: computational and cognitive approaches to the communication of knowledge.* San Fransisco, CA: Morgan Kaufmann.

Wenger, E., & Lave, J. (1991). *Situated Learning: Legitimate peripheral participation.* Cambridge, New York: Cambridge University Press.

Wheeler, B. C. (1998). The state of business education: Preparation of the past? *Selections, 14*(2), 19–21.

Whitten, W. B., & Bjork, R. A. (1977). Learning from tests: Effects of spacing. *Journal of Verbal Learning and Verbal Behavior, 16*, 465–478. doi:10.1016/S0022-5371(77)80040-6

Wickens, T. D. (1999). Measuring the time course of retention. In C. Izawa (Ed.), *On human memory: Evolution,progress, and reflections on the 30th anniversary of the Atkinson-Shiffrin model* (pp. 245–266). Mahwah, NJ: Lawrence Erlbaum Associates, Inc.

Wijekumar, K. J., Meyer, B. J. F., Wagoner, D., & Ferguson, L. (2006). Technology affordances: the 'real story' in research with K-12 and undergraduate learners. *British Journal of Educational Technology, 37*(2), 191–209. doi:10.1111/j.1467-8535.2005.00528.x

Wills, G., Woukeu, A., Carr, L., & Hall, H. (2003). The need for semantic in web design. *Semantic Web workshop, Fourteenth Conference on Hypertext and Hypermedia Nottingham,* UK

Witt, P. L. (2000). *An experimental of teachers' verbal and nonverbal immediacy, student motivation, and cognitive learning in video instruction.* University of North Texas.

Witt, P. L., & Wheeless, L. R. (2001). An experimental study of teachers' verbal and nonverbal immediacy and students' affective and cognitive learning. *Communication Education, 50*(4), 327–342.

Witt, P. L., Wheeless, L. R., & Allen, M. (2004). A meta-analytical review of the relationship between teacher immediacy and student learning. *Communication Monographs, 71*(2), 184–207.doi:doi:10.1080/036452042000228054

WML. Wireless Markup Language and WAP: Wireless Application. Retrieved November 17, 2008, from http://www.wapforum.org

Wong, W. K., Chan, T. W., Chou, C. Y., Heh, J. S., & Tung, S. H. (2003). Reciprocal tutoring using cognitive tools. *Journal of Computer Assisted Learning, 19*, 416–428. doi:10.1046/j.0266-4909.2003.00046.x

Woodgate, D., & Stanton Fraser, D. (2005) *eScience and Education: A Review.* (Review report commissioned by JISC). http://jisc.ac.uk/uploaded_documents/ACF2B4.pdf

Woodgate, D., Stanton Fraser, D., Paxton, M., Crellin, D., Woolard, A., & Dillon, T. (2008, March 23-26). Bringing school science to life: personalization, contextualization and reflection of self collected data. *Proceedings of the Fifth IEEE International Conference on Wireless, Mobile and Ubiquitous Technologies in Education, WMUTE 2008*, Beijing, China. IEEE Computer Society.

Woods, R. H. Jr, & Baker, J. D. (2004). Interaction and immediacy in online learning. *International Review of Research in Open and Distance Learning, 5*(2).

Wozniak, P. A. *(1990).* Optimization of learning, Master's Thesis, University of Technology in Poznan, 1990 and adapted for publishing as an independent article on the web.

Wurman, R. S. (1989). *Information Anxiety.* New York: Doubleday.

Xie, X., Liu, H., Ma, W.-Y., & Zhang, H.-J. (2006). Browsing large pictures under limited display sizes. *IEEE Transactions on Multimedia, 8*(4), 707–715. doi:10.1109/TMM.2006.876294

Yahoo! Answers (2005). Retrieved 24 July, 2008, from http://answers.yahoo.com

Yamada, T., Shingu, J., Churchill, E., Nelson, L., Helfman, J., & Murphy, P. (2004) Who Cares? Reflecting who is reading what on distributed community bulletin boards. In *Proceedings of the 17th annual ACM symposium on User interface software and technology* (pp. 109-118). ACM press.

Yang, J., Chiu, C., Tsai, C., & Wu, T. (2004). Visualized online simple sequencing authoring tool for SCORM-compliant content package. In *Proc. of the IEEE International Conference on Advanced Learning Technologies 2004* (pp. 609–613), IEEE Computer Society.

Yang, S. J. H. (2006). Context aware ubiquitous learning environments for peer-to-peer collaborative learning. *Educational Technology & Society, 9*(1), 188–201.

Yingling, M. (2006). Mobile Moodle. *Journal of Computing Sciences in Colleges, 21*(6), 280–281.

Yusop, N., Lowe, D., & Zowghi, D. (2005). Impacts of Web systems on their domain. *Journal of Web Engineering, 4*(4).

Zagal, J. P., Piper, A. M., & Brukman, A. S. (2004). *Kids telling fables through 3D animation: Georgia Institute of Technology.* Retrieved, from http://hdl.handle.net/1853/3732

Zimmermann, A., Lorenz, A., & Oppermann, R. (2007). An Operational Definition of Context. *Proceedings of Context 07.*

Zurita, G., Nussbaum, M., & Salinas, R. (2005). Dynamic grouping in collaborative learning supported by wireless handhelds. Educational Technology & Society, 8(3), 149–161.

About the Contributors

Dr. **Tiong-T. Goh** holds a BSc and an MSc in Electrical Engineering from the Ohio State University, Columbus Ohio, USA, an MBA with distinction from Manchester Business School, UK, and a PhD in Information Systems from Massey University, New Zealand. He has worked in the polytechnics and universities, research laboratory, commercial organisation and defence organisation as research and development engineer and project engineer. He is currently a Lecturer with the school Information Management, at Victoria University, New Zealand. He has published various book chapters and Journals in e-learning. His areas of research interest include multiplatform e-learning systems, database system development, enterprise business intelligence and health informatics.

* * *

Marc Alier (1971) received an engineering degree in computer science and a PhD in Sciences in the Politecnical University of Catalonia. The last 15 years has worked in research and development related to the e-learning industry. He has participated in the development of several LMS and authoring tools, and has been an online teacher. Since 2001, he has taught project management and computing ethics as a lecturer in the Universitat Politecnica de Catalunya (http://www.upc.edu) and other universities such as UOC (http://www.uoc.edu), in computer science studies. He is director of a master's program in software for organization management. He is the technical advisor and teacher in the Sciences of Education Institute, UPC (http://www.ice.upc.edu) in the implantation of Moodle. Since early 2004, he has been a developer of the http://moodle.org community contributing with third part modules and core functionalities such as the Wiki module and the Webservices layer. Alier is the representant of Moodle in the IMS Common Cartridge discussion forum, and is a collaborator of the Open Knowledge initiative.

Jaime Muñoz Arteaga, is Professor in Computer Science at Universidad Autónoma de Aguascalientes.. Universidad Autónoma de Aguascalientes. He is a researcher in Human-Computer Interaction, and web technologies. He holds a PhD in Computer Science, Human-Computer Interaction (University Toulouse 1 (UT1), Toulouse, France, 2000). He has a number of collaborations with very important researches in prestigious universities around the world. He has reported his work in journals, book chapters, conference papers, and technical reports, and so far has supervised several students at the graduate and undergraduate level. His current research interests include several topics on: Human-Computer Interaction, Mobile Technologies, Software Engineering, and Artificial Intelligence.

Jonathan Bishop is a Chartered IT Professional and a director of Glamorgan Blended Learning Ltd and Chair of its Centre for Research into Online Communities and E-Learning Systems. A Town

Councillor in Wales, he is committed to finding technological solutions to social, economic and cultural problems of his community, including tackling the 'digital divide'. Jonathan has developed and researched online communities since the 1990s and is noted for inventing the Circle of Friends social networking technology and the PARLE e-learning system for which he was a finalist in the Innovation category of the New Statesman New Media Awards in 2004. He holds a BSc(Hons) in Multimedia Studies, specialising in online communities for the Web and TV, an MSc in E-Learning, specialising in persuasive technology and an LLM in European Union Law, specialising in the e-learning industry. In his spare time Jonathan enjoys taking part in debating competitions, swimming, listening to music, spending time in his garden, and playing video games.

Peter Byrne is a PhD research student in the Centre for Research in IT in Education (CRITE), at the School of Computer Science & Statistics, Trinity College. His research concerns the use of mobile technology to support collaborative and constructionist applications to for digital narrative and animation production. He has studied for a B.A. (Mod) in Information and Communications Technology (1998-2002) and for an M.Sc in Networks and Distributed Systems (2003-2004) at Dublin University, Trinity College. He has worked as a research assistant in the Machine Learning Group at Trinity College investigating an application Case Based Reasoning to physiotherapy treatment.

Juan Manuel González Calleros is a Researcher assistant of the University catholic of Louvain (UCL). He is member of the Belgian Lab of Human Computer Interaction and of the UsiXML consortium. Having received a scholarship award, he came to Belgium and completed his DEA in computer Science from UCL. He received a master degree in Computer Sciences at National Institute of Astrophysics, Optics and Electronic (Mexico). Juan Manuel Gonzalez Calleros is pursuing a PhD in Computer Sciences with the thesis model Driven engineering of 3D User Interfaces at UCL, Louvain School of Management (IAG-LSM). Her research interests include workflow models, Model-Driven engineering for developing User Interfaces, in general, and 3D User Interfaces, in particular.

Maria Jose Casany (1973) received an engineering degree in computer science in the Politecnical University of Catalonia. She has participated in the development of several LMS and authoring tools, and has been an online teacher. Since 2003, she has taught software development, and databases design and administration as a lecturer in the Universitat Politecnica de Catalunya (http://www.upc.edu) and other universities such as UOC (http://www.uoc.edu). Maria Jose has developed several Open source projects such as the J2MEMicroDB (a database engine for Mobile devices) and a migration of the authoring software JClic to the OLPC X0 platform (http://laptop.org). She is writing her phd dissertation about Ubiquitous learning for sustainable development. Her current main project is the development of an open source mobile client for Moodle.

Dr. **Alan Chamberlain** is an internationally published research fellow, working in the Mixed Reality Lab at the University of Nottingham. He currently sits on the editorial board of five international journals and has reviewed for numerous conferences and workshops in the field of Human Computer Interaction. His interests are mobile/wearable human computer interaction, pervasive computing, art and user evaluation and critical interpretations of new technologies/HCI-based frameworks. The Mixed Reality Laboratory (MRL)is an interdisciplinary research laboratory at the University of Nottingham .It brings together leading researchers from the Schools of Computer Science, Engineering and Psychology to research mixed reality - new technologies that merge the physical and digital worlds.

Dr. **Edward V. Chapel** has focused his professional efforts for most of his career on the strategic uses of technology in higher education environments. Currently the CIO and Vice President for Information Technology at Montclair State University in Northern New Jersey, where he has been for the past eight years, Ed has pursued a particular interest in the living and learning benefits of mobile computing, and broadband cellular services in particular, for contemporary college students. Prior to working in the information technology arena, he utilized his doctoral preparation in sociology to support his activities as an Institutional Research professional and a faculty member, teaching a broad range of courses in the discipline with a special focus on quantitative research methods and sampling design.

Simona Colucci (s.colucci@poliba.it) received the laurea "cum laude" degree in Management Engineering in 2002 and the Ph.D. degree in 2006 from Technical University of Bari. She is currently a Post Doc research student at Technical University of Bari. Her research interests include Knowledge Representation, Description Logics and their applications to Knowledge management. She is involved in various research projects and has authored various papers on journals and conferences on these topics.

David Crellin was educated at Rugby, Bristol, and Cambridge. He joined PA's, Technology division where he was responsible for managing a number of major projects. From 1982 David has run his own business (Abington Partners). Abington originated data-logging for schools. The company has won three SMART awards from the DTI. In 2002 we introduced the ScienceScope Datalogging products. Over the last 4 years the range has grown to be the most innovative and comprehensive available. These products are being used extensively in innovative developments in science teaching including: 'Participate', A £3 million collaborative research and development project which aims to demonstrate and evaluate how technologies can help people create and share digital content. Participate is supported through the Technology programme with grant funding from the Technology Strategy Board and the EPSRC. David lives with his wife and four children in Bath. He is 55.

Dr. **Teresa Dillon** is an artist-researcher and director of the intermedia company Polar Produce. Through Polar Produce she creates and produces location-based intermedia work, which combines performance, visual and sound arts, interactive design, new and old media. Alongside her work in Polar Produce, Teresa also creates her own installations and performances, and works with Kathy Hinde in the sound art duo BOP. Her work has shown internationally at various festivals and conferences and she publishes on creative collaboration, music technology, new media design, open source technology and educational media. In 2007, she set up New Interfaces for Performance, a distributed research and touring network of artists from across Europe. She also directs the UM Intermedia Festival, Lisbon, Portugal and the OFFLOAD-Systems for Survival art-research programme. Teresa also works as a freelance producer and researcher, (including for the BBC) and teaches on the Arts, Culture and Education MA at Cambridge University.

Daniel C. Doolan, is a lecturer in the School of Computing, Robert Gordon University, Aberdeen. His main research interest is on Mobile and Parallel Computing. He has published over 40 articles in the areas of mobile multimedia, computer graphics and parallel processing.

Dr. **Danaë Stanton Fraser** is a Reader in Psychology at the University of Bath where she leads the CREATE group evaluating novel technologies and studying their effects on collaborative learning. Dr. Stanton Fraser has led a number of technology enhanced learning projects including the JISC-funded SENSE project exploring children's use of pollution monitoring sensors for understanding environmental impact and 'the mobile phones in schools' project. She is currently an investigator on the EPSRC Cityware programme, evaluating collaborative impacts of mobile technologies on children and adults across heritage environments and an investigator on the TSB/EPSRC Participate project exploring pervasive computing for mass participation in environmental monitoring. Working in an interdisciplinary environment, she has published over 65 papers in peer-reviewed journals and conferences.

Alba Fuertes graduated in Industrial Engineering at the UPC Technical University of Catalonia (Spain), School of Industrial and Aeronautical Engineering of Terrassa (ETSEIAT), in 2007. Since then, she has been working in the research group "Group of Research and Innovation for Construction - GRIC" at UPC. Currently, she is a PhD student in Construction Engineering and her research interests are Innovative Information and Computing Technology applications for knowledge management, decision-making, and processes improvement in construction projects. She has also been working at UPC as an assistant lecturer since 2007. Currently, she is involved in different research projects focusing on "collaborative e-learning systems" and "knowledge management in construction research projects", one of them the European eContentplus project MACE.

Maxine Glancy is a Behavioural Scientist and User Experience Designer at the BBC (Research and Innovation) in London, England. Her research explores behavioural and design factors that shape the design and use of broadcast content and technologies. This includes subjects such as location-based computing, targeted help & accessibility, public displays, mobile collaborative 'learning-games', user generated content, pervasive media, broadcast entertainment services, and broadcast system design. She has degrees in Psychology, Computing, Design, and Environmental Science.

Amanda Gower is a principal researcher in the Media Interfaces Group in the Broadband Applications Research Centre, BT Innovate. Amanda trained as an animator, and received a BA (Hons) degree in Design from Edinburgh College of Art and a Masters degree in Computer Graphics and Animation from the National Centre for Computer Animation, Bournemouth University. Since joining BT in 1994, Amanda has developed virtual reality and pervasive media prototypes and applications, for learning, collaborative working, the arts and entertainment. Amanda has worked on collaborative research projects for some time, as a designer for eRENA (Tools for the Art of Tomorrow), and as the BT project lead for TOWER (Theatre of Work Enabling Relationships) and E-drama (an interactive role-play tool combining drama, ICT and education). She is the project manager for the Participate project, which explores how convergence in pervasive, online and broadcast media can support mass-participatory campaigns.

Andy Gower leads the Media Interfaces Group in the Broadband Applications Research Centre, BT Innovate. Andy's team is focused on creating new ways for consumers to access, interact, create and share media using Broadband enabled services and devices. Considering how emerging interfaces and technologies can enable people to break free from the desktop PC and gain new experiences supported by both fixed and wireless Broadband. Andy received a BA (Hons) degree in Three Dimensional Design from Leicester Polytechnic. Since joining BT in 1990, Andy has worked on the design of consumer and capital plant products and a variety of vision setting research projects. Andy's current research work is focused on 'located' media, leveraging User Generated Content and Character Mediated Communications.

Josefina Guerrero García is doing her PhD research in the domain of Management Sciences, Option "Information Systems" at Université catholique de Louvain (Belgium), Louvain School of Management (IAG-LSM). She received a DEA (Diplôme d'Etudes Approfondies) in Management Sciences from the same university in 2006 and a Master's Degree in Management at Instituto de Estudios Universitarios (Mexico) in 2001. She is a member of the Belgian Laboratory of Computer Human Interaction (BCHI) and the UsiXML Consortium. Her research interests include workflow models, computer supported cooperative work, and information systems.

Lester Gilbert is a Lecturer in Information Technology at the University of Southampton. He has published a textbook, Principles of e Learning Systems Engineering, co-authored with Veronica Gale, integrating his business-oriented practical experience of Systems development with Multimedia and Computer Aided Instruction to form the basis of his focus on e learning and the use of technology in learning and teaching. Lester is the Principal Investigator on the JISC-funded REAQ and EFSCE projects, and a Co-Investigator on a number of other JISC-funded projects including mPLAT, Remora, AsDel, FREMA, CORE, and EASiHE. He has published a number of papers on e-learning and e-assessment. Lester has a background in experimental and cognitive psychology, in particular learning theory and the statistical analysis of data and psychometric measures of test quality.

Paul Hayes is a Lecturer and Researcher with the School of Computing at National College of Ireland. He graduated in 1990 with an honours degree in Electronic Engineering from the University of Limerick. He subsequently completed a Master's degree by full-time research in the area of wireless networks. He was then employed by Queen's University Belfast as a Research Associate to work on a collaborative research project in telecommunications between the University and Nortel. After working in the telecommunications industry as a senior engineer for a number of years in 2000 he joined the National College of Ireland as a lecturer in telecommunications. His research interests include mobile learning, data communications and multimedia.

Shinichi Hisamatsu is a researcher of interfaculty initiative in information studies, the university of Tokyo in Japan. He received Master of Media and Governance from Keio university. His speciality is Educational technology and media art. He is interested in how to construct user interface in education context. He belongs to Japan society for educational technology, Human Interface Society, Information Processing Society of Japan, the institute of electronics, information and communication engineers(IECE),Association for the Advancement of Computing in Education(AACE).

Lars Erik Holmquist is a senior researcher at the Swedish Institute of Computer Science in Kista, Sweden, a research leader in the Mobile Life Center at Stockholm University, and leader of the Future Applications Lab at the Viktoria Institute in Göteborg, Sweden. He previously founded and led the PLAY research group at the Interactive Institute. He received his master's degree in Computer Science in 1996 and his PhD in Informatics in 2000, both at the Göteborg University, and became an assoiate professor at the Göteborg IT University in 2004. His research interests include human-computer interaction, information visualization and ubiquitous computing. He has been member of many international conference committees and published extensively in these research fields. He was chair of the UbiComp 2002 conference and started the Mobile Music Technology workshop series in 2005. He is an associate editor of the Springer journal Personal and Ubiquitous Computing and writes a regular column in ACM's interactions magazine, "On the Edge".

Dr. **Yvonne Howard** is a Senior Research Fellow within the Learning Societies Lab at the University of Southampton. She holds an Honours degree in Computer Science and a PhD in modelling software development processes. Her initial research lay in process models for software development, particularly investigating dynamic processes where evolution is driven by feedback in the domain. More recently this has focused on Agile and social methods of user participation in the development process. She has been involved in many UK projects: the EPSRC funded ABCD project, the Open Middleware Infrastructure Institute (OMII), and JISC funded FREMA, r2q2, CLAReT, mPLAT, Remora, FAROES and InnovationBase projects. In all of these projects, a particular interest has been how information systems can provide social, knowledge and organisational capabilities to encourage emergent innovation in practice.

Tim de Jong is a PhD student at the Open University of the Netherlands. He received his Master's degree in Knowledge Engineering from the University of Maastricht in 2005. Since 2006, he has been working at the CELSTEC institute of the Open University of the Netherlands mainly researching mobile learning, contextualized computing, and knowledge engineering. The topic of his PhD focuses at the support of authentic learning processes with mobile technology using mobile media delivery, mobile media creation, and contextualised notifications. Moreover, he is currently working in the European eContentplus project MACE, which focuses on the metadata enrichment of architectural content.

Dr. **Patricia Kahn** has been employed at Montclair for eight years supporting faculty and providing leadership in developing a community around teaching and learning using technology. Through her leadership as Director of Technology Training and Integration, faculty are able to explore innovative teaching methodologies that promote a learning environment that encourages collaboration, critical thinking, and exploration in response to a diverse community of learners. Patty completed her doctoral degree in Education, where her research focused on learning styles and instructional design and has recently applied her research with academic uses of mobile technology in higher education.

Dr. **Siu Cheung Kong** is an Associate Professor in the Department of Mathematics, Science, Social Sciences and Technology at the Hong Kong Institute of Education. Dr. Kong received a Doctor of Philosophy from the Department of Computer Science of the City University of Hong Kong. Dr. Kong publishes in the fields of information technology in education, information literacy education, collaboration in mobile learning and cognition and technology in mathematics education. He is currently a member of the Steering Committee on the Strategic Development of Information Technology in Education that was set up by the Government of the Hong Kong SAR.

Professor Dr. **Rob Koper** is the dean of the Centre for Learning Sciences and Technologies (celstec.org), a Research centre of the Open University of the Netherlands (formerly OTEC) that aims to research, develop, test and provide sustainable and evidence-based solutions for the advancement of education, training, professional development and learning experiences at work, at school, at home and on the move. His personal research concentrates on professional development in self-organised distributed learning networks. In the past, he developed the Educational Modelling Language (the predecessor of IMS Learning Design) and led the OUNL's contribution to the IMS LD specification activities. He is furthermore coordinating the European Integrated Project TENCompetence that aims to develop an infrastructure for lifelong competence development.

John Lenarcic is a Physicist and Applied Mathematician by training, an IT academic by fortunate accident and an Armchair Philosopher by conscious choice. He is currently a Lecturer in the School of Business Information Technology at RMIT University in Melbourne, Australia.

Dr. **Seng W. Loke** is a Senior Lecturer at the Department of Computer Science and Computer Engineering in La Trobe University. He leads the Pervasive Computing Interest Group at La Trobe, and has authored 'Context-Aware Pervasive Systems: Architectures for a New Breed of Applications' published by Auerbach (CRC Press), Dec 2006. He has (co-)authored more than 180 research publications including numerous work on context-aware computing, and mobile and pervasive computing. He has been on the program committee of numerous conferences/workshops in the area, including Pervasive'08, and PerEL (Pervasive E-Learning).

Tracey J. Mehigan is a researcher with the IDEAS Research group at the Department of Computer Science, University College Cork, Ireland. Her principle area of research focuses on the use of mobile computing for eLearning, to facilitate the inclusion of those with disabilities and special needs, into mainstream and ubiquitous collaborative learning environments.

Dr. **David Metcalf** is the director of the Mixed Emerging Technology Integration Lab at the University of Central Florida. Specific areas of focus include learning business strategy, performance measurement, operational excellence, outsourcing, blended learning and mobile learning. Dr. Metcalf was formerly the Chief Learning Technologist at RWD Technologies. There he was responsible for the analysis, design and strategic alignment of RWD's technology solutions for learning. He also led the development of several award-winning mobile learning innovations.Dr. Metcalf joined RWD with the sale of his NASA Kennedy Space Center laboratory spin-off company, Merrimac. He was responsible for the management and operations of the award-winning multimedia laboratory for various government contractors.Dr. Metcalf holds a B.A. in Computer Graphics from the University of Texas, an M.S. in Computer-Based Learning, and a Ph.D. in Information Systems from Nova Southeastern University and keeps current by continuing to hold appointments at several universities.

Dr. **David Millard** is Lecturer of Computer Science within the Learning Societies Lab at the University of Southampton. He holds an Honours degree in Computer Science and a PhD in Contextual Hypermedia Systems. David has been involved in Hypertext and Web research for over ten years, firstly in the area of Open, Adaptive and Contextual hypermedia and more recently in the area of Web 2.0, Semantic Web, Knowledge and Narrative interfaces, and the impact of Web Literacy on e-learning and m-learning. He has been the Principle and Co-Investigator on many projects in the UK, including the JISC funded FREMA, MPLAT, Remora, Semtech, Synote, R2Q2, PeerPigeon, Faroes and EdShare projects, and has over one hundred international publications in these areas. He is interested in the ways that people use information systems in-the-wild, and how we can use emergent social, organisational and semantic structures to help them make sense of their world.

Dr. **Kiyoshi Nakabayashi** is a professor of R&D Department at National Institute of Multimedia Education. After receiving a M.Sc. degree from Tokyo Institute of Technology in 1982, he entered Electrical Communications Laboratory of Nippon Telegraph and Telephone Corp. where he was engaged in the research and development of parallel processing, character recognition system, and network-based learning support system. In 1999, he joined NTT-X Inc. where he has conducted development and operation of e-learning service system. He has also worked in the field of e-learning technology standardization serving as the ISO/IEC JTC1 SC36 (Sub-committee for Information Technology for Learning, Education and Training) Japan national body head of delegation and the vice president chairing Executive Committee of eLC (e-Learning Consortium Japan). He received 1994 JSAI (The Japanese Society for Artificial Intelligence) Research Encouragement Award and ICCE95 (International Conference on Computers in Education) Outstanding Paper Award. He received a Ph.D degree in human science from Waseda University in 2006.

Dr. **Wan Ng** is Senior Lecturer in the areas of science & technology education and gifted education in the Faculty of Education (Bundoora) at La Trobe University, Australia. Her major research interest lies in the use of innovative and integrated curriculum in learning, and the use of technology, including handheld and mobile technologies, in education at both school and tertiary levels.

Dr. **Howard Nicholas** is Senior Lecturer in Language Education. His major focus is on the relationship between interaction patterns and the development of language and thinking with handheld devices viewed as a tool to support innovative pedagogies that promote divergent thinking and open-ended educational practices.

Tommaso Di Noia (t.dinoia@poliba.it) is an assistant professor in Information Technology Engineering at Technical University of Bari (Politecnico di Bari). He got his Ph.D. from Technical University of Bari. His main scientific interests include: Description Logics - Theoretical and Practical Aspects; Resource Matchmaking; Knowledge Representation Systems for Electronic Commerce; Automatic (Web) Services Discovery and Composition; Knowledge Representation Systems and Applications for the Semantic Web. He co-authored papers which received the best paper award at conferences ICEC-2004 and IEEE CEC-EEE-2006.

Jordi Piguillem is a Computer Software Engineer by Universitat Politecnica de Catalunya. During last three years has been collaborating on several Moodle projects as a programmer and as a software designer. During Summer'08, he had been working in Google Summer of Code initiative, where he has been developing an IMS LTI compliant client for Moodle. Nowadays, he is working on his PhD about integration of information systems.

Agnese Pinto (agnese.pinto@doom-srl.it) received a laurea degree in Management Engineering from Technical University of Bari (Politecnico di Bari), Italy, in 2004. She is currently a project manager at D.O.O.M. s.r.l., where she leads the ontology design group. Since July 2004 she also collaborates with the Information Systems Research Group of Politecnico di Bari. Her main research interests are in ontology design, natural language ontology mapping, ontology design pattern and in e/m-learning field. She has co-authored various research papers in these fields.

Ian Pitt, lectures in Usability Engineering and Multimedia at University College, Cork, Ireland. He took his D.Phil at the University of York, UK, then spent a year as a post-doctoral research fellow at Otto-von-Guericke University, Magdeburg, Germany, before moving to Cork in 1997. He is the leader of the Interaction Design, E-learning and Speech (IDEAS) Research Group at UCC, and has published over 60 articles within the areas of HCI and the use of speech and non-speech sound in human-machine interfaces.

Dr. **Joan Richardson** is a Senior Lecturer in the School of Business Information Technology at RMIT University in Melbourne, Australia. She has led University projects designed to strategically support adoption and implementation of new technologies in the higher education sector for a number of years. Collaborations with industry and consultancies have ensured that she has maintained a leading position in the application of ICT in the higher education sector, shifting effective use of technology into the mainstream and delivering state of the art courses and programs. Dr Richardson has won University awards for curriculum development and construction and incremental development of suites of resources that utilise technology innovations for delivery. Since 2001 she has been the principal author of four editions of the text Computing For Business Success published by Pearson Education Australia. Each version of the text has included 'state-of-the-art' digital media and resources, in the form of companion CD-ROM's and Web sites, constructed in conjunction with Pearson Education Australia.

David Rogers is a researcher at the Institute for Simulation and Training, Mr. Rogers is investigating mobile learning applications in the developing world, and the use of digital repositories in education. Mr. Rogers served as a director of Aid & Development programs in East Africa for several years, where he was responsible for establishing and managing one of the only operating secondary schools in Southern Sudan during the civil war. He continues to manage a scholarship fund that sponsors promising African students to attend college and university programs. Mr. Rogers holds a B.S. in Basic Sciences from the United States Air Force Academy, and is currently a doctoral student at the University of Central Florida in the Texts and Technology program. His research areas include international development, cognitive psychology, and mobile learning.

Mattias Rost is a PhD student at the IT University of Gothenburg in Applied IT, and has a Masters of Science degree in Computer Science from Chalmers University of Technology. He has been a research assistant in The Future Applications Lab at the Viktoria Institute since 2005, and at the IT University of Göteborg since 2006. He is now conducting research at the Mobile Life Centre in Stockholm, Sweden, on mobile applications and services. His research domains include ubiquitous computing, human-computer interaction, as well as computer supported collaborative work technologies.

José Rouillard is an Associate Professor in Computer Science at the University of Lille 1 in the LIFL laboratory. The LIFL (Laboratoire d'Informatique Fondamentale de Lille) is a Research Laboratory in the Computer Science field of the University of Sciences and Technologies of Lille (USTL) linked to the CNRS, and in partnership with the INRIA Lille - Nord Europe. José Rouillard obtained his PhD in 2000 from the University of Grenoble (France) in the field of Human-Computer Interfaces. He is interested in Human-Computer Interaction (HCI), plasticity of user interfaces, multi-modality and multi-channel interfaces. He has written one book on VoiceXML in 2004, another one on Software Oriented Architecture (SOA) in 2007 and more than 70 scientific articles. He is now engaged in research on mobility and pervasive/ubiquitous computing.

Francisco Javier Martinez Ruiz is a professor of computer science at the University of Zacatecas (Mexico). Most of his research centers in the field of Human Computer Interaction. He is studying how to tackle the challenges of the User Interface modeling in the Web and the integration of these technologies to e-learning. He received a Bachelor's degree in Computer Sciences from the Institute of Technology of Zacatecas (Mexico), a Master Degree in Computer Sciences from The Instituto Tecnológico y de Estudios Superiores de Monterrey (ITESM,Mexico) and a DEA Degree from the Université catholique de Louvain (Belgium). From 2004 to 2006 he directed the Specialization in Computer Science in the University of Zacatecas. He is a researcher of the Belgian Laboratory in Computer Human Interaction (BCHI) at the Université catholique de Louvain. He is the recipient of an AlBan scholarship, the European Union Program (www.programalban.org) of High Level Scholarships for Latin America, under reference E06D101371MX.

Michele Ruta (m.ruta@poliba.it) received the laurea degree in Electronic Engineering from Politecnico di Bari (Technical University of Bari) in 2002 and his Ph.D. in Information Engineering from the same University in 2007. His research interests include pervasive computing and ubiquitous Web, mobile service discovery and composition, Knowledge Representation systems and applications for wireless ad-hoc contexts. On these topics he has co-authored various papers in international journals and conferences. He is involved in various research projects related to his research interests.

Tally Schmeits graduated in Computer Science at the Fontys Hoge school of Eindhoven (Netherlands), specialisation Technical Computer Science (TCK), in 1998. After that, he started to work for a company which provided multi-medial learning solutions for preschool and higher education as Senior Software Developer/System Architect. Currently, he is working at the Open University as a Software Developer on the European MACE project, focusing on the development of a Competence Catalog. His next challenge is starting up his own company Schmeits Websolutions which provides web-based Solutions for a wide range of clients.

Eugenio Di Sciascio (disciascio@poliba.it) received the "laurea" degree with honors from University of Bari, and the Ph.D. from Politecnico di Bari (Technical University of Bari). He is currently full professor of Information Technology Engineering at Technical University of Bari, and leads the research group of SisInfLab, the Information Systems Laboratory of Technical University of Bari. Formerly, he has been an assistant professor at University of Lecce and associate professor at Technical University of Bari. His research interests include multimedia information retrieval, knowledge representation and e-commerce. He is involved in several national and European research projects related to his research interests. He co-authored papers that received best paper awards at conferences ICEC-2004 and IEEE CEC-EEE-2006.

Floriano Scioscia (f.scioscia@poliba.it) received the Laurea Degree in Information Technology Engineering from Politecnico di Bari, Technical University of Bari in 2006. He is currently pursuing his PhD in Information Engineering at the same University. His research interests include pervasive computing, mobile service discovery and composition, knowledge representation systems and applications for wireless ad-hoc contexts.

Boon-Chong Seet obtained his PhD in Computer Engineering from Nanyang Technological University, Singapore, in 2005. After graduation, he was recruited as a Research Fellow under the Singapore-MIT (Boston) Alliance (SMA) computer science program at the National University of Singapore, School of Computing. In 2007, he was awarded a Visiting Scholarship by ETSIT Technical University of Madrid, Spain, to pursue research under an EU co-funded project on multi-disciplinary advanced research in user-centric wireless network enabling technologies (MADRINET). Since December 2007, he is with Auckland University of Technology (AUT) as a faculty member of its Electrical and Electronic Engineering Department. His research activities are generally in the areas of mobile networking, computing, and communications. Recently, he has also an interest in the research and development of mobile applications. To date, he has about thirty technical publications in refereed journals, books and conferences.

Marcus Specht is Professor for Advanced Learning Technologies at the Open University of the Netherlands and is currently involved in several national and international research projects on competence-based lifelong learning, personalised information support and contextualised learning. He received his Diploma in Psychology in 1995 and a dissertation from the University of Trier in 1998 on adaptive information technology. From 1998 until 2001, he worked as senior researcher at the GMD research centre on HCI and mobile information technology. From 2001, he headed the department "Mobile Knowledge" at the Fraunhofer Institute for Applied Information Technology (FIT). From 2005, he was Associated Professor at the Open University of the Netherlands and working on competence-based education, learning content engineering and management, and personalisation for learning. Currently, he is working on mobile and contextualised learning technologies, learning network services, and social and immersive media for learning.

Sabin Tabirca is a lecturer in Department of Computer Science of National University of Ireland, Cork. His main research interest is on Mobile and Parallel Computing for Scientific Problems. He has published more than 100 articles in the areas of mobile multimedia, parallel computation, number theory and combinatorial optimization.

Brendan Tangney is a Fellow of Trinity College Dublin and is a Senior Lecturer in the School of Computer Science & Statistics. He is co-director of Trinity's Centre for Research in IT in Education (www.cs.tcd.ie/crite). His research is concerned with the innovative use of ubiquitous technology to enhance the experience of learners in a variety of formal and informal settings.

Dr. **Torab Torabi** is Senior Lecturer at Department of Computer Science and Computer Engineering. He is the head of Software Engineering Interest Group. He has (co)authored more than forty journal and conference papers in Software Engineering and Mobile Computing. His research interests include Software Engineering, Process Modeling, and Context-Aware Mobile Services. Dr. Torabi serves in editorial board of number of journals, and he has served in program committee and organization of number of conferences and workshops. He has supervised and coordinated more than thirty major research and industry projects.

Jean Vanderdonckt is Full Professor in Computer Science at Université catholique de Louvain (Belgium), Louvain School of Management (IAG-LSM) where he leads the Belgian Laboratory of Computer-Human Interaction (BCHI). This laboratory is conducting research, development, and consulting services in the domain of user interface engineering, a domain that is located midway between software engineering, human-computer interaction, and usability engineering. Jean Vanderdonckt is the founder and the coordinator of the UsiXML Consortium that structures activities towards the definition and the usage of UsiXML (User Interface eXtensible Markup Language) as a common User Interface Description Language. He is the coordinator of HCI activities within the Similar network of excellence (The European research taskforce creating human-machine interfaces SIMILAR to human-human communication). He is also a member of the European COST n°294 Action MAUSE on usability engineering and of the SESAMI Working Group. He is a Senior member of IEEE, ACM, and SIGCHI. He is also co-editor in chief of Springer HCI Series of books.

Stephan Weibelzahl is a Lecturer and Researcher with the School of Computing at National College of Ireland. He is programme director of the MSc in Learning Technologies. Prior to that, Stephan headed a research group in learning technologies at Fraunhofer Institute Experimental Software Engineering, Kaiserslautern, Germany. He holds a Diplom (Masters) in Psychology and received a PhD from the University of Trier, Germany. Stephan's main areas of expertise are personalisation in e-Learning and the evaluation of learning technologies. Exploiting artificial intelligence techniques, Stephan and his team designed, developed and evaluated a number of e-Learning systems that monitor and adapt to the user in order to improve the learning process.

Dr. **Gary Wills** is a Senior Lecturer in Computer Science at the University of Southampton. He graduated from the University of Southampton with an Honours degree in electromechanical engineering, and then a PhD in Industrial hypermedia systems. He is a Chartered Engineer and a member of the Institute of Engineering Technology and a Fellow of the Higher Educational Academy. Gary's main research interests are in Personal Information Environments (PIEs) and their application to industry, medicine and education. PIE systems are underpinned by Service Oriented Architectures, adaptive systems and advanced knowledge technologies. In education he researches into adaptive assessment systems and the use of mobile technology. In Medicine he researches Virtual Research Environments enabling collaboration, discussion and dissemination of results form clinical trials with large datasets. For Industry his research, allows engineers to view federate knowledge, using knowledge technologies and ontologies to present appropriate information to the engineers.

Dr. **Dawn Woodgate** is a Research Fellow and member of the CREATE Group in the Department of Pyschology at the University of Bath, UK. She has an interdisciplinary background, with a first degree in Biology, a Masters in Research, and a PhD in Science and Technology Studies. She is a qualified and experienced former science teacher, and now researches and publishes in the areas of Education and Technology, Human Computer Interaction and Science and Technology Studies.

Kin Choong Yow obtained his B.Eng (Elect) with 1st Class Honours from the National University of Singapore in 1993 and his Ph.D. from the Cambridge University, UK in 1998. He joined the Nanyang Technological University, Singapore as a faculty member in May 1998, and he is presently an Associate Professor of Computer Engineering in the College of Engineering. His research interests include

Computer Vision, Wireless Communications and Computational Intelligence. He has published over 57 top quality international journal and conference papers and he has served as reviewer for a number of premier journals and conferences, including the IEEE Wireless Communications and the IEEE Transactions on Education. He was the winner of the NTU Excellence in Teaching Award 2005, and he also won the Most Popular SCE (School of Computer Engineering) Year 1 Lecturer, for 4 consecutive years (2004 – 2007).

Index

U

V

W

Z